A Handbook of Organic Farming

Arun K. Sharma
Scientist (Agronomy)
Central Arid Zone Research Institute
Jodhpur, India

Published by:
AGROBIOS (INDIA)
Agro House, Behind Nasrani Cinema
Chopasani Road, Jodhpur 342 003
Phone: 91-0291-2643993
Website: *www.agrobiosonline.com*
E. mail: *agrobiosindia@gmail.com*

AGROBIOS

ISBN No.: 978-81-7754-099-4

Published by: Dr. Updesh Purohit for Agrobios (India), Jodhpur
Laser Typeset at: Yashee Computers, Jodhpur
Cover Design by: Shyam
Printed at: Agrobios Digital, Jodhpur

Preface

The word '*Organic*' means origin from a living thing and farming with the philosophy of Organic is to make production system alive with long life. It is not just to replace fertilizers and pesticide with manure and predators but it is an ongoing dynamic process for making healthy soil, and ultimately a vital living system of the world.

Organic farming is similar to the other sustainable farming systems e.g., permaculture, eco-farming etc., which are based on harmony with nature or near to nature approach. The only distinguishing character is the *certification* of production in organic farming. Certification is a procedure in which certain rules and regulation have to be followed. This is necessary to obtain certificate from recognized agency. This agency certify that the product is produced strictly with organic methods.

In this handbook almost all the aspects related organic farming are covered with a balanced approach. The author demonstrates the newness of the organic concept for readers. This will help the readers to discover easily the philosophical and technical differences between organic and conventional (synthetic input dependent) farming systems.

Chapter 1 deals with the need and progress towards sustainable farming systems. In chapter 2 covers various rules and regulation related to soil and crop management for a standard organic farming. Chapter 3 gives basic information about soil environment in relation to plant growth. Various basic and applied aspects of input and crop health management are discussed in chapters 4 to 13. The author raised several issues and concerns, specially on adoption, marketing and research in organic farming in chapters 14 to 16.

Considering the fast development in technology and marketing in organic farming, detailed list of electronic sources of information is included in the end. References are not given in the text because the readers should feel that he or she has learned the subject at some depth depending upon the level of the writing. If references are allowed, then the tendency of the writing will be to keep telling the reader more about where the subject matter is available rather than presenting the same reasonably well on the spot. Such style makes the writing directive rather then instructive.

After reading the handbook, hopefully the readers will have a balanced view about organic farming and will get answers of most of the quarries with updated knowledge.

Arun K. Sharma

Contents

CHAPTER 1

Cycle for Survival

Increasing population and to some extent capitalism in the second half of the 20th century forced to develop shortcut and fast output system, accordingly modern agricultural production has developed in a direction contrary to the requirements of productive and sustainable ecosystems for cohesiveness and functional diversity. The trend is:

1. Rather than being cyclical, modern agricultural production tends toward an open flow of material. The growing use of off-farm production inputs (mineral fertilizers, plant protection agents, machinery etc.) means that largely *closed* systems are being increasingly opened up and, thus, losing their cohesiveness.
2. Instead of being complex and diverse, modern farming systems tend to be one-sided (specialized) in response to pressure toward economic rationalization. Production is becoming increasingly confined to only a few lines of activity, limited crop rotations and a small number of species and varieties, generally with a greatly restricted genetic base (e.g. hybrid varieties).

The aim of this development is production according to industrial considerations in order to maximize yields in the short to medium term. This is backed up by agricultural research which investigates the immediate, and not so much the long-term effects of isolated production factors (e.g. nitrogen or water) on crop yield. This mono-causal approach does not take account of the multifunctional nature and close inter-linkages of the individual factors in the farm ecosystem.

This has resulted in decreasing sustainability due to breaking of natural cycles of carbon and nutrient besides having several other deleterious effects on the environment which first affecting the agriculture production but ultimately will become question mark for the human survival.

THE GREEN REVOLUTION-UNSUSTAINABLE IMPACT

The population of developing countries has been growing rapidly and the demand for food has not always been able to keep up with population growth. This plus the failure of a monsoon to arrive over Asia in 1965 and 1966 made many people think that a famine maybe ahead. The Ford Foundation and the Rockerfeller Foundation in the US could find out the solution to be in HYV (High yielding varieties) crops, which had proved successful in Mexico. These two institutions encouraged the acceptance of this new technology and introduced in 1966 - *The Green Revolution*. These HYVs soon spread all over Asia. Traditional grains of rice were replaced by these *miracle grains*.

Yields at first were exceptionally high. In the past 30 years cereal yields have doubled and have kept pace with the need to feed a rising population. Then soil fertility began to decline due to the vast amounts of artificial fertilizers the grains needed. HYVs were susceptible to many pests that traditional types would have tolerated. To counteract these pests vast amounts of pesticides were used. Not only did this cause harm to the environment but the pests became tolerant to the most poisonous of pesticides. Water quality was also seriously effected, as chemicals leeched away into water sources. As Table 1.1 shows yields rose initially between 1949-1965 but an overall decline can be seen from 1967-1978.

TABLE 1.1: Compound rates of growth

Crop	Production		Area		Yield (% per annum)	
	(a) 1949-50 to 1964-65	(b) 1967-68 to 1977-79	(a) 1949-50 to 1964-65	(b) 1967-68 to 1977-78	(a) 1949-50 to 1964-65	(b) 1967-68 to 1977-78
Food grains	2.98	2.40	1.34	0.38	1.61	1.53
Non-Food grains	3.65	2.70	2.52	1.01	1.06	1.15
All Crops	3.20	2.50	1.60	0.55	1.60	1.40
Rice	3.37	2.21	1.26	0.74	2.09	1.46
Wheat	3.07	5.73	2.70	3.10	1.24	2.53
Pulses	1.62	0.20	1.87	0.75	-0.24	-0.42

Source: The Organic Farming Source Book 88 pp.

Traditional methods based on organic methods may lost and these methods would have been handed down from generation to generation but with the advent of conventional farming these may have been lost along with many varieties of traditional seeds.

One of the groups of these people who made these decisions have observed that these inputs were not in fact the right decisions and they are now seeing the mistakes of such practices. The best way they see now to improve the environment and increase yields of crop is to rely on traditional and ecological methods.

Our fathers and forefathers, who were traditional farmers, were the pioneers of organic farming. Today organic farming is looked upon as the answer to these problems posed by technical farming.

Many farmers are now returning to organic methods of farming that are similar to their traditional methods but now, with more knowledge attached, organic farming can prove to be as successful as, if not more as conventional framing.

IMPACT OF ADVANCED AGRICULTURAL TECHNOLOGY HAZARDS

Modern agriculture largely depends on the use of fossil fuel-based inputs, such as chemical fertilizers, pesticides, herbicides and labour saving but energy intensive farm machinery. While the applications of such high input technologies have undoubtedly increased production and labour efficiency. There is a growing concern over their adverse effects on soil productivity and environmental quality which is emerging to recognize that the farmer has a great social responsibility as a land owner than merely agribusiness considerations.

Adverse Effects of Agricultural Practices on the Farm

Although, high-input agriculture has certainly resulted in spectacular gains in productivity it has degraded some of the natural bases on which this system rests. The ultimate costs of this degradation on the farm are borne by the farmers themselves and by society at large, be they in lowered productivity, environmental cleanup, or further technology to solve a technologically induced problem. Some of these on-farm impacts are mentioned here.

Soil Erosion

During the late 1970s, the USDA informed that up to half of the cropland in USA showed serious erosion of its precious, nutrient-rich topsoil. The numbers are hard for a layperson to comprehend; over 15 tons per hectare of topsoil lost, on average. Global figures are just as alarming: the United Nations is using satellites to track soil erosion, and estimates that in northern India, losses per hectare exceed those of the United States. The effect of these losses of topsoil on the farm is devastating. As wind and water remove the finer particles, the subsoil that is left is coarser, with less organic matter to break up the soil, and more fertilizers are needed to make it support plant growth. Ultimately, eroded soils can become useless for crop production.

Water Availability

Water is an essential component for plant growth, and in arid and semiarid regions, the availability of water for agriculture must be improved through irrigation. According to the FAO, 270 million ha in the world are irrigated, and this consumes 70% of the world's total human-induced water consumption.

About half the water used in the United States for irrigation comes from underground aquifers (for example, the *Ogallala aquifer* in the Southwest), from which it is pumped to the surface. The same is true in the Indo-Gangetic plain of India. These aquifers are slowly and constantly recharged through rain percolation in the soil, but the rate at which they are being depleted is much faster than slow recharge, and the level of the water table is falling, mainly in the view of the demand of the intensive aquatic systems to water requirements.

Various crop plants have different needs for water. When this is expressed in economic terms, the cash value of crop per volume of water used is seen to vary greatly. Yet, governments often have *cheap water* policies, providing subsidized water to those farmers who grow crops that are water hungry in locations that are water proof. This promotes depletion of water resources.

Salinization

Irrigation of salt-rich arid and semi-arid lands has an added environmental cost. When water use is excessive, the underground water table rises and brings with it dissolved salts from the lower regions of the soil. This initially lowers crop yield, and if the water evaporates and leaves salts caked on the surface, ultimately renders the soil agronomically unless drastic and expensive measures are taken to remove the salts. The extent of salinization is more difficult to measure than simple soil erosion, because the effects of salinized soil are underground, at least initially. Estimates suggest that in Bangladesh river delta salinization is significantly reducing yields in about one-third of the cropland, while in the desert of the Southwest U.S. the figure approaches one-fourth.

Fertilizer and Pesticide Contamination

Applying nitrogen fertilizer in excess amount and removed each year with the crop can lead to loss of nitrogen from the soil environment. Because nitrate is negatively charged, it does not bind to negatively charged soil particles and therefore, is poorly stored in the soil. When soil water exceeds plant needs (through rain or irrigation), nitrate leaches into soil ground water. This can contaminate the wells, and nitrates in drinking water pose a health problem to the animals and people who use such wells.

The extensive use of pesticides poses a more direct problem for farm workers. Because these people are generally outside of the main-stream of health care estimates of the extent of pesticide induced health problems and are fraught with difficulty. But incidents of acute poisoning certainly occur with alarming frequency, and epidemiological data show that workers who handle pesticides more than 20 days a year have an increased risk of developing certain types of cancer.

Genetic Erosion

The demands of modern agriculture have led to a narrowing of the genetic base of crop plants. The initial success of the newly bred and selected strains has repeatedly led to their rapid adoption by all farmers, and abandonment of the richly diverse landraces and wild relatives. For example, in the United States six cultivars (out of tens of thousands worldwide) account for a third of the wheat planted. In Greece, virtually all landraces of wheat have disappeared over the past 40 years. The dangers of such a narrow genetic base are obvious in terms of susceptibility to pests (for example, the Tungro virus that devastated IR-8 rice in the Philippines during the 1970s).

Adverse Effects of Agricultural Practices Away from the Farm

In addition to the serious effects of high-intensity farming in degrading the resource base for future farming, the environment away from the farm can be adversely affected in the following ways:

Soil Erosion

Eroded soil particles, especially carried by water, often end up filling reservoirs, increasing the costs of water extraction, and clogging navigable waterways. According to the United States Department of Agriculture, about 1 billion tons of eroded soils are deposited in waterways each year in the USA. The removal of 10 per cent of the highly eroded lands from cultivation currently would reduce this deposition by 20 per cent. Most difficult to cultivate without erosion are hillsides. Hillside cultivation requires careful management if soil erosion is to be avoided.

Fertilizer and Pesticide Contamination

Farming is a major source of water pollution by fertilizers and pesticides dissolved or suspended in soil water. Excessive nitrogenous fertilization can contaminate drinking water. It is estimated that 25 per cent of the people living in Europe are drinking water with nitrates above recommended levels (25 mg per litre). Pesticide contamination of groundwater, and then drinking water, is also a potentially serious problem. Although the effects of DDT on certain bird populations are well known, the effects of low concentrations of other pesticides on people have not been established. In the USA, detectable levels of 39 pesticides have been observed in ground water in 34 states. Although in most cases the levels are below those deemed unsafe, long-term effects of this exposure are not known. Pesticides Residue in the vegetables

Since the qualities, importance and attributes of vegetables are seen unparalleled in resisting irrecoverable diseases is well known above spraying of pesticides in our crops should be avoided. Vegetables come to the market with poison applied in the form of the pesticides in the field. In spite of prohibition, some pesticides that are macabre to fauna, flora as well as environment come to the farmers door through black hand.

The indestructible nature of the poisonous residues causing environmental pollution enters into the human body through food chain. This poison makes the body poisonous day by day and brings a curse to human life through the creation of various irrecoverable diseases like the loss of memory, abortion, insanity, tumor, cancer etc. Every year, seven lakh people are falling under poisonous reaction of pesticide among which 14000 die of it. This is not the end, pesticide kills useful insects and germs living in the soil, ultimately makes it unfertile and creates environmental imbalance.

The pesticides were introduced to environment as pesticide after Second World War. Within the sixty's decade of twentieth country, the consciousness among the people of western world about the harmfulness effect of pesticide was developed to a great extent. Our farmers were getting this pesticide publicly free of cost up to 1974. The import and distribution management was given to public sector since 1979. Distributors (dealers) paid scant-attention towards the betterment of fauna, flora and environment.

The greedy producer after applying easily available pesticide in his vegetables field, brings to the market observing no waiting period of the poison. Thus the consumers swallow poison and suffer from unknown irrecoverable diseases. For favour of consumption, recently applied vegetables take enough time to get effacement position.

Socio-Economic Values

Until recently in human history, there was enough land and associated resources to meet the needs of rural populations around the world. But now the easily converted land and easily extractable resources are already being used for agriculture, and an increasing number of poor people lack the economic resources to exploit the available land for optimal crop growth.

The cycle of resource degradation, hunger, and poverty can best be illustrated by considering a rural family living in northern India. Conversion of forest lands to farming, and the needs of a growing population for wood for cooking and heating, have led to a serious decline in the availability of fuel wood. Animal manure and crop residues may be used instead of wood for fuel; but this eliminates them from the soil as fertilizer, with concomitant reductions in crop yields and increases in erosion. If the local market is used to buy fuel wood, the price may be high and this results in less money available to purchase agricultural inputs; once again, yields fall and the lands are not cared for.

Ultimately, the small farmer must make difficult decisions, and it is a *no-win situation*. In the struggle to survive, long-term considerations of land productivity are far from paramount. Many subsistence farmers in developing countries face a dilemma. Either they must eke out a living with dignity on marginal land, thereby contributing to deforestation and land degradation, or they must go to the rapidly growing cities where there are few jobs and where life itself is degrading for so many.

Economic and Political Issues

Two objectives have driven the agricultural policies of the developed countries:

1. To provide adequate food for the people
2. To provide adequate incomes for farmers

Left unstated but surely considered are such aspects as the preservation of the family farm, foreign trade balances, the incomes of ancillary industries that provide inputs, political voting blocks, and so on, and left almost un-thought are such things as nutrition and nutritional diseases, the distribution of food to all the population, and the environmental costs of agricultural production.

In the USA, the policy that has the greatest impact on farming is the Farm Law, a commodity programme under which the federal government guarantees the farmer a minimum price for crops produced on the farm. Two-thirds of all U.S. croplands are now enrolled in commodity programmes, with 80-95 per cent of the acreage planted to rice, maize, wheat, sorghum and cotton. Total payments are about $25 billion per year.

Farming in the United States has become a taxpayer-supported industry, where the government contracts for production, buys products, and indirectly pays workers. Because of the commodity programme, farmers cannot easily introduce crop rotation, because they have no *authorization* to grow the crops that might fit in a rotation schedule. The 1990 revision of the Farm Bill began to deal with this, and allows a supported farmer to plant up to 25 per cent of the acreage with other crops without losing acreage-based supports.

In Japan, the desire for local production of rice rests not only on tradition, but also on the historically strong tendency for self-sufficiency. With China as an unreliable trading partner and strong local political pressures to support rice farmers, the Government of Japan has the highest price supports of the industrialized world.

The subsidy programme also encourages the excessive use of fertilizers and pesticides, to attain high crop yields. If yields are high, the farmer safeguards future financial returns, even if it costs more to produce the crop than it is worth. According to the law of diminishing returns, it takes much more fertilizer and other inputs to raise production from 6 to 8 tons per hectare, than from 4 to 6 tons per hectare. Pushing production up to reach such, a high level may not be cost-effective but the taxpayers foot the bill. However, the large amounts of fertilizer and pesticide required reaching this production level put an unnecessarily heavy load on the environment.

The General Agreement on Tariffs and Trade (GATT) governs trade between many countries of this world. In the recent negotiations of the *Uruguay Round,* it is not surprising that a major stumbling block was agriculture. The United States put forward a bold proposal to eliminate tariffs and subsidies on agricultural products altogether. However, this was rejected by the European countries, for domestic political reasons. Nevertheless, these negotiations were the beginning of the process of removing government subsidies and allowing the free market to govern trade in agricultural products.

Although freeing the international agricultural markets would make the worldwide trade in these commodities more rational, it could have adverse effect on the developing countries. All over the developing world, except for China, a period of profound economic change is underway. Called structural adjustment, this is in its broadest sense a transition from socialism to free market economies. It is having, and will continue to have great implications for food production and consumption. Liberalization of economic policies in India has started yielding good results.

Environmental Accounting

As economics around the world undergoes these rapid changes, their impacts on the environment are not always considered. One reason is that the politicians, and the senior civil servants or the bureaucrats who advise them, are not aware of the real costs of soil erosion, depletion of nutrients, and pollution. Those who are aware of these factors do not know how to integrate them into a policy framework; or they may simply ignore them in responding to them for the moment.

Putting a value tag on an environment such as a piece of land or a body of water is difficult. An approach to doing this is to survey the resource (such as land) and quantify its possible economic uses (such as forest, crop growth, fallow etc.). Then some measure of ecological stress is developed (for example, nutrient depletion leads to a transition from crop growth to fallow) and an economic value is put on this change. This must include not only the reduced yields from the resource, but also the costs of cleanup (for pollution) and regulation. If this process is done for a region or a country, a monetary value can be put on the rate of resource depletion leading to lowered economic value.

Some countries, notably Norway and France, have attempted to make annual environmental balance sheets, reporting stocks of renewable resources (such as farmland), nonrenewable resources (such as minerals) and cyclical resources (such as air and water). These are then used as a guide to government policies in using these resources. These approaches require not only trained people to conduct the inventories of the natural environment and analyze them, but policy makers educated in these fields.

Food Security

When policy makers use the term *food security*, they are usually referring to grain stocks in excess of demand, left in storage against bad crop years. This has been a concern of planners, since agriculture began. But there is another aspect of food security, at the level of the individual household. True food security means that each person has a reliable, available, and affordable food supply. The United Nations sponsored World Commission on Environment and Development, set up to prepare for the U.N. conference on these subjects in Rio de Janeiro in 1992, used the term *equity* as a criterion for distributing the world's resources, including food, and defined the term in two ways.

1. Equity for people living now who do not have equal access to natural resources or to social and economic goods.
2. Equity for the human generations to come, whose interests are not represented by current economic analyses or by market forces that do not consider the future.

However, at the Rio de Janeiro conference, the third world countries demanded that in the name of equity, their people should share agricultural resources, such as genetic strains of crop plants, because much of the biological diversity is in the poorer countries. They could derive monetary benefit by ownership, although most countries agreed to this principle, the United States initially refused to sign the treaty, under pressure from some industries that felt that exploration would be made less economically viable.

SUSTAINABLE AGRICULTURE: NEED OF THE HOUR

There is an urgent need to develop farming techniques, which are sustainable from environmental, production, and socioeconomic points of view. The means to guarantee sufficient food production in the next decades and beyond is critical because modern agriculture production throughout the world does not appear to be sustainable in the long-term. The agricultural community is thus setting it hopes on *sustainable agriculture,* which will maintain the cycles of input-output and ecosystem balance. Definition form FAO (1988), seems appropriate in this contest.

> *Sustainable rural development is the management and conservation of the natural resource base, and the orientation of technological and institutional change in such a manner as to assure the attainment and continued satisfaction of human needs for the present and future generations. Such sustainable development, in the agriculture, forestry and fishers sectors, conserves land, water, plant and animal genetic resources, is environmentally non-degrading, technically appropriate, economically viable and socially acceptable.*

Evolution of Sustainable Agriculture

In the 1960s and 1970s, a growing environmental agriculture movement evolved in response to increasing soil erosion, pesticide use, and groundwater contamination. Simultaneously, economic conditions for farmers were becoming more stressful and the number of family farms declined.

In 1980s, Wes Jackson of The Land Institute in Salina, K. S., began using the term *sustainable agriculture* to describe an alternative system of agriculture based upon resource conservation and quality of rural life.

While sustainable agriculture has become the umbrella under which, many of the above-mentioned alternatives farming systems fall, it is important to note that sustainable agriculture is really a long-term goal, not a specific set of farming practices. In temperate zones sustainable agriculture was defined as such:

> *Sustainable agriculture is a philosophy based on human goals and on understanding the long-term impact of our activities on the environment and on other species. Use of this philosophy guides our application of prior experience and the latest scientific advances to create integrated, resource-conserving, equitable farming systems. These systems reduce environmental degradation, maintain agricultural productivity, promote economic viability in both the short and long term and maintain stable rural communities and quality of life.*

In this context, sustainable agriculture embraces all agricultural systems, striving to meet these criteria. Many aspects of modern conventional agriculture are included in sustainable agriculture, just as are many aspects of alternative farming systems.

One aspect of modern agriculture receiving a lot of attention in the sustainable agriculture discussion is the use of chemical inputs to supply fertility and pest control. White agriculture chemicals will continue to play an important role in American agriculture. Many farmers are looking at alternatives due to environmental, economical, or regulatory reasons. In a transition to farming systems more reliant on biological methods of production, low-input farming serves as in intermediary step

Sustainable agriculture emphasizes the conservation of its own resources. For a farm to be sustainable, it must produce adequate amounts of high-quality foods, be environmentally safe, and where appropriate, be profitable. Sustainable farms minimize their purchased inputs (fertilizers, energy and equipment) and rely, as much as possible on the renewable resources of the farm itself. This is especially important in the 90 per cent of farms that exist in the third world, where these inputs are often not available or affordable.

Agricultural ecosystems, unlike the natural ecosystems, are human manipulated ecosystems and regular and sometimes intense disturbances are a major part of economically viable agriculture management systems.

Sustainable agriculture is a complex issue associated with producing food while maintaining our biophysical resources including soil, water and air with no adverse impacts on the wider environment. It should:

1. Maintain or improve the production of 'clean' food
2. Maintain or improve the quality of landscapes, which includes soils, water, biota and aesthetics.
3. Have minimal impact on the wide environment.
4. Be economically viable
5. Be acceptable to society.

There are different concepts of sustainable agriculture, but none is generally accepted. It embraces several forms of non-conventional agriculture that are often called organic, alternative, ecological or low input. However, from economic and ecological perspectives, two basic criteria must be met if agriculture is to be sustainable in the long term. These are:

Sustainable farming uses some form of integrated pest management for pest control, and this can include the use of chemical pesticides that are not used by organic farmers. Thus, sustainable agriculture does not mean a return to the farming methods of the late 1800's. Rather, it combines traditional techniques that stress conservation with modern technologies, such as improved seed, modern equipment for low-tillage practices, integrated pest management that relies heavily on biological control principles, and weed control that depends on crop rotations. Sustainable farms try to use wind or solar energy instead of purchased energy, and use organic animal manure and nitrogen-fixing legumes as green manure to maintain soil fertility, as much as possible, thereby minimizing the need to purchase inputs from outside the farm. The use of genetically engineered crop strains is certainly not excluded by sustainable farming. The emphasis is on maintaining the environment, not on rules about what can or cannot be done. Profits from sustainable farms can exceed those of conventional farms.

Innovative farmers have developed many alternative farming methods and systems. These systems consist of a wide variety of integrated practices and methods suited to the specific needs, limitations, resource bases and economic conditions of different farms. To make wider adoption possible, however, farmers need to receive information and technical assistance in developing new management skills.

Sustainable Livelihood

Scientists, the world over, have developed new technologies for application in the field. Over the years they have come to realize that these technologies must be integrated into existing farming systems, and have been going out into the fields to try to understand the problems and to see how the solutions are working. The large-scale adoption of technology-based agriculture can lead to a loss of knowledge of traditional farming methods, which were developed over many centuries and are adapted to the physical, biological and social environments of the people. The knowledge may be useful in devising strategies for sustaining agriculture in the developing regions. Traditional agricultural methods may be applicable for designing sustainable systems in the developed countries. In other words, the rich and sophisticated can learn a lot from the poor and unsophisticated.

During the past one-decade, there have been important changes in the thinking of agricultural scientists and policy makers concerned with agricultural research in developing countries. It is clear that the type of research that led to the Green Revolution has benefited the resource-rich farmers, whereas resource-poor families have been left behind.

> *It is increasingly recognized that who produces food and where it is produced may actually be as or more important than how much is produced?*

If resource-rich farmers produce food that subsistence farmers cannot purchase, there is no improvement in the food situation. Population projections show that many rural areas with fragile ecologies will have to continue supporting large populations.

Food can be produced in a more sustainable way that protects the resource base. The techniques are known, available and used by some farmers. Although more researches are needed, however, these techniques could find more widespread use given the proper incentives. Currently, the incentives (price supports, tax policies and export subsidies) favour industrial agriculture. Society needs to replace them with an incentive that favours sustainable agriculture. Because sustainable agriculture uses fewer inputs, this may cause some dislocation in the industrial sector and will be strongly resisted. In the Third World, much research on indigenous farming methods is needed, and the top down approach to research needs to be replaced and/or supplemented by on-farm research.

According to Dr. M.S. Swaminathan, *the greening of agriculture requires the greening of both technology and public policy*. Producing more food and agricultural commodities from less land, water and energy is a task that will call for the integration of the best in modern technology, with the ecological strengths of traditional farming practices.

OTHER ECO-FRIENDLY FARMING SYSTEMS

There are several established approaches to ecofriendly farming systems. A common thread in all schools is an emphasis on biological systems to supply fertility and pest control rather than chemical inputs.

Organic Farming

The most widely recognized alternative farming system. Modern organic farming evolved as an alternative to chemical agriculture in the 1940s, largely in response to the publications of J.I. Rodale in the U.S., Lady Eve Balfour in England, and Sir Albert Howard in India.

In 1980, U.S.D.A. released a landmark report on organic farming. The report defined organic farming as:

> *Organic farming is a production system, which avoids or largely excludes the use of synthetically compounded fertilizers, pesticides, growth regulators, and livestock feed additives. To the maximum extent feasible, organic farming systems rely upon crop rotations, crop residues, animal manures, legumes, green manures, off-farm organic wastes, mechanical cultivation, mineral-bearing rocks, and aspects of biological pest control to maintain soil productivity and tilth, to supply plant nutrients, and to control insects, weeds, and other pests.*

Biological Farming

Biological farming has become synonymous with farmers using the Reams fertility system as the basis for crop production. *Eco-agriculture* is the term used to describe this system by the monthly Acres, U.S.A. The Reams system is based on the LaMotte-Morgan soil test and the use or rock phosphate, calcium carbonate, and compost to achieve nutrient ratios of 7:1 calcium to magnesium, 2:1 phosphorus to potassium, and so on. Biological farming allows the use of selected chemical fertilizers (avoiding disruptive materials such as anhydrous ammonia and potassium chloride) and adopts low-inputs approaches to use of herbicides and insecticides.

Diagnostic instruments to monitor plant and soil conditions are frequently used in biological farming. These include refractometer to monitor sugar content (Brix) in plant tissue sap; electrical conductivity meters to monitor ERGS (or energy released per gram of soil); ORPS meters (or oxygen reduction potential of soil); and radionics. Based on data gathered, foliar sprays containing biostimulants and soluble nutrients are applied. The Pandol Brothers, a large commercial fruit and vegetable operation in California, reduced their annual pesticide bill from US $500,000 to US $50,000 per year after adopting a Biological Fertility Program.

Nature Farming

Nature Farming was developed in Japan in the 1930s by Mokichi Okada, who later formed the Mokichi Okada Association (MOA). Nature Farming parallels organic farming in many ways but includes special emphasis on soil health through composts rather than organic fertilizer, when possible. Kyusei Nature Farming, a branch group, emphasizes use of microbial preparations in addition to traditional Nature Farming is most active in the

Pacific rim, California and Hawaii.

In addition to these methods-based approaches to sustainable farming, regenerative agriculture and permaculture are widely recognized. However, these latter systems, like sustainable agriculture, are more conceptually oriented than methods-based.

Regenerative Agriculture

In regenerative agriculture bunds on nature's own inherent capacity to cope with pests, enhance soil fertility, and increase productivity. It implies a continuing ability to re-create the resources that the system requires. In practice, regenerative agriculture uses low-input and organic farming systems as a framework to achieve these goals.

Permaculture

Permaculture is a contraction of *permanent agriculture.* The term was coined by Bill Mollison, an Australian forest ecologist, in 1978. Permaculture is concerned with designing ecological human habitats and food production systems, and follows specific guidelines and principles in the design of these systems. To the extent that permaculture is not a production system, per se, but rather a land use planning philosophy, it is not limited to a specific method of production. Thus, practically any site-specific ecological farming system is amenable to permaculture.

Some of the Eco-friendly farming systems are described here in detail:

Alternate Agriculture

Plant growth and crop production are complex processes that depend on many interactions between organisms. The trend of modern agriculture has been:

1. To substitute monoculture and continuous culture for crop rotation and diversified agriculture
2. To use genetically more uniform plants, that have a narrow genetic base
3. To use inorganic fertilizers, rather than the more difficult-to-use organic manures, in combination with green manures
4. To use herbicides and pesticides to combat weeds and pests, rather than more complex biological control mechanisms
5. To make bigger fields by eliminating all vegetation between them

Recently, there has been a resurgence of interest in alternative agriculture, organic farming and sustainable agriculture. In 1989, the Board on Agriculture of the National Research Council of the United States lent credence to this movement with the publication of a major study called *Alternative Agriculture*. Alternative agriculture is not a single system of organic, low input, regenerative, or sustainable farming. All these systems share an emphasis on management practices and on biological relationships between organisms.

Alternative agriculture recognizes that a piece of land on which crop plants are grown is first and foremost an ecosystem, and not a factory. An ecosystem has many interacting organisms that must remain in balance. Many natural processes occur in such an ecosystem and farmers should take advantage of these natural processes, rather than try to circumvent them or destroy them with chemicals. Alternative agriculture rejects certain practices (such as heavy use of inorganic fertilizers), but most important are the practices it favours:

1. Tillage that minimizes soil erosion even if it is more expensive
2. Reliance on animal manures and green manures with minimal input of inorganic fertilizers
3. Integrated pest management for pest control
4. Management systems, such as crop rotations, that help control weeds and disease organisms
5. These techniques are not new of course, although some of them have been refined in recent years.

Organic farming is defined as production system, which avoids or largely excludes the use of synthetically compounded fertilizers, pesticides, growth regulators, and livestock feed additives. To the maximum extent feasible, organic farming systems rely upon crop rotations, crop residues, animal manures, legumes, green manures, off-farm organic wastes, mechanical cultivation mineral-bearing rocks, and aspects of biological pest control to maintain soil productivity and to supply plant nutrients, and to control insects, weeds and other pests (USDA, 1980).

In my view the following definition of organic farming seems to be more appropriate as it considers the important aspects like sustainability of natural resources and environment. Organic farming is a production system which favours maximum use of organic material (crop residues, animal excreta, legumes, on and off farm organic wastes, growth regulators, bio-pesticides etc.) and discourages use of synthetically produced agro-inputs, for maintaining soil productivity and fertility and pest management under conditions of sustainable natural resources and healthy environment.

Ecological Agriculture

The growing concern about environmental degradation, dwindling natural resources and urgency to meet the food needs of the increasing population are compelling farm scientists and policy makers to seriously examine alternatives to chemical agriculture. A sustainable agriculture backed-up by *green technologies* in an integrated farming system has been considered a promising and potential pathway. The twin problems confronting agricultural production are all pervasive erosion of natural resources such as land, water and biodiversity, fast declining soil fertility and use efficiency of inputs such as water, fertilizer and energy. Demographic pressures accelerate the first and agronomic deficiencies the second.

The gravity of the environmental degradation, arising from faulty practices, has set several experts in the field to focus attention on ecologically sound, viable and sustainable farming systems. Though there has been a series of scientific and policy conferences on this issue, the sustainable farming methods are yet to get the approval of most agricultural scientists.

There are no serious attempts to develop a package of practices for large-scale adoption of sustainable agriculture in India. Only a few enterprising farmers and private institutions stand testimony to the utility of safe and highly productive natural farming methods.

The adverse effect of pesticides on the environment has been well documented, and their residues in the food chain have endangered the whole life sustaining systems in many regions. Chemical fertilizers have also jeopardized the environment through nitrate poisoning and exterminating the beneficial soil microflora and micro-fauna by adversely altering the chemical and physical structures of the soil.

Ecological Farming Systems

The shift from chemical to ecological agriculture should, however, be gradual. A sudden switchover can spell disaster and discourage farmers from taking to this course. At least two to three seasons will be needed for the transition, and during the interim years the farmers could build up a sufficient organic base to fertilize the fields and improve the productivity of the soil. Once a state of ecological balance is attained and there are a good number of beneficial organisms to check the explosion of pests and pathogens, crop yields of a high order can be ensured. The cost of cultivation can be brought down substantially as many farm-grown inputs can be integrated efficiently in the farming system.

Ecological farming can be sustainable and profitable too. It is a self-reliant method and has to be an integrated system. Since it is a knowledge-intensive practice, one has to keep pace with the dynamics of nature to increase the biological productivity of the soil.

Since eco-farming uses several farm grown inputs, and less dependent on market purchased inputs, it is economically attractive to the growers. Case studies show that when a chemical far incurred about Rs. 11,250 towards the cost of cultivation per hectare of rice, an organic farm spent Rs. 10,590. Alternatively, if the biofertilizers and organic nutrient supplements such as neem cake are subsidized to extent the same monetary advantage as chemical fertilizers, the cost of ecological farming will come down significantly.

It has been observed that in chemical farming, fertilizers and plant protection chemicals accounted for nearly 30 per cent of the total cost. The cost of market purchased inputs in ecological farming was far less, and bulk of it went towards the purchase of neem cake.

In both the cases, the yield remained almost the same at 5,625 kg paddy per hectare. The straw yield was eight tonnes. The net returns from the ecological farming system at the current cost of Rs. 3. 34 per kg paddy is Rs. 8,197. 50. The chemical farming, the net profit is Rs. 7,500. If one gets a premium price for the poison-free, organically grown rice, the economic returns from the ecological returns from the ecological farming system will be highly encouraging. While the modern system of farming expended nearly 31,000 MJ of energy, the ecological farming method accounted for only 23,400 MJ. High energy inputs such as fertilizer, plant protection chemicals, herbicides and high speed diesel are substituted with biological inputs and low-energy human and cattle power. The output/input ratio in the case of modern farming is 3. 76, while in the case of ecological farming it is more than 4. 95.

Objectives of Ecological Farming

Bringing to match the crop, soil and climate of a region the ecology and farming and gaining from the economy and efficiency of inputs are the objectives of eco-farming. According to Sankaram (1996), this technology may be grouped into four categories.

1. Those that reduce the environmental burden of greenhouse gases (CO_2 and CH_4), global warming, ozone depletion. All attention to promote renewable sources of energy (draught animal power), electrical energy from garbage disposal and biogas from organic wastes.
2. Those that reduce the demand on land, water and biodiversity without adverse effect on agricultural production and nutritive value of food. Nurse the soil back to health, change cropping patterns to maximize ecological productive efficiency, improve, water use efficiency- through conjunctive use of rain, tank, underground, well and river waters; reduce conveyance losses; phase out subsidies.
3. Those that continue to improve crop productivity, under shrinking land resources (genetic and agronomic) hybrid vigour, gene pyramiding; multiple cropping patterns, integrated nutrient management and integrated pest management.
4. Reduce hunger and poverty, adopt, cost effective farming to bring equity of food price and wage, encourage job promoted growth right in the village to arrest migration as ecological refugees.

For each specific location, eco-farming involves the search for the optimum rather than the maximum in cohesiveness and functional diversity. The fairly general nature of these properties, makes it clear that the designation *ecological* cannot be applied in an absolute sense nor can it be expressed in quantitative terms. It relates, rather, to a development process in which changes in the ecosystem must be continuously assessed to determine whether or not specific measures are of value. This means that farming techniques are not good or bad per se; their value depends on their applicability within a particular system; e.g. whereas minimum tillage may be recommendable in one case, ploughing may be necessary in another. There are scarcely any objective and generally valid criteria for assessing the extent to which a technique is environmentally appropriate for a given farm situation.

Prospects

In contrast, eco-farming systems and techniques have been developed on the basis of a holistic view of man within the biosphere and the awareness of man's dependence on scarce natural resources. In some cases, recognition may have been reached that the use of certain external production inputs is undesirable from the ecological point of view or is, under certain circumstances, even unprofitable. In other cases, as in some developing countries, the inputs may not be available on account of supply constraints where this level of consciousness of these economic conditions prevail. Intensification of farming is sought through more productive use of available resources such as soil nutrients, rain water and local energy, together with the knowledge, labour and initiative of the people.

The term *eco-farming* implies that farming regions and individual farms must be treated as ecological systems. Ecology is a science, which deals with the relationships between organisms and their environment. But in this context the environment is not confined to the natural conditions (e.g. soil, climate); it encompasses the entire complex of physical, economic, social and cultural conditions which affect the growth and development of an organism or organic system.

Eco-farming can be sub-stainable and profitable too. It is a self-reliant method and has to be an integrated system. Since it is a knowledge-intensive practice, one has to keep pace with the dynamics of nature to increase the biological productivity of the soil.

Since eco-farming uses several farm grown inputs, and is less dependent on market purchased inputs, it is economically attractive to the growers. When a chemical farm incurred about Rs. 11,250 towards the cost of cultivation per hectare of rice, an organic farm spent Rs. 10,590. Alternatively, if the biofertilizers and organic nutrient supplements such as *neem* cake are subsidized to extent the same monetary advantage as chemical fertilizers, the cost of ecological farming will come down significantly. It has been observed that in chemical farming, fertilizers and plant protection chemicals accounted for nearly 30 per cent of the total cost. The cost of market purchased inputs in ecological farming was far less.

In both the cases, the yield remained almost the same at 5,625 kg paddy per hectare. The straw yield was eight tonnes. The net returns from the ecological farming system at the current cost of Rs. 3.34 per kg paddy is Rs. 8,197.50. In chemical farming, the net profit is Rs. 7,500. If one gets a premium price for the poison-free, organically grown rice, the economic returns from the ecological returns from the ecological farming system will be highly encouraging.

While the modern system of farming expended nearly 31,000 MJ of energy, the ecological farming method accounted for only 23,400 MJ. High energy inputs such as fertilizer, plant protection chemicals, herbicides and high speed diesel are substituted with biological inputs and low-energy human and cattle power. The output/input ratio in the case of modern farming is 3.76, while in the case of ecological farming it is more than 4.95.

Ecological farming is more labour intensive, and several of its components such as vermicompositing, vermiwash, and generation of botanical pesticides and rearing beneficial insects such as *Trichogramma* parasites, ladybird beetles, and spiders for release in the field as bio-control agents offer scope for unemployed and landless youth.

Group and community endeavor holds the key for successful implementation of ecological farming over larger areas. Community participation in pest control and rodent eradication will be rewarding. Even in the case of developing biofertilizers, such as blue green algae and *Azolla*, group action will pay rich dividends.

Instead of giving the labeled *ecological* or *un-ecological approach*, it would be more correct to refer to the degree of appropriateness of a farming techniques for a specific site and time.

Integrated Intensive Farming System (IIFS)

The Integrated Intensive Farming Systems (IIFS) methodology provides the pathway to achieving an evergreen revolution in agriculture. According to Prof. M. S. Swaminathan, Chairman, M. S. Swaminathan Research Foundation (MSSRF), Madras, the IFS involves agricultural intensification, diversification and value-addition. It helps improve physical and economic access to food, there by fostering sustainable food security at the level of each individual in a house hold. Emphasis on food security at the individual level is important in view of the growing feminisation of poverty, according to him.

As the name indicates, IIFS involves intensive use of farm resources. To be ecologically sustainable, such intensification should be based on techniques which are knowledge-intensive rather than capital-intensive, and which replace, to the extent possible, market-purchased chemical inputs with farm-grown biological inputs.

Such a shift in the nature of inputs used is brought about through integrated farming involving animal husbandry, fishery and agroforestry. Such integrated farming practices which constitute the second element of the IIFS strategy provide scope for organic recycling. The third element of the strategy is value-addition to every part of the plant and animal biomass through the establishment of bio-refineries. On-farm and off-farm employment can then be linked in a symbiotic manner. In the case of farm-women, who are invariably over worked because of their multiple roles in a household, IIFS aims to reduce the number of hours of work and add economic value to each hour of their work, according to Dr. Swaminathan.

With an ever-expanding urban population, the demand for a wide range of processed and pre-cooked foods will expand, as urbanisation leads to diversification of foods habits. The IIFS with its emphasis on crop-live-stock-fish integration responds to this challenge. Also, the emphasis placed on post-harvest technology in IIFS helps both producers and consumers. To sum up, IIFS involves a pro-nature, pro-employment, pro-women and pro-poor orientation to technology and dissemination. It leads to labour diversification and not displacement. It leads to resource-based, in contrast to the widely prevalent commodity-based, agricultural development planning, according to him.

Dr. Swaminathan has prescribed *seven pillars of IFFS'* for successful implementation of such systems. Soil health care is fundamental to sustainable intensification. In addition, vermiculture constitutes an essential component of IIFS. IFFS farmers should maintain a *soul health card* to monitor the impact of farming systems on the physical, chemical and microbiological components of the soil.

Water harvesting and management is the next crucial component. Crop and pest management including integrated nutrient supply (INS) and integrated pest management (IPM) systems form important components as well. Energy management and post-harvest component, choice of crops, other components of farming system and information technology are also important. These will help to ensure economic viability, environmental sustainability, social and gender equity in IIFS villages.

It should be emphasized that IIFS will succeed only it is a human-centered rather than a mere technology-drive programmer. It essence is the symbiotic partnership between farming families and their natural resource endowments of land, water forests, flora, fauna and sunlight. Without appropriate public policy support in areas such as, land reform, security of tenure, rural infrastructure. Input and output pricing and marketing small farming will find it difficult to adopt IIFS according to him. The eco-technologies and public policy measures needed to make IIFS a mass movement should receive concurrent attention. The programme will fail if it is based solely on a technological quick-fix approach.

On the other hand, it can trigger an evergreen revolution, if mutually reinforcing packages of technology, training. Techo-infrastructure and trade are introduced, according to Dr. Swaminathan.

Ecological agriculture that effectively combines use efficiency of Input and economic yield maximisation will begin to prevent the abuse of natural resources. On its progress depends the achievement of food security and sustainability of farming systems. Ecological agriculture will reactivate the fatigued green revolution to move forward through promotion of soil health, package of agronomic practices with each component a blend of tradition and modernity, and a policy to value each input based on its intrinsic worth free from subsidies.

There is the need to recapture local knowledge about framing systems. A synthesis of traditional wisdom and ecological prudence of topical agriculture with modern advances in technology will help higher use efficiency of inputs and sustainable crop yield maximisation.

This symbolic of low input sustainable agriculture of (LISA) and sustainable agricultural research and education programme (SAREP) widely promoted in the U. S.

Low External Input Supply Agriculture (LEISA)

Low-Input Agriculture

The term low-input agriculture has been defined as a production activity that uses synthetic fertilizers or pesticides below rates commonly recommended by the Extension Service. It does not mean elimination of these materials. Yields are maintained through greater emphasis on cultural practices, IPM, and utilization of on-farm resources and management.

The LEISA concept seeks to optimize the use of locally available resources by maximizing the complementary and synergistic effects of different components of the farming systems. External inputs are used in a complementary way.

Although, the term *low-input farming* has often been used to describe any system of alternative agriculture, if can be seen that it is distinctly different from organic farming etc. Nevertheless, any system that reduces purchased chemical inputs can be called low-input farming, some examples are:

1. Poultry litter can replace nitrogen fertilizers in the production of watermelons.
2. Legume cover crops can supply the total nitrogen requirements of pecan trees.
3. Compost amended potting mixes produce superior vegetable transplants than traditional soilless mixes.
4. No-till vegetable systems are feasible using reduced herbicide rates to kill cover crops.
5. Subterranean clover living mulches supply nitrogen and weed control in peach orchards.

Integrated pest management is probably the oldest and most widely recognized Extension Service program devoted to low-input agriculture. However, only recently have the "non-chemical" approaches-such as cultural, mechanical, and biological-within the IPM framework been emphasized over the chemical component. Some programs, in fact, are now termed *biologically-intensive IPM.*

The intentions of the LEISA concept are obvious, but in practice the way it differs from integrated agriculture (understood to be a gradual minimization of external inputs) is not too clear. Defining what constitutes *low* inputs will always be difficult, therefore, LEISA remains an ambiguous concept.

Important characteristics of LEISA systems are that thy are based on a preventive approach whereby the problem is tackled at its roots, as opposed to the more symptom-curing nature of *modern agriculture*. Ecological and biological principles are the basis of the farm system. Nature works for the farmer, the farmer does not have to work against it. They are often based on local knowledge and production systems adapted to modern requirements and much less on external expertise. They are generally more labor-intensive compared to the mechanization-and petrochemical-intensive character of the agriculture envisioned by the Green Revolution and what it has become.

Criteria for LEISA

Ecological Criteria

1. Balanced use of nutrients and organic matter
2. Efficient use of water resources
3. Diversity of genetic resources
4. Efficient of genetic resources
5. Efficient use of energy sources

6. Minimal negative environmental effects
7. Minimal use of external inputs

Economic Criteria

1. Sustained farmer livelihood systems
2. Competitiveness
3. Efficient use of production factors
4. Low relative value of external inputs

Social Criteria

1. Wide-spread and equitable adoption potential, especially among small farmers
2. Reduced dependency on external institutions
3. Enhanced food security at the family and national level
4. Respecting and building on indigenous knowledge, beliefs and value systems
5. Contribution to employment generation

Biodynamic Agriculture

Biodynamic farming means biological dynamics. It is a method of organic agriculture, which considers farm as a living system and where one activity affects the other

Biodynamic farming evolved in Europe in the 1920s following lectures on agriculture by the Austrian anthroposophist Rudolf Steiner. Biodynamic farming parallels organic farming in many ways but places greater emphasis on the integration of animals to create a closed nutrient cycle, effect of crop planting dates in relation to the calendar, and awareness of spiritual forces in nature. A unique feature of this system is the use of eight specific preparation derived from cow manure, silica, and herbal extracts to treat compost piles, soils, and crops.

Demeter (tm) is a certification program for and feed produced by strictly biodynamic farming methods. The Community Supported Agriculture (CSA) marketing programs, gathering popularity as an innovative method of subscription farming, were largely introduced into the U.S. by the biodynamic movement. An article on soil quality and financial performance of biodynamic and conventional farms in New Zealand in the April 16, 1993 issue of Science. In a comparison of 16 adjacent farms, the biodynamic farms exhibited superior soil physical, biological and chemical properties and were just as financially viable as their counterparts.

Organic Farming vs. Biodynamic Farming

The biodynamic farming is more than just another organic method. It stands for a truly scientific way of producing humus. It does not involve in the application of organic matter in a more or less decomposed form but the use of completely digested from of crude organic matter known as stabilized and stable humus. In this aim, the method differs from organic farming. IN the case of biodynamic farming the organic material to be used as basis for compost is transformed either by means of the biodynamic compost preparations or by means of the biodynamic compost starter.

Principles of Biodynamic Farming

1. Biodynamic farming aims to restore humus status of the soil ecosystem to hold its fertility and productivity.
2. Also it helps to restore the soil for a balanced functioning of flora and fauna. Because soil is a living system wherein the microbes can be fully established and maintained.
3. The biodynamic farming does not deny the role and importance of mineral nutrients of the soil like nitrogen, phosphate, potash, calcium, magnesium etc., and it considers the skilful use of organic matter as the factor for soil life.
4. It involves skilful application of all the factors contributing to soil life and health. Because a plant grows under the influence of abiotic factors like temperature, oxygen, CO_2 light, water etc. And these energies are transformed in the plant system into chemically active energies by way of photosynthesis.
5. Biodynamic farming considers a plant as living entity which consists not only of mineral elements (like N, P, K, Ca, Mg, Mn, Cl, Fe, etc.,) but also of organic matter such as protein, carbohydrates, cellulose, and starch.
6. Biodynamic farming gives importance to enzymes and growth substances.
7. To restore the soil fertility, biodynamic farming gives priority to proper crop rotation Soil exhausting crops should be cultivated alternatively with fertility restoring crops.
8. It also provides recognition to green manuring and cover cropping.
9. To restore the soil environmental conditions, forests, wind protection and water regulation are important.
10. Biodynamic farming also gives importance to maintain the soil structure (physical characters like bulk density, pore space, water holding capacity and texture).

Rules for using Biodynamic Agriculture

1. Use of biodynamic sprays to stimulate and retain soil fertility
2. Following crop rotation to combat pests and disease and to increase the beneficial micro flora
3. Recycling on farm wastes
4. Following prevention strategies against pest and disease by strengthening plant and animals in the farm.

The produce of this farming is guaranteed by presence of international certification trademark Demeter this assures for a buyer that produce is biodynamically grown. The Demeter appeared first in Germany around 1928. It is certification trademark of biodynamic farming.

Few Biodynamic Preparations

When a living organism/entity, leaves behind the dead mineral of some form, that is subjected to laws of decay. The life force held the living creature from decay. To sustain and foster these forces biodynamic preparations were developed. The concept behind this is, these preparations/mediates influence the planets of the solar system and livings in the ethnic formative forces into the heap.

TABLE 1.2: Few biodynamic preparations

Preparation	Herb/material	Elements	Results
502	Yarrow Flower *Achillea millefolium*	Sulphur (S) Potassium (K)	Permits plants to attract trace elements in extremely dilute Quantities for their best nutrition.
503	Chamomile Flower	Calcium (Ca)	Stabilizes Nitrogen (N) within the compost and increase soil life so as to stimulate plant growth.
504	Stinging Nettle *Urtica dioeca*	Sulphur (S) Potassium (K) Calcium (Ca) Iron (Fe)	Stimulate soil health, by providing plants with the individual nutrition
505	Oak Bark	Calcium (Ca)	Provides healing forces (or qualities) to combat harmful plant diseases.
506	Dandelion Flower *Taraxacum officinale*	Silicon (Si) or Silicic Acid Potassium (K)	Stimulates relation between Si and K so that the Si can attract cosmic forces to the soil.
507	Valerian Flower *Valeriana officinalis*	Potassium (P)	Stimulates compost so that Phosphorous component is properly used by the soil

Research Achievements in Biodynamic Farming

It has been proved in experiments carried out in Germany, America and Sweden, that the biodynamic organic spray preparations have several positive effects with increase in yield in number of crops like carrot and beetroot. There is also increase in quality parameters like percentage of carbohydrate, protein, nitrate, soluble amino acids and vitamins (Gielen, 1973).

Samaras (1971) stated that the use of biodynamic organic preparations showed an improvement in the keeping quality of carrots. These sprays applied during the growth reduced the losses during the storage. The reduced the production of carbon dioxide, enzyme activity and the number of epiphytic bacteria. Alela (1973) stated that several research projects carried out in Germany and Sweden have shown that the use of biodynamic organic spray increased the yield of sugar beet by 8-14 per cent, stimulating the growth of leaves by 8-26 per cent. Spiegs (1979) stated significant increase in yield of wheat grains was achieved over four years in the bio-dynamic organic farming systems than the conventional systems.

An experiment conducted by the Swedish Biodynamic Institute together with the University of Uppsala, showed higher relative content of protein, high vitamin C content, less darkening and better preservation of taste in potato (Peeterson, 1977).

A field experiment conducted by the Biodynamic Institute in Jarna, Sweden, over a period of forty years showed that the biodynamically treated organic plots showed improvement in the organic content and bulk density of the sub soil. And also it encouraged the activities of earthworms, which in turn aerated the soil to a deeper level.

At Horticultural Research Station, research work on biodynamic farming has already been initiated in vegetable crops like potato, cabbage and carrot by using various biodynamic preparations. By biodynamic farming system, there is a reduction in cyst nematode population in potato crop. In cabbage crop under biodynamic cropping system, yield was two and half times more than the conventional method of cultivation. Encouraging results have been recorded in other crops also. Further research programmes are under progress.

Microbes are working round the clock as nutrient stabilizers, absorbers, retainers and transporters as well as producers of plant growth promoters and other biologicals. It is possible to capitalize on their natural capabilities through use of microbes within the existing cropping practices. By adopting suitable management practices, soil nitrogen can be manipulated to augment the contributions of nitrogen-fixing bacteria. The VAM is more effective under low phosphorus situations, nitrogen-fixers perform better under low soil nitrogen situations. Organic amendments can be used to augment the benefits of microbes. More systematic studies are required to exactly know the role of microbes in various cropping systems under different agro-climatic zones. Mass awareness on the conservation of organic sources, recycling of nutrients, bio-fertilizers and bio-pesticides need to be built.

Permaculture Farming

Meaning and Objectives of Permaculture Farming

A Permaculture Farming System is adapted to Ecological, Social and Economic Conditions

- adapted to ecological conditions means: farming methods, which reduce erosion and destruction of the soils;
- adapted to social conditions means: the farming methods are acceptable to the farmers and appreciated by the farmers;
- adapted to economic conditions means: the farmers can afford the new farming methods and can increase their income by using the new farming methods.

A Permaculture Farming System is Mechanized by the use of Draft Animals in order to:

- reduce labour burden,
- accelerate farm work,
- increase the farm area, production and income of the farmers

Permaculture Farming Methods should improve the Living Conditions of Farmers. Improvement of living conditions means:

- having less and easier work,
- having more time for themselves,
- having more and better food, and
- having more income for school fees, health care, etc.

The Elements of the Permaculture Farming System

The new permaculture farming system has a lot of elements, which are working hand in glove to improve farming.

1. Making Contour Bonds (big ridges) along contour lines prevents erosion;
2. Planting of Permanent and seasonal crops reinforces the contour bonds and produces a lot of food and income;
3. Small ridges prepared by oxen between the contour bonds prevent erosion and provide optimum growing conditions for seasonal crops;
4. Mixed cropping and crop rotation of seasonal crops conserves the soil, protects against diseases and guarantees good yields;
5. Planting of legumes improves the soil fertility for higher yields;
6. Draft oxen, used for farm work reduce the labour burden and accelerate farm work;
7. With the draft oxen, you can increase the size of your farm, your production and your income.
8. Keeping of draft oxen in the cowshed during the night protects the oxen and allows the production of cow dung;
9. Use of cow dung and green manure reduces the expense of mineral fertilizer, improves the soil and increases the yields;
10. Vegetable gardening improves the family diet and produces additional income by marketing of surplus vegetables.
11. Working one big farm with the family reduces the labour burden for the individual family members and guarantees continuation of farm activities when one family member is sick.

Organic Agriculture System

Practicing organic agriculture is managing the agro-ecosystem as an autonomous system, based on the primary production capacity of the soil under the local conditions.

Although there are some differences between these approaches, the common understanding is that practicing organic agriculture is managing the agro-ecosystem as an autonomous system, based on the primary production capacity of the soil under local climatic conditions. Agro-ecosystem management implies treating the system, on any scale, as a living organism supporting its own vital potential for biomass and animal production, along with biological mechanisms for mineral balancing, soil improvement, and pest control. Instead of the tendency towards mechanical thinking characteristic of industrial agriculture, agro-ecosystem management requires a nursing, cultivating attitude, a professional feeling for the organism Farmers, their families, and rural communities, are an integral part of this agro-ecosystem. Both sexes are involved on equal terms.

We thought it an advantage to present a drastic alternative to modern high-tech farming, one totally without synthetic chemicals and fertilizers (nitrogen!!). In this way the reader must think in a different way, and not just adapt to a trend of *using a little bit less*, which is what many so-called sustainable agriculture systems are based on. We advocate the development of consistent sustainable agriculture systems by starting from perhaps a drastic alternative, determine its impossibilities, and develop feasible systems from there.

A hypothetical *High External Input Farmer* (HEIF), who wants (has) to become more *environmentally friendly* will reduce inputs. This is what many would call *integrated agriculture*. There are different grades, usually starting with *Integrated Pest Management* (IPM), adding *Integrated Nutrient Management* (INM), and making a real change when the farm becomes an *Integrated Farm System* (IFS). Continuing the quest to become more environmentally friendly and more sustainable, the farmer will move forward to LEISA when optimizing the IFS. The farmer will end up with an Organic Agriculture System.

The Major Aims of Organic Farming

Many outsiders perceive organic agriculture as a craze, not a method, in which all agrochemicals are dogmatically banned because of private and environmental health concerns. Others believe that yields are much lower in organic agriculture, or that the cost of labor is much higher. In short, it is not an option to solve the world's agricultural problems.

In reality, organic agriculture is a consistent systems approach based on the perception that tomorrow's ecology is more important than today's economy. Its aim is to stop degradation and reestablish natural balances. The economy must readjust to the primary production factors, and not the other way around. Without ecology, there is no economy. If conventional agriculture had been made to pay for the degradation and environmental damage it is causing, the move towards organic farming system would have been made long ago.

Besides this other possible aims of organic agriculture are-

1. To work as much as possible within a closed system and draw upon local resources
2. To maintain the long term fertility of the soil

3. To avoid all forms of pollution caused by agricultural techniques
4. To provide a foodstuff of high nutritional quality in sufficient quantity
5. To reduce the use of fossil energy in agricultural practices to the minimum tending of zero
6. To give to all livestock the conditions of life that conform to their physiological needs
7. To make it possible for agricultural families to earn a living through their work and develop their potentialities as human beings
8. To maintain the rural environment and also preserve nonagricultural ecological habitats

Organic agriculture recognizes that everything effects everything else, and that one component cannot be taken out of the system, without positively or negatively affecting other things.

Concept of Organic Farming

Organic refers to the agricultural systems used to produce food and fiber. Organic farming systems do not use toxic chemical pesticides or fertilizers. Instead, they are based on the development of biological diversity and the maintenance and replenishment of soil fertility. Organic foods are minimally processed to maintain the integrity of the food without artificial ingredients, preservatives, or irradiation. Organic farming describes two major aspects of alternative agriculture:

1. The substitution of manures and other organic matter for inorganic fertilizers.
2. The use of biological pest control instead of chemical pest control.

Approaching organic agriculture from the vantage point of a scientist trained in conventional farming techniques, Harwood (1984) outlines the following assumptions:

1. Organic agriculture is a reversion to the agriculture of the 1940's.
2. Organic agriculture is merely conventional agriculture without the use of chemicals. This assumption often leads to the use of a no treatment check plot in a conventional experiment as an organic treatment.
3. Organic agriculture is relevant only to backyard gardens.

Organic agriculture is based on the assumption of a difference between chemicals produced by biological organisms and the same chemicals, which have been synthesized in an industrial process.

Difference between Organic and Conventional Farming

Organic and conventional agriculture belonged to two different paradigms. The fundamental difference between the two competing agricultural paradigms as follows:

Conventional	Organic
Centralization	Decentralization
Dependence	Independence
Competition	Community
Domination of nature	Harmony with nature
Specialization	Diversity
Exploitation	Restraint

In contrast, several agro-ecologically based researchers stress more the fluid transition between conventional, integrated and organic farming, as an outcome of different assessments of economic, ecological and social goals. Consequently, technical strategies such as integrated pest management of balanced nutrient supply might improve conventional agriculture to such an extent that it may appear unnecessary to strictly ban pesticides and mineral fertilizers as required by organic standards.

However, there is scientific evidence that organic agriculture differs from conventional agriculture not only gradually but fundamentally. Implementing organic methods consequently seems to provide a new quality in how the agro-ecosystem works. This functioning cannot be explained by summing up single ecological measures. Organic farming seems to improve soil fertility in a way and to an extent which cannot be achieved by conventional farming even if the later consistently respects some ecologically principles.

Organic agriculture is one of several to sustainable agriculture and many of the techniques used (e.g., inter-cropping, rotation of crops, double-digging, mulching, integration of crops and livestock) are practiced under various agricultural systems. What makes organic agriculture unique, as regulated under various laws and certification programmes, is that: (1) almost all synthetic inputs are prohibited. And (2) *soil building* crop rotations are mandated.

The basic rules of organic production are that natural inputs are approved and synthetic inputs are prohibited, but there are exceptions in both cases. Certain natural inputs determined by the various certification programmes to be harmful to human health or the environment are prohibited (e.g. arsenic). As well, certain synthetic inputs determined to be essential and consistent with organic farming philosophy, are allowed (e.g. insect pheromones). Lists of specific approved synthetic inputs and prohibited natural inputs are maintained by all certification programmes and such a list is under negotiation in Codex. Many certification programmes require additional environmental protection measures in adoption to these two requirements. While many farmers in the developing world do not use synthetic inputs, this alone is not sufficient to classify their operations as organic.

History of Organic Farming

Organic Agriculture Systems are not a repudiation of the assets of modern agricultural technology, neither are they defined by the simple elimination of synthetic fertilizers or pesticides. Methods in organic agriculture are less intensive in terms of synthetic and other external inputs compared with conventional farming methods, but are much more intensive from a biological point of view.

The term *organic farming* was first used by Lord North Bourne in the book, *Look to the Land* Lord North Bourne, who embraced the teachings of Rudolph Steiner and biodynamic farming, had a vision of the farm as a sustainable, ecologically stable, self-contained unit, biologically complete and balanced-a dynamic living organic whole. The term thus did not refer solely to the use of living materials (organic manures, etc) in agriculture although obviously it included them, but with its emphasis on *wholeness* is encompassed best by the definition' of, pertaining to, or characterized by systematic connection or coordination of part of the one whole.'

As defined by a USDA Study Team on Organic Farming, Organic farming is a production system which avoids or largely excludes the use of synthetically compounded fertilizers, pesticides, growth regulators, and livestock feed additives. To the maximum extent feasible, organic farming systems rely upon crop rotations, crop residues, animal manures, legumes, green manures, off-farm organic wastes, mechanical cultivation, mineral-bearing rocks, and aspects of biological pest control to maintain soil productivity and tilth, to supply plant nutrients, and to control insects, weeds and other pests.

> *According to the produced Codex definition (FAO), Organic agriculture is production management system which promotes and enhances agro-ecosystem health, including biodiversity, biological cycles, and soil biological activity. It emphasizes the use of management practices in preference to the use of off-farm inputs, taking into account that regional conditions require locally adapted systems. This is accomplished by using, where possible, agronomic, biological, and mechanical methods, as opposed to using synthetic materials, to fulfill any specific function within the system.*

Organic agriculture systems include approaches and methods like organic/UK), biodynamic/D Rusch-Muller/CH, Lemaire-Boucher/F, regenerative/USA, nature farming/JPN (Fukuoka 1985) and permaculture/AUS (Mollison 1990). These were all developed during this century, most of them the last 50 years.

In the Western world, organic agriculture is fairly well developed. There is a growing market for these products (including imports from developing countries) based on consumers' willingness to pay a substantial premium for reliably labeled, organically grown foods. Various types and methods of organic agriculture have been developed in the Northern Hemisphere, such as the biological-organic and biodynamic method

Needs of Organic Farming

Needs for Organic Inputs

Organic farming is the pathway that leads us to live in harmony with nature. Organic agriculture is the key to a sound development and a sustainable environment. It minimize environment environmental pollution and the use of non-renewable natural resources. It conserves soil fertility and soil erosion through implementation of appropriate conservation principles.

Several reasons have been emphasized for the need of organic agriculture, like limited land holdings, poor socio-economic conditions of farmers, rise in input cost etc. The broadest view shows two major reasons viz., population and environment, emphasized the ultimate need for eco-friendly technologies. FAO estimates that by the year 2000 the global population will be 25% higher than in the mid-1980s and that 90% of this population increase will be in the developing countries like India. It will be necessary to increase the supply of food and other agricultural products to meet their needs. At the moment, many of the methods being used to increase production are damaging to natural resources and the environment and farmers are supposed to invest heavily into inputs to improve yields and productivity. Further, the produce from organic farming has become inevitable. The food containing pesticides and other chemicals are increasingly made obvious by many research studies revealing the presence of pesticide residues in eatables. About one million people suffered from pesticides poisoning and 20,000 die every year due to the toxic effects of these chemicals worldwide but the crop loss due to pest is still 15,000 crores, apart from killing a portion of the human population by the same pesticide use. This perspective is also one of the needs for organic approach to farming. In order to avoid the deleterious effects of synthetic chemical fertilizers and pesticides, organic agriculture is needed as an alternative to provide ecologically safe methods of farming.

Although the concept of organic agriculture has received much elaboration at different levels, the description offered by has been found to be the most comprehensive covering all essential factors.

Organic agriculture is a production system, which avoids or largely excludes the use of inorganic fertilizers, pesticides, growth regulators and livestock feed-additives. To the maximum extent feasible, organic farming system rely on crop rotations, crop residues, animal manures, legumes, green manures, off farm organic wastes and aspects of biological pest control to maintain soil productivity and tilth, to supply plant nutrients and to control insects, weeds and other pests.

Sustainable agriculture, by definition, is that form of farming which produces sufficient food to meet the needs of present generation without eroding ecological assets and productivity of life supporting system of the future generation. The key characterization of organic farming in relation to sustainable soil fertility and organic farming include:

1. Protecting the long-term fertility of soil by maintaining organic matter levels, fostering soil biological activity and careful mechanical intervention.
2. Providing crop nutrients indirectly by using relatively insoluble nutrient sources which are made available to the plants by the action of soil microorganisms.
3. Nitrogen self-sufficiency through use of legumes and biological nitrogen

fixation as well as effective recycling of organic materials, including crop residues and livestock wastes.

4. Weed, disease and pest control relying primarily on crop rotations, natural predators, diversity, organic maturing resistant varieties and limited thermal, biological and chemical intervention.
5. The extensive management of livestock, paying full regard to their evolutionary adaptations, behavioural needs and animal welfare issues with respect to nutrition, housing health, breeding and rearing.
6. Careful attention to the impact of farming system on the wider environment and the conservation of wild life and natural habitats.

Organic agriculture is viable alternative to conventional agriculture. It protects the soil from erosion, strengthens natural resources base and sustains biological production at levels commensurate with the carrying capacity of managed agroecosystem because of reduced dependence of fertilizers and plant protection chemicals; problems of environmental pollution are greatly reduced if not totally avoided. The practice leads to regeneration of ecosystem. There is now a demand worldwide for organically grown foods, which command a premium in export markets. The demand for such safe foods is increasing annually and this opportunity needs to be exploited.

Organic agriculture is often associated with low yields. The demands for products of organic agriculture are growing in India. The European Union (EU) and the International Federation of Organic Agriculture Movement (IFOAM) have defined rules for organic agriculture producers. India too needs to establish its own certification scheme to facilitate exports of organic agriculture products and an accreditation agency to certify the produce of the farm. The essential elements of an organic quality assurance scheme are as follows:

1. Development of standards
2. Inspection and verification
3. Certification
4. Accreditation

As low yields are expected, the market's ability to pay a price premium is a key determinant of sustainable growth of organic agriculture. The European Union is currently world's leading market for organic products with various countries within the EU having deficits of fresh organic products. The legislation on organic agriculture varies across countries. At present certification is compulsory for the organic foods exported from India. To encourage mass production and consumption of organic products, there should be clear-cut policies and incentives to promote organic agriculture and necessary infrastructure and market.

National level efforts are needed to facilitate the successful adoption of organic agricultural by the farmers. The approach shall be farmer centered and programmes developed shall create conditions for the conservation and efficient use of locally available resources as inputs in agriculture. Research has also to play to the extent of developing appropriate agro-techniques in conformity with the basic principles of organic agriculture.

Increasing Demand of Organic Foods

The concept of organic farming has been getting very popular of late with many advocating a return to traditional methods of agriculture. The debate on the advantages and disadvantages of organic farming is now taking center stage in Europe and other parts of the world.

A recent opinion poll conducted in U.K revealed that almost one in two people believe that foods produced by traditional methods are more nutritious than conventional produce. Most people consider it better for the environment.

The organic food business is growing and is fast becoming a very lucrative one. German consumers spent nearly £ 1.2 billion in 1997 on organic food and the market is growing by nearly 10% a year.

Organically grown foods are being produced using techniques, pre-dating modern agro chemicals and intensive farming. The European Union regulates organic farming by setting standards, which prevent the use of artificial pesticides. Fertilizers and growth hormones are also banned.

The main reason for the popularity of organically produced food is that it has been grown on land certified as pesticide free for many years. However, residues from old chemicals may still turn up in organic food certified by the authorities. According to UK's Institute of Food Science and Technology, *Organic food can never be defined as pesticide free* but in comparison with produce derived from farms using conventional farming methods, it definitely contains lower pesticide levels.

A recent study by the U.S Food and Drug Administration had indicated that over 60% of foods have no detectible pesticide residues with less than 1% having residues exceeding health tolerance levels. However, a 1987 study conducted by the National Academy of Sciences found that the average Americans consume, 45 micro grams of potentially carcinogenic pesticide residues in their food on a daily basis, which suggests that there is some risk of cancer from the foods that we normally eat.

The exact risks associated with consumption of pesticide residues are still unclear. Long-term health problems cannot be ruled out, as the data available at present is not consistent. Bacteria pose a greater threat to human health, as food borne microorganisms are responsible for the death of thousands of people each year. This becomes a matter of concern as far as organic farming is concerned, as the fertilizers commonly used are animal dung, which is known to harbor lethal bacteria. There have been claims that conventionally grown crops are more likely to turn out produce contaminated by the potential killer *Escherichia coli* 0157:H7 but these claims are yet to be proved. Studies have shown the presence of other bacteria on organic produce.

Farmers using organic methods of farming must compost manure first if the risks associated with their produce are to be reduced. This however requires knowledge of critical times and temperatures needed to make composted manure safe microbiologically. The presence of pathogenic microorganisms known to produce toxins in food has been a source of concern in recent years. The major source of this form of contamination has been attributed to the manure added by farmers to these crops.

A study conducted in the US compared apples grown through both conventional and organic methods for their properties. The study concentrated on a single chemical: a mycotoxin named patulin, which occurs in varying concentrations in apples. Mycotoxins are a series of fungal toxins which when ingested cause a variety of reactions in human beings and animals. The conventionally grown apple yielded juice with a patulin concentration ranging from 244 micrograms per liter to 3993 micrograms per litre. On the other hand, the organic apple juice had patulin at rates up to 45,000 micrograms per litre. This study suggests that the use of artificial pesticides and concentration of patulin are inversely correlated.

Companies are misleading consumers. Some of them claim that organically grown food implies that the food is safe. Scientists agree that current farming practices need to change the environmental effects of high intensity farming increasingly haunt us, there is a great need to find new approaches.

Meanwhile the sales of organic produce have touched a record $2.8 billion per year. According to the Organic Market Overview Report, sales of organic produce have risen to a record $2.8 billion per year. The report claims that sales have increased by more than 20 percent per year for six years in a row. Farmers bringing their produce to the weekly farmer's market in Napa, California, say business has never been better. *It's definitely more business ever year* said one seller. *It's an under saturated market.*

CONCLUSION

The fast development of technology for increasing production without given due importance to the agro-ecosystem balance resulted in disturbed *natural cycles* of carbon, nutrient and food chain of flora and fauna. The results are visible in terms of decreasing yield and increasing unsustainability in agro-ecosystem. The need is to revive natural balance with sustainable farming concept. There are several approaches has been recommended by various scientist, farmers, and forums. Organic farming is one such approach which emphasis on maintaining the cycle of input–output with ecofriendly methods, is becoming widely recognised by researchers and farmers and gaining popularity in the international market.

CHAPTER 2

Towards Organic Farming

Change from conventional to organic farming is the change of philosophy and approach and it takes time. The Key Principle of the philosophy Organic farming system is the autonomous ecosystem management which is based on three strongly interrelated principles, which take place under: mixed farming, crop rotations, and organic cycle optimization.

MANAGEMENT OF AUTONOMOUS ECOSYSTEM

In all types of organic agriculture production understanding with production capacity of an agro-ecosystem is necessary. The process of extreme specialization propagated by the Green Revolution led to the destruction of mixed and diversified farming and ecological buffer systems. The disregard of biochemical and ecological feedback mechanisms of the agro-ecosystems resulted in, for example: a decline in soil organic matter content and soil activity, the elimination of pest-predator insects, cultural practices replaced by agrochemical inputs, and the deregulation of the animal autoimmune systems by prolonged use of antibiotics. These effects can be reversed by basing the farm system on diversification and optimum use of internal inputs. The function of this autonomous ecosystem management is to meet the need for food and fibers based on the local ecological carrying capacity.

Mixed Farming

For appropriate diversification in Organic Agriculture Systems, which ideally means mixed farming, or the integration of crop and livestock production on the farm. In this way, cyclic processes and interactions in the agro-ecosystem can be optimized, for example, using crop residue in animal husbandry and manure for crop production. This does not mean that some specialization is not possible; there is often a focus on a certain activity. This enables the farmer to acquire a high level of knowledge, take advantage of a market opportunity, and to concentrate investments. Mixed farming is not a dogma because it is often not feasible for every individual farm. In such situations, one should look for an appropriate scale in the relevant region or watershed to connect animal with plant production. Various types of cooperation offer a longer-term solution for building organic agroecosystems between several farmers, each with appropriate expertise and specialization.

Diversification of species, biotopes, and land use as a means to optimize the stability of the agro-ecosystem is another way to indicate the mixed farming concept. The synergistic interactions between plants, animals, soils, and the biosphere support this idea.

Plants

Plants (including shrubs and trees) are mainly valued as primary producers of photosynthetic carbohydrate and related protein production, as well as for nitrogen fixation (mainly by leguminous plants). Like the soils and animals, plants are multifunctional on the farm. They produce human food, herbs and medicines, non-food products (fuel, shelter, flowers, fibers), animal feed, the edaphon (mineral mobilization, structure), and the landscape. The main types of plants are gramineous food and feed grains, feed grasses, and food and feed legumes complemented by others such as fresh food plans. The main principle of plant growing in organic agriculture is that food and feed production is also soil improvement. This implies a structural, preventive approach to land degradation, based on improved soil life and structure, together with improved infiltration and retention of rainwater.

Animals

Domestic animals within these beneficial ecosystem interactions also have multiple functions (Baldwin 1980). They can produce protein (dairy, meat, hair, wool), feed on roughage (optimal use of biomass of which only part is fit for human consumption), produce manure (nutrient recycling and redistribution, stabilization of soil aggregates), and provide intelligent traction (low external energy input). From this point of view, ruminants are most essential. Non-ruminants can easily be a disadvantage when not fed on wastes.

A key issue in Organic Agriculture Systems is that domestic animals are linked with the soil, resulting in feed and manure production for the farm or region leading to increased food production for human consumption. Livestock health is dependent primarily on good husbandry, which means appropriate breeds, proper housing, feed, and hygiene. Breeding policy is not aimed exclusively at short-term maximum production, but at long-term production, including animal health and feed utilization. Similar to human medicine, a whole range of alternative methods are developed to control parasites and cure diseases.

Soils

Living soils (mineral-clay-humus complexes including a wide range of microbial, microfaunal, and microfloral species called the edaphon) are a source of and storage mechanism for nutrients for crops, a biotope for crop residue feeders, and a regulatory mechanism which makes nutrients available for plant growth.

Deep-rooting plants are appreciated for their capacity to bring mineral deposits from deeper soil layers into the nutrient cycle. Nodulation and N-fixation by legumes, N-balance contributions of N-fixing and denitrificating bacteria, P-solubilizing bacteria, and root symbiosis with mycorrhiza are all highly valued for their energy efficiency and crop-growth compatibility. Many problems of soil degradation and subsequent soil erosion, which are conceptually consistent with the physical-chemical plant nutrition model, are structurally prevented by the living soil concept. Issues like efficiency of resource use and energy, and quality of food further legitimate these basic considerations.

The main question is how a maximum bio-ecological effect can be triggered by minimal external inputs. However, where the subsoil cannot deliver the required nutrients by plant-initiated and edaphon-supported weathering, and when animal manure is not available, the application of rock dusts is widely accepted in organic agriculture, although locally available industrial organic by-products and wastes are preferred.

Calculating a mineral balance by estimating or measuring nutrient inputs and outflow is a technique in Organic Agriculture Systems to estimate the nutrient volume in the system and correct imbalances over a longer period of time.

Biosphere

The local biosphere provides the context within which, with minimal external inputs and maximal use of the ecological potential, optimal production conditions are created. This includes the appropriate tuning of the interactive synergy of crops and animals, rotations, and manuring to comply with local and seasonal conditions. Microclimatic improvements by means of tree and shrub planting, use of raised beds, moderate irrigation and drainage, and adaptations of varieties are also accepted strategies. Direct radiation from the sun on the soil surface is prevented by using soil covers and mulches, which also stimulates edaphon formation by creating the food and environmental conditions needed for development. Depending on the situation, tillage techniques are minimized.

Crop Rotation

Within the mixed farm setting, crop rotation takes place as the second principle of organic agriculture. This means cultivating a variety of different crops in a specific sequence of optimize their mutual interactions. Besides the classical rotation (one crop per field), intercropping, mixed cropping, and under sowings are other options to optimize interactions. Benefits from combinations of species and varieties should be considered.

Benefits of Crop Diversification

1. Increased yield stability
2. Reduction in disease incidence and severity
3. Reduced pest incidence
4. Increased arthropod diversity
5. Improved weed control
6. Reduced soil erosion
7. Recycling of nutrient reserves from depth in soil
8. Transfer of N from N fixing species

In addition to plant functions, important effects such as weed suppression, suppression of soil-borne insect pests and diseases, complementary in nutrient demand, nutrient catching, and soil covering can also be mentioned. The main point is to find synergistic interactions between crops in space and time, producing increased yields in terms of direct and indirect harvest (improved conditions for a next crop, N-fixation, maintenance of soil fertility). Crop rotation and multiple cropping systems have the structural advantage of spreading the risks of weather and pests. Sowing at staggered intervals, crop maintenance, and harvests reduce labor peaks. Evidence shows that most problems with insect pests, diseases, and weeds are prevented by good rotations. Advances in biological control have shown the potential for non-chemical pest management. In this system, pests do occur, but, generally to a degree that does not affect the economic feasibility. Breeding for resistance or tolerance, combined with rotation and use of organic fertilizers, would avoid most of the disease problems. In addition to a wide rotation, mechanical and manual weed control guarantee sufficient weed management.

Organic Cycle Optimization

Each field, farm, or region contains a given quantity of nutrients. Nutrients should be used in such a way that optimal use is made of this finite amount. This means that nutrients should be recycled and used various times in different forms. Second, it means that care should be taken that a minimum amount of nutrients leave the system, so that in effect, the "import" of nutrients can also be restricted. Third, the quantity of nutrients available to plants and animals can be increased within the system by activating the edaphon, resulting in increased weathering of mother material.

The larger the quantity of biomass produced, the faster the cycling and decay processes take place. A healthy, biologically active soil is the basis of this recycling process, requiring the replenishment of organic matter and nutrients removed by crop harvesting. A nutrient deficiency problem should therefore be solved by assessing and amending the physical and biological condition of the soil, the balance of the rotation, and the suitability of farming practices before "imported" fertilizers, whether mineral or organic, are used. The foremost principle in Organic Farming Systems is that the farmer continuously strives to improve the efficiency of the primary production system.

In Partnership with Nature

In Organic Agriculture Systems, the interactions described here can be used to develop and manage a holistic farming system which is an autonomous organic agro-ecosystem based on a partnership with nature. Such a system is not static; it should be adapted continuously to changing conditions using practical experience and new knowledge. There is no quick and easy recipe. Redirecting agriculture away from the unsustainable situation of the present requires more than just another set of technologies. It requires an attitude, which emphasizes human and ecosystem capacities, and cultivates them in a structurally sustainable way.

Mobilizing indigenous knowledge and inspiring farmers to grow their crops autonomously in the local agroecosystem places the farmers in the center of farming activities. This is a prerequisite for human resource development of farmers.

The principles mentioned above provide a basis for the development of agricultural methods which fit well in the concept of SARD.

BASIC STANDARDS AND GENERAL PRINCIPLES FOR ORGANIC AGRICULTURE

Organic farming system is based on the dynamic interaction between the soil, the plants, animals, humans, the ecosystem and the environment. The system is directed towards enhancing natural life cycles rather than suppressing nature. It relies largely on locally available resources. Organic farmers have proven to the world that their farming system is distinguishable from other agricultural systems, and above all, is competitive and able to provide agricultural products of good quality while minimizing negative side-effects.

Crop and Soil Management

Organic farming systems encourage the use of rotations and manures to maintain soil fertility. Carefully managed soils with a high proportion of humus offer essential advantages with respect to water retention, ion exchange, soil erosion, and animal life in the soil. A high proportion of humus in the soil gives uniform distribution of nutrients and also promotes plant hygiene.

Green manuring and intercropping of legumes is another important aspect for biological farming systems not only in regard to weed control, but also in reducing the leaching of nutrients and in reducing soil erosion. A green cover throughout most of the year is one of the main goals of such farming methods. Depending on the green manure mixture or the legumes used for under sowing, there may be an increase in soil organic matter and soil N as well as in other nutrients.

Choice of Crops and Varieties

1. All seeds and plant material should be certified organic.
2. Species and varieties cultivated should, as far as possible, be adapted to the soil and climatic conditions and be resistant to pests and diseases.
3. A moratorium on all transgenic food and agricultural applications should be imposed.
4. In the choice of varieties genetic diversity should be taken into consideration.

Crop Rotations

Rotations should be as varied as possible and aim to:

1. maintain soil fertility,

2. reduce nitrate leaching,
3. reduce weed, pest and disease problems.

It is recommended that specific rotations including legumes are insisted upon by the certification programmes.

Fertilization Policy

Increasing prices of chemical fertilizers have enabled organic wastes to regain an important role in the fertilizer practices on the farm. Good manure management means improved fertilizer value of manure and slurry and less nutrient losses. Composting of all organic waste in general, and of farm yard manure or feedlot manure in particular, is important in organic farming with the following guidelines:

1. Sufficient quantities of organic material should be returned to the soil to increase or at least maintain its humus content on a long term basis.
2. Accumulation of heavy metals and other pollutants should be limited.
3. Organic material produced on organic farms should form the basis of the manurial programme.
4. Management and handling of manure and compost should minimize nutrient losses.
5. Soil pH-values, which are appropriate to the soil type and the crops cultivated, should be maintained.
6. Use of human faeces is restricted. Consideration should be paid to the absence of pollutants. Steps should be taken to prevent the transmission of pests, parasites and other infectious agents e. g. by high temperature composting.
7. Non synthetic mineral fertilizers and brought in fertilizers of biological origin should be regarded as supplementary and not a replacement for nutrient recycling.
8. Minimizing the downward movement of nutrients in the soil profile:
   a) The rotation with deep-rooted crops,
   b) The avoidance of high-solubility nutrient sources,
   c) The avoidance of mould board ploughing in favour of chisel ploughing,
   d) The insertion of nutrients into the rotation onto a sod crop, if possible, to maximize uptake, and
   e) The seasonal use of cover crops in and around the major cash crops.

There are differences in nutrient profile with depth between organic and conventional systems, if the assumption is correct and if any of the above practices have a major effect in vertical nutrient movement.

Management of Pests, Diseases and Weeds

Organic farmers' primary strategy in controlling pests and diseases is the prevention. Organic farmers build healthy soils--fertilizing and building soil organic matter through the use of cover crops, compost, and biologically based soil amendments. This produces healthy plants, which are better able to resist disease and insect predation. Organic farmers also rely on a diverse population of soil organisms, insects, birds, and other organisms to keep pest problems in check. When pest populations get out of balance, growers will implement a variety of strategies such as the use of insect predators, mating disruption, traps, and barriers. As a last resort, botanical or other non-toxic pesticides may be applied under restricted conditions. Weeds are controlled through increased cultivation, as well as through cover crops, mulches, flame weeding, crop rotation and similar management methods.

1. Organic farming systems should be carried out in a way which ensures that losses from pests, diseases and weeds are minimized. Emphasis is placed on the use of varieties well-adapted to the environment, a balanced manurial programme, fertile soils of high biological activity, correct rotations, companion planting, green manures, etc.
2. The natural enemies of pests and diseases should be protected and, through proper habitat management, encouraged such as hedges, nesting sites, etc.
3. Weeds are controlled by a number of preventive cultural techniques limiting their development, e. g. suitable rotations, green manures, a balanced manurial programme, early seedbed preparations and predrilling, mulching and by mechanical control. Physical and thermic methods are permitted. Growth Regulations, etc.

Wild Products

The collection should positively contribute to the maintenance of natural areas.

Pollution Control

1. All relevant measures should be taken to minimize pesticide contamination, from outside and within the farm by wind drift, drainage and irrigation. This includes a clearly defined demarcation between conventionally and organically farmed fields.
2. In case of risk or reasonable suspicion of risk of pollution, the certification programme shall set limits for the maximum yearly addition of heavy metals and other pollutants.
3. Accumulation of heavy metals and other pollutants should be limited.

Soil and Water Conservation

The collection should positively contribute to the maintenance of natural areas.

Landscape

1. The certification programmes should, where necessary, set standards for a minimum percentage of the farm area to facilitate biodiversity.

These areas should be managed properly and be linked.

2. Areas which might be included in this percentage are:
   a) extensive grassland such as moorlands, reed land or dry land
   b) in general all areas which are not under rotation and are not heavily manured:
   c) extensive pastures, meadows, extensive grassland, extensive orchards, hedges, hedgerows, groups of trees and/or bushes and forest lines
   d) ecologically rich fallow land arable land (with no inputs),
   e) ecologically diversified (extensive) field margins,
   f) waterways, pools, springs, ditches, wetlands, swamps and other water rich areas which are not used for intensive agriculture or aqua production
   g) areas with ruderal flora.

PRINCIPLE REQUIREMENTS AND PRE-CONDITIONS

The Organic Agricultural Movement has adopted certain techniques that respect natural ecological balances to avoid products and methods which are contrary to the principle aims.

The basis for crop production in gardening, farming and forestry is the consideration for the structure and fertility of the soil and surrounding ecosystem, as well providing a diversity of species. This is achieved by a combination of:

1. A versatile crop rotation
2. The recycling of organic material, and includes
3. A wide range of methods for the control of pests, diseases and weeds which avoiding the use of synthetic fertilizers, pesticides and herbicides.
4. The basis for animal husbandry is the respect for physiological and ethological needs of the animals. This is achieved by a combination of:
5. Providing sufficient amounts of good quality organic fodder
6. Providing husbandry systems appropriate to behavioural needs
7. Proper veterinary treatment
8. Animals are an important part of organic farming system because:
9. They contribute to closing the nutrient cycles,
10. Animals convert organic matter and are thus major contributors to soil fertility
11. Some animal species can utilize agricultural areas which cannot otherwise be used
12. Growing forage crops improves the crop rotation, the diversification and balance of the farming system
13. They may be for draught purposes
14. They can utilize by-products from agricultural production
15. They contribute to higher yields

The basis for natural ecological balances in livestock production is a harmonious relationship between the crop production and animal husbandry. This is achieved by a combination of:

Establishing as much self-self-sufficiency of manure and animal feed on the farm unit as possible through specifying maximum stocking rates.

Converting plant protein and energy into animal protein and energy brings about losses during the metabolic process. For this reason crop, production for human nutrition and that for animal nutrition should generally be balanced. The basis of processing organic products is that its vital qualities are maintained throughout each step of the process. This is achieved by a combination of:

1. Choosing and developing methods, which are adequate to the specifics of the ingredients.
2. Developing standards, which emphasize careful processing methods, limited refining, energy saving technologies, minimal use of additives and processing aids etc.
3. The production and handling of organic products should seek to minimize the environmental degradation.

CONVERSION FROM CONVENTIONAL TO ORGANIC FARMING

The approach is to fit an organic system into any given local situation. It will thus be clear that organic farming systems cannot be introduced in one day. The period needed to develop an organic agriculture system is called the conversion period. Conversion Requirements

When traditional agricultural methods fulfill the principles of these standards no conversion period is required. When claiming virgin land for organic agriculture, no conversion period is required.

The whole farm, including livestock, should be converted according to the standards over a period of time. This time is defined by the certification programme.

If a farm is not converted all at once, it should be done on a field by field basis, whereby the full standards are followed from the start of conversion on the relevant fields. The area of landing being managed to the full standards will therefore progressively increase.

The composition of a farm unit can vary widely according to geographical conditions, ownership structure, time span, etc.

Farms with Plant Production and Livestock

The entire operation should be seen as a whole. It is not recommended to separate the conversion of individual fields or individual livestock productions, unless this is on the basis of imposed public restrictions or can be justified according to local conditions.

Farming with plant production/livestock and using remote areas (outlaying fields, alpine pastures, etc). When organic farmland and remote areas are operated under the same management, the remote areas in general should be included in the conversion plan as well. Areas which:

1. Requires disproportionately long transportation of farmyard manure,
2. Have very different climatic conditions,
3. In any other way have difficult access in relation to the other operations,
4. May, with restrictions, be operated conventionally, provided that the associated prohibited aids are not kept within the operation area for which certification is applied.
5. Operation of individual, separate farms. If a farmer of farming community operates two or more farms within a local area, all the farms should be converted according to the organic standards.
6. The time taken for conversion of the whole farm should not exceed one complete rotation.

Farmers will probably experience some loss in yields when converting their operation to organic production. There is period of time between the discarding of synthetic inputs and sufficient biological insect populations, nitrogen fixation from legumes) during which pest suppression and fertility problems are typical. The degree of yield loss varies, however, and depends on factors such as the inherent biological attributes of the farm, farmer expertise, and the extent to which synthetic inputs were used under the previous management system. Where soil fertility is low and biological processes have been seriously disrupted, it may take years to restore the ecosystem to the point where organic production is possible. In such cases other sustainable approaches, which allow judicious use of synthetic chemicals, may be more suitable start-up solutions. One strategy to survive the difficult transition period involves converting farms to organic production in partial installments so that the entire operation is not at risk. This is the period in which organic methods are introduced and the organic system is developed. Techniques have to be learned and labor reorganized. Experience shows that such a period takes from three to five years, depending on the situation. Often, a conversion plan is made, but the conversion can be a difficult time for farmers. Banks often deny credit to farmers who convert to a system where they might have lower yields. It might take a few years before pest predators are at full strength again. Farmers have to learn to observe more than they used to. After conversion, the system should be flexible enough to adapt itself from within to any changes in circumstances. From this point of view, it should be an autonomous system.

Limitations

Although this concept fully fits in the goals and criteria of organic agriculture systems are not proposed as a miracle solution for all the problems mentioned in Chapter 1. One should not set high hopes on a single concept, especially after experience with the Green Revolution and biotechnology, which were understood by many as miracle solutions.

These are assumptions throughout the organic literature of differences between organic and conventional systems with respect to their effects on soil physical properties, soil insect fauna, nutrient flow within the soil, crop health, and nutritional value of the harvested crop. One finds many principles commonly held by organic farmers such as feed the soil and not the crop, and the need to rotate with deep rooted crops to bring nutrients from deep in the soil profile. There are other perplexing observations such as the decline in yields during the process of conversion from conventional to organic practices. These assumptions, principles and observations cover virtually every aspect of crop and animal management in organic systems. Some are readily understandable, but many deal with physical and biological interactions which have not been researched and are little understood.

Initiating Organic Farming

Drylands are the potential place where organic farming can be started first because in drylands-

1. Less effect of high input agriculture thus least residue of pesticide and less time required for conversion
2. Organic manure improve the fertility and water retention capacity of poor soils of drylands
3. Economic conditions of drylands farmers are also comparatively poor and they are not able to purchase high cost input on the other side they can do the labour intensive operations and both are requirements of organic farming.

Medicinal Plants-the First Crops for Organic Farming

Organic farming in drylands can be started with medicinal plants because

1. The forest resource of medicinal plants is decreasing but demand is increasing thus cultivation is the only solution to fill this gap.
2. Medicinal plants are for the curing of disease and any residue of pesticide can convert it into poison. Hence medicinal plants should only be cultivated in organic farming
3. Use of high dose of inputs like fertilizers, irrigation etc may change the composition and quality of medicinal plants. Growing near to natural conditions is the best way to maintain the quality which is possible in the organic farming.

MANAGEMENT OF PERMACULTURE FARM

Although organic farming and permaculture both are the types of sustainable farming yet organic farming has certain man made standards and rules while permaculture has only one principle i.e. work in synergism with nature. Thus the farmers from the marginal areas or having less interest in export of certified produce to developed countries, but would like to make a self-sustainable production system with least monitory input, can adopt permaculture first and in future they can move to organic production when they will have funds for organic certification. This makes conversion convenient and a better understanding with nature.

Permaculture Farm

A Permaculture farm is a farm, where crops are grown every year continuously and forever without shifting in another farm.

Use of Draft Animal

The traditional way of farming with the hard-hoe is tedious and not very efficient. With the help of draft animals, you can mechanize nearly all farm works, like clearing, plowing, harrowing, planting, ridging, weeding, moulding and transportation. With the draft animals you:

- reduce your labour burden: *farm work becomes easier;*
- accelerate working operations; *farm work becomes faster,*
- can increase the size of your farm land: *production increases;*
- can produce manure and increase yields without buying expensive mineral fertilizer,
- can do contract labour to help other farmers and increase your income.

Making Permanent Farm

When the traditional farmer reduces his fallow period or starts to crop permanently on one farm, he will soon recognize, that the soil becomes less fertile and yields decrease.

In addition, bush fires or the farming method of bury and burn as well as the heavy rains, which carry away the soil, reduce the soil fertility.

To make a farm permanent, we therefore have to protect, conserve and improve our soil, to have a good yield every year.

Conservation of Soil

The soil is a living thing like man, animal and plant. *Conservation of soil means, keeping its power active to produce food.* Without adding soil food, it will have no power to produce and will "die". The most important enemies of the soil are fire and erosion.

Protection of the Soil against Fires

Destruction of soil by fire can be caused by the farming method of bury and burn of organic matter or by wild bush fire.

Protection form Water Erosion

Only running water can carry away soil. Running water can be found on sloping land. The steeper the slope, the more power the water develops and the more soil it washes away. The power of running water needs to be stopped to prevent erosion.

To stop running water on a slope, cultivation along contour lines and building dams or ridges along the contour lines is necessary.

Protection from Wind Erosion

Wind can attack the soil only when:

- the wind is powerful;
- the soil is bare; and
- the soil structure is damaged.
- To break the power of the wind, wind-breaks. need to be established

A well-established bund with permanent crops can reduce the power of the wind. In addition, wind-breaks can be established at the corners of farms with trees and hedges to break and wind. Trees and hedges can also serve as a living fence for farms, to protect them against destruction by cattle.

Practice mixed cropping with a high plant population and double cropping to have a good soil protection.

A soil with a poor structure is easily attacked by wind erosion. Therefore to protect and improve soil structure:

- avoiding bush fire;
- covering the soil with plants or mulch to protect it against sun;
- adding cowdung, plant residues (corn stalks etc.) and green manure to the soil; and reduce soil preparation (ridging instead of plowing and harrowing).

Improvement of the Soil

To allow permanent farming activities on a farm, with good yields, the need is to improve the soil. The use of manure, organic farming mixed cropping, crop rotation and use of mineral fertilizer will help to improve soil for steady and high yields.

For a draft animal farmer, dung is the most important manure. It is free of charge because, it is produced by his draft animals. When applied sufficiently on the farm, dung will improve the soil and increase the yields of the crops. The crops will produce more food for man and animals.\

How to Bury Organic Matter

To avoid loss and improve the decomposition of organic matter from plant drop, this need to mix with the soil. The oxen farmer uses the cart, the roller cutter and the ridger for this exercise.

With the cart, he transports the manure to the farm with the roller cutter, he can clear plant residues. With the ridger, he buries the organic matter into the soil.

The buried organic matter gives humus to the soil. And improves soil and the following crop will produce a lot of food.

Mixed Cropping

To conserve and improve our soil, practicing mixed cropping, is necessary mixed cropping is the planting of two or more crops on the same field at the same time.

To get all possible advantages, combine mixed cropping with crop rotation.

Permaculture for Wastelands

The rehabilitation of the wastelands is thought of as being a Herculean task, if not nearly impossible. Usually one thinks of such desolate places as naturally and inevitably inhospitable and hostile. The soil is either too dry or too wet, too stony or too steep, etc.

Many are the places where regeneration attempts have been made, more often than not with very little or results. Definitely, it is a huge labour, but when properly done, it can be of enormous benefit to the economy of the country. After many years of practical experience in this field, going often to the depths of despair, there are not wastelands, but lots of wasted land.

The techniques for regeneration are very simple. Though financial inputs are necessary, it can be done at a much lesser cost than at first imagined. The vast majority of the wastelands are community lands. Only if the communities concerned are involved in the regeneration process can there be a lasting result. This needs to be worked out so that benefits occurring from the regeneration will go to all parties concerned, land owner and landless alike and will thus not be a cause of discontent. These initial steps of care and protection alone can eventually change the appearance of community lands in a very positive way.

The actual method for reclaiming the wastelands includes three essential points:

- soil and water conservation
- the introduction of pioneer species
- the introduction of secondary species

Soil and Water Conservation

This consists of creating bunds with earth or stones to hold the rain water in place where it falls, preventing any further soil erosion, or to slow down the flow of water downhill. It is clear that such work starts always on the highest point of a given area or watershed, never at the bottom or anywhere in the middle of the slope. These bunds can be of different sizes and shapes. The can follow different patterns, such as contour bunding, crescent of half-moon bunding, netting etc. Instead of bunds, pits and trenches can also be made. The advantage of pitting for soil and water conservation is that those pits will collect organic materials like leaves brought by the wind and that they will gradually fill up with silt. Once that has happened one can successfully plant trees there.

A second step in the soil and water conservation measure is the creation of small water harvests. Here one takes advantages of the places where erosion has already carved out the landscape. By building earth or stone dams across a gully, for instance, one obtains immediately a small water reservoir. Care should be taken that an adequate overflow is provided for each such water harvester. If worked out correctly a whole network of small water harvesting structures can be laid out. From these ponds water percolates into the soil and the effects of this will be immediately visible. Remember that water is the driving force of all Nature. It is preferred to keep all works small and easy to maintain. Better to make two smaller ponds than a bigger one, unless one is very sure about one's skills.

Pioneers

A second major point in the regeneration process is the introduction of pioneer plants. These are mainly small trees and shrubs, but include also weeds and grasses. It should be possible to just sow seeds of these pioneer plants around bunds, in trenches, behind stone crescents, etc. Nature has provided a wide variety of plants. Whatever is the initial condition of the wasteland it should be possible to find a few species that will anyhow thrive there. It is only up to use to identify such suitable pioneer plants for our area.

Pioneer Trees and Plants

For the regeneration of the barren wastelands it is obviously necessary to select plants that can grow in harsh and very adverse conditions. Nature has provided for many kinds of plants, even those that can grow on the poorest soils and in the most hostile environments.

Such plants are called pioneers. These can be weeds, grasses, shrubs, creepers, and trees.

The criteria for identifying good pioneers

- ability to grow fast in very poor soils under adverse conditions
- ability to produce large amount of biomass in a short time
- easy to grow from seed or easy to plant without need for pitting or any other care
- drought or flood resistant
- resilient, capable of recovery after stress
- powerful deep penetrating root system
- nitrogen fixing
- ability to unlock or accumulate nutrients form the vast reserves of the sub soil
- unpalatable to livestock
- economically valuable, even if only minor

An ideal pioneer should have at least five of these characteristics. Furthermore, pioneers should be good colonizers. Their seeds should sprout in full sun, they should preferably be intolerant to shade and have a short life span. It is not necessary that pioneers have a major economical value.

The function of the pioneers is to cover the shade the sol as fast as possible. They will improve the soil by bringing about porosity and by adding organic matter. They will also fix nitrogen in the soil. Also, they will create a favorable environment needed for the secondary species to grow and thrive.

Once established, pioneers can be pruned and the pruning used to mulch the young saplings of secondary species.

Pioneer plants are very invasive and care should be taken that they don't become a nuisance in agriculture fields. It might be necessary to cut back the flowers to prevent seeding or even uproot whole plants at a later stage. Once their duty is done, pioneers retire and move on to other areas to work.

It might sometimes be necessary to raise such pioneer plants in the nursery and plant them out carefully. As they are early and prolific seeders such pioneer plants should come up over the whole area within a very short time.

For instance, for dry lateritic soil *Acacia holosericea* and for saline clays acacia ampliceps are excellent pioneer trees readily colonizing whole areas. *Stylosanthes* varieties are excellent soil binders and soil improves and can yield quality fodder.

When the pioneers have grown, providing wind protection and improved soil conditions, one can start introducing secondary species.

At first one can undertake trials with a wide variety of trees and plant species. This is to identify those ones which might be successfully introduced later on.

Secondary Species

A third step in the regeneration process is the planting of the secondary species. These will be for timber, fodder, minor fruits, non-edible oil seeds, construction materials, etc. The valuable and multi-purpose indigenous trees can offer us all these multiple benefits. The growth performance of the secondary species will be enhanced by using effective biofertilizers (*Azospirillum*, phosphobacteria, vermicompost, rock powders).

It is very important to go for a mixed plantation. No single species should cover more than 10 percent of the given area. Diversity is the golden rule of Nature. Along with the secondary species, fodder grasses and legumes can also be introduced. *Cenchrus* and *Stylosanthes* species and *Desmanthus virgatus* are excellent plants for this purpose.

Now the pioneer trees can be pruned for mulching the secondary species. In no way clear cut the area before planting. This would undo all the work that the pioneers have done. Mulching will protect the soil, conserve moisture and enhance soil microbial activity.

When these three simple rules are followed there should be no big problem to reclaim even the most degraded watersheds.

As an outcome of environmental improvement, wildlife will develop in great numbers (hares and partridges) providing a valuable source of protein. Also beekeeping or the collection and processing of medicinal herbs can further be valuable and remunerative activities.

Care and protection of the wastelands by the local communities can of itself regenerate the land if it is not too degraded. In other places the simple method described above can be followed. Plant trees and shrubs that the soil wants to grow. This might not be necessarily the ones that we think should grow. Those can be planted eventually much later on amidst the lush greenery of the regenerated "washed" land.

CONCLUSION

Among the sustainable production systems organic farming has one distinguish feature of certification. For certification certain rules and standards has to be followed although it need both time and money but ultimately get due recognition by the *Mother Nature* and in the market.

CHAPTER 3

Soils Environment and Plant Growth

Soil, in its traditional meaning, is the natural medium for the growth of land plants, whether or not it has developed discernible soil horizons, in this sense soil has a thickness that is determined by the depth of rooting of plants. Soil is defined as the unconsolidated mineral material on the immediate surface of the earth that serves as a natural medium for the growth of land plants. Soils are that portion of the earth's crust in which land plants can grow, if water and temperature are adequate, at least the minimum nutrients are available, and toxic substances are in low concentration.

Soils are often deposits of weathered, loosened, and transported particles. These materials are called parent materials. They are deposited by water, glaciers (ice), wind, or gravity into land areas called landforms. Many landforms are named for their method of transport and final shape: Alluvial is from flowing water; moraines are from glacier movement and melting; dunes are from wind movement of sands; and colluvium is from gravity's action. All of these various deposits, and others, become soils when exposed to weathering at the earth's surface. The soils, and their landform materials, vary enormously in the sizes of particles, slopes, uniformity of properties, and extensiveness across the landscape. The soils vary widely in quality and productivity.

Sand dunes do not ordinarily support plant life, but with stabilization (retarded mobility) and water, if needed, they could do so, and so can be considered soils. Organic deposits, such as are found as peat or muck bogs, are very different from the majority of soils because they sometimes contain almost no mineral materials, which are the major components of most soils; nevertheless, they are soils.

MINERAL VS ORGANIC SOILS

Because soils range from almost no organic matter to nearly 100 per cent organic substances, an arbitrary division of soils into mineral soils and organic soils is made. Organic soils have more than half of the upper 80 cm (32 in.) as organic material. Mineral soils are all other soils.

SOIL FACTORS FOR PLANT GROWTH

Soils are the medium in which crops grow and provide food and cloth to the mankind. The basic needs of crop production are to maintain soil fertility and soil productivity.

Soil Fertility

It is the inherent capacity of soil to provide essential chemical elements for plant growth. Soil fertility is vital to a productive soil. But a fertile soil is not necessarily a productive soil. These are so many factors e.g. poor drainage, insects, drought etc. that can limit production, even when fertility is adequate. Soil fertility in modern-day agriculture is a part of a dynamic system. Nutrients are constantly being 'exported' in the forms of plant and animal products. Unfortunately, others can be lost by leaching or erosion. Still others, like phosphorus (P) and potassium (K), can be tied up by certain soil clays. Organic matter and soil organisms immobilize, then release, nutrients throughout time. If production agriculture were a closed system, nutrient balance might be relatively stable.

Soil Productivity

Productivity emphasizes the capacity of soil to produce crops and it is expressed in terms of **yields**. To understand soil productivity, one must recognize existing soil-plant relationships. Certain external factors control plant growth: air, heat (temperature), light, mechanical support, nutrients and water. The plant depends on the soil (at least partly) for all these factors, except light. Each directly affects plant growth. Each is linked to the others. Since water and air occupy the pore spaces in the soil, factors that affect water relations necessarily influence soil air. In turn, moisture changes affect soil temperature. Nutrient availability is influenced by soil and water balance as well as by soil temperatures.

Essential Plant Nutrients

Sixteen chemical elements are known to be essential for plant growth. They are divided into two main groups: non-mineral and mineral.

Non-Mineral Nutrients are carbon (C), hydrogen (H), and oxygen (O). These nutrients are found in the atmosphere and water. They are used in photosynthesis. Products of photosynthesis account for most of the increased plant growth.

The 13 mineral nutrients, which are available in soil are divided into three groups: major, secondary and micronutrients:

Major Nutrients:	**Nitrogen (N), Phosphorus (P), Potassium (K). The major nutrients usually become deficient in the soil first, because plants use them relatively in large amounts.**
Secondary Nutrients:	Calcium (Ca), Magnesium (Mg), Sulfur (S).
Micronutrients:	Boron (B), Chloride (Cl), Copper (Cu), Iron (Fe), Manganese (Mn), Molybdenum (Mo), Zinc (Zn).

Five additional nutrients-sodium (Na), cobalt (Co), vanadium (V), nickel (Ni), and silicon (Si)-have also been established as essential nutrients in some plants. These five micronutrients are almost never deficient in soils. The secondary nutrients and Micronutrients are usually deficient less often and smaller amounts are used. But they are just as important as the major nutrients. Plants must have them when and where they need them.

Soil and Root Environment

The growing tips of the root wriggles through the growing medium following the line of least resistance. Roots are able to enter cracks, which are, or can be readily opened up to about 0.2 mm in diameter. Compacted soils severely restrict root exploration but once into these narrow channels the root is able to overcome great resistance to increase its diameter. Anything, which reduces root exploration and activity, can limit plant growth. When this happens action must be taken to remove the obstruction to root growth or to supply adequate air, water and nutrients through the restricted root volume.

In order to grow and take up water and nutrients the root must have an energy supply. A constant supply of energy is only possible so long as oxygen is brought to the site of uptake (see respiration). Consequently the soil spaces around the root must contain air as well as water. A lack of oxygen or a build-up of carbon dioxide will reduce the root's activity. Furthermore, in these conditions anaerobic bacteria will proliferate and many produce toxins such as ethylene. In warm summer conditions roots can be killed back after one or two days in waterlogged soils.

Soil Physical Properties

The physical properties of soils-texture, structure, density, porosity, water content, strength (consistency), temperature, and colour are dominant factors affecting the use of a soil. These properties determine the availability of oxygen in soil, the mobility of water into or through soils, and the ease of root penetration.

Soil Texture

Natural soils are comprised of soil particles of varying sizes. The soil-size groups, called soil separates, are sands (the coarsest), silts, and clays (the smallest). The relative proportions of soil separates in a particular soil determine its soil texture.

Texture is an important soil characteristic because it will, in part, determine water intake rates (infiltration), water storage in the soil, the ease of tilling the soil, the amount of aeration (vital to root growth), and will influence soil fertility. For instance, a coarse sandy soil is ease it till, has rapidly and easily loses plant nutrients, which are drained away in the rapidly lost water. High-clay soils (over 30 per cent clay) have very small particles that fit tight together, leaving little open pore space, which means there is little room for water to flow into the soil. This makes high-clay soils difficult to wet, difficult to drain, and difficult to till.

Soil Separate Sizes

The U.S. Department of Agriculture (USDA) has established limits of variation for the soil separates and has assigned a name to each size class (Table 3.1). This system has been approved by the Soil Society of America and is the one used in this book. Other particle-size classification systems are used in the United States and throughout the world.

Soil Textural Classes

Textural names are given to soils based on the relative proportions of each of the three soil separates are sand, silt, and clay. Soils that are preponderantly clay are called clay (textural class); those with high silt content are silt (textural class); those with a high sand percentage are sand (textural class). A soil that does not exhibit the dominant physical properties of any of these three groups (such as a soil with 40 percent sand, 40 percent silt, and 20 percent clay) is called loam. Note that loam does not contain equal percentages of sand, silt, and clay. It does, however, exhibit approximately equal properties of sand, silt, and clay.

The textural triangle is used to determine the soil textural name after the percentages of sand, silt, and clay are determined from a laboratory analysis. Since the soil's textural classification includes only mineral particles and those of less than 2 mm diameter, the sand plus silt plus clay percentages equal 100 per cent. (Organic matter is not included.).

TABLE 3.1: Soil Separates and Their Diameter Ranges

Soil Separate Name	Diameter Range (mm)	Visual Size Comparison of Maximum Size
Very coarse sand	2.0-1.0	House key thickness
Coarse sand	1.0-0.5	Small pinhead
Medium sand	0.5-0.25	Sugar or salt crystals
Fine sand	0.10-0.10	Thickness of book page
Very fine sand	0.10-0.05	Invisible to the eye
Silt	0.05-0.002	Visible under microscope
Clay	Less than 0.002	Most are not visible even with a microscope

TABLE 3.2: Textural classes of soil

Textural Class	% Sand	% Silt	% Clay
Clay (see center)	20	20	60
Silt loam	20	70	10
Sandy loam	65	25	10
Loam	40	40	20

Separation of sand into various sand sizes and the U.S. standard sieve numbers used for these separations. A 10-mesh sieve has 10 wire divisions per inch in the two directions making the screen.

Soil Structure

Soil texture is the relative proportion of sands, silt, and clay in a soil; how these particles are grouped together into stable collections or aggregates is soil structure. Natural aggregates are called peds. They vary in their water stability; the word clod is used for a coherent mass of soil broken into any shape by artificial means such as by tillage.

The formation and stability of soil aggregates is a very important phenomenon. Although there is still some uncertainty about the exact mechanism, which lead to formation of granules, several factors are responsible for their geneses. These include:

1. Soil tillage
2. Wetting and drying
3. Freezing and thawing
4. Activities of plant roots and soil fauna
5. Decomposition of organic matter
6. The modifying effects of adsorbed cations.

The stability of aggregates is of great physical importance. Three major factors influence aggregate stability- I) binding action of soil microorganisms, ii) cementing action of intermediate products of microbial synthesis and decay, such as gums and certain polysaccharides, and iii) cementing action of the more resistant stable humus components, which contributes to long term stability of the soil aggregates. Aggregate stability is not entirely an organic phenomenon as interaction does take place between organic and inorganic components. Polyvalent inorganic cations (e.g., Ca^{++}, Mg^{++}, Fe^{++}, and Al^{3+}) are thought to act as bridges between the organic matter and soil-clays, thus encouraging development of clay-organic matter complexes. Soils rich in kaolinite and hydrons oxide clays have aggregates with highest stability. The larger the aggregates in a soil, the lower is its stability.

Soil Structural Classes

Soil structural units (peds) are described by three characteristics: type (shape), class (size), and grade (strength of cohesion). Types of structure describe the ped shape with the terms angular blocky, subangular blocky, columnar, granular, platy, and prismatic.

Structureless

Soils have no noticeable peds. It might be an unconsolidated mass such as noncoherent sand and is called single grain, or it might be a cohesive mass, such as could occur in some loams or clayey soils, and is called massive. Soil structure influences the rate of infiltration of water in soil.

Soil stability-resistance to erosion by wind or water-is determined partly by the particle sizes in the soil (soil texture). More often, soil structure-the bonding of individual mineral particles into larger stable configurations called aggregates-resists erosion even further.

Natural or machinery-induced hardened soil layers (hardpans) are common to many soils. Such hardpans may inhibit root penetration and thereby reduce the effective soil volume from which plants can absorb nutrients and water. Subsoiling, a process of ripping the hardpan with tillage equipment, may be needed. Water may be trapped above a hardpan, causing a waterlogged layer of soil. Oxygen needed by the roots cannot pass through these water filled pores even for short distances of a few centimeters (an inch or so).

The magnitude of plant growth reflects a composite of many favorable and unfavorable factors. Favorable growth factors include adequate aeration, water, nutrients, adequate soil depth, and proper soil temperature. Unfavorable growth factors are many, among which are toxic levels of certain elements, diseases, harmful insects, adverse temperatures, and inadequate or excess sunlight. Any accurate estimate of the plant productivity of a soil must include all favorable and unfavorable factors.

Soil Porosity and Permeability

Pore spaces (also called voids) in a soil consists of that portion of the soil volume not occupied by solids, either mineral or organic. Pores in soil are the result of irregular shapes of primary particles and their aggregation; the forces of penetrating roots, worms, and insects; and of expanding gases entrapped by water. Under field conditions, pore spaces are occupied at all times by air and water.

Pores are officially described according to their average diameter in millimeters are follows:

Very fine .... *< 0.5*
Fine *0.5-2*
Medium *2-5*
Coarse .... *> 5*

Water drains by gravitational force from pores larger than about 30-60 microns (μm) (0.0012-0.0024 in.). In comparison, root hairs, the smallest plant roots, are between 8 and 12 μm (0.00032-0.00048 in.) in diameter. In soils where most pores are smaller than 30 µm (0.0012 in.), attraction forces in the soil retain the water within the fine pores, which results in a waterlogged soil and poor aeration. Thus, to the growing plant, pore sizes are of more importance than total pore space. The best balance of water retention (smaller pores) plus adequate air and water movement (larger pores) is in medium-textured soils such as loams. Aggregation, or the lack of it, can modify this balance of large and small pores, which results from the soil texture. Soluble bicarbonates, silicates, and iron during wet periods will move in the soil and precipitate when drying occurs. Over many decades and centuries, pores become filled with the precipitates and often are cemented into hard layers.

The relative amounts of air and water in the pore space fluctuate continuously. During a rain, water drives air from the pores, but as soon as soil water disappears by deep percolation (downward movement), evaporation, and transpiration (evaporation from plant leaf openings), air gradually replaces the water as it is lost from the pore spaces.

Soil Air

To survive, all living organisms require exchanges. The most important of these in soils are for respiration by plant roots and for organic matter decomposition by microorganisms. The most desired condition is well-aerated soil, a condition in which oxygen exchange between soil air and atmospheric air is rapid. The factors influencing the rate of gaseous exchange include soil pore sizes and continuity, temperature, depth in soil, wetting and drying of soil, and coverings (mulches) on the soil surface.

Composition of Soil Air

Atmospheric air has about the following composition of the gases important in soils:

Dinitrogen (N_2).. 79 %
Oxygen (O_2).. 20.9 %
Carbon dioxide (CO_2).. 0.03 %
Water vapor (relative humidity).. 20-90 %

The most important of these of plant growth is the oxygen content. Oxygen is required in respiration and organic material oxidation (decomposition). As roots and microbes use oxygen, they give off carbon dioxide into the soil air.

Rates of Oxygen Exchange: The rate at which in the soil exchanges with oxygen in the atmosphere is called the oxygen diffusion rate (ODR). Many large soil pores would speed air exchange (diffusion). Small pores, with bottleneck portions filled with water, would decrease exchange rates.

The Oxidation-Reduction Potential (Eh or Redox Potential): In most soils, plants grow best in oxidized soil (rice and marsh plants are exceptions). The major reason is the need for oxygen. Oxygen is the primary acceptor of electrons released in the oxidation of carbon in organic materials. If the oxygen concentration is too low, certain elements or irons become the electron acceptors and are, in turn, reduced. Mostly undesirable things happen when the ODR is too low and redox value drops: toxic organic acids form, nitrate is denitrified and nitrogen is lost to the atmosphere, available sulfate is reduced to sulfide, and most plant roots cannot respire. Thus, plant root growth slows or stops.

Aeration and Plant Growth: Reduced aeration causing deficient oxygen concentrations can be caused by various factors. Some of these are: (1) waterlogging (ponding); (2) soil compaction of loams and clays' (3) very high contents of clay that swell when wet, thereby closing large pores; and (4) organic matter decomposition by soil microorganisms that uses oxygen in soils already low in ODR.

Soil Strength

Soil strength is the degree of resistance of a soil mass to crushing or breaking when force is applied. A non-cemented soil mass is evaluated under two moisture conditions; air-dry and field moisture capacity. When a soil mass is cemented, soil strength is evaluated when air dry and again after it has soaked in water for 1 hour.

Soil Color

Soil color indicates many soil features. A change in soil color from the adjacent soils indicates a difference in the soil's mineral origin (parent material) or in soil development. White colors are common when salts or carbonate (lime) deposits exist in the soil.

Dark soil absorbs more than do light-colored ones. Some black coal mining wastes and dark-colored oil-shale residues reach temperatures of 65.6-70.0°C (150-158 °F), which are lethal to many plants that could otherwise grow in those soils. Although black soils having high humus content absorb more than light-colored soils, they also frequently hold more water. The water requires relative larger amounts of heat than the soil minerals to raise its temperature; water also requires considerable heat to evaporate it.

Within geographic regions, darker colors usually indicate higher organic matter contents. Soil color determination is standardized and determined by the comparison of the soil color to Munsell color charts. These color charts are similar to books of color chips found in paint stores where many gradations of different color groups are shown.

Soil Temperature

Soil temperature has important role in soil biological activities, nutrients recycling, water movement etc. major regulatory factors of soil temperature are soil texture, structure, depth, organic matter content and vegetation cover. Soil temperatures resulting from the sun's radiation change with depth and with the time of day. For example, the maximum daily temperature at deep soil depths is delayed, even by several hours, after the time when the air temperature reaches a maximum.

Soil Water

Water is required for all life. In soils, water is supplied to plants through the roots, it lubricates the soil allowing root penetration, it is necessary for microbial mobility and action, and it allows nutrient mobility. In a dry portion of soil, water uptake stops, nutrient absorption essentially ceases, and the root growth practically stops. A typical wetted loam soil will contain water in about 25 per cent of its volume, with only about half of that water being available to plants.

The properties of soil water are usually most favorable in the medium-textured soils (loams, sandy loams, silty clay loams). Water entry into soils (infiltration) is too fast in sands and too slow in clays for optimum crop production. Inadequate air exchange will occur in the small and water filled pores in clayey soils. Sandy soils hold low plant available amounts of water following wetting. Thus, sands will require frequent rains or irrigations to supply adequate water. Clayey soils will have slow drainage of excess water; sandy soils will be well drained, if drainage is possible.

Water Retention Forces: Soil serves as a water reservoir but a leaky one. When too much water is added, the excess runs off over the surface or into deeper layers. Water is held in soils because of the attraction between unlike charges- a positive ion attracted to a negatively charged ion. The positively charged hydrogens of water are attracted to nearby negatively charged ions, such as oxygen, even to the oxygen of another adjacent water molecule.

Water potential is defined as the work the water can do when it moves from its present state to a pool of water (at a specified elevation, temperature, soluble salt content, etc.). Because it refers to work, it is an energy term, not a pressure term. The soil water potential is a combination of the effects of the surface area of soil particles and small soil pores that adsorb water (matric potential), the effects of dissolved substances (solute or osmotic potential), and the atmospheric or gas pressure effects (pressure potential). In non-salty well-drained soil, the matric potential is almost equal to the water potential.

Soil Water Classification for Water Management: The most useful classifications of water are those relating water plants can use and to water which they cannot use. Terms include gravitational (drainage) water, plant-available water, field capacity, and permanent wilting point.

Ecological Classification of Soil Water: Plant-available water is the portion of stored soil water than can be absorbed fast enough by plant roots to sustain life. It is defined as the weight percentage of total soil water held with a water potential between -33 and -1500 kPa (for sands, -10 to -1500 kPa). The water held within these potentials makes up most of the storage water used by plants. (Gravitational water also is equally available as it flows past the plant roots if adequate is maintained. Since gravitational water is present only for short periods of time permeable soils, it cannot be depended upon in calculating plant water requirements).

The **wilting point** (sometimes called permanent wilting point percentage) is the percentage of soil water held with water potential less (more negative) than -1500 kPa; it is held so strongly that plants are not able to absorb it fast enough for their needs.

Field capacity is the percentage of soil water that is held with water potential less (more negative) than -33 kPa and is a measure of the greatest amount of water that a soil can hold, or store, under conditions of complete wetting followed by free drainage. Field capacity values are used to determine the amount of irrigation water needed and the amount of stored soil water available to plants. Plant-available water is equal to the difference of water percentage at field capacity and at permanent wilting point.

Capillary water moves through the soil pores because of a water potential gradient. Water potential gradient is the difference in total water potential between two locations in the soil.

The saturation percentage is the water content of soil when all pores are filled with water. It is approximately double the field capacity value. The saturation percentage is approximated by the water in a soil.

Soil Water Reservoirs: Water in soil is held as films on particle surfaces and in small pores. Large pores (as in sands and in between large aggregates) allow water to drain by gravitational flow. Small pores (clays) retain water by capillary force. Generally, the more clayey the soil and the higher the humus content, the larger the amount of water the soil can store not only does the quantity of water retained in clay soils remain large, but much of it is tightly held to the large clay surface area. This means that the clay soil will hold a large amount of water at the permanent wilting point percentage.

When Plants Most Need Water?: Plants are in need of water when they begin to wilt, but by the time wilting is visible, plant growth has already been reduced significantly. Wilting usually is due to insufficient soil moisture. Most plants have critical periods of growth during which a lack of water is most damaging. For plants producing seed, the period from flowering to seed set (fertilization) is the most critical. If drought occurs during this period, grain or seed yield will be greatly reduced.

Other Soil Physical Properties

Additional soil physical properties include plasticity, stickiness, smeariness, and fluidity.

Plasticity is the degree to which soil is permanently deformed, without rupture, by a force applied continuously in any direction. Plasticity is determined in the field by rolling wet soil between the hands until a 3-mm wire will break when the soil has dried to approximately the plastic limit.

Stickiness means the property of wet soil to adhere to another object.

Smeariness is a field term for clays that are thixotropic. Thixotropy is the property exhibited by certain fine-textured clay gels of being solid upon standing but turning liquid when shaken or mixed.

Fluidity refers to non-thixotropic clay soils that flow under hand pressure.

Tilth

The soil surface or seedbed should be carefully managed to produce the required tilth: the structure of the top 50 mm of the soil. Sandy soils are easily broken down to the right size with cultivation equipment. Heavier soils are less easy to cultivate and benefit from weathering to produce a frost mould.

The fineness of a seedbed should be related to the size of seeds but ideally consists of aggregate or crumbs between 0.5 and 5 mm in diameter. Cloddy surfaces lead to poor germination as well as poor results from soil herbicide treatments. Tilth is broken down by the rain on the soil surface. As soil crumbs break up, the particles fill in the gaps; this reduces infiltration rates. As the surface dries, a cap or pan so formed. Thus fine "dusty" tilths should be avoided and the soil crumbs should be stable be that they can withstand the effect of rain until the crop is established. This is particularly important on fine sandly and silty soils which tends to have poor structural stability.

Soil Colloids and Chemical Properties

The chemical properties of soils are determined by the colloidal portion of soils. A colloid is any solid substance whose particles are very small; thus, its surface properties are relatively more important than its mass (weight). Most colloids are smaller than a few micrometers (microns) in diameter. Because of their large surface-to-mass ratio, soil colloids settle slowly from suspension and have many reactions at their surfaces that determine soil properties. The predominant soil colloids are soil clays and humus.

Humus is a mixture of residues left after partial decay of organic substances in and on top of soils. Chemical properties of soils are more influenced by the clays and humus than by equal weights of the larger silt and sand particles.

The chemical properties of soils include mineral solubility, nutrient availability, soil reaction (pH), cation exchange, and buffering action.

Clays, the major mineral cation exchange particles, have negatively charged sites in their lattices and attract and hold positively charged ions (cations) at the clay surface. The quantity of these cations that can be held, or exchanged, by a given amount of soil is the cation exchange capacity of that soil. The exchangeable cations are not easily lost by leaching action until they are replaced (exchanged) by other cations. Plants often excrete hydrogen ions (H^+) to the soil solution, which can replace nutrient cations from their exchange sites, thus allowing the nutrients to move to the plant root and be absorbed.

If the adsorbed ions on the colloids are mostly $Al(OH)_2^+$ and some H^+, the soils are acidic. The pH is a measure of acidity or alkalinity: soil pH ranges from about pH 3.5 (extremely acidic) to pH 7 (neutral) to about pH 11 [extremely alkaline (or basic)]. Plants grow well between pH 5 and pH 8.5. Strongly acidic soils are undesirable because soluble aluminum and manganese can reach toxic levels and microbial activity is greatly reduced. Strongly alkaline (basic) soils have lower micronutrient availability, except for molybdenum; iron, zinc, manganese, and macronutrient phosphorus may be deficient. Truly, the colloids in soils are very important to the properties of soils. In fact, the colloids have dominant influences on the soil's physical and chemical properties.

Organic Colloids (Humus)

Humus is a temporary intermediate product left after considerable decomposition of plant and animal remain-temporary because the organic substances remaining continue to decompose slowly. They humus is often referred to as an organic colloid and consists of various chains and loops of linked carbon atoms.

Humus is amorphous, dark brown to black, nearly insoluble in water, but mostly soluble in dilute alkali (NaOH or KOH) solutions. It contains about 30 per cent each of the nitrogen-rich proteins, the slow-to-decompose lignins, and complex sugars (polyuronides). The polyuronides comprise much of the varied organic substances that cement soil aggregates together. Humus has about 50 per cent carbon, some oxygen, 5 per cent nitrogen, and lesser amounts of sulfur, phosphorus, and other elements. It also has a cation exchange capacity on a dry weight basis many times greater than that of clay colloids. On a weight basis, a few percent humus exerts much greater influence than does several times more percent clay, especially for properties such as the cation exchange capacity.

Cation Exchange

Plant growth is altered by modification of chemical properties, such as soil acidity, availability of nutrients, and toxicities. These concerns involve cation exchange, soil acidification, and nutrient balance. Any electric charged colloidal surface area will attract mobile substances of opposite charge in the soil water. The dominant residual charge on most soil colloids is negative. These negative charged sites attract positively charged ions in the soil water. Positively charged ions are called cations. Soil acts as a cation exchanger.

Cation exchange is an important reaction in soil fertility, in causing and correcting soil acidity and basicity, in changes altering soil physical properties, and as a mechanism in purifying or altering percolating waters. The plant nutrients calcium, magnesium, and potassium are supplied to plants in large measure from exchangeable forms. Cation exchange is very important in soils because of the following relationships:

1. The exchangeable K is a major source of plant K.
2. The exchangeable Mg is often a major source of plant Mg.
3. The amount of lime required to raise the pH of an acidic soil is greater as the CEC is greater.
4. Cation exchange sites hold Ca^{2+}, Mg^{2+}, K^+, Na^+, and NH_4 ions and slow their losses by leaching.
5. Cation exchange sites adsorb many metals (Cd^{2+}, Zn^{2+}, Ni^{2+}, Pb^{2+}) that might be present in waste waters. Adsorption removes them from the percolating water, thereby cleansing the water that drains into groundwaters or surface waters.

The amounts of cations in the soil solution are intimately related to the exchangeable ions. Any change in concentration of a cation in the solution (fertilizer or lime) forces a change in proportions of all exchangeable ions.

The cation exchange sites immobilize cations but also keep them exchangeable, and, thus, available to plant roots. Any water moving through the soil (exchanger) will lose many of its soluble cations to the soil (exchanger) and pick up those cations that have been replaced from exchange sites.

Cation Exchange Capacity (CEC): Cation exchange capacity (CEC) is the amount of exchangeable cations per unit weight of dry soil. It is measured in centimoles (+) of cations per kilogram of soil [cmol(+)kg]. The term centimoles (+) is used because the number of negative sites in a given soil sample does not change, but the weight of the cations that may be adsorbed to those sites at one time do change. One centimole (+) of cation X occupies the same number of cation exchange sites as one centimole (+) of cation Y (a different cation).

Anion Exchange and Adsorption: Colloids that have an appreciable anion exchange are those that also have low cation exchange. Multi charged positive ions, such as iron and aluminum, have hydroxyls (OH-) that can be exchanged with sulfate (SO_4^{2-}), phosphate ($H_2PO_4^-$ or HPO_4^{2-}), molybdate (MoO_4^{2-}), and some other anions. The highest anion exchange capacities (AEC) occur in amorphous silicate clays (volcanic ash soils) and in iron and aluminum hydrous oxide clays; lesser exchange occurs in kaolinite. Phosphates particularly can be held firmly on anion exchange sites.

The Soil Reaction (pH)

The term pH is from the French *pouvoir hydrogene* or hydrogen power. Soil reaction (pH) is an indication of the acidity or basicity of the soil and is measured in pH units. The scale ranges from 0 to 14 with pH 7 as the neutral point. At pH 7, hydrogen ion concentration (H+) equals the hydroxyl ion concentration (OH^-). From pH 7 to 0 the soil is increasingly more acidic; from pH 7 to 14 the soil is increasingly more alkaline (basic). The H^+ concentration, which is the substance measured when determining pH, has a tenfold change between each whole pH number. Thus, a soil of pH 5 has 100 times more H^+ in solution than a soil solution with a pH of 7.

Importances of Soil pH: The soil pH is easily determined and provides clues about other soil properties. The soil pH greatly affects the solubility of minerals. Strongly acidic soils (pH 4 5) usually have high and toxic concentrations of soluble aluminum and manganese. Azaleas, tea, rhododendrons, cranberries, pineapple, blueberries, and several conifer timber species tolerate a strong acidity and grow well. In contrast, alfalfa, beans, barley, and sugar beets only do well in slightly acidic to moderately basic soils because of a high calcium demand or inability to tolerate soluble aluminum. Most minerals are more soluble in acid soils than in neutral or slightly basic solutions.

The soil pH also influences plant growth by the pH effect on activity of beneficial microorganisms. Most nitrogen-fixing legume bacteria are not very active in strongly acidic soils. Bacteria that decompose soil organic matter and thus release nitrogen and other nutrients for plant use are also hindered by strong acidity. Fungi usually tolerate acidity better than do other microbes.

On mineral soils most agricultural crops do best in slightly acidic soils (pH 6.5); on organic soils, about pH 5.5.

Soils become acidic, even from basic parent materials, by the leaching away by rainwater of the basic cations (Ca^{2+}, Mg^{2+}, K^+, Na^+) and replacement of many of them by H^+ from carbonic acid (H_2CO_3) formed from water and dissolved carbon dioxide. Eventually, $Al(OH)_n^+$ replaces H^+.

Soil basically, although more difficult to alter than soil acidity, may be just as undesirable for plants. Non-leached soils or those high in calcium (low rainfall areas) may have pH values to 8.5 With increased exchangeable sodium, soils may reach values of over pH 10. Plants on soils of pH greater than about 9 usually have reduced growth, or even die. Some plants (halophytes) are tolerant of high salt or pH. Soluble $Al(OH)_4^-$ may exist at these higher pH values.

Most micronutrient problems caused by high soil pH are solved by adding special fertilizers such as water soluble chelates (key'-lates), which are "stable" but soluble complexers of the metal ions (although susceptible to microbial decomposition). If excess exchangeable sodium exists in the soil, it is usually removed by adding gypsum, followed by irrigation to leach the exchanged Na^+ ions.

Buffering in Soils: Most soils can resist appreciable pH changes when large amounts of a material either strongly acidic or basic are added, such as an acid-forming or base-forming fertilizer. This ability to resist a change in pH is the buffering capacity of the soil. The buffering capacity increases as the cation exchange capacity increases. To exhibit buffering, the soil must remove hydrogen ions (H^+) of added acids or neutralize the hydroxyls (OH^-) of added bases. The concept of buffering can be broadened to include a resistance to change in the concentration of any ion in the solution that is also adsorbed to the colloid.

Buffering action should be effective in controlling soluble (and sometimes toxic) concentrations of H^+, Al^{3+}, Ca^{2+}, Mg^{2+}, K^+, and Na^+. In all instances the larger the CEC of the soil, the greater should be the buffering capacity. Soils high in humans and/or clay, particularly montmorillonite or vermiculite clays, will have high buffering. Organic soils and clays have higher CEC values and are more strongly buffered than are sandy soils of the same geographic area.

Origin of Organic Soils

Organic soils are derived from plants growing in environments that decompose dead residues slowly. Usually, stagnant waters (swamps, marshes, moors, bogs) in temperate climates allow good growth of adapted plants, which die or shed leaves into the water. The decomposition of plant material in stagnant water lacks enough oxygen (is anaerobic) and so is very slow. Over centuries, such accumulations may reach depths of several meters. Examples of these are California's Stockton delta, Florida's everglades, and the many ponded waters in depressions in northern Michigan, Minnesota and Canada.

Common organic soil parent materials include many mosses (such as sphagnum), pondweed, cattails, sedges, reeds, grasses, and various "water-loving" deciduous and coniferous shrubs and trees. The term fibric means low extent of decomposition; more than two-thirds of the material can be recognized as to its plant part or source.

Soil Biology and Ecology

A small portion of the seemingly inert soil mass consists of live organisms-insects, rodents, worms, plants and their roots, and millions of microorganisms (microbes). The scientific study of these living organisms in and on soils is called soil biology. When interactions are included it is **soil ecology**. As detailed knowledge has grown, the classification of organisms has been modified. Previous classifications of all organisms into either animals (faunae) or plants (florae) is now expanded into new categories, all of them not always accepted by all scientists. In this book we use the system of five kingdom, which are Animalia (animals, worms, insects), Plantae (plants, algae), fungi (molds, yeasts, rusts), protista (protozoa, slime molds), and monera (bacteria, actinomycetes).

Some soil microbes and Animalia (macroorganisms) have little obvious effect on the soil one way or another. Some are absolutely essential to plant yields and by spreading animal infections and disease. Although the detrimental soil organisms are many, the beneficial ones exceed them in effect. Animalia aerate and aggregate soils. Some bacteria convert toxic carbon monoxide to carbon dioxide, fix inert atmospheric dinitrogen into useful forms, decompose organic materials, and reduce excess nitrates to relatively inert dinitrogen gas (N_2). Actinomycetes are a major source of antibiotics used for disease control and some also fix dinitrogen gas to usable forms. Even the lowly algae contribute to soil productivity by adding organic matter and fixing atmospheric dinitrogen. Microbes are major decomposers of organic materials; they continually work away at the resistant soil organic residues called humus.

Humus, the residual parts of one-living matter after initial rapid decomposition of organic materials, results primarily from the actions of bacteria, fungi, and actinomycetes and their enzymes. Fungi and bacteria are the most important organic matter decomposers. Both groups form symbiotic relationships in which two different organisms live in association, with both benefiting. Mycorrhizae, the symbiotic root-fungi association is found most dramatically on roots of trees but now is known to be possible on most plants. Mycorrhiza aid nutrient and water uptake. Rhizobia (N_2-fixing bacteria) form nitrogen-fixing nodules most commonly on legumes; similar microbe-plant relationships exist on many other plants, even on corn and other grasses. Certain bacteria called autotrophs get their energy from oxidizing important soil elements, including nitrogen, sulfur, iron, manganese, and carbon monoxide. Some algae fix atmospheric nitrogen and become part of the organic matter. Actinomycetes fix N_2 and help decomposition of organic materials. Protozoa are also decomposers and act as predators on bacteria.

Soil Formation

Soil formation comprises two different processes. First, the changes from a consolidated mass (rock) not capable of growing plants to the development of an unconsolidated (loose) layer of material that can support plants if climate is suitable and water is available; Second, the changes occuring within the loose material as time passes. This latter process is also called soil development. Actually, the change from a solid mass to loose soil material and other changes within the soil profile occur simultaneously. Soil formation is used to mean both the production of unconsolidated material by weathering processes and soil profile development, which are the changes involved in development of horizons.

Horizons are soil layers which differ from adjacent layers. They tell much about the characteristics of a soil. They include information about hardpans, depth of organic matter accumulation, soil denseness from clay deposition, and the extent of leaching.

Changes during Soil Formation

Soil formation-the processes of weathering and of change within a soil mass-is caused by many factors. Additions, losses, transformations, and translocations occur constantly.

Additions are made by water (rainfall, irrigation), nitrogen from bacterial fixation, energy as sunlight, sediment from wind and water, salts, organic residues, and fertilizers, and in other ways.

Losses result from chemicals soluble in soil water, eroded small-sized fractions, nutrients in grazed and harvested plants, water losses, carbon losses as carbon dioxide, and denitrification loss of N_2.

Transformations happen because of the many chemical and biological reactions that decompose organic matter, form insoluble materials from soluble substances, and alter or dissolve some minerals. As water and organisms move within the soil, many substances are translocated to different depths-for example, clays to deeper layers, and soluble salts into the groundwater or to the soil surface as water evaporates. Even localized sorting is active, such as forming deposits of carbonates or iron oxides in portions of the profile.

Physical Weathering

The processes of soil formation begin with the weathering of rocks and minerals. Physical abrasion of particles and chemical reactions constantly cause changes in the mineral and organic soil mass. Generally, the chemical processes are more dominant than physical processes in warm wet climates, but the centuries during which weathering is active allow physical weathering to exert even dominant effects, especially in cold arid zones.

Climate and Soil Formation

Climate is an increasingly dominant factor in soil formation with increased time, mainly because of the effects of precipitation and temperature. Some direct effects of climate on soil formation include the following:

1. A shallow accumulation or retention of lime (carbonates) in areas having low rainfall occurs because calcium bicarbonates (from dissolving carbon dioxide, minerals, and lime) are not leached if sufficient water is not present. Such soils are usually alkaline.
2. Acidic soils form in humid areas due to intense weathering and leaching out of basic cations (calcium, sodium, magnesium, potassium).
3. Erosion of soils on sloping lands constantly removes developing soil layers.
4. Deposition of soil materials downslope covers developing soils.
5. Weathering, leaching, and erosion are more intense and of longer duration in warm and humid regions, as in Hawaii, where the soil does not freeze. The reverse is true in cold climates, as in Alaska.

Moisture is important in soil formation. A soil is said to be developed when it has detectable layers (horizons), such as of accumulated clays, organic colloids, carbonates, or soluble salts that have been moved downward by water. The extent of colloid movement and the depth of deposition are determined partly by the amount and pattern of precipitation, which produce the leaching action.

Climate also influences soil formation indirectly through its action on vegetation. Semiarid climates encourage only scattered shrubs and grasses. 13 Arid climates supply only enough moisture for spars, short grasses or shrubs, which may not be dense enough to protect the soil against wind and water erosion. Many arid soils exhibit only slight organic matter accumulation and small amounts of profile development.

Soils resulting from year-around hot and humid weathering, such as occurs in some parts of Hawaii on old, and well-drained surfaces, are typically deep (often more than 3 m; 10 ft.), reddish in color (due to the presence of oxidized iron), with well-decomposed organic matter, and low in essential elements because of the leaching by the high rainfall.

Biota and Soil Formation

The activity of living plants and animals and the decomposition of their organic wastes and residues (the living environment, the biota) have marked influences on soil development. Differences in soils that have resulted primarily from differences in vegetation are especially noticeable in the transition where trees and grasses meet. Minnesota, Illinois, Missouri, Oklahoma, Texas, and most western states are places where such differences can be observed readily.

Some soils beneath humid forest vegetation may develop many horizons, are leached (washed, eluviated) in the surface layers, and have slowly decomposing organic matter layers on the surface. In contrast, some grassland soils near the transition zone of forests are rich in well-decomposed organic matter, frequently to depths of 30 cm or more (1 ft. or more) into the mineral soil.

Burrowing animals, such as moles, gophers, prairie dogs, earthworms, ants, and termites, are highly important in soil formation when they exist in large numbers. Soils that harbor many burrowing animals have fewer but deeper horizons because of the constant mixing within the profile, which nullifies the organic colloid and clay movements downward.

Microorganisms help soil development by slowly decomposing organic matter and forming weak acids that dissolve minerals faster than pure water. Some of the first plants to grow on weathering rocks are crustlike lichens, which are a beneficial (symbiotic) combination of algae and fungi.

Topography and Soil Formation

The earth's surface contour is called its topography (sometimes called relief). Topography influences soil formation primarily through its associated water and temperature relations. Soils within the same general climatic area developing from similar parent material and on steep hillsides typically have thin A and B horizons because less water moves down through the profile as a result of rapid surface runoff and because the surface erodes quite rapidly. Similar materials on gently sloping hillsides have more water passing vertically through them than do materials on steeper slopes. The profile on gentle slopes generally is deeper, the vegetation more luxuriant, and the organic matter level higher than in similar materials on steep topography.

Materials lying in landlocked depressions receive runoff waters from surrounding higher areas. Such conditions favor greater production of vegetation but slower decomposition of dead plant remains because of oxygen deficiency in waterlogged (saturated) soil; this results in soils with large amounts of organic accumulations. If the area above the soil surface is wet for many months of the year, organic (peat or muck) soils develop. If the accumulating waters dissolve salts from surrounding soils, the depression may become a salt marsh with unique tolerant plants growing, or it may develop toxic salt conditions with no plants at all. When soils on the watershed are strongly acid, iron may leach from them and be deposited in depressions to form bog iron (limonite). Alkaline soils on sloping topography in humid regions may result in lime being eroded into depressions and lead to the formation of marl.

Time and Soil Formation

The length of time required for a soil to develop the distinct layers called genetic horizons depends upon many interrelated factors of climate, nature of the parent material, the organisms, and topography. Horizons tend to develop most rapidly under warm, humid, forested conditions where there is adequate water to move colloids (subvisible-sized clays, humus, and other materials). Acid sandy loams lying on gently sloping topography appear to be the soils most conducive to rapid soil profile development.

Under ideal conditions, a recognizable soil profile may develop within 200 years; under less favorable circumstances, the time may be extended to several thousand years.

Soil development proceeds at a rate that is a combination of the effects of time plus the intensities of climate and biota (organisms), further modified by the effect of land relief (topography) on which the soil is situated and the kind of parent material from which it is developing.

Interactions of Soil-Forming Factors

The formation of soil is a diverse and complex process, with effects from the five major factors working in combination. For instance, soil with good drainage (topography or parent material) and mild temperature and high rainfall (climate) will probably support ample plant life (biota) because favorable drainage provides an aerated location for the plant roots. The plants, in turn, decompose, producing carbon dioxide (CO_2), which combines with water from rainfall to form carbonic acid (H^+ and HCO_3^-). Acidity increases the solubility of parent minerals; sodium, potassium, calcium, and magnesium are dissolved and leach away in the draining soil water, and the soil becomes ever more acidic. High rainfall also moves some clays and small organic matter particles deeper into the soil's B horizon.

The same climate (or clayey parent material) but different topography (a low-lying valley or depression instead of a gentle slope) might produce water-logged soil (which has poor drainage). Poor drainage results in stagnant water and non-aerated soil, often with resultant poor plant growth and decreased rates of organic matter decomposition. The accumulating drainage waters also mean the accumulation of dissolved salts. The horizons of these adjacent soils developed from common factors but with differing topography will become very different.

Soil Taxonomy

So far we have discussed soils as a product of continual natural change. Soil horizons are usually altered slowly, so in soil we see now few changes are expected during the next decades or even centuries. Thus, soils can be grouped (classified) into categories based on their present properties, and those classifications will remain unchanged for decades.

Taxonomy (derived from the Greek taxis, meaning *arrangement* or *order*) is the systematic grouping of similar things; soil taxonomy is the scientific grouping of similar soils. According to The U.S. Soil Classification System the most general soil category is called order. All world soils are placed into 11 orders; in the United States these are subdivided into 47 suborders, 185 great groups, 970 subgroups, 4500 families, and 12,620 soil series. As a brief orientation, simple descriptions of the 11 orders.

TABLE 3.3: Brief Characterizations of the 11 Soil Orders of the United States

Soil Orders	General Features
Entisols	Entisols have no profile development except perhaps a shallow marginal A. Many recent river floodplains, volcanic ash deposits, unconsolidated deposits with horizons eroded away, and sands are Entisols.
Inceptisols	These soils, especially in humid regions, have weak to moderate horizon development. Horizon development has been retarded because of cold climates, waterlogged soils, or lack of time for stronger development.
Andisols	A tentative soil order. Andisols are soils with over 60 per cent volcanic ejecta (ash, cinders, pumice, basalt) with bulk densities below 900 kg/m^3. They have enough weathering to produce dark A horizons and early-stage secondary minerals (allophane, imogolite, ferrihydrite clays) b. Andisols have high adsorption and immobilization of phosphorus and very high cation exchange capacities.
Histosols	Histosols are organic soils (peats and mucks) consisting of variable depths of accumulated plant remains in bogs, marshes, and swaps
Aridisols	Aridisols exist in dry climates and some have developed horizons of lime or gypsum accumulations, salty layers, and/or A and Bt horizons.
Mollisols	Mostly these are grassland, but with some broadleaf forest-covered soils with relatively deep dark A horizons; they may have B horizons and lime accumulation.
Vertisols	Vertisols have a high content of clays that swell when wetted. Vertisols require distinct wet and dry seasons to devolop because deep wide cracks when the soil is dry are a necessary feature. Usually, Vertisols have only deep self-mixed A horizons (top soil falls into cracks seasonally, gradually mixing the soil to the depth of cracking). These soils exist in most temperate to tropical climates with distinct wet and dry seasons.

Soil Orders	General Features
Alfisols	Alfisols develop in humid and subhumid climates, have precipitation of 500-1300 mm (about 20-50 in.), and are frequently under forest vegetation. Clay accumulation in a Bt horizon and available water much of the growing season are characteristic features. A thick E horizon is also common. They are slightly to moderately acid.
Spodosols	Spodosols are typically the sandy, leached soils of cool coniferous forests. Usually, O horizons, strongly acidic profiles, and well-leached E's are expected. The most characteristic feature is a Bh or Bs or accumulated organic matter plus iron and aluminum oxides.
Ultisols	Ultisols are strongly acid, extensively weathered soils of tropical and subtropical climates. A thick E and clay accumulation in a Bt are the most characteristic features.
Oxisols	Oxisols are excessively weathered; few original minerals are left unweathered. Often Oxisols are over 3 m (10 ft.) deep, have low fertility, have dominantly iron and aluminum oxide clays, and are acid. Oxisols develop only in tropical and subtropical climates.

The 11 soil order (which includes the tentative order called Andisols) are categorized by the nature of the developed pedogenic horizons they contain, by the extensive mineral weathering (Oxisols and to some extent Ultisols), by the importance of swelling clay contents (Vertisols), and by being organic soils (Histosols).

CONCLUSION

In the organic farming, soil is considered to be live and knowledge of origin (birth) and various factor affecting its behaviour and health, are necessary for holistic management of soil in sustainable organic production system.

CHAPTER 4

Soil Organic Matter and Humus

Since ancient times, farmers have recognized significant benefits of soil organic matter to crop productivity. These benefits have been the subject of controversy for centuries and some are still debated today. The following list includes many of the recognized benefits of soil organic matter.

1. It serves as a slow-release source of N, P, and S for plant nutrition and microbial growth.
2. It possesses considerable water-holding capacity, and thereby helps to maintain the water regime of the soil.
3. It acts as a buffer against changes in pH of the soil.
4. Its dark colour contributes to absorption of energy from the sun and heating of the soil.
5. It acts as *cement* for holding clay and silt particles together, thus contributing to the crumb structure of the soil, and to resistance against soil erosion.
6. It binds micronutrient metal ions in the soil that otherwise might be leached out of surface soils.
7. Organic constituents in the humic substances may act as plant-growth stimulants.

The term **soil organic matter** is generally used to represent the organic constituents in the soil, excluding undecayed plant and animal tissue, their partial decomposition products, and the soil biomass. Thus, this term includes:

1. Identifiable high-molecular-weight organic materials such as polysaccharides and proteins.
2. Simpler substances such as sugars, amino acids, and other small molecules, and (iii) humic substances.

Soil organic matter is frequently said to consist of humic substances and Non humic substances. **Non humic substances** are all those materials that can be placed in one of the categories of discrete compounds such as polysaccharides and sugars, proteins and amino acids, fats, simple organic acids, and so on; humic substances are the other, unidentifiable components.

The term **humic substance** refers to an operationally defined, heterogenous mixture of naturally occurring organic materials. These substances are ubiquitous in nature and arise from the decay of plant and animal residues in the environment. Humic substances are generally classified into humic acid, fulvic acid, and humin on the basis of their solubility in water as a function of pH.

INTERACTIONS BETWEEN SOIL BIOTA, SOIL ORGANIC MATTER, AND SOIL STRUCTURE

Physical protection mechanisms are important determinants of the stability of organic matter in soils. It is necessary to integrate the different concepts of physical protection of soil. This is partly based on the observation that the turnover rates of easily decomposable compounds are much higher in liquid microbial cultures than in soils and on electron microscopy studies that have demonstrated sites of accumulation of organic residues of clearly cellular origin. The phenomenon that drying of soil samples and disruption of soil aggregates can increase C and N mineralization is indirect evidence of the existence of physical protection in soil. It has been established that there is more physical protection in fine-textured soils than in coarse-textured soils, leading to higher carbon (C) contents and larger amounts of microbial biomass.

Several following mechanisms have been suggested to explain the physical protection of organic matter in soil against decomposition:

1. The adsorption to or coating by clay and silt particles is the main mechanism of physical protection of organic matter and microbes in coarse-textured soils. The clay silt fraction remains highly enriched with organic C and microbial biomass. In fine-textured soils this mechanism does not seem to be the main mechanism of physical protection.
2. The clay and silt fraction does not enriched with organic C and microbial biomass. Here, the location in small pores in aggregates seems to be an important mechanism of protection. Physical disruption of fine-textured soils released some of the protected organic compounds.
3. The positive correlation between clay content and total organic C and microbial biomass indicates that the capacity to physically protect microbes and organics is greater in fine-textured soils than in coarse-textured soils.
4. The entrapment of organics in small pores in or between aggregates, which renders them inaccessible to the decomposing microbial community.
5. The biomass of nematodes and bacteria are regulated by pore size distribution and clay content, leading to a higher biomass of nematodes in coarse-textured soils and a higher biomass of bacteria in fine-textured soils.
6. Soil structure may also control decomposition processes by its effect on the grazing intensity of the soil fauna on microbes. The soil fauna may stimulate microbial growth rates through grazing, and predation of microbes by protozoa and nematodes has been suggested as an important mechanism of nutrient turnover in soil. A large proportion of bacteria may occupy pores <3 μm, while protozoa and nematodes are restricted to larger pores. This means that a large part of the bacterial population will be physically separated from the protozoa and nematodes in soil.

7. Organic matter turnover and microbial activity have been studied in different soils. It is generally found that the turnover of organic matter and microbial activity is less in fine-textured soils than in coarse-textured soils.
8. The apparent higher grazing pressure on the bacteria in coarse textured soils in comparison with fine-textured soils, however, did not lead to a more active bacterial population.
9. Most of the bacteria were protected from being grazed in both coarse- and fine-textured soils.

To understand how structure can affect soil organic matter turnover it is important to know how primary soil particles and organic matter interact. Tisdall and Oades presented a soil structure model describing the association of organic matter with free primary particles (i.e., sand, silt, and clay), microaggregates, and macroaggregates. The basic structural units in soils are considered to be microaggregates (clay-polyvalent, metal-organic matter complexes, <250 μm in diameter) that are water stable and not affected by agricultural practices. These microaggregates are linked together to form macroaggregates, which are defined as being larger than 250 μm. Macro- and microaggregates contain primary particles, organics, and pores of different size.

Fungal and Bacterial Pathways of Organic Matter Decomposition and Nitrogen Mineralization in Arable Soils

The development of sustainable agricultural practices depends largely on promoting the long-term fertility and productivity of soils at economically viable levels. Efforts to achieve these goals have focused on:

1. Lowering fertilizer inputs in exchange for a higher dependence on biologically fixed and recycled nutrients.
2. Reducing pesticide uses while relying more on crop rotations and biocontrol agents.
3. Decreasing the frequency and intensity of tillage.
4. Increasing the return of crop residues and animal wastes to land.

The principal objectives of these approaches are to match the supply of soil nutrients with the fertility demands of the crops, to maintain acceptable pest tolerance levels, and to develop soil physical properties that optimize oxygen supply, water infiltration, and water-holding capacity at levels that minimize the losses of nutrients by leaching and gaseous export. Determining the suitability of these sustainable practices to a broad range of crops, soil types, and climatic regimes requires an understanding of their effects on the physical, chemical, and biological properties of soils.

The importance of soil biota as causal mechanisms for sustaining the fertility and productivity of soils has been the focus of several major programs on the ecology of arable farming systems. These include, but are not restricted to;

1. The long-term conventional (CT) and no-tillage (NT) trails (Horseshoe

Bend site) of the Georgia Agroecosystems Project in the United States.

2. The conventional and integrated farming trials (Lovinkhoeve site) of the Dutch Programme on soil Ecology of Arable Farming Systems.
3. The barley, grass ley, and lucerne ley trials (Kjettslinge site) of the Swedish project on the Ecology of Arable Lands.
4. The long-term stubble mulch and no-tillage trials (Akron site) at the Central Great Plains Research Station in the United States.
5. The cultivated barley trials (Ellerslie and Breton sites) of the University of Alberta, Canada.

In all the above programmes, much attention has been directed at understanding the contributions of fungal- and bacterial-based food webs to the accumulation and loss of soil organic matter (SOM) and to nutrient cycling. The importance of distinguishing these two primary pathway is based on the theses that:

1. Bacteria have lower C assimilation efficiencies and faster turnover rates than fungi, factors that are likely to increase rate of nutrient mineralization and organic matter decomposition in bacterially dominated soils.
2. The mycelia growth form is more conservative of energy and nutrients, enhancing organic matter storage and nutrient retention in fungal dominated soils.

TABLE 4.1: Estimates of N Flux (kg N ha^{-1} 10 cm^{-1} yr^{-1}). Through bacteria and fungi and the n passing through microbial consumers and predators in bacterial and fungal channels

Dept (cm)	Biomass N		Energy channel	
	CF	IF	CF	IF
Bacteria				
0-10	32	49	30	55
10-25	36	40	30	25
Fungi				
0-10	0.60	1.10	0.70	0.85
10-25	0.70	0.90	0.70	0.55

Source: Moore, J. C. and de Ruiter, P. C. 1991. *Agric*, Ecosystems *Environ.,* 34: 371-397.

When bacterial production is greater, bacterial-feeding fauna are expected to dominate. The most common bacterial-feeding fauna are protozoa and many non stylet-bearing nematodes, which require water films for locomotion and feeding. They are generally believed to increase organic matter loss and nutrient mineralization due to their relatively large biomass and high turnover rates and to their feeding on bacteria. The predominance of fungal- and bacterial-based food webs in NT and CT agroecosystems, respectively, could account for many of their observed differences in organic matter turnover and nutrient cycling.

Soil Structure and Soil Organic Matter Storage

Microbial Influence

Microbes (fungi and bacteria) indirectly influence the storage and release of nutrients through their effects on soil physical properties and the protection of **soil organic matter** (SOM). Mycelial fungi and bacteria contribute directly to the formation and stabilization of soil aggregates through their deposition of extracellular polysaccharides and hyphal entanglement. Where aggregates remain intact for relatively long periods of time (as in no-tillage soils), clay surfaces and micropores (<1 μm diameter) become occluded with the extra-cellular byproducts of microorganisms, restricting the access of microorganisms and extracellular enzymes to physically isolated SOM. As such, the formation and stabilization of soil aggregates represent potentially important mechanisms for the storage and protection of SOM.

Polysaccharides and humic substances are the dominant-binding agents produced by several species of bacteria and Streptomyces. In contest, humic and lignin-like compounds are found to be the principal binding agents produced by most species of fungi. The aggregates bound by filamentous structures (and/or their associated binding agents) are more resistant to physical disruption than those bound by bacterial polysaccharides. The stabilization of macroaggregates (>250 μm) is most often attributed to fungi, while that of microaggregates is usually associated with the production of adhesive metabolites by bacteria and the fungal polysaccharides are more stable to degradation than those derived from bacteria.

The fungal-based food webs often support large populations of microarthropods (especially mites and Collembola) and earthworms. In some soils, especially those of temperature regions, the feeding activity of litter-dwelling microarthropods may be responsible for significant accumulations of fecal pellets in the surface soil.

Earthworms also influence the stabilization of aggregates both directly, through the rearrangement of particles and the deposition of mucus, and indirectly, through the stimulation of microbial, especially fungal activity. With proper conditioning the fecal pellets and casts of these organisms form stable aggregates that contain their excretory byproducts, as well as fragmented plant and microbial debris. Once stabilized, these fecal aggregates can function as anaerobic microsites where conditions are suitable for high levels of denitrification. In bacterial-based food webs, the feeding of protozoa on bacteria in the micropores of aggregates may contribute to increased organic matter turnover and reduce aggregate stability.

The changes in soil structure that result from intensive cultivation are often associated with significant losses of SOM. The magnitude of these effects depends greatly on the intensity of cultivation and the quantity and quality of fertilizers and organic residues returned to the soil.

Minimum and no-tillage soils often support higher standing stocks of SOM and greater soil aggregation than conventionally tilled soils. The formation and stabilization of macroaggregates in no-tillage soils represent important mechanisms for the protection and maintenance of SOM that may otherwise be lost under conventionally tilled practices. This mechanism may be explained in nearly 18 per cent higher standing stock of soil organic matter in no-tillage (30.7 Mg C ha^{-1}) as compared conventionally tilled (26.1 Mg C ha^{-1}) soils at this site.

Various effects of conventionally tilled and no-tillage management on C and N dynamics may be attributed to differences in the composition of their decomposer communities.

The mycelial fungi appear to play a primary role in regulating the decomposition of surface-applied crop residues in no-tillage, while bacteria are more important to the decomposition of incorporated residues in conventionally tilled. Recent studies indicate that mycelial fungi also contribute more significantly to the formation and stabilization of soil aggregates in no-tillage than in conventionally tilled soils.

TABLE 4.2: Effects of Fungicide on Water-Stable Aggregate Distributions in Surface Soils (0-5 cm) of Conventional and No-Tillage Plots at HSB

% Water-Stable aggregates[a,b]					
Aggregate	**No-tillage**		**Conventional tillage**		**Tillage**
Size-class	Control	Fungicide	Control	Fungicide	(p <0.05)
>2,000 μm	65.5	42.3*	43.2*	38.7	**
250-2,000 μm	17.2	31.2*	20.1	23.8	
106-250 μm	5.9	10.0*	9.9	10.4	**
53-106 μm	2.5	5.8*	3.5	2.9	
<53 μm	8.9	10.6	23.4	24.2	**

[a]Values are normalized to a sand-free basis.
[b]Asterisks indicate significant (ANOVA/LSD; p <0.05) effects of the fungicide within tillage and aggregate size-class (*) and the main effects within size-class (**).

The above results are based on studies using the fungicide Captan to manipulate fungi in fields. In the later experiment, a 42% reduction in fungal hyphae in the fungicide-treated soils of no-tillage resulted in a 40% decline in the largest macroaggregates and redistribution of particles into smaller size classes. Despite similar reductions in hyphal densities, there were no significant effects of the fungicide on Water-Stable Aggregate (WSA) distributions in conventionally tilled.

Food Web in Soil

Understanding how agricultural management practices contribute to the sustained fertility and productivity of arable soils requires knowledge of both the structure and function of belowground food webs with particular attention to their spatial and temporal variances. When soils are disturbed by ploughing and the crop residues incorporated, there is much less spatial and temporal differentiation of food web structure and function. Under conventionally tilled practices, bacterial-based food webs play a relatively greater role in regulating SOM dynamics and nutrient cycling. Consistent with their higher rates of decay, incorporated residues retain larger populations of both primary and secondary decomposers than do surface applied residues, yet the population dynamics of microbivous fauna in conventionally tilled appear to be much less tightly coupled to that of the microbial population. In conventionally tilled, the soil microbial biomass also makes up a much higher proportion of the whole soil C than in no-tillage. Furthermore, a greater proportion of this biomass is bacterial in origin and concentrated in the plough layer where it is susceptible to the flushes of mineralization imposed by tillage events and wet/dry cycles. Overall, these properties of cultivated soils appear to contribute to greater organic matter losses and lower nutrient retention.

When cultivation is minimized and crop residues are retained on the soil surface, there is a much greater spatial and temporal differentiation of belowground food webs and the processes that they mediate as compared with conventionally cultivated soils. In no-tillage systems, for example, fungi-based food webs often develop on surface-applied crop residues. In this environment, mycelial fungi play a key role in the immobilization of mineral N; however, it is their interactions with fungal-feeding fauna that determine the amount and timing of N releases from decaying residues. The vertical stratification of belowground food webs is also much more pronounced under no-tillage, shifting from a fungal-based food web near the soil surface to a more bacterial-based food web deeper in the profile. Where fungi are concentrated near the mineral soil surface, they play an important role in the stabilization of aggregates formed by the casting activities of meso- and macrofauna (e.g., microarthropods, earthworms). This and other mechanisms of aggregate stabilization can significantly enhance the physical protection of organic matter, yielding greater SOM storage in no-tillage soils.

Determining the extent to which differences in the structure and function of soil food webs can be used to predict changes in the sustainability of agricultural management practices remains an important area for future research.

Formation Mechanisms of Complex Organic Structures in Soil Habitats

Soil contains natural reservoir of biotic debris. It consists of plant remains, dead animals, and microorganisms that are subject to continuous turnover. With time, dead biomass is either mineralized or transformed to yield diverse complex organic substances. Transformation proceeds in two stages (Hayes, 1991):

1. The **first stage** involves degradative processes that lead to the formation of substances. The more recalcitrant lignin components are selectively preserved (Kogel-Knabner, 1993).
2. In the **second stage**, the substrates and the preserved products are further transformed by synthetic processes to result in the formation of humus.

Formation of humus is a biochemical process. The degradation of biotic debris is catalyzed mainly by microbial enzymes. The subsequent synthesis of the humic material can be mediated by both enzymes and abiotic catalysts that are present in the soil. Soil enzymes originate mostly from microorganisms and to a lesser extent from plants, and clay minerals or metal oxides constitute abiotic catalysts. Poly-condensation reactions between reducing sugars and amino acids play a role in humus formation. One of the most important catalytic reactions in soil is oxidative coupling that links phenolic products of the degradative processes into humic polymers through enzymatic reaction and mineral catalysis. Oxidative coupling can also result in the binding of xenobiotics to humus. This reaction is usually accompanied by a detoxification effect (Bollag, 1983; Calderbank, 1989) and may be catalyzed by either enzymes or minerals. Organic chemicals can be bound to soil through various mechanisms involving either physical sorption or covalent binding. By definition, the adsorption process is reversible; in practice, however, xenobiotics are increasingly difficult to desorb as they become gradually sequestered in inaccessible microsites within the soil matrix. The compounds that are bound through stable covalent linkages are believed to the an integral part of humus, thus representing little or no threat to the environment.

Organisms Involved in Organic Matter Formation

Soil contains various living and non-living components, without this the organic matter would not exhibit such richness. Soil has following components:

Living organisms	less than 1% of the total volume of the soil
Minerals	about 50%
Organic matter	3-6%
Air the water	up to 50%

Trophic groups of living organisms in soil

1. Photoautotrophic higher plants, algae and photosynthetic bacteria
2. Photoheterotrophic bacteria
3. Chemoautotrophic bacteria
4. Chemoheterotrophic organisms

They include all soil animals, both vertebrates and invertebrates, most bacteria, and most fungi. These organisms are responsible for the biological and biochemical processes of humus formation. Chemoheterotrophs are the only organisms that can perform the two critical functions necessary to the humification process:

1. An active role of transforming raw materials into humus.
2. A passive one of serving as a source of biomass after death.

Chemoheterotrophs (microorganisms) play the most important active role in humification. Plants, however, are most important in passive processes because they serve as the major source of biomass. The tree other groups of organisms play only a passive role in the humification process.

Chemoheterotrophic Aminals

These include beetle larvae, centipedes, millipedes, and termites in tropical soils. They are primary decomposers and break up plant litter into smaller pieces.

Saprophytic Bacteria, Fungi and Protozoa

They are also important in the primary decomposition process because these organisms produce extracellular enzymes that catalyze the dissolution of outer protective tissues. Secondary decomposition is performed exclusively by microorganisms, which include most of the soil saprophytic bacteria and fungi. These microorganisms specialize in selectively degrading various large organic molecules such as lignin and cellulose.

The products of microbial degradation can also be linked into humic polymers by the action of the same microorganisms. The microorganisms involved in humification are those that synthesize the oxidoreductive enzymes required to mediate the processes of synthesis.

Oxidoreductive Enzymes

They are designated as either peroxidases or mono-phenol monooxygenases. Peroxidases are found in most of the higher plants and numerous bacteria and fungi include two distinct classes:

Classes of Mono-phenol monooxygenases	
Laccases	Laccase activity was first detected in the Japanese lacquer tree *Rhus vernicifera.* Microbial laccase were subsequently found in many fungi (e.g., Agaricales, *Aspergillus fumigatus, Rhizoctonia praticola, Trametes versicolor,* and others) and in actinomycetes like *Actinomyces galbus, and Streptomyces* spp.
Tyrosinase	Tyrosinase activity has been detected in fungi such as *Agaricales (Aspergillus nidulans)*, brown-rot fungi, *Psalliota arvense*, and *Russula nigricans* and in various actinomycetes.

Both peroxidase and laccase activity have also been detected in soil extracts, apparently as a result of the production of the extracellular oxidoreductases.

Synthetic Reactions in Soil

The microorganisms responsible for the humification process are ubiquitous, therefore, the synthetic reactions are common in soil. The extent of these reactions is limited only by proliferation of the oxidoreductase producers, and the ability of these microorganisms to proliferate appears to be practically limitless. Various studies have shown that essentially all microbial species can be found in most soils, differing only in numbers and proportions. Even in deep layers of soils (up to 500 m) where only anaerobic conditions are expected to prevail, the composition of the microflora is essentially the same as that observed on the surface. Also, the transformation of organic compounds in both types of environments differs little except in the extent of the reactions.

DEGRADATIVE PROCESSES

Humus formation is a slow decomposition process. It undergoes a constant slow at a rate of about 2-5% a year. Though this process is slow, even than the content of humus in soil remains fairly constant over considerable periods of time due to its systematic replenishment. Plant materials decompose fairly and rapidly, with half-lives varying from days to months. Only a small portion of the total plant input, however, is converted to humic substances. During the first year, about 70% of the plant material is degraded to CO_2, and about 5-10% forms new biomass. Thus, 75-80% of plant substances undergo an initial rapid phase of plant litter decomposition. The material remaining is decomposed more slowly, and only the products of this second phase are incorporated into soil humus.

Types of Plant Material for Decomposition

Alexander (1977) divided these substances into six broad categories:

TABLE 4.3: Plant materials, which decompose in soil. The type of material with their availability

Plant material	Availability in %
Cellulose	15-60% of the dry weight
Hemicelluloses	10-30%
Lignins	5-30%
Water-soluble sugars, amino acids, and aliphatic acids	5-30%
Ether and methanol-soluble fats, oils, waxes, and pigments	
Nitrogen- and sulfur-containing proteins	

The decomposition in the soil may be the *preferential decomposition* (decomposition of easily mineralizable materials) or the *selective preservation* (decomposition of refractory plant components).

Plant components (cellulose, hemicelluloses, and proteins) are easily mineralized while lignins and lipids are refractory to decomposition. During the period of rapid decomposition, the former are quickly mineralized to CO, by microorganisms as a source of carbon and energy. Nitrogen contained in proteins is mineralized to ammonium and nitrates.

The Pathway of Lignin Decomposition

Lignin degradation, however, is critical for humification to occur because the process generates phenols and phenolic acids, which are fundamental constituents of the core humus polymers. The pathway of lignin decomposition to soil is not yet fully understood. Lignin has a complicated structure of phenylpropanoid units connected irregularly by C- O-C and C-C side chain and direct ring-to-ring linkages. The CO_2 released during lignin degradation is the product is the product of cometabolism rather than the result of direct microbial mineralization. Compared to cellulose, very little (less than 1%) lignin carbon is incorporated into new biomass.

Microorganisms for Lignin Biodegradation

White-rot fungi are well known of the lignin biodegraders, because they contain peroxidases that remove electrons from lignin aromatic rings with completely etherified hydroxyl groups. This reaction yields cationic radicals that are then stabilized by the cleavage of C-C bonds in the lignin side chains. The radicals then react with O_2 or water and become stabilized by the formation of hydroxy or keto derivatives.

Because lignin is highly resistant to biodegradation, it may stabilize reactive components like cellulose, hemicellulose, or proteins by forming chemical or physical linkages to these molecules. A similar protective effect is produced by soil minerals and humus polymers that anchor unstable plant constituents by various adsorption forces or chemical binding.

Highly degradable proteins, for example, may be protected against rapid biodegradation by their nucleophilic addition to aromatic polymers through free NH_2 or SH groups and/or by adsorption on humus or clay.

Acquired Stability

The plant components may not be completely mineralized during the initial rapid decomposition phase. Therefore, many relatively unstable plant constituents, such as polysaccharides, proteins, soluble sugars, ad amino acids, can survive in the soil for a sufficient length of time to participate directly in humus formation.

SYNTHETIC PROCESSES FOR HUMUS FORMATION

The Role of Oxidative Coupling in Humus Formation

Humus (humic acid, fulvic acid, and humin) is a complicated structure. All main fractions of humus have fundamentally similar structures. They have a large aromatic polymer that in nature occurs as micelles. Incorporated into such polymers are various phenols, quinines, nitrogen heterocycles, and benzoic acids having reactive carboxyl, hydroxyl, carbonyl, and thiol groups that can act as binding sites for carbohydrates and amino acids. Humus also contains aliphatic moieties, often hydrophobic in nature, and abounds in stable free radicals that give oxidative properties to humus and enable the attraction of ionized or protonated compounds

Theories of Humus Formation

1. One of the oldest theories proposes that humic polymers are lignins or other resistant plant materials that have undergone minor modifications under soil conditions
2. Plant materials serve exclusively as a source of energy suggests that the role of microorganisms is limited to the production of various amino compounds and phenols that, once discharged into soil, polymerize form humus. It seems possible that each of these processes can contribute in the humus formation.

Modern Theories of Humification

This theory is based on the fact that a central role is played by the process of polymerization. The phenolic compounds derived from lignins (ferulic acid, pyrogallol, orcinol, syringic acid, and catechol) are considered to be the most important substrates for these polymerization reactions. This hypothesis was confirmed by many researchers. They have successfully synthesized humic-like substances by polymerizing one or more phenol-derived compounds. The synthetic polymers produced in this way are quite similar to natural humus, particularly with respect to elemental analysis, exchange capacity, total acidity, resistance to microbial degradation.

Phenolic compounds may be polymerized during the process of humus formation through oxidative coupling reactions that, as with most biological processes, require catalysts. This function is fulfilled by a series of biotic catalysts (the oxidoreductive enzymes) produced by soil microorganisms and abiotic catalysts (clay minerals and metal oxides) present in soils. Humus itself also exhibits catalytic properties due to the presence of free radical components.

The mechanism of coupling requires the oxidation of phenolic compounds involving the loss of a proton and an electron from the hydroxyl group.

This loss results in the production of free radicals or reactive quinines that covalently couple to form chains of humic polymers. Such coupling can be shown) in a simple model involving one of the components of humus, namely syringic acid, to demonstrate the formation of various oligomers, ranging from dimers to hexamers, by a laccase isolated from the fungus *Rhizoctonia praticola*. Mass spectroscopy and nuclear magnetic resonance (NMR) techniques were used to identify the products of an enzymatic reaction in which horseradish peroxidase was used to catalyze the oligomerization of guaiacol. Such studies have contributed to an understanding of mechanisms involved in humus formation.

The syringic acid and guaiacol (polymeric forms) are similar to the structure particularly with regard to the oxygen bridges and the direct carbon-carbon linkages between aromatic rings. It is possible to incorporate a per tide unit similar to that depicted in Stevenson's structure into a simplified synthetic model of humic acid. This incorporation could also be mediated by oxidoreductive enzymes, such as the laccase isolated from *Rhizoctonia praticola*.

NUTRIENT DYNAMICS

Biological management of soil fertility depends on the manipulation of organic inputs to the soil more than on any other practice. The organic matter supplies energy and nutrients to the soil biota to stimulate their activities and therefore promote soil fertility.

Organic inputs are processed by a complex web of soil organisms, but biological management is based on the assumption that the outcomes are relatively consistent. These outcomes are multiple and include the generation of inorganic nutrients such as NH_4, NO_3, PO_4, and SO_4 from organic compounds of these elements; the synthesis of soil organic matter (SOM) from precursor compounds in the organic inputs; and modification of the physical structure of the soil as a result of stimulation of biological activities. Quantitative variation in these outcomes occurs as result of following variety of potential aspects:

1. Organic inputs
2. The diversity of organisms involved in the conversion processes.
3. The complexity of the interactions between them.
4. The biological activities are regulated by the quantity and quality of the organic manure added.
5. The environmental conditions under which the processing takes place (including climate and soil type).
6. The status of the soil community.

Increasing insight into the ways in which these factors regulate soil biological activity constitutes the main progress in achieving a predictive capacity for biological management of soil fertility. Most attention has been given to the use of organic mulches to supply nutrients such as nitrogen and to the possibilities of utilizing *resource quality control* as a tool for managing nutrient availability.

TABLE 4.4: Chemical Composition, Decomposition Rate, and N-Release Rate for Various Agricultural Residues

Material	Lignin (%)	Polyphenol (%)	C/N	kDM[a]	kNa
Gliricidia	12	1.62	13	0.127	0.152
Rice straw	5	0.55	42	0.106	0.052
Maize stover	7	0.56	43	0.085	0.077
Leucaena	13	5.02	13	0.062	0.055
Acioa	48	4.09	28	0.010	0.001

[a] Decomposition constant (week^{-1}) for dry matter (DM) and nitrogen (N), respectively. From Tian, G. *et al.*, 1992. *Soil Biol. Biochem.,*24: 1051-1060.

Resource quality refers to the regulatory effect of the chemical composition of an organic resource on its rate and pattern of decomposition and nutrient release.

Early studies of decomposition processes demonstrated the importance of N concentration and C:N ration as determinants of the N supplying capability of plant residues.

Studies in natural ecosystems subsequently demonstrated that other indices such as the lignin concentration or the lignin: N ratio provided a better predictor of N release patterns. N is not released from forest litters until decomposition of lignin commences, and others have also stressed the importance of the more recalcitrant components as regulators of decomposition rate and nutrient release.

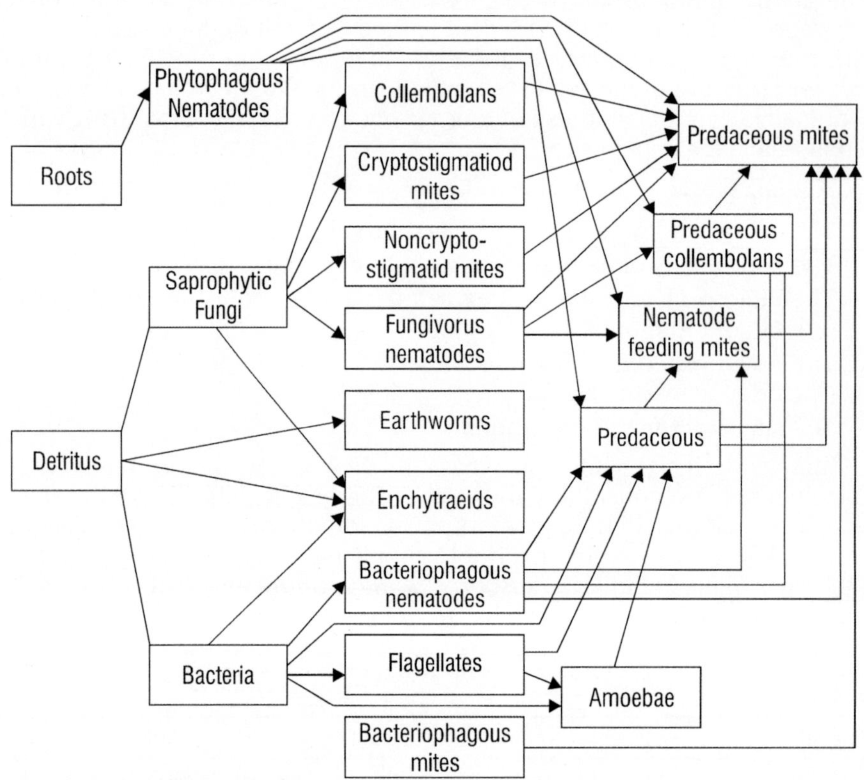

FIG 4.1: A crop residue (detritus)-based soil food web from a calcareous soil under arable cropping in The Netherlands. Under conventional tillage the bacteria-based food web is dominant; under reduced tillage the fungal and earthworm paths are more prominent (Verhoef and Brussaard, 1990).

Polyphenols have also been implicated as regulators of N release in materials that, despite being nutrient-rich and low in lignin, are slow to release N. It has been proposed that the polyphenols interact with N to form stable polymers that are resistant to breakdown and therefore delay N release.

The above theories have been used to evaluate organic residues as sources and suppliers of nutrients within cropping systems. Bulky Organic Manures/Animal Manures/Crop Residues/Composts

Environmental Regulation of Organic Matter Management

Organic matter chemistry – resource quality –is not the only factor needed to make reliable predictions of the effects of organic matter inputs. The physicochemical environment has control over biological processes from a higher level in the hierarchy than resource quality. Much of this regulation such as that deriving from the climate lies outside the control of the farmer. There are practices, however, that can be manipulated so as to *environmentally tune* the influence of organic inputs on soil biological processes. These include the location and timing of the application of organic matter. It is well established, for instance, that incorporation of residues in the soil, as opposed to surface mulching, can accelerate decomposition processes significantly and thus alter the dynamics of nutrient availability. The breaking up of the residues during tillage may be a factor in this effect, as well as the transfer of the organic matter into the more favorable environment below the soil surface.

It is well established that cultivation changes the composition, diversity, abundance, activity of the soil community and a variety of farming practices involve the management of organic inputs. For instance, there is now a considerable body of data relating to differences in soil biological populations and processes in minimum-tillage systems as compared with conventional tillage. These effects have been shown to be dramatic in relation to many groups of soil fauna, including such keystone groups as earthworms, with subsequent benefits gained in low-till systems to the physical structure and water regime of the soil. The potential exists to utilize these practices to manipulate the biological community and thence to determine the outcome of decomposition processes. A major challenge is to determine pragmatically the extent to which these *fine-tuning* factors of microenvironment and community structure need to be taken into account when developing management systems.

Sustainability of soil fertility and a steady increase in crop productivity would have to be the priority in future for avoiding food scarcity. Exhausted soils with poor or low amounts of nutrient content, impeded drainage causing salinity and alkalinity, erosion due to both water and wind which takes away the nutrients of the soil and increases soil acidity in highly leached condition and continuous fragmentation of holdings making mechanical operation difficult make yield increase difficult.

Soil fertility and productivity have to be restored and maintained at a high level to ensure continuous growth in agriculture. Poor soils are becoming a limiting factor more particularly due to the deficiencies of micronutrients.

Soil amendment/reclamation measures and use of micronutrients have to be simultaneously popularized. A check on further deterioration of soil fertility is critical and improvement of physical and biological properties of the soil coupled with a balance in the nutrient content in the soil alone can ensure sustained crop production. Integrated Nutrient Supply System (IPNS) and Integrated Pest Management (IPM) system have become an accepted strategy to bring about improvement in soil fertility and protecting environment

A high proportion of organic sources of nutrients, as a renewable source replacement to fertilizers, is considered environment friendly in terms of protecting the quality of underground waters, soil property and the environment in general.

Methods of Organic Matter Recycling

Organic matter is the life of soil and recycling of organic matter is essential for maintaining long term soil fertility.

Organic manures, including animal manures, crop residues, green manures, and composts were traditionally and preferentially used as a major source of organic matter in developing countries until the 1960's when inorganic chemical fertilisers began to gain popularity.

The shift away from organic recycling practices also served to reemphasize the value of, and need for, organic amendments for short- and long-term improvement of cultivated soils and maintenance of soil productivity. Without regular additions of adequate amounts of organic materials to soils, there is often a concomitant increase in nutrient runoff, erosion, and gradual deterioration of soil physical properties.

There are three ways for marinating and recycling the organic matter in the soil in other words these are the methods of applying manure in the soils.

In-Situ Manuring

In this method the manure is provided by animal or plant sources at site. Thus it is the direct recycling of the organic matter available in the soil and useful for making nutrient in the available form.

Ex-Situ Manuring

The organic matter and plant nutrients taken out in terms of biological yield need to be returned to the soil for sustaining the soil fertility. Incorporation of decomposed plant and animal wastes are the major source for this type of manuring.

Biological Manuring

Besides the above two methods there are several types of fauna and flora which either help in mobilizing available organic matter and nutrient or act as nutrient provider form other sources e.g., Nitrogen fixing microbes.

All the three methods are describes in details in the next chapters.

CHAPTER 5

In Situ Manuring

This is the method of manuring to the soil *at the site.* On the basis of origin of manure there are two types of *in-situ* manuring

1. Manuring by animals at site
2. Manuring with plant material grown at site, popularly known as **green manuring**.

Manure and compost have been the major traditional means sustaining plant nutrients in the soil throughout history, and they are equally as important today. However, in addition, the other methods of fertility maintenance are *in-situ* manuring, tapping flood water, cutting and carrying natural green manure species, slicing of weeds and soils from terrace risers, tree leaf litter decomposition in agroforestry and burying of crop residues on the land.

IN-SITU MANURING BY ANIMAL

This is a widespread practice in areas where herders traditionally migrate their flocks each year for the search of better pastures. There is a mutual advantage to both herders – who need a place to settle their sheep each night – and farmers, who get urine and manure right in their fields for a small fee The manuring of fields by tethering the animals directly in the fields, is also an important strategy that farmers developed to suit their conditions.

The System

There are two types of *in-situ* manuring.

1. One is keeping of cattle and buffalos in the open lands (unirrigated, unbunded terraces) just after harvesting of the crops. This is commonly done in October-November (after the harvest of finger millet, and before planting winter crops, like wheat or barley).

   The second period occurs in March-April after harvest of wheat or barley, and before planting maize or millet. Animals are moved from one peg to another after every 2-3 nights so as to completely manure the field. Certain pieces of lands are intensively manured, by keeping animals for a longer period where land is available, for about 2-3 month after the harvest of first crop. This is done by constructing temporary sheds in the fallow lands, and the animals are kept in these sheds for a few weeks at a time. Such temporary sheds are moved from time to time so as to manure the complete follow area. In such intensively manured lands, nursery raising of finger millet and rice, and planting of maize or potatoes are the practices subsequently carried out. Some farmers also keep animals under the canopy of fruit trees, in order manure the orchard. Generally, these animals are kept for a period of one to three nights at a site depending upon the number of animals, or the size or type of land. All these systems, assist the farmer to feed crop residues at the same spot, and at the same time reduces the transportation cost of manure from the cattle shed to the fields. Depending on the season and location, some farmers provide temporary shade, but in most cases, animals are kept in the open overnight. Farmers have reported that urine and dung of a buffalo tethered for three days in one spot of land is equal to heavy application of compost.

2. The other system of *in-situ* manuring is carried out by using migratory flocks of sheep and goats kept under a transhumance system. In hilly areas, these flocks graze on the high hill alpine pasture (above 2500m), where almost no crops are grown (from May to October/November) while in hot arid areas the time starts just after harvest in the one and only rainy season crop i.e. from November to June. The sheep and goats are kept overnight, preferably in dry lands.

When the flocks are kept overnight on a person's land, other than the sheep owners, the land owner has to pay the Shepherd in cash or kind usually food grain, for every night of the stay on the land. There is a very high demand for such flocks.

One method of farm yard manure application is the herding of sheeps into the field (in some parts, donkeys are used). The sheeps are allowed to remain in the field overnight. Urine and excreta deposits are available right where they are needed.

The droppings are later ploughed directly into the soil. This practice is undertaken when there are no crops in the field.

Advantages

3. Useful to small and marginal farmers who do not own cattle and have limited land.
4. Useful to farmers having limited farm yard manure.
5. Transportation and labour charges are avoided.
6. Pastorals can generate income by charging a fee for this service.
7. Manure decomposes easily.
8. Release nutrient into the soil through decomposition.
9. Promotes easy infiltration in vertisols.
10. Utilizes urine which is normally wasted.

Limitations

1. It is difficult for those areas where sheep population is low.
2. Farmers have to depend on migratory practices of the herders.

Verification of Farmers Experiences

Farmers report that sheep and goat manure has a greater immediate and lesser residual effect on crops than that of cattle and buffalo. They have also experienced that once fields have been manured by sheep and goats, they should continue to be manured by them thereafter. Farmers also believe that *in-situ* manuring makes the soils more fertile than using farm yard manure (FYM). The results of the experiment for verifying this fact showed that *in-situ* manuring by sheep was superior to cattle. *In-situ* manuring by sheep produced 28.7% higher maize yield over the direct application of an equivalent amount of sheep manure. This can be attributed better utilization of urine under *in-situ* manuring. This result fully agrees with the farmers' experiences, and shows that farmers gained experiences are through their own experimentation and observation.

Sustainability

The most common practice is used by large proportion of farmers in the hills and hot arid areas, and is a system which has been developed by the farmers through their ingenuity and has been sustained since long time. The system seems quite sustainable as it is a very old and traditional system developed by the farmers themselves. Farmers' have already recognized its importance and value.

Labour is the key component for management of soil fertility. Limitations in the more widespread of compost and FYM are created because of labour constraints. The *in-situ* system is an efficient method of recycling organic residues, since crop(s) residues and animal waste are properly utilized directly into the soils, with no nutrient loss, and minimal involvement of human labour. Neither does the system involve any external inputs. Further, as a traditional practice, it does not suffer any communication barrier, as many new technology does.

Prospects

Declining soil fertility is an acknowledged production constraint in rainfed areas. For subsistence farming, use of external purchased inputs such as chemical fertilizers is very difficult if not impossible. In the remote areas, the wide scale use of chemical fertilizer is restricted because of lack of transportation facilities, the lack of awareness of farmers and their low purchasing power. Perhaps most importantly, it is constrained by unavailability of fertilizer of the right type, in the right quality, at the right time. Under such a situation, traditional practices should be encouraged.

Crop intensification and afforestation programmes cause have limitations to *in-situ* manuring by the migratory sheep and goats, because of lack of land for grazing. However, for the system of manuring by cattle and buffalo, *in-situ* manuring can be fitted in during the turn round period of few a days between two crops. This practice is more attractive in terms of labour requirements and the urine is also completely utilized. Formal research has shown that this practice is more useful than the application of FYM. Traditional practices are often neglected by development workers where energy and time is used in recommending the apparent benefit of low technology. Tradition is, however, the language of the farmers rainfed areas. Better knowledge of traditional agricultural system is required in order to develop sustainable alternative technologies under these circumstances.

INSITU MANURING WITH PLANTS (GREEN MANURES)

Green manuring can be defined as a practice of ploughing or turning into the soil undecomposed green plant tissues for improving physical structure as well as soil fertility. From the time immemorial, the turning in a green crop for improvement of the conditions of the soil has been a popular farming practice. Green manuring, wherever feasible, is the principal supplementary means of adding organic matter to the soil. It consists in the growing of a quick-growing crop and ploughing it under to incorporate it into the soil. The green-manure crop supplies organic matter as well as additional nitrogen, particularly if it is a legume crop, which has the ability to fix nitrogen from the air with the help of its root nodule bacteria. The green-manure crops also exercise a protective action against erosion and leaching.

Growing a green manure is not the same as simply growing a legume crop, such as beans, in a rotation. Green manures are usually dug into the soil when the plants are still young, before they produce any crop and often before they flower. They are grown for their green leafy material, which is high in nutrients and protects the soil.

If food is in very short supply, it may be better to grow a legume from which a bean crop can be harvested and then dig the plant remains into the soil. These plant remains will not break down in to the soil so quickly and will not be as good for the soil as younger plants but they will still add some nutrients to the soil for the next crop.

Benefits of Using Green Manures

Green manuring offers an inexpensive way of improving crop yields and it takes little extra effort. Green manures are especially important on farms where there is not enough animal manure available, and when it is not possible to bring in natural fertilizers from elsewhere. Although the use of green manures may seem to create extra work, they provide following benefits:

Enhance Soil Fertility

Green manures recycle nutrients and add organic matter to the soil. They help prevent nutrients being washed out of the soil. The nutrients are taken up by the green manure and held inside the plant. When the nutrients are needed for the next crop, the plants are dug into the soil or used as a mulch on top of the soil. This helps to increase crop yields. Legumes and other nitrogen fixing plants, which take nitrogen from the air to the soil are particularly beneficial.

Supplement for Nutrients

Different green manures and grain-legumes are used to increase the nitrogen content and texture of the soil. Among the green manures *Sesbania aculeata* accumulated the largest amount of bio-mass and nitrogen contribution and among the grain legumes, cowpea ranks first both in terms of grain yield and biomass addition. The available green manure crops and utilization pattern are as follows:

TABLE 5.1: Nutrient potential of green manures

Green manure	Biomass (tonnes)	N accumulobase (Kg/ha)
Sesbania aculeata	22.50	145.00
S. rostrata, Crotalaria	20.06	146.00
Juncea, Tephrosia	18.40	113.00
Purpurea	6.80	6.00
Green gram	6.50	60.20
Black gram	5.12	51.20
Cow pea	7.17	63.30

Improved Soil Structure

Green manures improve soil structure, letting more air into the soil and improving drainage. Green manures help sandy sol hold more water and not drain so quickly.

Prevention of Soil Erosion

Green manures help to stop the soil being carried away by wind and rain. The roots penetrate the soil and hold it in place.

Weed Control

Green manures help to control weeds. Bare soil can become quickly overgrown with weeds, which can be difficult to remove. Green manures cover the ground well and stop weeds growing beneath them, by competing for nutrients, space and light.

Method of Use

Farmers often observe the benefits of green manures but many do not use them because they do not know which species to use and how to include them in their own farming system. It is, therefore, important to plan in advance where and when they are to be grown.

The adoption of green manuring depends upon the agroclimatic conditions. Broadly, the following two types of green manuring can be thought of:

Green Manuring In Situ

In this system, green manure crops are grown and buried in the same field which is to be green-manured, either as a pure crop or as an intercrop with the main crop. In India, various methods of growing green manure crops *in situ* are followed to suit local conditions. For the proper decomposition of the green manure, it is necessary that the green material should be succulent and there should be adequate moisture in the soil. Plants at the flowering stage, contain the greatest bulk of succulent organic matter with a low carbon/nitrogen ratio. The incorporation of the green-manure crop into the soil at this stage allows a quick liberation of nitrogen in the available form.

Green manuring *in situ* is followed in northern India while green leaf manuring is common in eastern and central India. Green manuring is a practice of ploughing or turning into the soil undecomposed green plant material for improving the physical conditions of the soil or for adding nutrients. Any crop or plant (generally leguminous) grown and ploughed *in situ* is called a green manure, for example, dhaincha, Sesbania, sunn hemp, wild indigo and pillipesara. A leguminous crop producing 10-25 tonnes of green matter per ha will add about 60 to 90 kg N. *green leaf manure* refers to turning under of green leaves and tender green twigs collected from shrubs and trees grown on the bunds, wastelands and nearly forest areas. Common shrubs and trees useful for this purpose are *karanj*, *Glyricidia*, *neem*, *subabul*, etc.

1. **Dhaincha (*Sesbania aculeata*):** It is suitable for loamy and clayey soils. Dhaincha is an ideal green-manure crop for rice based cropping system. It is fairly resistant to drought as well as stagnation of water. The green manure grows well even in alkaline soils and corrects alkalinity if grown repeatedly for 4-5 years. The roots have plenty of nodules. It yields about 10-15 tonnes of green manure per ha and requires a seed rate of 30-40 kg/ha.
2. ***Sesbania speciosa:*** It is a valuable green manure for wetlands and can be grown in a wide range of soils. Seed production is prolific; however, pods are frequently attacked by insects. The green manure can be raised on the field borders along the bunds. *Sesbania* seedling (21 days) can be planted in a single line at 5-10 cm apart in the borders of the fields close to the bunds. In about 90 days it produces about 2-4 tonnes of green manure per ha. It does not affect the rice yield by shading or root effect. If second rice crop is planted immediately after the first crop, the manure can be incorporated into the field. About 300-400 g of seeds are sufficient to raise nursery and plant the seedlings along the bunds of one hectare.
3. ***Sesbania rostrata:*** One of the important features of this green manure is that in addition to the root nodules, it produces nodules in the stem. The stem nodulation is an adaptation for waterlogged situation since flooding limits growth of green manures and may reduce root nodulation. Under normal condition, both root and stem nodules are effective in N fixation. It can produce about 40-50 tonnes of fresh biomass from one ha in about 50-60 days. It has higher N content of 3.56 % on dry weight basis. Biomass production is higher during summer (April – June) than in winter (Dec.-Jan.) season. This green manure can also be produced by raising seedlings (30 days old) and planted in the paddy field along

the bunds or as intercrop with rice.

4. **Sunn hemp (*Crotalaria juncea*):** It is a quick growing green manure crop and gets ready for incorporation in about 45 days after sowing. It does not withstand heavy irrigation leading to flooding. The crop is at times subject to complete damage by leaf eating caterpillars. The crop can produce about 8-12 tonnes of green biomass per ha. The seed requirement is 30 kg/ha.
5. **Wild indigo (*Tephrosia purpurea*):** This is a slow growing green manure crop and cattle do not prefer to graze it. The green manure is suitable for light textured soils, particularly in single crop wetlands. It establishes itself as a self-sown crop and the seeds remain viable till the harvest of rice. On an average about 3-4 tonnes of green manure is obtained in one ha. The seed rate is 30 kg/ha. The seeds have a waxy impermeable seed coat and hence scarification is required to induce germination. Soaking seeds in boiling water for 2-3 minutes is also equally effective in promoting germination.
6. **Indigo (*Indigofera tinctoria*):** It resembles wild indigo and is a long duration crop with more leafy growth. It comes up well in clayey soils with one or two irrigations.
7. **Pillipesara (*Phaseolus trilobus*):** This is a dual purpose crop yielding good fodder for the cattle and green manure. Pillipesara comes up well in hot season with sufficient soil moisture. Loamy or clayey soils are best suited. After taking one or two cuttings for fodder or light grazing by animals, the crop can be incorporated into the soil. About 5-8 tonnes of manure can be obtained from one ha.

Green Leaf Manuring

Green-leaf manuring refers to turning into the soil green leaves and tender green twigs collected from shrubs and trees grown on bunds, waste lands and nearby forest areas. The common shrubs and trees used are Glyricidia (*Glyricidia maculata*), *Sesbania speciosa*, Karanj (*Pongamia pinnata*), etc.

A number of leguminous and non-leguminous plants are grown on bunds or wastelands with the prime objective of utilizing their foliage as green manure.

1. **Glyricidia (*Glyricidia maculata*):** This is a shrubby plant that comes up well in moist situations. Under favourable conditions, it grows well like a tree. It can be easily grown in waste lands, farm road sides, field bunds, etc. The crop can be established by stem cuttings or seedlings planted in the field borders. It can be pruned for its tender loppings and compound leaves for green leaf manuring at the time of puddling rice. On an average, a well-established plant yields 12-15 kg green matter. About 400 plants on the peripheral bunds yields 5-6 tonnes green manure/ha.
2. **Karanj (*Pongamia glabra*):** It is a leguminous tree grown in wastelands. On an average, a tree can yield 100-120 kg of green matter. The leaves contain about 3.7 % N (on dry weight basis).
3. **Neem (*Azadirachta indica*):** It is an indigenous tree with wide range

of adaptability. Almost all parts of the tree have been used for several millennia for preparation of medicine and its leaves and leaf–extracts to repeal insect pests of crops and stored grains. The leaves are pruned at the time of planting rice and incorporated as green leaf manure.

4. **Calotropis (*Calotropis gigantea*):** On roadsides and fallow lands, the plant grows wild under different soil and climatic conditions. The leaves are more succulent and a plant can produce about 4-5 kg of green matter. Besides it also helps in controlling soil born pests like termite.

The nutrient content of green manure and green leaf manure crops are given in the Table 5.2.

Table 5.2: Nutrient content of important green manure and green leaf manure crops

Crop	Nutrient content (% on dry weight basis)		
	N	P	K
Green manure			
Sesbania aculeate	3.3	0.7	1.3
Crotalaria juncea	2.6	0.6	2.0
Sesbania speciosa	2.7	0.5	2.2
Tephrosia purpurea	2.4	0.3	0.8
Phaseolus trilobus	2.1	0.5	-
Green leaf manure			
Pongamia glabra	3.2	0.3	1.3
Glyricidia maculata	2.9	0.5	2.8
Azadirachta indica	2.8	0.3	0.4
Calotropis gigantea	2.1	0.7	3.6

Place in Farming System

Green Manures in Rotation

Growing green manures as part of a crop rotation is an important part of an organic farming system. These help to build soil fertility and are particularly useful when grown before crops, which need a lot of nutrients.

Green manures can be used in rotation:

1. Whenever there is no crop in the ground, rather than leaving the land bare and allowing weeds to grow and nutrients to leach out of the soil.
2. As break crops, when there is only a short time between main crops.

Timing of sowing is important. The green manure must be ready to dig in before the crop next is sown. There should not be a long gap between digging in the green manure and planting the next crop. This is to prevent nutrients from the green manure leaching out of the soil, before being taken up by the next crop.

Green Manures and Under Sowing

Under sowing involves growing a green manure at the same time as a crop, among the crop plants. Sometimes they are sown with the crop or slightly later when the crops are already growing. This reduces competition between the green manure and the crop.

For example, under sowing is sometimes used with maize crops where a green manure is sown under the young maize plants. The green manure seeds are broadcast sown when the second weeding of the maize is carried out. In this way when the maize is harvested the green manure is already established and ready to grow quickly. This method means that no extra time is spent preparing the land and sowing the green manure.

Long Term Green Manures

Green manures can be grown for more than one season and used in the following ways:

1. Long term green manures restore poor soil. Using them over a long time has a greater benefit on soil fertility and structure of poor soil.
2. Long term green manures can be used when new land is being prepared for use, especially to help control difficult perennial weeds.
3. Long term green manures are used where land is to have a long fallow period. They can be sown at the beginning of the fallow of bush-fallow systems. They help to quickly build up the fertility of the soil and reduce the length of time before the land can be used to grow crops again.
4. Long term manures provide green material which can be cut and carried to other fields. Green material can be harvested from perennial species such as alfalfa (*Medicago sativa*), for digging in, mulching, composting or feeding to livestock.

Green Manures as Mulch

Green manure plants can be cut and left on the soil surface as a mulch. Mulching releases nutrients slowly but has some advantages:

1. Mulching helps to prevent weed growth
2. Mulching protects the soil from erosion
3. Mulching keeps the soil moist by reducing evaporation

Green Manures in Agroforestry

Agroforestry is the practice of growing trees and/or shrubs together, with crops and/or animals. The trees/shrubs act as long term green manures and the leaves can be used for digging in or as a mulch.

The regular pruning of agroforestry trees such as Leucaena (*Leucaena leucocephala*), Mother of cocoa (*Gliricidia sepium*) and Calliandra (*Calliandra calothyrsus*) during the crop growing period provides large amounts of green material for digging into the soil and reduces competition with the main crop. The material can also be used as a mulch. It is spread on the top soil, usually between crop rows or before a crop has been planted.

As well as improving the soil in the ways described above, trees and shrubs also provide food, fodder, fuelwood, erosion control and other benefits.

Management of Green Manuring

Time of Sowing and Seed Rate

March is the best suited period for raising green manure. Studies at Coimbatore (Tamil Nadu) reveal that summer season is the ideal for seed production of *S. rostrata*. Pre-season showers can be very well used to raise a green manure crop in single crop wetlands (August-January).

A seed rate of 40-50 kg/ha is required to get higher biomass of green manure. If the population is dense, the stem will be very thin and the crop will not seed early (e.g., dhaincha).

Seed Treatment

Germination is a problem for the green manure seeds, like *Tephrosia purpurea* because of their hard seed coat. Scarification of the seeds with sand and pre-soaking in hot water (50°C) for 5-minutes are found to improve seed germination upto 98%. Acid seed treatment is also effective.

The seeds can be treated with concentrated commercial grade sulphuric acid (400 ml/kg of seed) for 30 minutes. The seeds are washed thoroughly with water, shade-dried and used for sowing.

Stem Cuttings

Glyricidia is one of the green leaf manure tree crops propagated through seedlings and stem cuttings. A large percentage of the cuttings fail to root readily. The cut ends have been kept immersed in 0.005% IAA and 0.1% seradix A for one day and planted in small beds and irrigated for six days. The hormonal treatment aided root formation to a remarkable degree (90%).

The green manure crops can be established using cutting of soft wood (from freshly formed branches) and hard wood (leaving top and bottom 30 cm in the stem and older branches). Soft wood cuttings of *S. aculeate* and hard wood cuttings of *S. rostrata* are superior. The establishment of the cuttings can be improved by dipping in fresh cowdung (10% suspension).

Mixed Cropping

It has been found feasible to cultivate green manure mixed with rice fallow pulses. Green manure crops like indigo and pulses (black gram) can be mixed with the ration of 1:3. Pillipesara can also be sown in the same proportion. By adopting such practice in rice, fallow pulses no yield reduction has been noticed. Growth of green manure crop is also not spoiled. By the time pulse is harvested, the green manure put forth growth only to a few cm after the harvest of pulses, the green manure put on growth.

Inter Cropping

Field investigations at Coimbatore reveal the possibility of intercropping green manure in transplanted rice. Seedlings of *S. rostrata* were planted at an interval of 1.5 m. On 30 days after planting, the green manure is cut at a height of 15 cm from the ground level and incorporated as manure. The ratoon is allowed and the refresh incorporated while planting second crop of rice.

Border Planting

Border planting of green manure, particularly *Sesbania speciosa* in the rice field bunds was advocated even during 1950s in Thanjavur (Tamil Nadu). The seedlings of the green manure were planted at 60-90 cm apart on the bunds at the time of planting kuruvai crop during June. About 70% of the crop was used as green manure for the thaladi crop planted in October. About 1.5-2.0 tonnes of green leaf was obtained from the bund crops and incorporated. The remaining was left standing for producing seeds. It was also possible to produce in one ha, the seeds sufficient for raising green manure for 2.5 ha.

Phosphorus Response

Incorporation of green manure with single superphosphate or rock phosphate facilitates higher response to added phosphorus. Recommended dose of phosphorus for rice can be very well applied to green manure crop and incorporated for greater efficiency

Digging in Green Manures

Before a crop is sown, the green manure is dug back into the soil. Here it decomposes and the nutrients held inside green manure plants are released.

1. The plants take a short time, usually about two weeks, to rot down into the soil before the next crop is sown.
2. Green manures should not be ploughed in as this buries the plants and the nutrients too deep. They should be turned in just under the soil surface.
3. Digging is easier if the plants have been chopped into small pieces before digging. This also helps to prevent the problem of regrowth, if this occurs.

Younger green manure plants are easier to dig into the soil then older ones and land can be ready to use more quickly after they have been dug in. Therefore, over a long period, two short term green manures may be better than one longer term green manure. However, this may involve more of time and effort.

For most green manure plants, the best time to dig in is just before flowering begins, but this is different for some species.

If plants become too old and tough, it is more difficult to dig them. Soil organisms will find it difficult to break down and decompose old, tough plants. If this happens, green manures can be cut and composted instead.

The Choice of Green Manure

When choosing which green manure plant to use, you should consider the following points:

1. A green manure must suit the local climate, and the soil that it is to be sown in. This will help to keep the green manure healthy and to keep pests and diseases to a minimum.
2. Fast growing and leafy green manures are often preferred as they provide more nutrients when dug in.
3. Green manures should not be closely related to the following crop as they could attract pests and diseases which may affect the following crop.
4. It is important to know whether seed is easily available and affordable.
5. The length of time that land is free and how long the green manure will take to grow.
6. Plants which can be grown as a green manure include legumes and non-legumes. Legumes have nodules on their roots which contain bacteria. These bacteria take nitrogen from the air. This is known as nitrogen-fixation. Plants use this to grow, but this extra nitrogen is also made available to future crops when the legumes are dug into the soil. The ability of legumes to fix nitrogen makes them very good green manures. However, they do have limitations and non-legumes can sometimes be more suitable.

   Legumes will only fix nitrogen if the right type of bacteria, for example *Rhizobium,* are in the soil. This is especially relevant if the legume is not a local plant. A product containing these bacteria can be bought but it may be expensive. Even if the bacteria are provided, other non-legume plants can be better in some situations. They may produce more organic matter and have a better root system. They may also survive better and grow faster and may be able to tolerate extreme weather conditions or poor soils.

Other Important Considerations

It is possible that some green manures may grow too vigorously and become weedy. This is especially true for plants which are not found locally. They may grow among the following crop or spread into new areas. The green manure should be chosen carefully to avoid this. If a green manure is to be used for the first time in an area, it should be tried on a small plot and checked to see that it does not become a weed. Things to check for are:

1. Light seeds being blown about by the wind.
2. The green manure plant growing in places where it was not planted.
3. Long stems that grow from the plant and spread along the surface of the soil. New roots then grow at intervals along these stems.

Checks should be carried out for at least two years, if possible, before the green manure is accepted for general use.

Growing perennial green manures as annuals, it will prevent them from taking over and growing in areas where they are not wanted.

Achieving Sustainability in the Use of Green Manures

The present widespread use and rapid adoption of green manure and cover crops has taken many people. To some extent, this is because little attention has been given to the extent to which green manures and cover crops have always been used in traditional systems. Gene Wilken, for example, in his otherwise excellent book, *Good farmers: traditional agricultural resource management in Mexico and Central America*, stated that, *cover cropping is not widespread in traditional Middle America*,. Many scientists believed the technology is inappropriate for village farmers. As late as 1989, Anthony Young in the classic work on *Agroforestry for soil conservation* dismissed green manuring as, *a form of non-productive improved fallow which has rarely found favour with farmers* (Young 1989).

Sustainability

For more than a decade it has been accepted that green manures and cover crops would only be accepted by small farmers if they could be grown on land that had no opportunity cost, could be intercropped with other produce, grown under tree crops or on fallow land and be cultivated in periods of expected drought or extreme cold. They would also be favoured if they involved no extra labour or out-of-pocket cash expense.

Whilst these assumptions have proved correct, recent experience has shown that the sustainability of green manure and cover crops is more likely to be guaranteed when they provide farmers with some other benefit besides fertile soil. This condition is consistent with the observation that village farmers generally prefer multiple use technologies.

Experiences Worldwide

Experience from many parts of the world confirms the value farmers' attribute to green manures and cover crops that have multiple uses. In most known, traditional systems legumes are appreciated not only because they maintain soil fertility, but because the seeds or pods can also be eaten. Examples include the *Vigna* spp. which is intercropped in Southern Honduras, El Salvador and South-east Mexico and the high-altitude scarlet runner bean (*Phaseolus coccineus*), which is widely used from upstate New York (Seneca bean) to Mexico (ayocote) and from Guatemala (piloy) to Honduras (chinapopo) and Northern Chile. The velvet bean (*Mucuna* spp) is the most popular of all the green manures/cover crops used today and was initially used and spread by farmers along the southern border of the Himalayas in Nagaland partly because it was such a valued source of food. In Central Honduras, where World Neighbours and COSECHA have taught farmers to intercrop velvet bean with maize, there has been a disappointing failure (35%+) to continue this technology except in those villages where it is consumed as a major component of coffee, hot chocolate, bread and tortillas. In fact, the value of green manures and cover crops as human food seems to be the strongest factor motivating their sustained adoption.

Perhaps the second most common use of green manures and cover crops is weed control. The South-east Asia, a perennial species of the velvet bean is used to improve fallow and to control weeds. Modern practices, include using jack bean (*Canavalia ensiformis*), tropical kudzu (*Pueraria phaseoloides*) and perennial peanuts (*Arachis pinto*) under a variety of plantation crops, including coffee, citrus, and African oil palm. The velvet bean is also used to control Imperata grass (*Imperata* spp.) and this practice is spreading rapidly throughout Benin, Togo, and Columbia. Velvet bean and jack bean are used to control paja blanca (*Saccharum* spp.) in panama and to combat nutgrass and several other countries.

A third practice, which is now more widespread but which is still under appreciated, is the use of green manures and cover crops to stabilize Sweden agriculture. Since decreased fertility and weed infestation are the two most important reasons where farmers abandon their fields today, and since green manures and cover crops can to some extent, often solve both these problems they have proved to be an effective way of stabilizing shifting cultivation in many countries.

One dramatic example can be drawn from the work of the Centro Maya in Guatemala's northern Peten region. In this humid forest area, farmers could only grow maize for one or two years and then the ground had to be left to regenerate. Now hundreds of farmers are growing velvet bean intercropped with maize on the same fields year after year. Those who initially adopted this system have been growing maize on the same land for eleven consecutive years and productivity has improved over time. Another interesting example is that of Central Ghana, where village farmers are inventing their own ways of stabilizing their agriculture, including one system in which 30,000 Leucaena trees (*Leucaena* spp.) on one hectare are intercropped with maize and burned very lightly each year. This practice has allowed maize to be planted on the same land for 20 years in succession.

A fourth potential benefit and one that will probably acquire more significance as experience increases, is the use of green manures as animal feed. Most green manures and cover crop species, with the major exception of *Melilotus albus* cannot be grazed well, but many can be used for cutting and carrying even after months of drought. The most notable examples of this type are *Lathyrus neurivalvis* and lablab bean (*Dolichos lablab*). Seeds also provide fodder. One good example is the seed of the velvet bean which in Campeche, Mexico is cooked for a half hour, mixed with an equivalent amount of maize and then ground into pig feed. The University of Yucatan calculated that this velvet bean feed cost less than commercial feeds per unit of weight gained.

Green manure and cover crops can be used in other ways as well. In Southern Brazil, thousands of farmers regularly use some 25 different species of green manure and cover crops for soil improvement partly because this allows them to increase the amount of organic matter in their soil to the point where tilling is no longer necessary. The financial as well as ecological advantages of zero-till systems are tremendously attractive.

CONCLUSIONS

A number of conclusions can be drawn from the examples given above.

- The variety of sustainable green manure and crop cover systems already established in traditional as well as more recently introduced agricultural system is remarkably diverse.
- Green manures and cover crops have been adopted on a wide scale despite the seemingly prohibitive conditions. The fact that virtually every system referred to have some elements of these conditions confirms their predictive value. Thus, programmes to introduce new green manure and cover crop systems should teach farmers not only how these species can be used to improve their soil but that they have other uses as well.
- Tremendous potential still exists for the development of new green manure and cover crop systems.
- Scope of potential systems for using green manure and cover crops still need to be investigated, most notably the major possibilities of using them for animal feed; the potential latent in new as yet untried species, including trees and non-legumes, and the value of combining of green manures and cover crops rather than using individual species. Experience leads us to believe that, with the possible exception of very intensive farming systems such as irrigated vegetable and rice, green manure and cover crops systems can probably be introduced into many, if not most of the world's, small-scale farming systems.

CHAPTER 6

Ex Situ Manuring

Many a times *in-situ* manuring not able to fulfill the nutrient requirement of the soil on the other side plenty of organic waste from agriculture, animals and human being is available which is rich in nutrient and needs safe disposal and crop field are the best place for this. In the organic farming the nutrient cycle has to be maintain in which the nutrient taken out in terms of economic yield has to be returned to soil and use of agriculture waste is the appropriate solution for maintaining the cycle. Organic manures from agriculture/animal/human waste, contain essential plant nutrients and other growth promoting agents like enzymes and hormones, while no synthetic chemical fertilizer can supply all together thus they are indispensable from the manural schedule for any crop production. The organic manure through the process of decomposition and humification gives humus which helps to improve the physical, chemical and biological properties of soil. The organic acid released during decomposition of organic manure controls certain fungal pathogens and nematode infestations. Commonly available organic manures like FYM, crop residues, poultry and sheep manure, oil cakes and other farm wastes will effectively be utilized for nutrient supplement.

AVAILABLE POTENTIAL OF ORGANIC MATERIALS FOR EX-SITU MANURING

Agricultural wastes (animal manures and crop residues or mixtures of the two) are now considered quite an important component for horticultural crop production especially that based on organic systems. Agricultural wastes can now be considered to include crop after harvest and primary processing, tree residues organic/plant residues from social forestry, animal excreta and processing left-overs from the slaughter houses and agro-industrial wastes. Thus, agricultural wastes comprise all organic wastes produced and disposed of or used in primary agricultural production. Of course, these materials are essentially by-products of production rather than wastes in the strict sense of the word.

Organic Resources and Potential

There are several types of material mainly form plant or animal origin which can be used for *ex-situ* manuring, directly or in decomposed form -

Livestock and Human Wastes

1. Cattle-shed wastes such as cattle and buffalo dung, and urine.
2. Other livestock and human excreta.

3. Byproducts of slaughter-houses and animal carcasses: Blood and meat wastes, bones, horns and hooves, leather and hair wastes.

Crop Residues, Tree Wastes and Aquatic Weeds

1. Crop wastes of cereals, pulses and oilseeds (wheat, paddy, *bajra, jowar,* gram, *moong, urad,* cowpea, *arhar, masoor,* ground-nut, linseed, etc.)
2. Stalks of corn, cotton, tobacco, sugar-cane trash, leaves of cotton, jute, tapioca, arecanut, tree leaves, water hyacinth, forest litter, etc.

Urban and Rural Wastes

1. Rural and urban solid wastes.
2. Urban liquid wastes – sewage and sullage.

Agro-Industries Byproducts

1. Oil-cakes
2. Paddy husk and bran
3. Bagasse and pressmud
4. Sawdust
5. Fruit and vegetable wastes
6. Cotton, wool and silk wastes
7. Tea and tobacco wastes.

Marine Wastes

Fish meal and aquatic/seaweeds.

TABLE 6.1: Average nutrient content of bulky manures

Manure	Percentage content		
	Nitrogen (N)	Phosphoric acid (P_2O_5)	Potash (K_2O)
Animal refuse			
Cattle dung, fresh	0.3-0.4	0.1-0.2	0.1-0.3
Horse dung, fresh	0.4-0.5	0.3-0.4	0.3-0.4
Sheep dung, fresh	0.5-0.7	0.4-0.6	0.3-1.0
Night soil, fresh	1.0-1.6	0.8-1.2	0.2-0.6
Poultry manure, fresh	1.0-1.8	1.4-1.8	0.8-0.9
Raw sewage, fresh	2.0-3.0	-	
Sewage sludge, dry	2.0-3.5	1.0-5.0	0.2-0.5
Sewage sludge, activated dry	4.0-7.0	2.1-4.2	0.5-0.7
Cattle urine	0.9-1.2	tr.	0.5-1.0
Horse urine	1.2-1.5	tr.	1.3-1.5
Human urine	0.6-1.0	0.1-0.2	0.2-0.3
Sheep urine	1.5-1.7	tr.	1.8-2.0
Wood ashes			

Manure	Percentage content		
	Nitrogen (N)	Phosphoric acid (P_2O_5)	Potash (K_2O)
Ash, coal	0.73	0.45	0.53
Ash, household	0.5-1.9	1.6-4.2	2.3-12.0
Ash, wood	0.1-0.2	0.8-5.9	1.5-36.0
Farm, factory and habitation wastes			
Rural compost, dry	0.5-1.0	0.4-0.8	0.8-1.2
Urban compost, dry	0.7-2.0	0.9-3.0	1.0-2.0
Farmyard manure, dry	0.4-1.5	0.3-0.9	0.3-1.9
Filter-press cake	1.0-1.5	4.0-5.0	2.0-7.0
Plant residues			
Rice hulls	0.3-0.5	0.2-0.5	0.3-0.5
Groundnut husks	1.6-1.8	0.3-0.5	1.1-1.7
Straw and stalks			
Pearlmillet	0.65	0.75	2.50
Banana, dry	0.61	0.12	1.00
Cotton	0.44	0.10	0.66
Sorghum	0.40	0.23	2.17
Maize	0.42	1.57	1.65
Paddy	0.36	0.08	0.71
Tobacco	1.12	0.84	0.80
Pigeonpea	1.10	0.58	1.28
Wheat	0.53	0.10	1.10
Sugarcane trash	0.35	0.10	0.60
Tobacco dust	1.10	0.31	0.93
Tree leaves, dry			
Calotropis gigantea	0.35	0.12	0.36
Careya arborea	1.67	0.40	2.20
Cassia auriculata	0.98	0.12	0.67
Dillenia pentagyna	1.34	0.50	3.20
Madhuca indica	1.66	0.50	2.00
Pongamia pinnata	3.69	2.41	2.42
Pterocarpus marsupium	1.97	0.40	2.90
Terminalia chebula	1.46	0.35	1.35
Terminalia paniculata	1.70	0.40	1.60
Terminalia tomentosa	1.39	0.40	1.80
Xylia dolabriformis	1.37	0.30	1.61
Green manures, fresh			
Cowpea (*Vigna catjang*)	0.71	0.15	0.58
Sesbania aculeata	0.62	-	-
Cluster-bean (*Cyamopsis tetragonoloba*)	0.34	-	-
Horse-gram (*Dolichos biflorus*)	0.33	-	-
Mothbean (*Vigna aconitifolia*)	0.80	-	-
Greengram (*Vigna radiata*)	0.72	0.18	0.53

Manure	Percentage content		
	Nitrogen (N)	Phosphoric acid (P_2O_5)	Potash (K_2O)
Sunn hemp (*Crotalaria juncea*)	0.75	0.12	0.51
Blackgram (*Vigna mungo*)	0.85	0.18	0.53

Agricultural Waste

Crop Residues

Residues left out after the harvest of the economic portions are called crop residues/straw. In the developing countries like India, they are mostly used as cattle feed. In the developed countries, harvesting is done using combine harvester and hence the straw cannot be used as cattle feed. They are generally burnt in the field itself. Straw has good manurial value since it contains appreciable amount of plant nutrients. On an average, cereal straw and residues contain about 0.5 % N, 0.6 % P_2O_5 and 1.5 % K_2O. The crop residues can be recycled by way of incorporation, compost making or mulch material.

Agro-Industrial Wastes

Agro-industrial wastes are available in substantial quantities at processing sites and can be effectively utilized as manure.

Rice Husk

It is the major by-product of the rice milling industry. Unhulled paddy grain constitutes 20-25 % of husk. It is a poor source of manure and the nutrient content is very low (0.3–0.4% N, 0.2-O.3% P_2O_5 and 0.3–0.5% K_2O). Rice husk should be incorporated into the wet soil and can be used in saline and alkaline soils to improve the physical conditions. It can also be used as a bedding material for animals.

Bagasse

The most important by-product of sugar industry is bagasse. It is mainly used as fuel in boilers of sugar factories. It can be used as manure raw or after composting. It contains 0.25 % N and 0.12 % P_2O_5.

Pressmud

It is a by-product obtained during the process of sugar manufacturing. It contains about 1.25 % N, 2.0 % P_2O_5 and 20-25 % organic matter. Addition of pressmud is highly useful to acidic soils since it contains high amount of lime (upto 45 %)

Tea Wastes

In the tea industry, tea wastes are available during the course of tea production, processing and storage. Tea wastes are used for extraction of caffeine. The decaffeinated tea wastes can be used as a manure. Nutrient content of the spent tea waste is 0.3-0.35 % N, 0.4 % P_2O_5 and 1.5 % K_2O.

Coir Waste

It is a waste product from the coir industry and mostly dumped near the road sides. Coir pith contains high amount of lignin (30 %), cellulose (26 %) and wide C:N ratio (112:1). To reduce the bulk and C:N ratio, composting is recommended. The composted coir pith contains 1.26 % N, 0.06 % P, 1.20 % K with C/N ratio 24:1 the lignin content is reduced to 4.8 % due to composting.

Characteristics of Agricultural Wastes

Knowing the characteristics of agricultural/animal wastes is essential for designing biological waste treatment systems. These characteristics can be expressed in terms of the physical, chemical, and biological properties of the waste material. Qualitative and quantitative measurements provide the engineer with data to properly select and design the system best suited to the livestock operation.

The characteristics of organic wastes are expressed by many variables. The variables most commonly used are shown in Table 6.1.

Solids: The determination of solids is important for evaluating the strength of wastes and for determining the type of treatment needed. It is also used as an index for evaluating the efficiency and effectiveness of waste treatment system components.

TABLE 6.2: Variable used to characterize wastes

Variable	Units of Measurement
Quantity produced	Kg/day
Moisture content	%
Strength, 5-day BOD	mg/liter
Solids	
Total	Kg/day/unit
Volatile	Kg/day/unit
Suspended	Kg/day/unit
Nitrogen	
Total	mg/liter as N
Ammonia	mg/liters
Phosphorus, total	
pH	mg/liter as P

Volatile and Fixed: Total solids may be separated into volatile matter and fixed matter. Generally, the volatile component represents the organic matter present in the waste. Thus, volatile solids are indicative of strength since it is this portion that may putrefy. The fixed solids have little significance in terms of designing and sizing waste treatment systems.

Suspended and Dissolved: Solids can be further divided into suspended and dissolved, each of which can be separated into volatile and fixed portions. The breakdown of solids into their separate components.

The suspended solid determination is extremely important in evaluating characteristics of wastes. Not only is it used to determine the strength of the waste but it is also used to evaluate the efficiency of treatment units. All suspended solids will eventually settle out of solution due to biological and chemical flocculation.

Settleable Solids: Settleable solids is a term applied to solids in suspension that will settle under quiescent conditions due to the effects of gravity. In raw sewage, it is generally a measure of the grit and dirt that can be removed by primary settling tanks before the waste enters into biological treatment processes. Settleable solids are also used to determine the effectiveness of biological treatment processes. It is important in evaluating the performance of oxidation ditches and extended aeration systems.

Nitrogen

Nitrogen in liquid waste can exist in 4 forms, all of which are of interest to the agricultural sanitary engineer. The 4 forms are organic nitrogen, ammonia nitrogen, nitrite nitrogen, and nitrate nitrogen. The total of these 4 forms constitute total nitrogen.

Organic Nitrogen: All nitrogen present in organic compounds is considered to be organic nitrogen. The nitrogen-containing organic compounds are derivations of ammonia, the oxidation of which forms ammonia nitrogen.

Ammonia Nitrogen: The ammonia nitrogen is a result of bacterial decomposition of organic matter. Fresh sewage is generally high in organic nitrogen and low in ammonia nitrogen. The sum of organic and ammonia nitrogen should remain constant for the same liquid wastes, unless ammonia is allowed to escape to the atmosphere because of septic action. The total concentration of the two serves as a valuable index for evaluating the strength of liquid waste and for determining the type of treatment process to select.

Nitrite Nitrogen: Nitrite nitrogen is formed by bacterial oxidation of ammonia nitrogen. It is not present in fresh wastes but appears after bacterial activity has taken place. The presence of nitrite nitrogen indicate that the waste has undergone partial decomposition and is unstable. Nitrites can either be reduced back to ammonia or oxidized to nitrates.

Nitrate Nitrogen: Nitrate nitrogen is formed by the oxidation of nitrites and represents the most stable form of nitrogen. It is an indication of stability and is a determination of the completeness of the biological decomposition process.

Phosphorus

Phosphorus plays an important role in the life processes and thus is important in waste treatment facilities. Phosphorus is especially important in assessing the potential biological activity of aerobic treatment processes. Excessive amounts of phosphorus in surface waters can stimulate algal bloom and eventually eutrophy the water.

pH

pH determinations have very little sanitary significance. It has no direct relationship to the strength of liquid waste or on the method of treatment to be used. Basically, pH is used to control the operation of various biological treatment processes. It is especially valuable in the design and control of anaerobic digestion.

Bio Plant Growth Promoters

The decayed plant extracts are also used liquid manure for promoting plant growth. The Eupatorium weed, the stinging nettle, *Glyricidia* and *Khaki* fruit can be used as liquid manure for growth promotion and observed to be good growth inhibitors of paddy crop.

Livestock Wastes

Generally speaking, livestock wastes can be characterized as high strength, low volume wastes in contrast to municipal wastes that are low strength and high volume. Basically, livestock wastes are solids containing some water whereas municipal wastes are liquids containing some solids. Consequently, approaches different from those used by sanitary engineers must be employed for handling and treating livestock wastes.

Type of Production Unit

Livestock waste can occur as solids, semisolids, or liquid depending upon the design and operation of the production unit. In solid floor buildings and paved feedlots where manure is allowed to accumulate, its moisture content and consistency at any given time are a function of the age of the manure and its exposure to environmental conditions. If it is under, roof, the moisture of the manure will be dictated by the slope of the floor, ventilation rate, temperature, and relative humidity. Manure exposed to adequate drainage, good ventilation and a hot, dry solid that can be readily scraped, scooped, stacked and transported. On the other hand, if the accumulated manure on a solid floor is exposed to poor drainage, moist conditions, and occasional rainfalls, the results will be entirely different. Under these conditions the manure will have a high moisture content and appear as a semisolid. In many cases, the excess moisture will leach out many of the solids and nutrients and drain from the solid floor as a liquid waste.

In some production units, the manure is allowed to drop through slotted floors or screened cages and accumulate in a pit of water beneath the animals. These units can be operated as either aerobic or anaerobic systems. In either case, the manure is mixed with a great deal of water and is treated as a liquid. This is also the case for production units using flushing gutter systems that either continuously or periodically discharge large volumes of waste to flush the manures from the building. Thus, it is apparent that the final consistency of animal waste may appear either as a solid, a semisolid, or a liquid. Its final form is dependent upon the type of production unit chosen.

Species and Age of Animal

The amount of manure produced along with its chemical and physical characteristics is dependent upon the age and species of the animal. For example, wastes from mature stock and milk animals will be digested to a greater degree than that from younger, growing livestock.

Waste from ruminants has a different composition than waste from simple-stomached species. Ruminants, because of the bacteria in their stomachs, and utilize feeds high in cellulose which are difficult to digest. Swine and poultry on the other hand consume diets that are highly digestible. As a result, the characteristic of waste from these two classifications of animals varies, requiring different design conditions for treatment.

Other Factors

Other factors that may affect the characteristics of waste are bedding materials and spilled feed. Bedding material may consist of straw, wood shavings, sawdust, or shredded paper. It may be used for the comfort of the animal or simply to absorb the excess moisture in the waste. In either case, it will affect the characteristics of the waste and must be taken into account in the design of the treatment facility.

Spilled feed can also alter the characteristics of livestock wastes. It is a good practice to eliminate this source of wastage. Not only does it increase production costs but it puts an additional load on treatment facilities.

Beef Cattle

Feedlots.-Most of the beef cattle raised for meat production are produced in feedlots that are either paved or unpaved. In either case, the manure is allowed to accumulate over a period of time. During this time, the manure is exposed to wetting and drying action caused by the weather, microbial activity, and mixing and compaction due to the animal activity. The quantity and characteristics of the manure will vary significantly with the rations. The studies conducted by Grub *et al.* (1969), cattle fed on three different diets consisting of a high concentrate ration, cottonseed hulls, and sweet sorghum silage produced 1.1, 3.0, and 0.73 kg of dry solids per day per animal, respectively. Cattle fed a predominantly high roughage diet will produce more waste than cattle on high concentrate rations.

Under these conditions, manure handling generally consists of periodic scraping and field spreading. Estimates by (Madden and Dornbush, 1971) indicate that approximately 95% of the total waste produced is either removed by cleaning operations or decomposed on the feedlot surface. Potentially 5% of the total waste generated may leave the feedlot as surface runoff. The characteristics of the runoff from feedlots are given in Table 6.3. A common method for treating the runoff is by use of aerobic or anaerobic lagoons.

TABLE 6.3: Runoff from Beef feedlots

Variable	Units of Measurement (Kg/Day/1000 Kg)
Quantity produced	5% of total
Moisture content	-
Strength, 5-day BOD	1.67
Solids	
total	0.10
volatile	0.05
suspended	-
Nitrogen	
total	0.33
ammonia	0.11
Phosphorus, total	0.11
pH	-

Slotted Floors

Some beef cattle are confined within building sufficiently insulated and ventilated. Under this management practice, the animals are placed on slotted floors and the manure is allowed to accumulate beneath the floor in manure pits. Using this system, a manager has two options. He may choose to let the manure in the pit accumulate and undergo anaerobic decomposition. Then when conditions permit, he can dispose of the waste by field spreading.

The other option may be to treat it beneath the floor aerobically and allow the excess to run into a polishing lagoon. In either case, the manure under this management practice will be different from that of feedlot operations. The characteristics of these wastes are presented in Table 4.3. Work by McCalla and Viets (1969) indicates that the fertilizer value of manure slurry contains 2.64, 1.64, and 3.00 kg of nitrogen (N), phosphorus (P_2O_5), and potassium (K_2O), respectively.

Dairy Cattle

Stanchions.-Dairy cattle are housed in buildings using either stanchions or free-stalls. Manure from stanchion barns is allowed to collect in gutters where it is manually or mechanically scraped and stacked in storage areas until it can be hauled to fields for spreading and utilization. Farmers employing stanchion barns generally use bedding such as sawdust, straw, or wood shavings for the animals.

While the manure and bedding are stacked, liquid wastes will seep from the stored manure. The seepage liquid has a high pollution potential and must be controlled in order to avoid contamination of surface waters.

The liquid wastes can be collected in storage tanks where they can be used as liquid fertilizer or they can be allowed to flow to a lagoon for subsequent treatment. Characteristics stacked stanchion barn manure depend upon the time of storage, environmental conditions, and the type and amount of bedding used.

Free-stall.-Manure produced from dairy cows housed in free-stall barns can be scraped by a front-end loader and stacked in a storage area for later utilization. In many of the newer setups, the manure is flushed by large volumes of water discharged a few times a day. The characteristics of fresh dairy wastes are shown in Table 6.4. The liquid waste from the flushing aisle is discharged to a series of lagoons for treatment. Normally the effluent from the lagoon is used as the flush water. In other setups, the liquid waste is passed through a separator where the majority of the solids are removed from the liquid waste. The solid portion is transported and stacked for further utilization while the liquid part is allowed to flow to a lagoon for treatment.

Fresh manure contains approximately 5.0 kg of nitrogen, 2.0 kg of phosphorus, and 4.5 kg of potassium per each 1000 kg fresh, wet manure. Much of this value is lost, however, due to seepage and the volatilization of nitrogen during decomposition.

TABLE 6.4: Dairy cattle manure, stacked and stored

Variable	Units of Measurement
Quantity produced	50 Kg/day/animal
Strength, COD	160,000-268,000 mg/liter
Solids	
Total	18-30%
Volatile	75-89% of TS
Suspended	-
Nitrogen	
Total	4100-6100 mg/liter
Ammonia	700-2500 mg/liter
Phosphorus, Total	3800-6900 mg/liter
pH	6.6-8.6

TS = Total solids

TABLE 6.5: Liquid seepage from stacked manure

Variable	Units of Measurement
Quantity produced	4.5-11.0 litre/day/cow
Strength, 5-day BOD	4200-31,000 mg/liter
Solids	
Total	1.8-4.3%
Volatile	0.9-2.6%
Suspended	0.2-0.8%
Nitrogen	
Total	1200-2900 mg/liter as N
Ammonia	780-2200 mg/liter as N
Phosphorus, Total	64-500 mg/liter as P
pH	6.4-7.8

Utilization of Agricultural Organic Waste

Chemical constituents like cellulose, hemi-cellulose and lignin lead to establish the potential of agricultural wastes as food, feed, fodder, fibre and chemical products of economic importance. The **calorific value** (CV) indicates the fuel utility of agricultural waste. The CV for various agricultural waste materials varies between 14 to 19 MJ/kg. Higher C content favours calorific value of the waste material. The volatile water in plant residues may be around 80 per cent and fixed carbon is 20 per cent. Volatile matter in plant residues contribute towards oil and tar content. The characteristics of common agricultural wastes available in India are mentioned in Table 6.7.

TABLE 6.6: Dairy manure characteristics

Variable	Units of Measurement (Kg/Day/Animal)
Quantity produced	36-45
Moisture content	85%
Strength, 5-day BOD	0.6-0.78
Solids	
Total	3.0-4.7
Volatile	2.6
Suspended	2.0
Nitrogen	
Total	0.17
Ammonia	0.10
Phosphorus, Total	0.05
pH	-

TABLE 6.7: Characteristics of common agricultural wastes
(a) Physical and thermal characteristics

Agricultural waste (biomass)	Density (kg/m^3)	EMC at 80%	Lower calorific value (MJ/kg)	Proximate analysis (%)		
				FC	VM	Ash
Pigeonpea stalks	180	20.58	17.60	15.12	82.90	1.98
Bagasse	70	34.86	18.30	15.86	79.20	4.94
Cotton sticks	160	27.05	16.00	15.30	81.40	3.30
Groundnut shell	100	-	18.60	11.67	83.90	4.43
Maize cobs	100	28.00	15.30	15.16	83.01	1.84
Maize stalks	50	38.00	16.10	17.10	79.60	3.40
Rice husk	105	29.40	14.10	12.50	71.00	16.50
Rice straw	30	36.70	13.50	11.10	69.70	19.20
Wheat straw	60	34.00	15.70	17.93	73.60	8.47

(b) Chemical characteristics

Agricultural waste (biomass)	Cellulose (%)	Pentosan (%)	Lignin (%)	Al/BZ extractives (%)
Pigeonpea stalks	41.90	10.30	39.30	2.20
Bagasse	40.00	22.60	14.80	15.90
Cotton sticks	41.90	19.00	27.20	9.30
Groundnut shell	36.50	13.90	31.30	12.20

Agricultural waste (biomass)	Cellulose (%)	Pentosan (%)	Lignin (%)	Al/BZ extractives (%)
Maize cobs	36.80	27.90	11.20	14.60
Maize stalks	43.40	19.60	22.80	9.30
Rice husk	44.10	17.80	17.20	3.40
Rice straw	41.40	20.4	12.10	5.60
Wheat straw	39.60	24.10	17.00	7.30

(c) Elemental composition

Agricultural waste (biomass)	Nutrient Content (%)								
	C	H	N	Na	K	P	Ca	Mg	SiO_2
Pigeonpea stalks	53.30	4.70	0.60	0.05	0.57	0.08	0.11	0.41	0.68
Bagasse	48.20	6.10	0.20	0.06	0.51	0.04	0.14	0.36	1.30
Cotton sticks	51.00	4.90	1.00	0.09	0.61	0.08	0.12	0.43	1.33
Groundnut shell	41.10	4.80	1.60	0.05	1.20	0.12	0.10	0.40	2.52
Maize cobs	46.20	4.90	0.60	0.03	0.54	0.07	0.08	0.28	0.90
Maize stalks	41.10	4.20	0.60	0.04	0.42	0.05	0.09	0.45	2.00
Rice husk	37.80	5.00	0.30	0.02	0.30	0.03	0.08	0.17	15.60
Rice straw	36.80	5.00	0.59	0.09	2.50	0.06	0.10	0.53	16.77
Wheat straw	43.80	5.40	0.40	0.06	0.78	0.04	0.10	0.35	7.08

Source: Pathak *et al.* (1990)

The characteristics of the agricultural waste materials make them important in respect of their use and economic value. Agricultural wastes can be put to use because of:

1. Farm development (improvement of soil health, soil fertility, soil physical conditions and plant protection).
2. It is the source of energy and power.
3. It acts as animal feed and fodder.
4. It is the human food.

Recycling of Organic Materials for Fertilization

Organic wastes and surplus crop residues can be recycled in soil by different methods such as *in situ* incorporation of organic matter or by organic mulch besides composting which will eventually improve the biological, chemical and physical properties of soil resulting in higher crop productivity. Each season/year substantial quantities of plant nutrients are removed from soil by crop harvesting. On an average, one tonne of grain alongwith straw of wheat, paddy, *jowar*, *bajra*, maize and barley are likely to remove 33 kg N, 12 kg of P_2O_5 and 58 kg of K_2O. Therefore, estimated 125 million tonnes of food grains produced in the country will remove about 12.9 million tonnes of plant nutrients. The nutrient supply through chemical fertilizers during 1977-78 was of the order of 4.29 million tonnes only. Thus the balance of 8.61 million tonnes has to be met through organic sources. In this context recycling of organic wastes/by-product becomes vital for returning back at least a part of plant nutrients drawn from soil for maintaining humus and the soil productivity.

Direct Incorporation of Organic Materials in the Soil

Organic materials of different sorts can be ploughed in the soil (0-20 cm layer) about 3-4 weeks prior to sowing of crop with optimum moisture level. The prior decomposition of organic materials is important to ensure benefit to the crop and without nitrogen immobilization by microorganisms. The rate of decomposition will depend on the nature of organics used. Leguminous residues and non-edible cakes will mineralize faster as compared to cereal residues/straw). Addition of nitrogen in the form of chemical fertilizers or non-edible cakes will accelerate the pace of mineralisation of cereal residues/ straw poor in nitrogen. On the other hand, mineralisation of humified FYM or compost is slow and steady process and acts as a slow release fertilizer. It can be incorporated in soil even just before sowing of the crops. Application of farmyard manure or compost is the best source for maintenance of soil organic matter. The next in order is cereal straw/ residues for the maintenance of organic matter in soil. Legume residues are not good for maintenance of soil organic matter in soil. Legume residues are not good for maintenance of soil organic matter in tropics but are good sources of plant nutrients, specially nitrogen and they may add as much as 40 k N/ha in one season.

Organic Mulch

This is an unique and simple method particularly for recycling of dry and nitrogen-poor organic materials since its incorporation in soil is not required at this stage and the costs of ploughing, water and inputs of chemical fertilizer for their favourable utilisation is not involved. Although decomposition of mulch is a slow process, its biomass and C/N ratio during the course of one crop season is appreciably reduced and this facilitates its incorporation in next season at a low cost. The experiments have shown that straw mulch besides controlling the weeds and conserving moisture and thereby saving irrigation, also increases crop yield and population of beneficial soil microflora. Mulch is discussed in detail in chapter 13

CONCENTRATED ORGANIC MANURES

Concentrated organic manures are those materials that are organic in nature and contain higher percentages of essential plant nutrients such as nitrogen, phosphorus and potash, as compared to bulky organic manures. These concentrated manures are made from raw materials of animal or plant origin. The concentrated organic manures commonly used are aquatic weeds, oilcakes, blood-meal, fish meal, meat meal and horn and hoof meal.

TABLE 6.8: Nutrient content of aquatic weeds

Weed	N (%)	P (%)	K (%)
Azolla pinnata	2.33-3.83	0.196	0.153
Eichhornia crassipes (small)	1.50-2.20	0.37-0.56	2.68-3.45
E. crassipes (large)	1.70-2.15	0.38-0.56	2.68-3.45
Lemna minor	3.56-3.75	0.23	0.18
Monochoria vaginalis	3.12-3.75	1.84-1.97	4.34-4.91
Pistia stratiotes	2.12-2.21	0.75-0.89	2.36-2.39
Salvinia sp.	1.12-2.05	0.34-0.61	2.10-3.30
Spirodela polyrhiza	2.76-4.54	0.37	0.59
Spirodela and *Lemna*	4.06	0.78	1.96

Source: Majid, 1983.

Aquatic Weeds

Aquatic weeds are still regarded by many people as a menace, because they are not yet aware of the great potential and economic value of these profusely growing, un-controllable plants. These highly productive plants are generally more productive than conventional field crops. Moreover, when grown on wastes, they do not compete with conventional crops for fertilizer, water or land. Their natural profuse growth in the humid tropical and subtropical areas of the world, requiring no intensive cultivation, make them a promising source of multipurpose raw material. As a matter of fact, these plants have proved to be good source of livestock feed, human food, fish feed, organic or biofertilizer, energy, fibre and paper. Furthermore, the aquatic weeds have the capacity to purify waste water through the uptake of dissolved nitrogen, phosphorus and undesirable excessive minerals, including heavy metals.

TABLE 6.9: Comparative effect of compost of aquatic weeds/algae, cow manure on crop yield

Crop	Comparison of Manure	Increase in Yield
Rice	Aquatic weed (*Eichhornia, Pistia, Spirodela* mixed) Vs. cow manure	28.57% more in cow manure
Rice	Aquatic weed (*Eichhornia crassipes*) Vs. cow manure	22% more in aquatic weed
Rice	Alga (*Rhizoclonium*) Vs. cow manure	40% more in alga
Rice	Aquatic weed (*Spirodela*) Vs. Cow manure	33.33% more in aquatic weed
Wheat	Aquatic weed (*Eichornia crassipes*) Vs. Cow manure	30.23% more in aquatic weed
Maize	Aquatic weed (*Eichhornia, Pistia, Spirodela* mixed) Vs. Cow manure	40% more in cow manure
Soybean	Alga (*Spirogyra*) Vs. Cow manure	55% more in the alga
Sesame	Aquatic weed (*Eichornia crassipes*) Vs. Cow manure	21.5% more in aquatic weed
Sesame	Aquatic weed (*Azolla*) Vs. Cow manure	50% more in aquatic weed
Sesame	Alga (*Spirogyra*) Vs. Cow manure	37% more in the alga

Crop	Comparison of Manure	Increase in Yield
Onion	Aquatic weed (Eichhornia, Pistia, Spirodela mixed) Vs. Cow manure	13% more in aquatic weed
Onion	Aquatic weed (*Eichornia crassipes*) Vs. Cow manure	25.5% more in aquatic weed
Basella rubra (Beng. Puin Shake)	Aquatic weed (*Eichhornia*, *Pistia*, *Spirodela* mixed) Vs. Cow manure	9% more in cow manure
Spinach	Aquatic weed (*Eichornia crassipes*, *Pistia* mixed) Vs. Cow manure	20% more in cow manure
Lettuce	Aquatic weed (*Eichornia crassipes*) Vs. Cow manure	23.65% more in cow manure
Cabbage	Aquatic weed (*Eichornia crassipes*) Vs. Cow manure	59% more in cow manure
Cauliflower	Aquatic weed (*Eichornia crassipes*) Vs. Cow manure	35.62% more in cow manure
Cauliflower	Drain algae (*Phacus*, *Chlamydomonas*, *Oscillatoria*, Diatoms) Vs. Cow manure	4.33% more in cow manure
Gourd	Aquatic weed (*Eichornia crassipes*) Vs. Cow manure	55% more in aquatic weed
Beet	Drain algae (*Phacus*, *Chlamydomonas*, *Oscillatoria*, Diatoms) Vs. Cow manure	7.34% more in cow manure
Turnip	Aquatic weed (*Eichornia crassipes*) Vs. Cow manure	7.5% in cow manure
Radish	Aquatic weed (*Eichornia crassipes*) Vs. Cow manure	3.70% more in aquatic weed
Brinjal (Eggplant)	Alga (*Spirogyra*) Vs. Cow manure	2.71% more in the alga

Source: Majid *et al.* (1980)

Whereas organic manure has been found to be absolutely essential for maintenance of soil productivity, continuous use of chemical fertilizer, without the supplement of organic manure, has been reported to adversely affect the physical properties of soil. Scientists have been urged to find organic and biological techniques to replace excessive use of chemical fertilisers and pesticides in view of the problems of ecological imbalance.

Aquatic Weeds as Source of Energy

Many developing nations like India, Bangladesh, Pakistan, Philippines and Thailand etc., have an inexhaustible supply of aquatic weeds which can form a potential source of energy. Fuel from aquatic weeds, even if in a small quantity has economic implications, especially for rural areas in developing countries. Aquatic weeds may be used for biogas production as an organic manure and soil conditioner.

Aquatic Weeds as Organic Manures

The beneficial role of organic manures in maintenance of soil fertility and crop productivity has already been discussed before and thus hardly needs any emphasis. Organic manures act in many ways in augmenting crop growth and soil productivity. The direct effect of organics relate to the uptake of humic substances or its decomposition products affecting favorably the growth and yield of plants. Indirectly, it augments the beneficial soil microorganisms and their activities and thus increases the availability of major and minor plant nutrients.

Aquatic weeds have been reported to be a good source of organic manure. Many aquatic weeds contain appreciable amounts of nitrogen, phosphorus and potassium. Composted aquatic weeds, when applied to farm land, have been reported to benefit the field crops. Aquatic weeds not only provide crops with nutrients, they also improve soil's physical properties like, texture and water holding capacity of soil, which are very important in sandy, lateritic and heavy clay soils widely distributed in developing countries. *Azolla*, a good source of biofertilizer, is capable of reducing nitrogen demand of soil. Residual effect is also appreciable when aquatic weed compost is added to the soil.

Oilcakes

These can be grouped into two classes, namely. (i) Edible oilcakes- suitable for feeding the cattle and (ii) non-edible oilcakes- not suitable for feeding the cattle. Edible oilcakes, are normally fed to cattle as concentrates. However, some of the edible oilcakes are also applied by the farmers to the soil. The quantity of nitrogen varies with the type of oilcake. It is from 2.5 per cent in *mahua* cake to 7.9 per cent in decorticated safflower cake. In addition to nitrogen, all oilcakes contain small percentages of P_2O_5 (0.8 to 2.9 per cent) and potash (1.2 to 2.2 per cent). Oilcakes are the quick-acting organic manures. Though they are insoluble in water, their nitrogen becomes quickly available to the plants in about 7 to 10 days after application. The *mahua* cake, however is an exception, as its nitrogen does not become available till about two months after application. As such *mahua* cake should be applied about two months before sowing, provided the soil is moist. This cake is suitable for application to fruit plants and plantation crops.

How to Use Oilcakes

These should be well-powdered before application, so that they can be spread evenly and are easily decomposed by microorganisms. They can be applied a few days prior to sowing or at sowing time. Oilcakes-especially groundnut cake, are also applied extensively in the form of a top-dressing to sugarcane. Depending on crop, oilcakes are applied broadcast, drilled or placed while earthing up near root zone.

Cattle, Pig, and Poultry Manures

In general, the higher the dry matter, the higher the nutrient content. For example, in Table 7.1 the higher nutrient content of the poultry manure is due mainly to the higher dry matter content. There is usually about the same level of nitrogen and potassium in cattle slurry, and phosphorus content is usually about 25% of the nitrogen and potassium level. Pig manure usually has about the same level of nitrogen and potassium in cattle slurry, and phosphorus content is usually about 25% of the nitrogen and potassium level.

TABLE 6.10: Composition of Cattle, Pig, and Poultry Manures

	Nutrients (kg per 10 tonnes fresh manure)				
	Dry Matter (%)	N	P	K	Mg
Cattle manure	4-23	24-65	4-18	20-58	2-6
Pig manure	5-25	16-68	6-21	17-36	3-7
Poultry manure	23-68	96-230	24-120	38-116	12-22

Pig manure usually has about the same level of nitrogen as does cattle manure, although pig manure is usually higher in phosphorus and lower in potassium than is cattle slurry. The difference in the phosphorus to potassium ratio between cattle and pig manure is principally a reflection of the diet of these animals. The relatively low potassium and high phosphorus in pig slurry reflects mainly a cereal diet, and the high potassium in cattle manure reflects the high potassium content of herbage, which often makes up a high proportion of the ruminant diet.

Poultry

The nutrient ratio of poultry manure is somewhat similar to pig manure and probably reflects the diets of pigs and poultry. The variation in slurry composition can be greater than for solid manures. Poultry water can be divided into two parts

Litter Grown

Most of the poultry produced is grown on litter which absorbs the moisture from the manure. The litter may be sawdust, wood shavings, vermiculite or other suitable material. The spent litter is removed from the buildings depending upon specific management practices and field spread. Waste management of poultry litter is strictly a solid waste handling proposition.

Cage Grown

Some broilers and layers are raised in screened cages, which allow the wastes to drop to a holding pit directly beneath the animals. In some cases, the pit contains water and serves as both a holding and treatment facility. The liquid waste can be pumped into a liquid spreader for field application or into a lagoon for additional treatment. In other systems, the pits are equipped with ventilation facilities to remove the moisture from the manure, leaving a dry material as the final end-product. The dried manure can be transported and field spread, or bagged and sold as commercial fertilizer. Manure from broilers contains approximately 17 kg of nitrogen 8 kg of phosphorus, and 12.5 kg of potassium per 1000 kg of manure. Manure from layers averages 13 kg of nitrogen, 12 kg of phosphorus, and 11 kg of potassium per 1000 kg.

Meat-meal

An adult animal can provide 35 to 45 kg of meat after slaughter or death. At present, meat meal is manufactured in India on a small scale. The process of manufacturing meat-meal is very simple. First the bones and the meat are cooked or digested in a special containers for two to three hours. The bones are then separated from the meat. This meat is dried on a sand bath till it is brittle and then it is powdered. The drying can be done in double-jacket trays worked by steam and the material dried over the steam for an hour. Mutton squeezer can also be used for removing water content of the meat, and then dried. Meat-meal is a quick-acting manure and is effective for all crops, on all soil types. Its application is similar to oilcakes. Meat-meal contains about 10.5 per cent N and 2.5 per cent P_2O_5.

Bloodmeal

An adult cattle gives about 14.0 kg, and a goat or sheep about 1.40 kg of blood. Slaughter-houses should be provided with *pucca* or concrete floor with a central drain leading into a blood storage tank. The blood is first treated with commercial copper sulphate @ 125 gm per 100 kg of blood. It is then evaporated to dryness on a sand bath. Next, it is spread on a concrete floor covered by a net, and allowed to dry in sun. When completely dried, it is powdered, bagged and marketed as bloodmeal. Bloodmeal is a quick-acting manure and is effective for all crops on all soil types. It should be applied like oil-cakes. Bloodmeal contains 10 to 12 per cent of N and 1 to 2 per cent of P_2O_5.

Fishmeal

Non-edible fish carcasses are used to prepare fishmeal. The fish is dried, crushed or powdered and filled in bags. Fish manure is available either as dried fish, or as fishmeal, or powder. The manurial constituents present in it vary with the type of fish. Fishmeal is quick-acting organic manure and is suitable for application to all crops on all soils. It should preferably be powdered before use. Fishmeal contains 4 to 10 per cent N, 3 to 9 per cent P_2O_5 and 0.3 to 1.5 per cent K_2O

Horn-and-Hoof Meal

A healthy animal gives about 3 to 4 kg of horn and hoof. These materials, are cooked dried and powdered before bagging. This meal contains about 13 per cent N.

COLLECTION AND STORAGE OF ORGANIC WASTE

Traditionally, animals were bedded with straw or other litter and the manure was mixture of urine, faeces, and straw. This farmyard manure was normally stored and spread as a solid. A high proportion of agricultural waste is still handled with straw, as farmyard manure. However, an increasing proportion is being stored and handled as a liquid or slurry where no bedding is used. This is particularly true on most new and large intensive animal enterprises.

Animal wastes can be divided into three main categories:

1. Farmyard manure, where straw or other bedding is used;
2. Slurry, which is a mixture of faeces and urine, often containing variable quantities of rainwater and wash water; and
3. Separated manure, where the waste is separated into two fractions-solid and liquid-which are handled separately.

The third category is receiving increased attention with the use of mechanical slurry separators, where the liquid fraction is spread or irrigated to land, and the solid fraction is handled separately as a solid. Other categories of manure in which non-mechanical separation takes place can be included here. In arid or semi-arid regions, manure quickly dries out by evaporation after it is produced in feedlots or corrals. This manure is handled as a solid and is sometimes stacked. With this method of storage, there is no liquid for disposal; however, there can be runoff problems after heavy rainfall. In temperate regions with relatively high rainfall, such as Northern Europe, slurry is sometime pushed into compounds and solid/liquid separation takes place during storage. A solid crust forms on the top of the compound and sludge on the bottom with liquid in the centre. This method is sometimes described as dung stead storage and is essentially non-mechanical solid/liquid separation. The liquid fraction can be drawn off and spread first, and then the solid fraction remaining can be spread.

Crop residues could be considered a source of agricultural wastes. Residues from most root crops, however, are returned directly to the soil during harvesting (for example, potato haulms). These residues could not be considered as wastes and are really in the same category as plant roots. The residues do, however, supply considerable quantities of essential nutrients for the succeeding crops after mineralization by soil microorganisms.

Straw on soil after harvesting cereal crops is in a somewhat different category, as it can interfere with subsequent mechanical cultivation. To avoid this problem, straw is often burned or mechanically chopped into small pieces after harvesting. Straw left on the soil in this manner supplies the small quantities of phosphorus, potassium, and other nutrients present for subsequent crops. Due to the high carbon to nitrogen ratio of straw, the mineralizing organisms may reduce the available soil nitrogen, and extra should usually be applied-about 30 kg N/ha, if straw is not burned or removed. In recent years, straw is increasingly being used as source of feed for cattle. Research has shown that the feeding value and digestibility of straw can be increased by treatment with an alkali, such as sodium hydroxide, or by mechanical treatment, such as very fine grinding

Dairy cattle are sometimes housed in straw bedded sheds. The manure is usually allowed to accumulate to a depth of about 1 metre before cleaning. The manure is then either spread directly to land or stored in a heap for subsequent spreading.

Increasingly, manure is being collected and stored as slurry. The concept of handling manure as slurry containing a mixture of faeces, urine, and water is not a new one. Different methods of slurry handling have been practiced over the past century.

Modern methods of handling animal wastes as slurry were greatly influenced by the development, in Scandinavia, of slatted floors for animal rearing. This development has revolutionized cattle and pig housing over the past three decades. Along the same lines is the "slope slot flush," which was developed and is currently studied in the United States. With this system, manure passes through slots into pipes in the floor and flows to storage tanks outside the building.

Free stall cubicle houses are a popular method of housing cattle, particularly dairy cows. With this system, the manure is usually cleaned by tractor scraper off concrete yards and passes into storage tanks. The storage tanks may be under ground or over ground and can be concrete or steel, glass lined, or made from other materials. Tractor-driven pumps are sometimes used to transfer the manure to an over ground storage tank.

Slatted floors are extensively used for beef animals, and in the past few years, slatted passages for dairy cows have been widely adopted. With the slatted floor house, the manure is usually stored in two metre deep tanks beneath the house. There is usually adequate manure storage for the winter feeding period, and there is no labour involved in manure handling while animals are in the house. It passes directly as it is produced through the floor into the storage tanks below.

For pigs, straw is still sometimes used for bedding, but the majority of intensive units collect and store manure as a slurry. Part of the floor area is normally slatted and the slurry flows in manure channels to outside storage.

In the case of poultry, broilers are usually reared on deep litter. The litter is usually wood shavings, chopped straw, or similar materials. The house is cleaned out after about 11 weeks, when the broilers are sold. Fresh litter is then put in for a new batch of chicks. This manure usually has about 75% dry matter and is either stored in a heap or spread straight from the house. Laying hens are now usually housed in cages and the droppings pass through the cage floor. The droppings are usually taken by conveyor belt to a spreader or storage outside the house. The hen droppings have about 25% dry matter. Of all agricultural wastes, poultry manure contains the highest level of plant nutrients and therefore has the highest fertilizer value.

Many machines for spreading manure have been designed for disposal, so that manure is often spread unevenly. To get maximum fertilizer value from manure it is important that machines should be capable of spreading evenly, as with fertilizer distributors. Most manure is spread on the soil surface. However, the possibility of injecting slurry into the soil is receiving attention. Slurry injection into cultivated soil has been used for many years, and more recently, injecting tines suitable for grasslands are being used. Injecting is now carried out primarily to reduce smell after spreading, but it can also increase the effectiveness of slurry nitrogen and reduce contamination of grass.

Treatment of slurry along the lines used for urban sewage has been studied extensively over the past two decades. The consensus of opinion at present is that application to land is the only economically practical method of disposing of animal manures. An intensive pig enterprise with 5.000 pig places would have a waste treatment problem equivalent to a town with about 12,000 people. Conventional treatment on this scale would make animal products very expensive. Veal calf wastes are very dilute, by comparison with other animal slurries and are being treated successfully in Holland. Some pretreatment of slurry prior to spreading on land may be desirable; for example, solid/liquid separation may reduce some of the problems involved in pumping high dry matter slurry. To avoid spread of disease, slurry can be treated with lime before spreading on land.

ECONOMIC VALUE OF ORGANIC WASTE

The economic value of an organic waste or residue to a farmer is the value of the increase in crop yield and/or crop quality that is derived from its use. Since crop yield response to an organic amendment follows the law of diminishing returns, the average yield increase is always larger than the yield increase attributable to the incremental unit of organic material. If the price of the farmer's product is the same regardless of the quantity produced, which is usually the case, then the revenue derived from the average unit of organic material will be greater than the revenue from the incremental unit. Thus, the value of the average unit of organic material will be higher than the value of the incremental unit of organic material.

Farmers can be expected to utilize organic materials to the point where the revenue from the incremental unit is equal to its price, assuming application is included in the price. They would certainly use no more than that quantity since additional application of the organic material would yield a loss. Likewise, they would not wish to use less since total profits would decline. Thus, an estimate of the value of the incremental waste is necessary by management agencies when analyzing the marketability of a recycled product. However, public decision-makers should also be interested in the estimates of the total economic benefits from resource recycling alternatives, such as composting, compared with ultimate of the value of the average unit of organic material would be the appropriate figure to use in public project analysis.

AVAILABILITY OF ORGANIC WASTE IN INDIA

There are various estimates of agricultural waste availability. These may be based on crop-residues, net availability on remote sensing techniques which is used in developed countries.

TABLE 6.11: Agricultural waste availability in India

Agricultural Waste	Quantity (mt.)
Crop waste (including industrial waste)	301.6
Animal waste (including fisheries and marine waste)	944.4
Crop residues (including industrial waste)	108.0
Crop waste (cereals only)	289.7
Crop residues	407.0

Agricultural Waste	Quantity (mt.)
Animal wastes	2018.0
Crop residues (including industrial waste)	347.2
Crop waste residues	287.3
Animal waste	633.0

Estimates of agricultural waste availability suggest that the average value for crop wastes is 350 Mt and that of animal wastes it is 650 mt. Hence, around 1000 mt. of agricultural wastes are available in the country.

PROCESSING OF AGRICULTURE WASTE

Before application in the field the agriculture waste need to be processed for-

1. Maintaining the low Carbon: Nitrogen ratio, otherwise it would compete with crop for the nitrogen required for decomposition in the field.
2. Making material hygienic so that pathogen should not be transmitted to crop/human beings.
3. Increase the nutrient availability by decomposition popularly known as composting and discussed in detail in chapter *Production of Compost*. Composting is one way in which some of the problems associated with the utilization of various organic wastes (e.g., odors, human pathogens, and storage and handling constraints) can be resolved. Composting is an ancient practice whereby farmers have converted organic wastes into resources that provide nutrients to crops and enhance soil tilth, fertility, and productivity. Through composting, organic wastes are decomposed, nutrients are made available to plants, pathogens are destroyed, and malodours are abated. Composts provide a more stabilized form of organic matter than do raw wastes, and can vastly improve the physical properties of soils. For example, addition of sludge compost to sandy soils will increase their ability to retain water and render them less droughtly. In heavy-textured clay soils, the added organic matter will increase permeability and water infiltration and clay soils have been shown to reduce soil compaction, lower the bulk density, and increase the rooting depth.
4. Reduced the possibility of attack of Cellulose loving insects e.g. termite, as the Cellulose is decomposed during composting.

CONCLUSION

Without regular application of organic manures and recycling of crop residues, we cannot hope to maintain and sustain productivity. We need policies and strategies to prepare and distribute for use in agriculture and using modern technology on reducing their bulk. This assumes greater significance in the coming years as with rapid urbanization the bulky organic wastes are increasing. On the other side, increasing interest in organic farming is creating immense possibility of their eco-friendly utilisation.

CHAPTER 7

Production of Compost

Production of compost is the heart of the organic method. Dying and decaying matter, whether the falling leaves of the forest or the decaying bodies of more complex life forms, is constantly being used as the raw material of new life. In nature, the forest floor is the *workshop* for a continuous compost production operation. In agriculture fields, this normal cycle is interrupted by taking out some of the organic material produced. In setting up a compost-making operation, we are restoring the cycle and returning to the soil the humus, which is the single most important element in its fertility. The process of compost or *life renewal* is achieved by a remarkably complex interaction of billions of microscopic life forms with biological materials.

COMPOSTING

Composting is a biological decomposition process that converts organic matter to a stable, humus-like product under controlled conditions. During the composting process, microorganisms utilize the decomposable microbial substrates present in the organic compost both as an energy source and for conversion to microbial substances.

The composting process of simply a means of converting raw organic compost, a potential source of odor and public health problems, into a safe product called humus. By definition, composting is an alternative system for solving some of the compost handling problems existing on today's farms.

> *Compost is the dark brown crumbly material that is produced when a collection of plant and animal material is decomposed into fine organic matter and humus. Once the compost has been mixed into the soil it will undergo the process of mineralisation in which the humus releases minerals into the soil, making them available to the plants.*

It is a natural process, which occurs in nature in which organic matter is decomposed by microorganisms forming a humus-like substance. The process itself is not new. It has been in practice for many centuries by farmers who have stacked animal manure into piles or gardeners who have placed garbage, leaves, grass cuttings, etc., into pits.

Despite its wide usage, no advances were made in the treatment process until 1925 when Sir Albert Howard, a British agronomist stationed in India, developed a systemized process for composting. The improved process was labeled the Indore Method after the stage in central India where it was conceived and tested. The method described by Howard (1935) involves the formation of a layered pile about 1.5m high using garbage, animal composts, sewage sludge, straw, and leaves. Initially the process was anaerobic and required 6 months for completion to occur. The process was later modified by turning the pile over twice, which reduced the composting time to 3 months.

During the next 30 years, various methods for mechanically composting large quantities of refuse were tested and patented in Europe. Research on composting did not begin in the United States until the 1950s when the University of California at Berkeley embarked on an extensive investigation. These studies were the first to scientifically describe the principles of composting and led to the development of a rational technology. In 1966, the United States Public Health Service supported a demonstration grant in cooperation with the Tennessee Valley Authority and the municipality of Johnson City, Tennessee, to study the feasibility of composting solid composts and sewage sludge. Since then a great deal of research has been conducted by many of the Land-Grant Schools to determine the feasibility of composting and its value as a means for treating agricultural composts.

IMPORTANCE OF COMPOSTING

A great deal of organic compost from the kitchen and farm can be recycled and given back to the soil in the form of compost. When compost is added to the soil the level of organic matter increased, which is beneficial in many ways:

1. The finished product obtained from composting may have some fertilizer value but it should be strongly emphasized that compose is an excellent soil conditioning agent.
2. Incorporating compost into the soil increases the organic content and improves the texture, the permeability, and the water-holding capacity of that soil.
3. An improvement in humus content, hygroscopic moisture, water retention capacity, and absorption capacity when organic matter is added to the soil.
4. Compost can be used for improving the fertility of marginal and arable land and also for restoration of land that has been severely eroded or strip mined.
5. Compost can also be used as a mulch for nurserymen and vegetable farmers. Compost used as a mulch has advantages over peat moss of bark in that the compost releases absorbed water more readily than does peat moss, and imposes a lesser demand on soil nitrogen than does bark.
6. Compost is an excellent material for litter or bedding. It is moisture absorbent, odorless, and eliminates the need to purchase bedding from an outside source.

Maximizing the Nutrients Availability form Agricultural Compost

Digested agricultural compost should be applied at rates to supply adequate nutrients for crop needs. Rates of application greatly in excess of crop needs are undesirable since that means nutrients are not being used effectively and the risk of soil and water pollution is increased. In the past, when manures were valued for their plant nutrients, the normal rate of application was between 25 and 50 tonnes/ha/annum or 5 to 10 tonnes of manure dry matter. Now manures are often applied in much higher quantities of disposal rates of up to ten times the traditional rates. However, such rates could not be recommended either from an agricultural or environmental viewpoint. Approximately 100 tonnes of 4% dry matter slurry would be required to supply the same quantity of nutrients as 50 tonnes 8% dry matter slurry.

It is clear that the potassium level at this rate is over double the maintenance requirement of most agricultural crops, while the nitrogen and phosphorus rates are high but not excessive. In fact, if only 50% of the nitrogen is available, supplementary fertilizer would be necessary for maximum yield of many crops. If 100 tonnes/ha of pig slurry were used, then 300 kg N, 90 kg P, and 150 kg K would be applied. In this case, the phosphorus level is in excess of the needs of most agricultural crops but the nitrogen and potassium levels are not excessive. The 100 tonnes of cattle slurry would supply 8 tonnes dry matter and the pig slurry would supply 4 tonnes dry matter per hectare based on dry matter.

TABLE 7.1: Materials used for the composting

Material	Conversion ratio	Comments
Urine	2:1	Excellent activator and source of potash; ideally mix it with water before putting on heap
Dried blood	4:1	Highly effective activator
Pig manure	5:1	
Poultry manure	10:1	Use sparingly; a powerful activator. Also source of phosphates
Comfrey	10:1	Rich in many nutrients, especially potash. Almost no fibre
Lawn trimmings	12:1	Too nitrogenous to use on their own, especially if soggy. Mix with damp straw, weeds or leaves
Kitchen compost	12:1	Ideal, balanced ingredients. Exclude meat and bones which may rot and attract rodents
Farmyard manure	14:1	e.g., animal excretions plus straw. Almost, a balanced compost in itself after decomposition. Used fresh, when the straw is still quite yellow, it makes a good activator
Seaweed	19:1	Superb source of trace elements; no fibre or structure
Garden compost	20:1	Pea, bean, tomato, potato haulms provide valuable nutrients, also useful for holding air in the heap
Coffee grounds	20:1	
Horse manure	25:1	See farmyard manure
Weeds	30:1	Can range from sappy to fibrous, good source of nutrients. Avoid pernicious perennials. Note: stinging nettles are high in nitrogen and make an effective activator. If used in quantity, best treated in same way as lawn trimmings

Material	Conversion ratio	Comments
Bracken	48:1	High in potash if used when green. Avoid handling when it is producing spores (health risk)
Straw	80:1	Must be damp before use and, preferably, part-rotted to avoid robbing too many nutrients from heap
Woody prunings	100:1	Shred first, or else use as mulch
Bark	100:1	See woody prunings
Newspaper, cardboard	200-500:1	Use sparingly, shredded or torn up. Dampen before use; avoid materials with coloured inks
Sawdust	500:1	Too demanding of nutrients in their breakdown and too slow, unless first stacked for two to three years

Note: C:N values should be regarded only as approximations and will vary according to age precise make-up, etc. They are best used for comparative purposes to balance high C:N and low C:N materials at opposite ends of the table. A wide range of ingredients is the safest bet and will ensure a nutrient-rich compost, especially if seaweed and comfrey are added, even in small amounts. Wood ash, supplying potash and lime, may also be used. Lime need not normally be added to the compost pile, unless you have an acid soil. Care should be taken with pet manure: cat and dog faeces can sometimes be hazardous if handled.

From the above values, it is evident that the amounts of nitrogen, phosphorus, and potassium in manure are not in the correct proportion for many crops. There is generally too much potassium in cattle slurry and too much phosphorus in pig slurry relative to the other nutrients. To maximize the value of nutrients in manure, cattle slurry should be applied at rates to give adequate potassium for crop need and he supplemented with fertilizer nitrogen and phosphorus. Pig slurry, on the other hand, should be applied at a rate to supply adequate phosphorus and be supplemented with fertilizer nitrogen and potassium. The same approach can be used to estimate the application rate of other composts so as to maximize the fertilizer benefit of the nutrients present.

Where more composts are available than are required for the crop nutrient needs, most crops will tolerate higher application rates. For example, cattle slurry could be applied at a rate to supply adequate phosphorus for the crop, and the maximum yield when relying on nitrogen from cattle slurry alone, as high rates of application may depress yield.

A dressing of 40 to 50 tonnes/ha of good slurry is suitable for most root crops. As for silage, cattle slurry normally needs to be supplemented with phosphorus and pig slurry with potassium fertilizer. Nitrogen fertilizer may also be necessary, and the rate will depend on crop requirements. Care is needed when using compost for cereals because of their sensitivity to excess nitrogen and the difficulty of applying accurate rates of nitrogen due to the variation in compost composition.

Effect on Soil and Crop

Composts can have considerable effects on the composition of soils and crops. Soil analyses on experimental plots indicated that land receiving cattle slurry had high soil potassium levels and land receiving pig slurry had high levels of phosphorus. The manganese content of grass from cattle slurry treated plots was significantly lower than with fertilizer treatment. Copper sulphate is added to the diet of fattening pigs and the slurry may contain several hundred parts per million on a dry matter basis. Crops receiving pig slurry can contain increased levels of copper. There can also be a higher total removal of phosphorus from pig slurry than from fertilizer or cattle manure treatments. The higher phosphorus removal is probably due to the higher soil phosphorus level on pig slurry treatments.

In addition, the slurry provides a micro-environment that can facilitate seedling growth in otherwise inhospitable conditions. It is important to ensure that the ammonia content of the slurry is not so high that it can inhibit seed germination. Composts can also be used to provide a micro-environment for seedling development in reclamation of land; it can be particularly valuable in the rehabilitation of mine heaps and other industrially disturbed soils. In these situations, the composts are often superior to inorganic fertilizers as they have the dual benefits of improving the physical environment of the plant root and also supplying essential nutrients.

Method of Spreading Compost

Composts cannot be spread at some stages of crop growth: it is usually not possible to spread composts on manure crops near harvesting time. It is also true that some crops are more suitable for compost application than are others. From a fertilizer point of view, spring application in USA and Europe and *kharif* (rainy season) application in India usually gives best results, supplying maximum benefit from the nitrogen in the compost. However, it is not always possible to spread all compost in the above sections. In pig production, for example, the manure is produced throughout the year, so adequate storage for twelve months would be necessary if all the manure was spread in spring. This would be very costly. In practice, it is customary that manure produced during winter is spread during spring and summer, and that produced in autumn be spread during late autumn. In some situations, where soil conditions are suitable, manure is sometimes spread throughout the year so that very little storage is required.

The pattern of agriculture determines when and where composts can be spread. Traditionally, most compost was spread on tilled land. With intensive agriculture, in areas like the British Isles- compost is spread on grassland. In countries such as Denmark, where most of the land is tilled, a high proportion of the compost is spread for tillage crops. In India, agricultural compost is incorporated in soil for crops.

Time of application of compost is often determined by when soil conditions are dry enough to carry manure spreading equipment. Under some soil and climatic conditions, there may be six months of the year when manure spreading equipment could cause serious damage to the soil surface. Under other conditions, it is possible to spread manure throughout the year without serious soil damage to grass than the same quantity applied in spring or summer. It appears that slurry can deplete the already low oxygen levels in the soil near the grass roots, killing the better yielding grasses, which will be replaced by weeds. It is therefore advisable that if slurry has to be spread on grassland in winter, low rates (say not more than 20 tonnes/ha) should be used.

Rate of Application

The quantities of plant nutrients required for different crop and soil conditions are reasonably well established for most countries, where soil fertility is adequate, maintenance dressings of phosphorus and potassium are applied annually. Maintenance dressings are rates that will maintain soil fertility over the years and should equal the difference between losses and the nutrient supplying power of the soil. The main nutrient losses are crop removal and loss in drainage water. Nitrogen is more complex because there are many sources of losses and inputs and it is the primary nutrient influencing yield in conditions of high soil fertility. With low soil fertility, extra rates of nutrients, above maintenance dressings, may have to be applied initially for maximum crop yield. Losses of phosphorus and potassium in water from agricultural land are small in comparison with crop and potassium in water from agricultural land are small in comparison with crop uptake and are usually in the region of a few kilograms per hectare per annum. Removal by the crop is related to crop yield and composition. Most agricultural soils can supply large quantities of potassium annually from soil minerals. Rates of nitrogen fertilization for agricultural land are normally less than 300 kg N/ ha/annum.

Mineralization of plant nutrients and their availability to plants is one of the major consequences of organic recycling in soil. The quantity released depends mainly on the content of a particular nutrient in organic residues undergoing decomposition. In general, there is immobilisation of nitrogen due to decomposition of cereal straw. However, with FYM/compost application no such effect is observed. Addition of FYM and cereal residues results in improvement of total soil nitrogen. The effect of organic matter in reducing the intensity of phosphate fixation by the soil sesquioxides and maintenance of soil fertility by use of organic manures alongwith superphosphate has also been established. FYM remains intermediate in building up available P status of soil. In sub-humid lateritic soils, P use efficiency under rice-rice sequence was increased tremendously with FYM application. Response of crops to organic manuring depends on degree of decomposition of organic residues, C:N ratio, time of application, soil characteristics etc.

Continuous application of organic manures improves the availability of Zn in soil and may not be sufficient to meet the requirements of the immediate crops. Contrary to macronutrients, the range between deficiency and toxicity limits of micronutrients is quite narrow for most of the crops. Inclusion of micronutrients in fertilisation schedule should, therefore be advocated after careful appraisal through soil and plant tests.

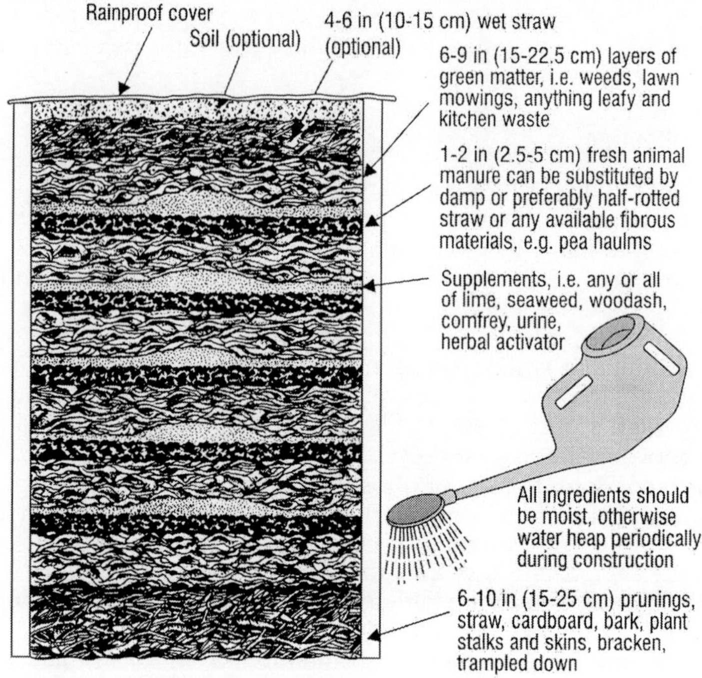

FIG. 7.1: Compost bin built up in layer system. Alternative layer can be mixed together.

Time of Application

To get a good response from compost application, good management practices should be followed. No more than 50 tonnes/ha should be applied in any one application, and there should be at least 30 days between application. Lower application rates would be advised for wet soil conditions or when applying in the wintertime on grassland. Heavy rates of composts, particularly slurry, can block soil pores. This can lead to anaerobic conditions and reduced infiltration capacity, leading in turn, to increased surface runoff during rainfall. The blocked soil pores will usually be cleared within 30 days by microbial flora after breaking down the organic matter. However, if compost is applied at too frequent intervals, soil microorganisms will not be able to clear the blocked soil pores.

Cattle or pig slurry at rates of 45 tonnes/ha applied six weeks before cutting did not reduce silage quality or animal performance (Tunney, 1976a). In order to reduce surface contamination and disease risks, it is advisable that composts should be applied at least four weeks before pasture is grazed.

Classification of Composting

It is classified into following general categories:

Oxygen Availability

Composting, categorized by the amount of oxygen available, is either **aerobic** or **anaerobic** in nature.

Aerobic Composting

The composting of municipal refuse and large-scale agricultural composts should be carried out under aerobic conditions. Aerobic composting is governed by the activity of aerobic microbes and hence required the availability of atmospheric oxygen during the period of decomposition. If the environmental factors are optimal, aerobic composting is characterized by high temperature, the absence of foul odors, and a short stabilization period. The high temperatures have a sterilizing effect by destroying weed seeds and pathogenic organisms. Hence, no malodors and its ability to destroy seeds and pathogens.

The Aerobic Heap: A heap should contain enough bulk of food for the heat-loving bacteria for their growth and activity. If the pile is too small, there will be insufficient heat build-up. To obtain a suitable C:N mix on a garden scale, the materials may first need to be assembled over a period of time. This can best be achieved by creating a preliminary stockpile. Garden debris, mowing, weeds and so forth can be loosely stacked and covered in one pile or placed in plastic bags; likewise kitchen compost.

Anaerobic Composting

It is governed by anaerobic bacteria that operate in the absence of atmospheric oxygen. The process is characterized by lower temperatures, the production of odorous gases, and longer stabilization times. Since anaerobic composting does not require atmospheric oxygen, the pile or bed can be sealed to prevent the escape of foul smelling gases and left alone. The major advantage of anaerobic composting is that the process can be carried on with a minimum of attention and as such requires little or no energy once the compost bed is established.

Temperature

The composting process can also be categorized by the operating temperature that exists within the pile. Temperature between ambient and 40° C support mesophilic organisms while temperatures between 40° C and 60° C support thermophilic organisms.

The mesophilic bacteria are more efficient than thermophilic bacteria and, therefore, decomposition occurs more rapidly in the mesophilic region. Those favoring thermophilic composting claim that decomposition proceeds more rapidly at the higher temperatures and in addition pathogens and weed seeds are destroyed.

The temperature in both aerobic and anaerobic composting gradually rises to well within the thermophilic range due to the excess heat energy generated by microbial activity. For this purpose it is essential to understand the effects of temperature and its relationship to microbial activity.

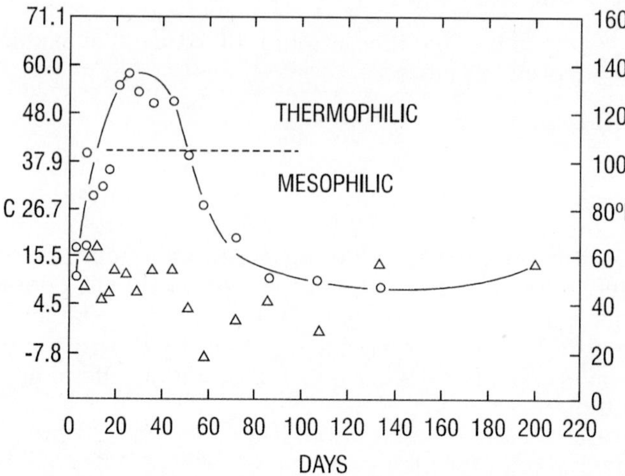

FIG. 7.2: A temperature profile of an un-shredded compost windrow constructed from leaves in the late fall.

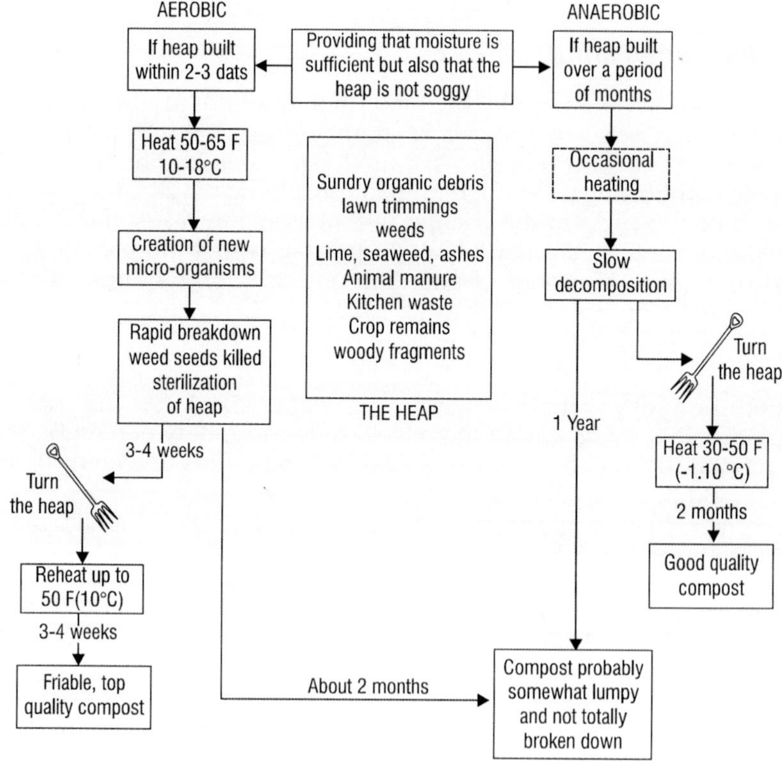

FIG. 7.3: The compost factory.

Sufficient air is required to sustain the temperature build-up. No n special air channels are required. Simply ensure that the material does not get compacted by forking through it and that grass mowing or the like, which are prone to settle into a solid, airless mass, are mixed with woody materials. Temperature is related to the weather, therefore composting is more reliable and fast in summertime. Compost, which may be ready in two summer months may require six months in winter.

Method of Operation

The composting processes are operated as either enclosed digesters or windrows.

Enclosed Digester

They are mechanized composters that provide aeration by some type of continued tumbling or stirring action. Some methods combine stirring with forced aeration. Such require a high capital investment and utilize a great deal of energy. Composting times for enclosed digesters are 10 to 15 days.

Windrow Composting

It is characterized by placing the organic composts in elongated piles called windrows. Depending upon the climatic conditions, the windrows may be placed in the open or covered to provide some protection against the elements. Aeration is provided by stirring and mixing the composter with a front-end loader or a specially designed rototiller type of implement. If the windrows are adequately mixed, the process will be aerobic and the composting time required is about 6 weeks. By way of contrast, anaerobic windrow composting takes 4 to 6 months.

Kinetics of Composting

The aerobic composting of all the municipal refuse plants are operated on aerobic process.

Biochemistry

The biochemistry or the rate at which organic matter decomposes is affected by the carbon-nitrogen relationship of the organic matter, moisture content, temperature, availability of oxygen, and the pH. The effects of these environmental factors on the decomposition of organic matter will help one to better understand the composting process.

Carbon-Nitrogen Balance

Carbon to nitrogen ratio controls the rate at which composting proceeds. As composting begins, the microorganisms require carbon as a source of energy for growth and nitrogen for protein synthesis. A C/N ratio above this range results in a slowdown of the composting process. If the carbon is present in a form highly resistant to bacterial attack, the C/N ratio can exceed the optimum level indicated. Examples of material that have a large percentage of carbon in a resistant form are paper, fiber, wood, and straw. The C/N ratio and hence the time required for composting can be lowered by adding a nitrogen source such as manure or activated sludge.

A low C/N ratio results in the loss of nitrogen as ammonia. The loss of nitrogen is not detrimental to the composting process, one should attempt to conserve the nitrogen as a soil nutrient.

Ideal C:N Ratio in Heap

The art of making successful compost is largely associated with achieving a correct C:N balance. 30 parts carbon to 1 part nitrogen, \ is ideal which reduces to 10-15:1 when turned into mature content. Fine grass mowing on their own will produce a treacly mass result, so straw, leaves or, possibly, cardboard would be good companions. If you include too much woody material such as bark and prunings without balancing this with sufficient green matter, the pile may not heat up at all.

A compost heap, in fact, is somewhere between a bonfire and nature's own method of decomposition, which is a slow rotting process culminating in incorporation by worms into the soil when breakdown is completed. By speeding up the decomposition process one can increase a soil's fertility very quickly, especially by making a well aerated heap where oxygen-loving organisms can flourish.

As composting proceeds, the C/N ratio continuously decreases with time, since the nitrogen remains relatively constant and the carbon is released as carbon dioxide gas. The compost is considered *ripened* once the C/N is lowered to a value between 12 and 20.

Moisture Content

Moisture is essential for microbial activity. In a dry summer the heap will definitely need some, preferably, added every so often while you are constructing it. Ingredients should be pretty damp. Too much moisture is unlikely at the time of making the heap, but it should be protected from rain with polythene on top.

The biochemical reactions during composting are ate also influenced by the moisture content of the organic matter. The optimum moisture content should be 100%, but this is impractical because the composting systems are based on the principle of dry handling. The maximum moisture content, therefore, is predicated on the ability of the composted material to be stacked. The optimum moisture content for greatest decomposition should be maintained between 50 and 60% (wet weight). Moisture content above 60% causes compaction of the material and also fills the voids with water, thus reducing the amount of air present. This may cause anaerobic conditions to occur, giving off foul odors and slowing down the decomposition process.

If this occurs, it is necessary to mix the compost in order to supply oxygen and restore the process to aerobic conditions.

Under aerobic conditions, it is also important for the composted material to form interstices which entrap air when the windrow is formed or the material is mixed. If these spaces are filled with water, no oxygen is available to the organisms. If composting is to be done anaerobically, a high moisture content is a desirable condition and can be obtained by saturating the windrow before sealing.

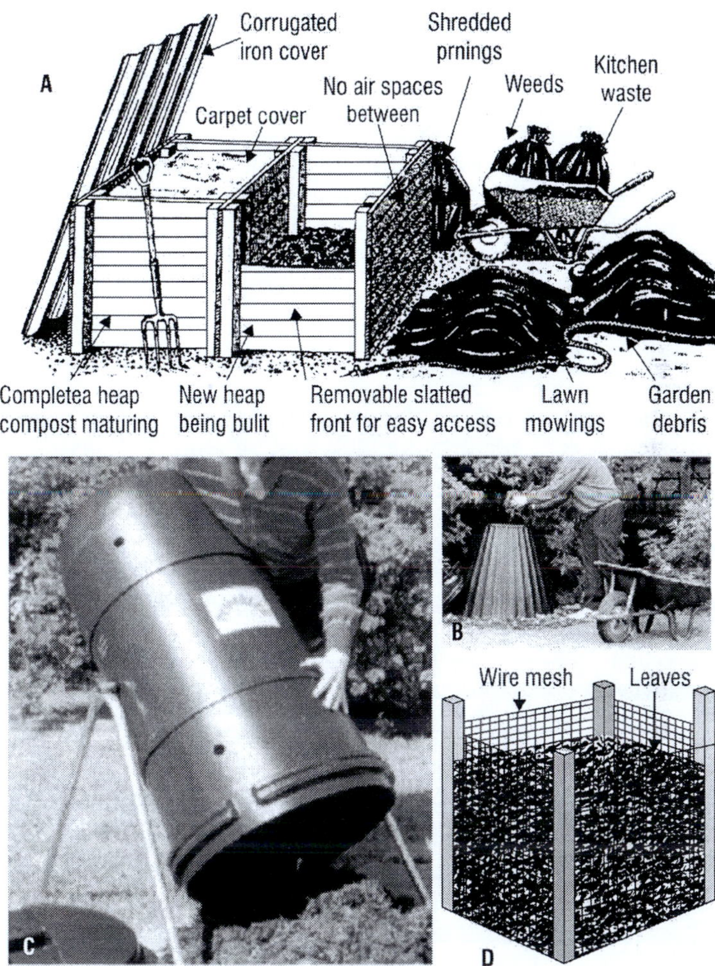

FIG. 7.4: Preparation of compost. (A) Initial stages for preparation of heap. (B&C) Classic, compost, conical shaped bin turns out useful quantities of compost. (D) Wire mesh cage, ideal for making leaf-mould.

If the moisture content falls below 50%, high temperatures occur in the center of the pile. The high temperatures begin to destroy the microorganisms, seriously curtailing the decomposition process. The problem can be corrected by adding water to the organic matter to raise its moisture content.

Ingredients to Avoid

Organic material that should not be included in a compost pile

1. residue from crops that have been sprayed with pesticides or herbicides;
2. material diseased with rusts and viruses;
3. meat scraps from the kitchen unless the compost heap is well protected from flies and vermin;
4. material with hard prickles or thorns;
5. persistent perennial weeds. To avoid any chance of these weed spreading they should be killed. Burning or spreading out in the sun so that they dry out completely are two methods for achieving this.

Paratoxics

Materials, which definitely should not find their way into your heap include all metals, rubbers, glass and plastics - even so - called *biodegradable* plastic will endure for many years. Large quantities of newspaper and cardboard will greatly slow the rotting process. Also inadvisable is diseased plant material, such as brassica stalks infected with club root, or white rot on onions, leeks or garlic. Potato blight, however, can be safely composted. If your heap does not meet these criteria it is safest to burn suspicious material.

Temperature

As indicated earlier, temperature is a key environmental factor and indicates the amount of biological activity taking place. The temperature within a compost pile is affected by moisture content, oxygen availability, and microbial activity.

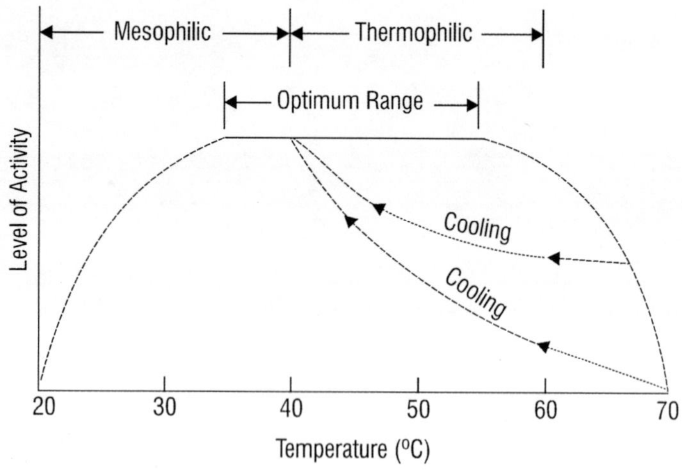

FIG. 7.5: The effects of temperature on microbial activity.

A drop in the temperature may indicate the material needs to be moistened or aerated or that the decomposition is in a late stage of activity. Figure 7.2 illustrates that the speed of the process increases with increasing temperature from ambient to 35° C. The process is at its peak efficiency between 35° C and 55° C. As the process exceeds 55° C, efficiency begins to drop abruptly and becomes negligible at temperatures in excess of 70°C. At temperatures above 65° C, spore formers begin to lose their vegetative ability and form spores, resulting in very little activity. As the process cools from 65° C or 70° C back to 40°C and lower, mesophilic organisms reappear in large numbers, establishing a high level of activity. Temperatures in anaerobic composting systems range from 38° C to 55° C, foul odors are given off, and pathogens may survive.

Oxygen Availability

Oxygen is essential to maintain aerobic conditions. An ample supply of atmospheric oxygen throughout the compost pile at all times is necessary. Oxygen can be incorporated into windrows by turning or by thoroughly mixing with mechanical means. The windrow should be turned once every 3 to 4 days to maintain aerobic conditions.

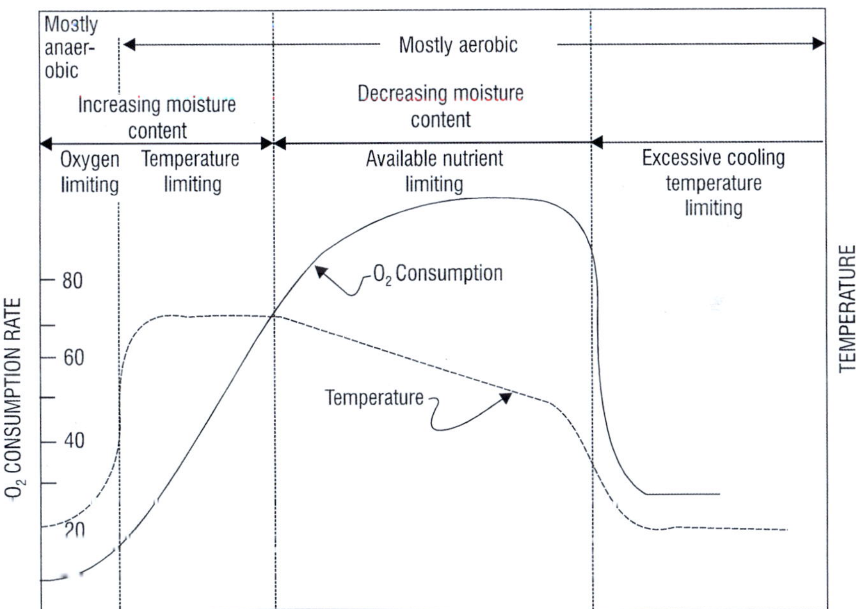

FIG. 7.6: Effects of aeration on temperature and O2 consumption in a high-rate enclosed digester.

Oxygen is added to enclosed digesters through continuous tumbling, stirring action, or through forced aeration. It is difficult to determine the true oxygen requirement because it is influenced by temperature, moisture content, and the bacterial population. It may be done by using the chemical oxygen demand (COD) as a means of measurement. Figure 7.3. The curve shows that at very low aeration rates, the oxygen supply is limiting and decomposition is anaerobic. Digestion under these conditions is relatively slow with little heat being produced.

With increased aeration, the process becomes anaerobic with a corresponding rise in temperature to a level, which restricts microbial activity. In this range, the capacity of the air to remove the excess heat is insufficient and the process becomes temperature limiting. As the compost begins to cool, water is removed faster than it is produced. In this range, decomposition is limited by the availability of nutrients.

If aeration rates are too high, heat will be removed faster than it is produced. This will lower the temperature below the thermophilic range, thus making the process under these conditions temperature limiting.

pH

During the decomposition process, changes in pH occur. At initial stage the material is slightly **acidic** because as composting proceeds, acid-forming bacteria cause the compost to become more acidic, thus lowering the pH.

The microbes in the compost then begin to metabolize the inorganic nitrogen to ammonium nitrogen, causing the pH to rise rapidly. At this stage the compost becomes **alkaline**. As decomposition continues, the ammonia may be released to the atmosphere or converted to nitrates. The nitrates are lost by leaching or by denitrifying bacteria, rendering the compost nearly neutral or slightly alkaline.

Microbes Involved in Composting

Activators of Biodegradation

Animal manure is used to enhance the heating process; it must be fresh rather then already rotted. Fresh manure *activates* by seeding the heap with bacteria. If enough is added it will also bring a lot of heat into the compost heap. This can be very useful in winter when it is cold and there is a shortage of green matter to engender heat.

The right bacteria can also be introduced by adding some of last time's compost as the heap is constructed. Commercial activators are unnecessary if the above conditions are met, but you might find it interesting to experiment with herbal products, which definitely make a difference. Packaged bacterial activators are also available but since the types of bacteria at work vary with the temperature, air conditions and other factors, these may be of limited use.

Composting is the conversion of biodegradable organic matter to a stable product called humus. The microbes involved include bacteria, fungi, and actinomycetes. During aerobic composting there is a continual change in the qualitative and quantitative nature of the microbial population. At first, fungi and acid-producing bacteria appear, causing the temperature to rise (mesophilic range). When the temperature rises above 40° C, these microbes are replaced by thermophilic bacteria, actinomycetes, and thermophilic fungi.

Transformation of biological material is carried out by microbial flora. They requires moisture, heat and, depending on the type of production operation chosen, air. A single gramme of compost will contain up to a billion bacteria, 100 million actinomycetes, a million fungi, algae, protozoae and others in their hundreds of thousands. Moreover, different bacteria with differing skills take over, depending on whether the heap is an aerobic (air-using) or anaerobic (airless) production line, the temperature, the pH level and the mineral content. Bacteria utilize nitrogen to break down the carbon materials and form compost.

Bacteria: Both mesophilic and thermophilic bacteria play an important role in the composting process. The mesophilic bacteria predominate during the initial and final phases of decomposition when temperatures are below 40°C. At initial stage, large numbers of aerobic mesophilic bacteria are present. They begin to multiply rapidly and sue to their increased biological activity they generates heat and temperature within the composting pile is increased. As the temperature rises above 40° C, the mesophilic bacteria (the exact role of mesophilic bacteria is not fully understood) are replaced by the thermophilic bacteria which continue to generate heat. Thermophilic bacteria, protein, lipids, and the non cellulose carbohydrate are the fractions of the compost. They are not capable of breaking down cellulose and lignin. At this stage, temperature is raised up to 70° C. The range of temperature from 40° to 70° C is known as the thermophilic range. The rise in temperature is influenced to a great extent by oxygen availability.

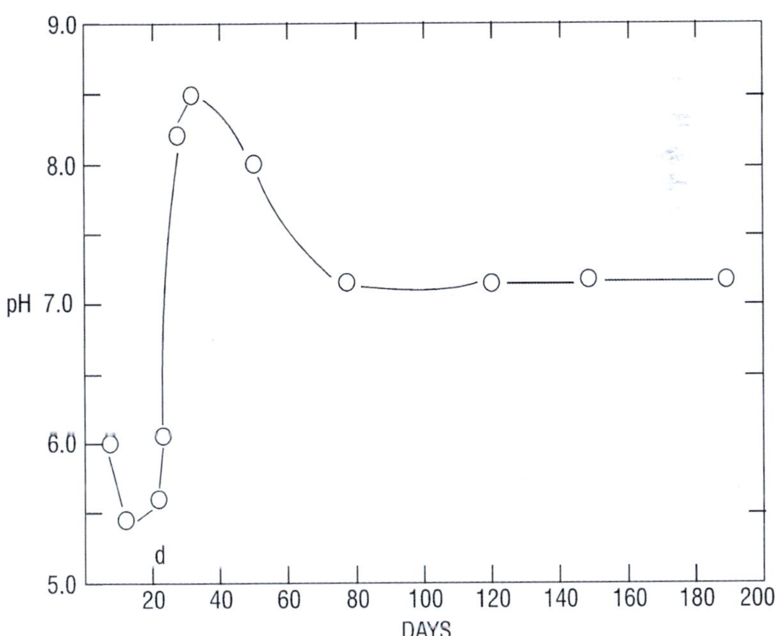

FIG. 7.7: Changes in pH during the decomposition process.

Actinomycetes: They exist in the thermophilic region and they utilize hemicellulose but not cellulose. Thermophilic actinomycetes are capable of decomposting cellulose. Thermophilic actinomycetes can grow at temperatures up to 72° C and, therefore, dominate the microbial population at the highest temperature level during the composting process and within the thermophilic range, the growth rate of action-mycetes increased 500-fold as compared to only a 5-fold increase for bacteria.

Fungi: They appear in both the mesophilic and the thermophilic stages of composting. The mesophilic fungi utilize the simple carbon substrates as their source of food. During the late stages of decomposition they utilize some cellulose and hemicellulose. Thermophilic fungi are less temperature tolerant than the thermophilic bacteria or actinomycetes. They operate in the range of 40° C to 60°C. Above 60°C, thermophilic fungi will die off. Mesophilic fungi (similar to like mesophilic bacteria) are active during the initial and final phases of decomposition. As the compost heats up above 40° C, the mesophilic fungi are replaced by thermophilic fungi. As the temperatures in the compost pile rise, both the mesophilic and the thermophilic fungi migrate to the outer edges of the compost pile where lower temperatures exist. When the temperatures drop, the fungi migrate throughout the pile and reappear in large numbers.

DESIGN CRITERIA

During designing there are a number of factors which influence the process like, the type of amount of compost to be treated, the climate, land availability and handling.

Type and Amount of Compost

The type and amount of compost to be treated is the first factor that must be considered. The type of compost will depend upon the livestock produced and the compost collection system employed.

In order to build a managed compost heap sufficient organic material must be collected together. It may be possible to collect some of the material needed for the compost heap, such as soil or crop residues, on the day of building the heap. However, the majority of the organic ingredients, eg kitchen compost, have to be collected on a regular or day-to-day basis. The materials collected before the day of composting need to be kept dry, cool and without ready air circulation. This will reduce decomposition. Putrescible material should be kept covered to prevent fly breeding. Suitable ways to keep the material dry would be to cover with banana leaves or a grass thatch.

The Climate

Since temperature and moisture affect the composting process, climatic conditions must be taken into account. For example, if a composting pile fails to heat up it may be due to either too much or not enough moisture. In extreme cases it may be desirable to add some sort of protective roof over the windrow. A protective cover will keep the rain and snow off of the windrow, prevent the hot sun from drying out the pile, and also retain much of the heat within the pile during cold weather.

Availability of Land

While designing for composting operation the factor of land availability should be considered. One should calculate the amount of land required for windrowing, curing the compost, and storing the finished product. The amount of land required, although relatively small, will depend upon the size of the operation and the size of the windrows constructed. Spacing of the piles may be determined from the proposed mixing procedures and the space required for maneuvering when turning and stacking. The amount of land needed should be doubled to account for contingencies, such as the inability to finish the compost on schedule due to equipment breakdown or inclement weather. The space is also needed for curing and storing the compost. Since compost is generally applied to the fields in the spring and the fall, at least six months of storage space should be provided.

Handling

Generally speaking, composting can be carried out without the use of specialized equipment. In most cases, a front-end loader can be used to collect the manure, transport it and spread it into layers to construct the windrow. Manure spreaders are also useful for spreading the manure to form windrows.

Many households and farms or small holdings will have a site where organic compost is piled up as it becomes available. This unmanaged heap will produce compost. However, the compost takes a very long time to develop and will be of inferior quality to compost produced in a managed heap.

A managed compost heap requires considerably more planning and labour. However, the compost produced under this controlled system will be available for use far sooner than that from an unmanaged heap; nutrient loss will be kept to a minimum and, due to the heat production during the composting process, many weed seeds and disease organisms will be killed. Consequently, for many people the production of a reliably high quality compost justifies the extra work.

TABLE 7.2: Managed vs unmanaged compost heaps

Managed	Unmanaged
Compost will be produced quickly	Compost takes a long time to develop
Nitrogen and other nutrient loss is minimized	Nutrient loss will occur reducing the value of the compost
Heat produced during the composting process will kill certain weed seeds and plant diseases	Weed seeds and plant diseases will not be killed
Heaps require a lot of labour and planning	Heaps require very little labour or planning

PRACTICAL METHOD OF MAKING COMPOST

Considerations in Building a Compost Heap

Size of the Compost Heap

A good basic size for a compost heap is 2 to 2.5 metres (about 6.5 to 8 feet) wide and 1.5 and 2 metres (about 5 to 6.5 feet) high. If the heap is too broad or too high aeration will be poor. The minimum size of a compost heap is usually recommended to be 1 metre^3 (about 1 yard^3).

Frequently more organic matter is needed to build a compost heap to the desired size than the inexperienced compost builder would expect. This can only be learnt through trial and error.

If it is difficult for an individual to collect sufficient organic material to build a compost heap it might be worth working with other farmers or householders, or even organizing the whole community into building their compost heaps together. Alternatively, extra material such as roadside cuttings or the organic by-products from local processing plants could be considered. The correct balance of slow and fast decomposable material, however, must be maintained.

Labour

Building a compost heap should be timed to fit in with the slack periods of labour. Some composting procedures e.g., regular turning, and some methods are more labour intensive than others.

Availability of Animal Feed

Much of the organic material suitable for composting is also suitable as animal feed. If animal feed is scarce it may be inappropriate to compost it. The farmer must make his decision based on his own circumstances.

If organic material is used as animal feed (e.g., kitchen compost for pigs, crop residues for cattle) it is likely there will still be other types of unwanted organic compost. Even if it takes longer to collect sufficient material it can still be worthwhile.

Water Availability

If water is scarce, it may be better to use available water directly for irrigation rather than for producing compost. However, compost added to the soil can improve its water holding capacity and, in the long term, would reduce the amount of the water required to irrigate the crops. Methods to reduce the water requirements of a compost heap include:

1. Covering the heap, thus reducing evaporation;
2. siting the heap in a shaded area;
3. enclosing the heap entirely within a wall of mud;
4. building the heap in a pit.

If a rainy season can be expected a good compromise is to concentrate on compost building when more water is available.

Pit or Heap

A heap is the most common form of composting. However, where water is scarce composts can be built in pits. This minimizes water loss, although watering is still required. Where there are heavy rains of a high water table compost built in a pit can easily become too wet.

Managing the Compost Heap

Water Content

In dry conditions, the heap will probably require water to be added about twice a week. The moisture content can be tested by placing a small bundle of straw in the compost heap. If the moisture is acceptable then the bundle will be damp when removed after about five minutes.

If the heap becomes too wet it should be opened up and mixed with more dry organic material or allowed to dry in the sun before being rebuilt.

Ingredients

Nearly all organic material, plant or animal can be used in a compost heap. However, organic material breaks down at different rates ranging from a rapidly rotting fruit to the slow breakdown of a dry maize stalk. When building a managed compost heap it is important to have a good balance between slow and fast rotting material.

These two main categories of organic material he to be balanced correctly because of the proportions of carbon and nitrogen they contain. Broadly speaking, young, living material that decomposes fast contains low levels of carbon relative to high levels of nitrogen. Tough, often dead material, for example straw, decomposes slowly and has a high ratio of carbon to nitrogen. This is called the C:N ratio; the higher the ratio the slower the rate of decomposition. If there is a shortage of easily decomposable material in the compost heap, the composting process will be very slow. If, on the other hand, there is an excess of the easily decomposable material the heap will become too acid and create a nasty smell, composting will occur slowly and the quality will be poor.

Building the Heap

In order to produce the conditions required by a compost heap it should be built in a particular way. First, a layer of course plant material such as stalks or twigs to ensure good air circulation and drainage is needed. Following that, organic material should be placed in layers, alternating between materials that is easily decomposed with material that decomposes slowly. A good thickness for each of these layers is 10cm (about 4 inches). If manure is used it should be applied in layers of 2cm (1 inch), and soil applied even thinner. Water should be applied after each sequence of layers.

With each layer of organic material, the edges should be laid down first, followed by the centre. This ensures that the heap edges will be firm and not collapse. Alternatively, a wire mesh or similar supporting structure that allows air to pass through can be used.

Air vents, made out of bamboo canes with holes cut in them and placed both vertically and horizontally throughout the heap, can improve aeration. A large pointed stick should be driven into the heap at a slant to monitor the temperature.

The heap should then be covered to protect it against moisture loss and/ or heavy rain. Sacking, grass thatch or banana leaves are all suitable.

Within three week the volume of the heap will have decreased considerably.

Digestion

Aerobic composting or digestion of livestock composts can be accomplished by placing the manure in windrows.

The Super digested compost: Application of super phosphate at 5 % over the raw compost materials during composting improves the manurial value of compost. Application of this compost to acidic, calcareous and heavy soils increases the P availability.

Windrows: They are elongated piles that are turned periodically to aerate the organic composts. They can be of any convenient length, about 2.5 to 4.0 m wide and 1.5 to 2 m high. The height of the windrow is critical and should be carefully managed in order to obtain good results. If piled too high, the material will be compressed by its own weight which will destroy the spore space and cause anaerobic conditions.

The height of the pile: It is also critical from the standpoint of controlling temperature. Windrows stacked too high may retain too much heat and develop temperatures above 70°C. Temperature in this range will destroy many of the microorganisms, slowing down the process and causing anaerobic conditions. Shallow piles may dissipate heat too rapidly, preventing optimum conditions for thermophilic organisms to be reached. Loss of moisture may also be excessive, causing the rate of composting to be retarded.

In drier climates, the cross-section of the windrow is usually trapezoidal, with the angle of repose being approximately 30°. In rainy climates, if no overhead protection is provided, the cross-section of the windrow should be semicircular in order to shed water.

Turning the pile frequently: It is essential for obtaining a rapid, nuisance-free composting action characteristic of the thermophilic aerobic process. In order to ensure a uniform and rapid decomposition, the compost must be thoroughly mixed. One method is to turn the outer edges into the center of the pile. This procedure causes weed-seeds, fly larvae, and pathogenic organisms which might survive near the cooler surface to be exposed to the lethal temperatures at the interior of the windrow. Turning is also useful for reducing the moisture content and aerating a pile that has become anaerobic. Turning can be accomplished by using a front-end loader or a clam-shell bucket. Commercially available machines are also available for mixing windrows. These machines are designed to straddle the windrow and mix the compost by means of long tines mounted on a rotating drum. Under normal conditions, when the moisture content is between 50 and 60%, the windrow should be turned at 3-day intervals. If the moisture level exceeds 60%, the windrow should be turned at 2-day intervals to prevent the occurrence an anaerobic conditions.

Curing

After the compost has gone through a period of active decomposition, the composted material must be removed from the windrow and allowed to cure. Curing is a process that allows the compost to continue its stabilization process at a slower rate under mesophilic conditions. Curing can be done in the open or placed under cover. After the compost is cured it is ready to be stored or utilized as a soil conditioner.

Maturation of the Compost

Once the compost heap has cooled down the heap should be left to mature. However, as long as most of the original material is no longer recognizable, and has turned into a brownish-black colour the compost can be used. If left to mature a finer product will be produced with a wider range of uses.

Maturation can take place on the ground, although the heap should be kept covered to protect it from the rain and sun. If the compost is kept for too long before use it may lose some nutrients and may also become a breeding place for unwanted insects.

Practical Applications Composting

Although composting is not widely used as a means of treating livestock composts, there are some farmers that are using it quite successfully. Whether or not composting fits into one's compost handling scheme is dependent upon the availability of labor, the degree of mechanization that can be afforded, whether the compost should be treated as a liquid or a solid, and how one plans to utilize the end products.

BIOGAS TECHNOLOGY

Biogas technology is anaerobic digestion of organic material producing clean fuel for cooking, lighting and running machinery and the bio-digest slurry besides providing plant nutrients both macro and micro to the crops, benefits the soil by bringing about physical, chemical and biological enrichment. This valuable organic manure not only releases plant nutrients slowly but also improves many of the soil properties like structure, water holding capacity, cation exchange capacity, resistance to soil erosion, and also provides energy for the activity of soil microflora, which enrich the soil and in turn increase productivity.

Raw Biomass
⇓
Bacteria Set I
⇓ ¬ Sugar etc.
Bacteria Set II
⇓ ¬ Hydrogen, CO_2, Organic Acids
Bacteria Set III
⇓
BIOGAS

FIG. 7.8: Schematic representation of the stages of methano-genesis.

The manurial value of the biogas slurry is superior to traditional FYM or compost in increasing crop productivity, and it enriches the soil with ammoniacal nitrogen and available phosphorus. While chemical fertilizers do increase crop yield over a short period, they do not maintain sustained productivity due to various reasons like deterioration of soil structure, deficiency of micronutrients and disintegration of soil microflora including N fixing and P solubilizing organisms.

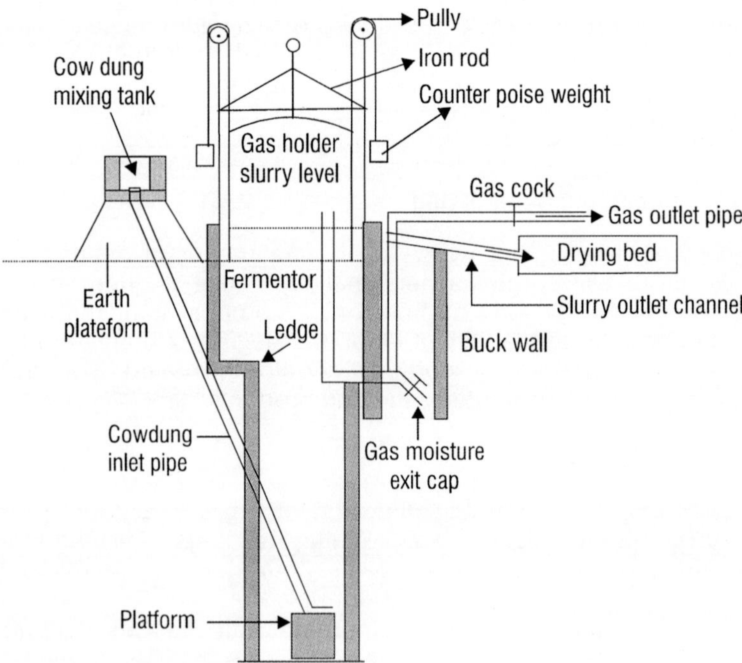

FIG. 7.9: Outline of gobar gas plant.

The slurry can be applied as liquid under wet, air dried and sundried conditions for enhancing the yield of agricultural and horticultural crops and is also found to be suitable as a seed coating media, fortified with *Rhizobium*, *Azospirillum*, Phosphobacteria and micro nutrients, which enhances seed germination, seedling establishment and induces vigour to crops. Biogas slurry enriched with inorganic fertilizers and biofertilizers such as a *Rhizobium* revealed higher crop productivity and sustained soil fertility.

A livestock unit (L.U.) is equivalent to a 540 kg the slurry production rate most often quoted for dairy cow. 500 kg beef cattle produces 40 kg. of undiluted slurry per animal per day. It has been estimated that about 3 tonnes of dung stead manure or 6 tonnes of farmyard manure would be produced by one livestock unit during a 140 day winter period.

The Gobar Gas Scheme, a development and popularization agency, provided producers of biogas with technical assistance and is responsible for the allocation of funds for construction of digesters, provide 3-5 head of cattle, with the view to meet energy needs of villages.

According to the New China News Agency, 7.15 million biogas installations have been made at the end of 1978, and current reports reveal that upto 1985 China have installed about 70 million of biogas plants. Government of India is also supporting in establishing regional and local offices which are responsible to establish biogas plants in every village. We are hopeful to meet out the demand of energy in near future through gobar gas.

Composition of Slurry

Cattle dung has been used practice in India since long in the form of dung cake as a domestic fuel and as compost manure. With the advancement of science, a technique of anaerobic digestion came into existence in late thirties, through which an efficient gaseous fuel i.e. biogas and digested slurry with higher manurial value can be obtained from the digestion of cattle dung. Besides, being rich in major plant nutrients i.e. N (1.5 per cent-2 per cent), P_2O_5(1 per cent), K_2O (1 per cent) biogas slurry is also a good source of micro nutrients like Zn, Fe, Mn and Cu. One tonne of biogas slurry on dry weight basis will provide approximately the following nutrients (Table 7.3).

In tropics, addition of organic manures to soils is very important from fertility and productivity point of view. Since organic manure is of biological origin, it contains nutrients and improves physical, chemical and biological properties of soil. There is, however, wide variation of nutrient availability in these manures due to differences in animal species, feed consumed and size and health of the animal.

TABLE 7.3: The equivalent cost of chemical fertilizer

Nutrient	One tonne biogas slurry (dry weight basis)	Equivalent chemical fertilizer	Cost Rs.
N	18 kg	40 kg urea	109.40
P_2O_5	10 kg	63 kg SSP	189.00
K_2O	8 kg	14kg MOP	66.90
Total			**365.30**

In the biogas plant, nutrient recovery represents an economic return to the farmer using this technology. The liquid effluent from the anaerobic digester is a relatively stable product in comparison to the influent. There is now a broader realization that recycled organic nutrients offer alternative to reliance on chemical fertilizers. In view of increasing prices of chemical fertilizers and withdrawal of subsidies on them, farmers should be provided with technology for enhancing them efficiency specially that of urea or some alternate source. Biogas slurry is one of the alternatives. The manure obtained from biogas plants has more nitrogen compared to conventional composting. It is reported that fermentation increases the ammonia content by 120 per cent and available phosphorus by 150 per cent. The output of sludge per year per plant (10 cm capacity) is roughly 10 cubic meter. The output of effluent is about 14 cubic meter per year.

Aerobic decomposition of organic materials for manure production results in a loss of 55 per cent of organic matter and 30 to 50 per cent of nitrogen. Whereas under anaerobic fermentation in the biogas plant, the loss of organic matter is 25 to 30 per cent and almost complete conservation of nitrogen: NH_4 - N is the main soluble end product in anaerobic unit and 15 to 18 per cent of the total nitrogen is converted to ammonia due to the alkaline nature of the slurry. In our country, the dung rarely contains more than one per cent of total nitrogen on dry weight basis, hence the ultimate nitrogen in slurry will not ordinarily increase 1.25 to 1.30 per cent unless other protein-rich organics are used as feed material in the process. It is directly linked to the nitrogen content initially present in such feed material. Freshly digested slurry has mineralized N to the extent of 21 per cent compared with 16 per cent of N in compost which shows the advantage of anaerobically fermented manure over aerobic decomposition.

Slurry for Agriculture

Several farmers in Gujarat having biogas plants on farm houses are incorporating the fresh slurry with irrigation water, which is to resulting in higher yields as compared to other modes of its application. This is particularly so with paddy nurseries. The application of such mixed slurry irrigation water resulted in higher yields on transplantation as compared to other methods of fertilization. But such method of disposal is possible only when the gas plant is located on a farm house. Location of the plant plays a significant role in efficient and effective utilization of the biogas slurry.

The effect of biogas slurry and gypsum enriched biogas slurry on rice-blackgram. Biogas slurry at 10 t/ha enriched with gypsum @ 250 kg/ha gave an additional grain yield of 1.80 t/ha compared to control. Gypsum enriched biogas slurry had an edge over slurry alone. The residual effect of FYM on the succeeding blackgram was comparatively better than that of biogas slurry. In another experiment, gypsum enriched biogas slurry in combination with 75 per cent recommended NPK resulted in higher seed yield under rice-blackgram cropping sequence.

Biogas slurry enriched with gypsum recorded maximum tiller number / hill (7.58) followed by enriched FYM and dried biogas slurry. Wet plain biogas slurry at 10 t/ha recorded increased tiller number by 0.27/hill over FYM alone. Muralikrishna Reddy (1987) had also observed an increase in tiller number through soil application of slurry enriched with gypsum. In respect of number of panicles, the enriched biogas slurry registered the maximum panicle number (412.5/m^2). These treatments resulted in an increase of panicles by 88.44 and 27.72 m^{-2} over control and FYM, respectively. FYM enriched with gypsum and dried biogas slurry at 10 t ha^{-1} and wet biogas slurry at 10 ha^{-1} gave identical panicle number. FYM alone produced more panicles than gypsum alone. Increased panicle number in rice due to application of biogas slurry was earlier reported by several researchers

In respect of seed and straw yields, the trend as observed in tiller and panicle number was almost reflected. Wet biogas slurry at 10 t/ha enriched with gypsum at 250 kg/ha proved to be distinctly superior to rest of the treatments. This treatment gave an additional seed yield of 1.80 t/ha compared to control. The response was well marked with FYM enriched with gypsum and dried biogas slurry at 10 t/ha recording increased yields by 21 and 18 per cent, respectively over control. Availability of soil ammoniacal nitrogen and phosphorus from slurry might have been responsible for fairly higher yields in slurry treatments.

TABLE 7. 4: Direct and residual effect of bio-digested slurry on rice and blackgram

Treatments	Rice				Blackgram	
	Tiller number / hill	Panicle number /m^2	Seed yield t/ha	Straw yield t/ ha	Pod number/ plant	Seed yield kg/ha
Wet bio-digested slurry @ 10 t/ha	6.83	384	7.46	13.03	9.21	422
Dried bio-digested slurry @ 10 t/ha	7.08	385	7.80	13.69	8.94	393
Wet bio-digested slurry @ 10 t/ha with gypsum 250 kg/ha (1:0.025)	7.58	413	8.41	15.25	9.12	402
Farmyard manure @ 10 t/ha	6.58	341	7.33	12.58	10.22	463
Farmyard manure @ 10 t/ ha with gypsum 250 kg/ha (1:0.025)	7.26	385	8.00	14.69	10.11	431
Gypsum 250 kg/ha	6.33	330	6.78	11.54	7.83	294
Control	6.17	324	6.61	11.03	7.70	292
CD (p=0.05)	0.22	0.06	0.19	0.49	1.24	51

The fermented compost had a superior effect on the response of maize in the early stages. At the termination of this series of experiments of 30 days, growth difference was less obvious. However, at the higher application rates, there was more protein per plant than with higher application of ammonia via chemical fertilizers.

The contribution of nitrogen from various organic manures in crop production, FYM. Compost, biogas slurry and sewage sludge performed equally well on equivalent N basis while poultry manure proved a better source. The residual effect of manure on succeeding crop also, the efficiency or FYM. As compared to that of fertilizer was noted to be 50 per cent. Due to severe loss of organic matter in sub-tropics the practice of applying heavy dose of manure only once in two years needs to be discouraged and it should be applied in small dose annually.

Biogas slurry combination with urea favorably influenced the growth, nodulation, dry matter, quality over other treatment and gave significantly higher pod yields over all the treatments. The treatments including slurry plus urea or slurry alone and PMC with urea gave 59.49, 50 and 49.83 percent increase in pod yield of groundnut over absolute control, respectively. However these treatments recorded an increase in pod yield of groundnut over chemical fertilizer treatment to the extent of 37.18, 22.53 and 22.27 per cent respectively.

Transfer of Biogas Technology

The saving of about 29 per cent of dung that is presently being burnt away as fuel in villages and its use as manure is a consideration of great practical importance in the context of the new strategy for increasing crop production. There is, no doubt, that in the present economy, the practice of burning cattle-dung as fuel to a limited extent is inescapable in rural areas and will continue at least for some more time. The question has, however, received attention and various solution have been suggested as follows:

1. The programme of installing and popularizing gas plants in the rural areas should be intensified.
2. Since the use of soft coke in rural areas is not going to be cheaper in view of tremendous cost of transport and other factors involved, the popularization of soft coke in urban and semi-urban areas may release wood to some extent for use as fuel in rural areas, thereby releasing cattle-dung for manurial purposes.
3. To augment the wood supply for fuel purposes in rural areas, stress must be laid on farm-forestry programme.
4. The quality of manure should be improved by adoption of scientific methods of composting and minimizing losses of plant nutrients. The programme of setting up biogas plants involves high cost of installation with KVIC plants whereas the Janata biogas plant is cheaper by 40 to 50 per cent. Hence, the biogas recovery in rural farm-sheds through the Janata biogas plant programme should be intensified.

In India, it is estimated that over 16 lakh biogas plants have been installed. Assuming 85 per cent plants in operation at a given time, 27.2 million tonnes of enriched biogas slurry is being produced every year, which is valued at over Rs. 225 crores. Thus in order to make the farmers fully aware of the benefits and utility of biogas slurry, there is a need for launching a mass awareness campaign. Ministry of Non-Conventional Energy Sources (MNES) has launched a countrywide demonstration programme in farmers field, through State Nodal Department, fertilizers, Gujarat State Fertilizer Corporation (GSFC), Indian Farmers Fertilizers Co-operative (IFFCO), Regional Biogas Development and Training Centres and Voluntary Organisations mainly FORRAD, New Delhi. Over 3000 such demonstrations have been conducted so far all over the country.

TABLE 7.5: Biogas plants installed up to March 31, 1992.

Type	Number of plants installed
Family size biogas plants	15,75,000
Community biogas plants	401
Institutional biogas plants	359
Night-soil biogas plants	59

The reports received from the State Nodal Department of Haryana, Himachal Pradesh, Kerala, Karnataka, Madhya Pradesh, Gujarat, Goa, Punjab, Uttar Pradesh and Tamil Nadu have been studied and analysed for different crops for the period 1984-95 to 1990-91 and the percentage increase in crop yield in demonstration plot.

TABLE 7.6: Summary of results of demonstrations conducted by State Nodal Departments / Agencies (1984-85 to 1990-91)

Crop	No. of Demonstration	Per cent increase in crop yield in slurry treated plot over untreated plot
Wheat	93	8-38
Paddy	67	18-36
Gram	11	17-11
Peas	10	15-50
Cotton	2	21-23
Potato	16	5-19
Tomato	18	38-42
Onion	6	10-11
Mustard	10	20-23
Groundnut	24	9-25
Watermelon	5	7-25
Sunflower	13	20-21
Cabbage	7	16-17
Banana	2	7-23

TABLE 7.7: Summary of results of demonstrations conducted by FORRAD (1989-90 and 1990-91)

Crop	No. of Demonstrations	Per cent increase in crop yield in slurry treated plot over untreated plot
Wheat	103	25
Paddy	79	32
Peas	14	60
Potato	3	43
Tomato	1	71
Groundnut	7	25
Ladies Finger	1	67
Chilies	1	49

Crop	No. of Demonstrations	Per cent increase in crop yield in slurry treated plot over untreated plot
Urad	1	1
Bajra	3	37
Maize	10	49
Ragi	8	27
Banana	3	9
Turmeric	1	27
Sugarcane	2	19
Deccan hemp	1	50
Mulbery	1	25
Mustard	2	23
Total	**257**	**35.5**

Besides increase in yield of various crops, various observations were made during the field demonstrations. These observations may be summarized as under:

1. Improvement in quality of produce. For example, the size of vegetables like cabbage, tomato, potato, brinjal, etc. were significantly increased.
2. Reduction in weed number in biogas slurry plots,
3. Biogas slurry does not attract insect breeding.
4. Reduces the usage of chemical fertilizers thus indirectly helps in reduction in unit cost of produce.
5. Improves the physical and chemical properties of soil.
6. Biogas slurry has high water holding capacity which helps rain-fed crops.
7. Seeds treated with biogas slurry have better germination rate.
8. Contains larger amount of organic matter and free ammonia than available in composted manure.
9. Biogas slurry is ready in shortest possible time.
10. If night soil and cattle urine is added, N and P_2O_5 availability in biogas slurry is enhanced.

In the context of reduction of subsidies on chemical fertilizers, biologically supplied or recycled organic nutrients such as biogas slurry offers an alternative to complete reliance on chemical fertilizers only. The manure obtained from biogas plants has higher nitrogen compared to conventional composting. The bio gas plant will prove more economical to the farmer only when both fuel and manure are used effectively and efficiently by him. Delinking these two aspects will make it uneconomical. It is well known that addition of organic manures to the soil improves its physical, chemical and biological properties. However, more research needs to be done to clearly indicate the benefits of the fermented animal compost on crop production.

For making the programme of organic farming/non-chemical farming more meaningful, effective and sustainable in terms of productivity also, following should be given due attention:

1. Establishing proper demonstration site in different agroclimatic conditions.
2. Developing efficient and suitable methods of slurry management and planning for its popularization.
3. Conservation of dung resource through the use of improved cooking devices.
4. Popularization of improved methods of composting such as 'NADP' technology alongwith biogas.
5. Documentation and scientific assessment of traditional farming systems, especially agricultural practices, cropping pattern, pest management, etc.

Maintenance unit for biogas plants in different locale.

GROWING OF MUSHROOMS

Spent compost from growing mushrooms, along with other composts, are spread on land and can have a significant fertilizer value. The quantities are small by comparison with other agricultural composts; however, they can be important in localized areas.

Conventional Types of Compost

For growing *Agaricus bisporus* a special type of compost is required which can be used for growing garden plants. But *Volvariella* spp and *A. bisporus* need different types of medium.

For the culture of *A. bisporus* some specific nutrients are required which are available in a good number of materials and they are used in making compost. Though the ingredients differ in the different mixtures the resultant nutritional value being almost the same. The availability of the ingredients are regional. Therefore, if only one mixture is recommended all of their ingredients may not be procurable locally. So for the sake of choice a number of mixtures are formulated and mentioned below.

Wheat straw .... *1000 kg.*
Poultry manure .... *400 kg.*
Wheat bran (Bhusi) .... *25 kg.*
Urea .... *14.5 kg.*
Gypsum (Calcium sulphate, $CaSO_4$*)* .... *30 kg.*

The mixture has been recommended by the National Centre for Mushroom Research and Training (Solan, Himachal Pradesh). The farmers of the village of Bhadana near Solan produce button mushroom using this compost.

Paddy straw .... *500 kg.*
Horse manure (stable manure) .... *100 kg.*
Poultry manure .... *300 kg.*
Wheat bran .... *12.5 kg.*

Urea .... *7 kg.*
Gypsum (plaster of paris) .... *30 kg.*

The mixture is recommended by the University of Horticulture and Forestry of Solan. By using this compost the many farmers of Chail, Simla, Manali, Khoradhara etc. of Himachal Pradesh are producing white button mushroom in quantities.

Wheat straw (1" - 4" long chips) .... *300 kg.*
Wheat bran .... *15 kg.*
Calcium ammonium nitrate (CAN) .... *6 kg.*
Urea .... *2.5 kg.*
Super phosphate .... *7.5 kg.*
Gypsum (calcium sulphate) .... *30 kg.*

This mixture is recommended by Indian Council of Agricultural Research.

Wheat straw (8-20 cm. long chips) .... *300 kg.*
Wheat bran .... *30 kg.*
Urea .... *4 kg.*
Ammonium sulphate .... *9 kg.*
Super phosphate .... *2 kg.*
Potassium sulphate .... *3 kg.*
Gypsum .... *30 kg.*

This quantity of compost can be sufficient for filling 20--22 trays. This has been also recommended by Indian Council of Agricultural Research.

Paddy straw .... *100 kg.*
Ammonium sulphate .... *18 kg.*
Urea .... *4.5 kg.*
Super phosphate .... *18 kg.*
Quick lime (CaO) .... *27 kg.*

This simple formula is innovated by the scientist of Taiwan (Ho).

Compost Making and Spawning

The compost making demands earnestness and carefulness on the part of the farmers. All details should be followed meticulously.

Compost

Ingredients..**Quantity**
1. *Wheat straw/paddy straw (1" - 4" long)* .... *300 kg.*
2. *Wheat bran* .... *15 kg.*
3. *Cal. Am. Nitrate (CAN)* .... *6 kg.*
4. *Urea* .... *2.5 kg.*
5. *Super phosphate* .... *7.5 kg.*
6. *Potassium sulphate* .... *3 kg.*
7. *Gypsum* .... *30 kg.*

Mix the ingredients of 3-6 nos. with 5 kg. of saw dust. Before mixing get the saw dust moist. Heap the mixture for a day so that the chemicals are mixed well with the saw dust. Next day soak the straw chips. First spread the chips on a cemented floor, then sprinkle water with a water can. Then spread the fertilizer mix and wheat bran on the soaked straw and mix them well.

Pile the mixture for decomposition. The width and height of the stack should be 150 m. The length is obviously dependent on the amount of the material. Cover the stack with a sheet of polythene.

Within 3-4 days the fermentation will start in the pile and consequently the temperature will rise to 60-80 °C.

Compost will be prepared when the process of decomposition will be completed which will take 28-30 days. In this period the pile has to be broken for aeration and piled again as before. This process has to be repeated at least thrice. Each time if some amount of extra manure is added the quality of the compost will be much improved. A mixture of the following ingredients may be used for this purpose.

Wheat bran.. 15 kg.
CAN (Calcium ammonium nitrate).. 3 kg.
Urea .... 1 kg.

Mix CAN and urea with the moist wheat barn and keep overnight. This can be applied when the stack is broken for the second time. Before piling, apply water by sprinkling to remove the shortage of moisture. For this use a water can fitted with its fine rose. Care should be taken to sprinkle water uniformly. Water in excess will be harmful. Have a pinch of compost and squeeze between the two fingers. If water oozes out take it for granted that watering has been more. Follow the work schedule given below.

First day...Mixing fertilizers with straw chips.
Sixth day.................... The stack is broken and oiled again for aeration.
Tenth day...The stack is broken.

Second time, some extra manure is mixed with it before piling again. Then on every third day the stack should be broken for aeration and piled again as before. If the moisture level goes down sprinkle water before piling. This process will be continued for 25 days. Whether the compost has been prepared that can be tasted easily. On the age of 28 days if the foul gas of ammonia does not evolve from the stack the compost has been ready for use. If not, a few days more might be required.

This is another process of compost making using the third mixture of the preceding chapter.

Soak the straw chips with water on a cemented floor. Then mix the full amount of wheat bran (15 kg.) and half of the fertilizer mixture together. Heap the mixture in a space of one square metre. Care should be taken to see that the pile should remain uniformly moist. The stack should not be disturbed for a week. Because the fermentation will start in this period and consequently the temperature will rise within the stack. After a week the stack should be broken, the rest of the manure should be mixed and piled again as before. The materials of the surface layer of the stack are subjected to less fermentation. For this reason it will be put in the middle of the stack at the time of next piling. To maintain the required moisture level sprinkling water may be needed. The full quantity of gypsum (30 kg) will be divided into three and each portion should be used in each time of breaking of the stack. When the compost will be ready its colour will be black, the condition will be moist and smell will not be foul, rather sweet like that of fresh hay. If not so, it is yet to be decomposed, and a few days more might be required.

The scientist P. K. Seth of the Agriculture University of Himachal Pradesh writes that there are processes for growing button mushroom-one is short and other is long. The short-termed process is suitable for the commercial purpose. Because this process involves the pasteurization of the compost that always ensures the high level of production. For sterilizing compost the equipment for applying steam at a high pressure is required. But for the domestic growers the compost that can be used without making germ-free has also been recommended. Now a long process of making compost will be discussed here. The ingredients are as follows.

Wheat straw (cuttings of 8--20 cm. long) .... *300 kg.*
Wheat bran .... *30 kg.*
Urea .... *5 kg.*
Ammonium sulphate .... *9 kg.*
Super phosphate .... *2 kg.*
Potassium sulphate /wood ash .... *3 kg.*
Gypsum .... *30 kg.*

Soak the straw chips in water for two days. To do this spread the materials with a uniform thickness on a cemented floor. Then sprinkle water with a water can at an interval of few hours.

Mix the ingredients of the numbers 3--6 with the half of wheat bran (15 kg.) and add a little water to moisten the mixture pile the mixture and cover the pile with a hessian cloth. Next day this mixture is to be mixed with wet straw cuttings. Stack the mixture as usual.

The Work Schedule

First Day (1)

Spread the wet straw cuttings on a cemented floor, scatter the mixture of wheat bran and other chemicals over it uniformly and all the ingredients will be mixed thoroughly. Then pile the mixture in a rectangular stack (1.8 m long, 1.5 m breadth and 1.5 m height). Within 48 hours the temperature inside the stack will rise above 60 °C due to fermentation. 6th day (2): Though the material inside the stack begins to decompose, that of the surface layer exposed to air remains intact. Separate the undigested material of 15 cm. deep of the surface layer and sprinkle water to soak it. Break the stack and mix the two lots together thoroughly, and add to it 30 kg. of CAN or ammonium sulphate, 3 kg. of urea and the remaining 15 kg. of wheat bran.

Of course, before mixing the chemical fertilizer and wheat bran they are mixed and kept in a moist condition for 24 hours. Then stack the mixture as before. This time a little insecticide (0.5 g) like Furadan 3G should be mixed with the compost material. On 10th day (3): Separate the undigested material of the surface layer and get it moistened. Break the stack and mix both the materials together thoroughly and pile again as before.

On 13th Days (4)

Follow the same as is done on the 10th day. But only exception is to add 30 kg. of gypsum to the decomposing material. Sprinkle water as needed. It is again said that the excess water is harmful. Take a pinch of the material and squeeze it between the fingers. If water oozes take it for granted that water has been added more than what is needed.

1. **On 16th day (5):** Break the pile and stack again.
2. **On 19th day (6):** The same as before.
3. **On 22th day (7):** The same as before.
4. **On 25th day (8):** The same as before.
5. **On 28th day (9):** Break the stack and check the smell. If ammonia gas is still generated stack again for a few days.

Watering the compost stack is to be done carefully. Water just to make it moist. On watering the mixture of straw cuttings and the fertilizer mix water should not spill out. Because with the spilling water certain amount of fertilizer is lost.

A simple method of compost making innovated in 1978 by a Taiwan scientist Ho is given below.

Paddy-straw cuttings........ 1000 kg.
Ammonium sulphate........ 18 kg.
Urea........ 4.5 kg.
Super phosphate........ 18 kg.
Quick lime (CaO)........ 27 kg.

Soak the straw cutting in water on a cemented floor. Then add the other ingredients and pile the mixture as usual. Cover the stack with a sheet of polythene. On the fourth day break the stack for aeration and pile it again as before. The process has to be repeated thrice on the 8th day, 12th day and the 16th day. After 20 days the trays will be filled with the compost. And it is then the compost is ready for spawning.

The quantity of water added to the compost greatly influences the mushroom production. If more water is added the nutrients will be leached out, and the quality of the compost will be poor. If the heap is too dry it will not complete decomposition and create problems for the mushroom growth.

The water content of the compost should necessarily be 68 per cent, which can be determined by a simple test. Squeeze a pinch of compost in-between the two fingers. On hard pressing if the droplets ooze it is be understood that the desired moisture level has been attained.

Spawning

Properly prepared compost is put in the trays and then spawn is added to it. The trays may be made of wood, plastic of fibre glass. The size and shape of the trays has been discussed in the relevant chapter.

Fill the trays with compost to the rim. The mycelium of mushroom is treated as mushroom seed or spawn. In the laboratory, the mycelium is grown in the medium of sorghum starch and put into bottles. The farmers have to procure these spawn bottles and the mycelium growing therein is taken out and added to the compost. Now mushroom spawn is available from the business firms, regional govt. laboratories, agricultural universities etc. Some voluntary organisations have come forward to extend their services for spreading mushroom culture. They distribute viable mushroom spawn among the farmers under their command area.

The spawn contained in a half-litre bottle is sufficient to spawn the compost of three trays. If the spawn bottles are not used preserve the bottles. At the 4°C of a refrigerator the spawn in bottles can be preserved well.

The spawn is taken out after breaking the bottle. Scatter the spawn uniformly over the compost of the tray. Then push them into the compost with the fingertips. It however should not go below the depth of 5 cm. Then cover the trays with newspaper. To soak the paper water is sprinkled over it. The trays are stacked in a well-ventilated room. How they will be stacked has been described and illustrated in the relevant chapter. The temperature of the room should be 24-25 °C. It will take 12-15 days to spread the mycelium in the entire mass of the compost. If the room temperature goes below 24° C the mycelium growth will be retarded. Care should be taken to see that the paper cover should always remain moist. Therefore sprinkling of water is needed at a regular interval. When the surface of the compost is covered with the mycelium growth remove the paper cover. Overspread a layer (2-3 cm. thickness) of clay paste. Pure clay should not be used as it will prevent aeration. Mix some leaf mould with light loam. To give it a paste-like consistency add water as needed. The porosity of the material is high, which will facilitate aeration and percolation of water applied by spraying. The covering of clay is termed as 'Casing'. The casing material should be sterilized before use. The one cubic metre of the soil mixture with three litre of one per cent solution of formalin. This quantity of soil is sufficient for casing 60-70 trays. Within a few days after casing the pin heads for fruit bodies will break the soil cover and raise their heads through it.

SUGGESTIONS

It is recognised that bulky organic manures, in particular, play their role in increasing crop yields through (a) supply of plant nutrients; and (b) improvement in several factors of soil fertility like the physical, chemical and biological properties of soils. Variable and often non-significant responses that are obtained through the use of these manures are now understood to be owing mainly to the low availability of nitrogen in them and consequently their inability to make good the primary deficiency of nitrogen in the dose applied. It is, therefore, recommended that following effective use of such manures should be made:

1. Use in adequate quantity, if available, on the basis of their content of plant utilizable nitrogen;
2. use in combination with fertilizer nitrogen in quantity to provide adequate nitrogen for the requirements of crops;
3. use after enrichment with nitrogen.

Since the quantity of manure available with farmers is much short of their requirements, the latter two alternatives provide a solution that would invariably give an assured response.

It must be realized that high crop yields induced by adequate fertilization with NPK salts are bound to project sooner or later the need for adjustment of other elements, from time to time to maintain the productive capacity of soils. It is in this role of supplying various secondary and trace elements that complex organic manures of this type can be relied upon to make good known and unknown deficiencies and rectify any adverse residual effects of fertilizers, particularly in their long-term use.

The urine component of cattle and buffalo excreta contains 50 to 70 per cent of total nitrogen and 90 to 95 per cent of the total potash excretion. The separate collection and utilization of urine absorbed and dehydrated in compost materials offers great scope for exploiting a potential source of highly effective manure available with every farmer. As an alternative to the recommended practice of adding urine-soaked litters to the manure heap, this approach would enable the limitation of fermented manures to be overcome and a high efficiency obtained from the liquid and solid excreta.

The technology on the method of composting available in the country is sound. However, the methods followed by the farmers are substandard. Composting of surplus and available farm composts such as straw, leaf fall, weeds, water hyacinth and animal dung should be practised by improved methods of composting available in the country for augmenting / supplementing the nutrients supply and humus substances for higher crop yields.

Preparation of rock phosphate digested compost is recommended. Low-grade rock phosphate @ 10 kg/tonne of organic material is recommended to apply while preparing the compost by pit or heap method. It has been found that rock phosphate application hastens the process of composting and improves the quality of compost.

The use of mesophilic cellulolytic (*Aspergillus* spp., *Penicillium* spp., *Trichurus spiralis*) inoculants is recommended for rapid composting of dry and high C/N ratio farm composts such as straw of different types, leaf fall, etc. It has been observed that by inoculating the compost heaps by an efficient and suitable culture, the composting period can be reduced with a net gain in nitrogen.

Mechanical composting has made progress in the country and semi-mechanized plans are more suitable to Indian conditions because of the advantage of low cost. The success of mechanical composting plants should not be judged purely on economic considerations because public health and checking of environmental pollution by systematic disposal of solid and liquid composts is an important factor.

Concentrated organic manures like oil-cakes, blood, meat, fish and bone-meals are sold in our country at abnormal premium as compared with chemical fertilizers of equal plant nutrient capacity. This does not appear to be justified on the basis of responses obtained with them which are at best equal to or slightly better than fertilizers. Considering also that the quantity of organic matter that get applied by the use of such organic manures is rather small and of little practical consequence, the prices paid for organic nitrogen from such sources are not commensurate with the returns obtained in crop yields, unless their prices are reduced. It should be realized that organic forms of nitrogen are first converted into mineral forms of ammonia and nitrate, before this element is assimilated by plants so that there is no particular superiority in effect that can be expected from this. Actually, organic manures do not nitrify completely and a part of their nitrogen always remains untransformed and, therefore, un-utilizable. Low-grade oil-cakes can be ammoniated or enriched with fertilizer for fuller and more effective use.

Sewage water, besides irrigation, adds plant nutrients to the soil and increases crop yields from 30 per cent to 50 per cent in food crops and 100 per cent in fodder crops as compared with ordinary irrigation water. The sewage water must, therefore, be fully exploited for increased crops production. Care must be taken for proper treatment of such water to avoid transmission of pathogens.

Organic materials rich in cellulose and lignin such as crop composts, straw @ 5 to 10 tonnes/ha can be recycled in the first instance as surface mulch and subsequently after one season partially degraded straw should be incorporated in the soil. This is a cheap method of utilisation of surplus crop residues instead of burning it. The results have shown that organic mulch is helpful in increasing the yield or crops significantly as well as the second crop when the left-over organic matter is ploughed in the soil.

Another promising method for direct disposal of crop composts such as paddy or wheat straw can be followed by incorporating the organic material @ 5 tonnes/ha and allowed it to decompose for a week followed by growing of a legume crops to advantage. The results have shown that the yield of legume crops can be increased by ploughing in straw.

It is recommended that adequate quantity of organic manure should be used in alkali soils in combination with gypsum as combined application interacts significantly in increasing the crop yields and also produces favourable effect on soil physical properties.

Further experimental evidence is needed on the comparative response and economics of various organic manures in relation to fertilizers and the effect of their combined application. Experiments are also necessary to study the effect of smaller applications of bulky organic manures, fortified with nitrogen and non-symbiotic nitrogen fixing bacteria (*Azotobacter*) and phosphorus solubilizing bacteria (*Pseudomonas striata, Bacillus polymyxa*).

Experience of users of organic manures has indicated that the quality of crops, particularly fruits and vegetables is superior to the quality obtained with chemical fertilizers. It would be useful to obtain some conclusive experimental evidence on this aspect relating to determinable quality factors like nutritive value, storage properties, colour, mineral and organic composition, etc.

It is often argued that even if dung is not at all used as fuel and all available organic composts with manurial value are carefully conserved and returned to the land, the supplies of plant nutrients cannot be adequate to improve and maintain optimum soil fertility. This may be true but in view of the various advantages of organic manures, it is not at all convincing that the improvement that can be brought about by the development of local manurial resources is not worth the trouble. The manurial policy of the farmers should, therefore, be to take every possible step to ensure the return to the land of all organic compost materials and supplementing such materials with chemical fertilizers adequately.

Economic Considerations

As the cost of fertilizers and energy continue to rise, new emphasis will be placed on the need to maintain the fertility of the soil and to reduce the amount of energy required for livestock production. Farmers may find that in the future the cost benefit ratio for composting may be lower than many of the more conventional compost handling systems used today. If that is the case, composting well may be a viable solution to our compost-handling problems.

Public Policy

Public policy can have a big impact on the future of composting. As available sites for compost burial become more difficult to find and as air quality standards for incineration become stricter, composting as a method for treating solid composts will become more attractive. In addition to this, the public will be more aware of the need to recycle some of our resource material, such as livestock composts and garbage, back into the soil.

CHAPTER 8

Vermicomposting

Soil fauna play a prominent role in regulating soil processes and among these the termites and the earthworms play a vital role in maintaining soil quality and managing efficient nutrient cycling. In organic farming practices the soil is considered to be a living component with is physical, chemical and biological characteristics.

Various aspects of earthworm activities in nature, and benefits that can be derived with mass culture, *viz.,* for cast production, vermicomposting and commercializing live material, indicate needs of developing mass earthworm culture. This is popularly referred as vermiculture. Vermiculture, as a subject, involve background of various aspects on biology etc. of earthworm. All these are scattered in literature.

Vermicompost is a method of making compost with the use of earthworms, which generally live in soil, eat bio-mass and excrete it in digested form. This compost is generally called vermicompost or wormicompost. It is estimated that 1800 worms which is an ideal population for one sq. meter can feed on 80 tonnes of humus per year.

Earthworms form a major component of the soil system and these organisms have been efficiently ploughing the land for millions of years and assisting in the recycling of organic nutrients for the efficient growth of plants. Advent of chemical fertilizers and biocides and their large scale application have changed the structure of soils and have, in most cases, eliminated soil organisms. Thus, human interference in natural ecosystems through modern practices of agriculture, silviculture and horticulture involving large scale application of chemicals in the form of fertilizers and pesticides has been responsible for the destruction of endemic soil fauna. For millions of years before the so called green revolution, earthworms have been ploughing the soil and fertilizing them.

Earthworms are one of the numerous range of burrowing organisms which improve soils. Fertile soils harbour bacteria, fungi, actinomycetes, protozoans, proturans symphylids, pauropods, spring-tails, pseudoscorpions, insects and their larvae, millipedes, centipedes, slugs, snails, snakes, rodents and the earthworms. These fauna along with the flora contribute to soil health. The presence of earthworms in soils is generally indicative of the presence of other faunal representatives contributing to soil fertility. It would therefore certainly be appropriate to call earthworms as indicators of soil health.

The greatest asset to our Country is that we have about 509 species of earthworms and that we need not depend on the exotic spices of earthworms. During the early nineties a New Zealand planarian flatworm *Artioposthia triangulata* had been reported to be eating up earthworms and spreading fast across England, Ireland and Scotland. This parasite had the potential to cause a plague of eliminating earthworms from soils. The chief cause attributed to the arrival of these flatworms into Scotland was by the inadvertent import of potted plants from New Zealand, the soils of which probably might have contained flatworm eggs. Thus transfer of organisms of any kind including the earthworms can directly or indirectly cause irreparable damage to local biodiversity as could be envisaged from this example. Advantages of Organic Manure and the role of Earthworms

The green revolution is the result of intensive agriculture. Presently, an extensive use of organic manures may boos agricultural production and can feed to the growing population of under developing countries.

With the intensive use of chemical fertilizers, farmers can harvest three crops in a year with good irrigation facilities, but there was no thought about its adverse effect in the long run on the soil condition in general and on the environment in particular for nearly four decades. To restore the soil composition, farmers are now showing inclination to revert to organic farming. With this changing trend, a comparative picture of the two practices gives better scope for adoption of organic farming to suit to the existing conditions. Along with that, it is necessary to produce the organic manure in large quantities from the available sources of organic matter to maintain the high level of produce in the available agricultural land.

Organic manure can be prepared from natural organic matter. These materials are the reservoirs of the nutrients that are required for the plant growth. These nutrients are in the form of complex, bound organic chemicals with strong bonds in most of these plant and animal debris. The entry of these chemicals in the form of simple organic or inorganic chemical into the plant system is the main basis of biogeochemical cycle. If these chemicals are judiciously made to enter the cycle of events in a regular systematic form in nature, then there will be no dearth for nutrient supply to the plants in a natural way. With this, there will be cleaning of the environment rather than polluting it.

Microbes are the **primary decomposers** while earthworms are the major **secondary decomposers**. Earthworm manure is formed from the dead tissues of plant and animals and thus is naturally the source of macro and micro nutrients in limited quantities. The weather conditions and the soil type in the tropical countries do not favour the restoration of carbon resource in the soils. The application of organic manure replenishes organic carbon to the impoverished soils. The presence of high level of oxidizable organic carbon helps in the slow release of nutrients from the manure and curbs the leaching of nutrients. The beneficial activity because of the nutrient availability.

EARTHWORMS AS INDICATORS OF SOIL FERTILITY

Earthworms can generally be called as biological indicators of soil fertility, for soil with earthworms most definitely support healthy populations of bacteria, fungi, actinomycetes, protozoans, insects, spiders, millipedes and a host of others that are essential for sustaining a healthy soil.

Earthworms act on litter and the absence of earthworms is characterized by the presence of litter mats on surface sol. Soils inhabited by earth worms have surface casts to a range of about 98 tons dry weight ha-1 yr-1 in woodlands to about 133 tons dry weight ha-1 yr-1 in grasslands, maximum production however occurring during the wet periods. Soil fertility results from the maintenance of soil conditions such the decomposition processes operate at a level adequate to release plant nutrients from the litter at a rate that will sustain optimum growth. Availability of earthworms in soils have always promoted plant growth. Scientists have reported the production of grass, whet and clover to increase many fold by the presence of earthworms, while increase in growth of maize, paddy, sugarcane, vegetables and ornamental plants have also been reported.

EARTHWORMS AND PLANT GROWTH

The effect by earthworms on plant growth may be due to several reasons apart from the presence of macronutrients and micronutrients in their secretions and in vermicasts in considerable quantities. Certain metabolites produced by earthworms may also be responsible to stimulate plant growth. It is considered that earthworms release into the soil certain vitamins and similar substances. Though the NPK value of the vermicasts is always lower than any standard chemical fertilizer, several experiments have proved that wormcasts can promote lush growth of plants. This probably may be due to plant growth promoters like, cytokinins and auxins present in casts.

Earthworms play in important role in low-input agricultural systems. Their casting activity involves up to 11.6% of the organic carbon and 12.9% of the total nitrogen of the 0 to 15 cm topsoil in undisturbed or recovering systems. Earthworms are very sensitive to changes in the ecosystem, expressed by strongly reduced surface casting activity. In improved cropping systems, earthworm activity can exceed that in the forest. This increase is due to the maintained groundcover and its reduced water consumption compared with forests. The negative impact of traditional cropping (lack of organic matter input through vegetation management) on casting becomes more important. The benefit of earthworm activity, apart from the effects on soil physical properties, is its concentration and deposition of large amounts of organic carbon and total nitrogen at the surface. Resources are placed at a location and in a form where (and in which) they are least likely to be lost. A firm conclusion on the effect of casts on plant nutrition is not yet possible.

In sustainable agricultural production, systems the resource-conserving aspect of earthworm activity might be the more important one. *H. africanus* is active over a broad range of soil qualities and shows that it has the potential to improve or mediate the buildup of favorable soil properties. Thus, in accordance with results from Ibadan show that earthworms are not merely a consequence of high soil fertility, but they contribute to its buildup and maintenance.

INTERACTION OF VERMICOMPOST-EARTHWORM-MULCH-PLANTROOT (VEMP)

VEMP interaction is essential for an effective utilization of the soil ecosystem in agro-practices. Use of green manures in organic farming is also essential. The possibility of interactions of the root hairs with nitrogen fixing organisms may be increased by the VEMP interaction. It is very well known that earthworms prefer lower levels of lignin and tannin, and in case of green manuring younger plants are recommended for ploughing-in for an effective release and uptake of nutrients.

VAM is effective in cases of wheat and other crops and the population of VAM is very well enhanced in the presence of earthworms, as earthworms are reported to be vectors of viable propagules of VAM. The uptake of phosphate, though facilitated by VAM, dispersal of these spores and conservation of such microbes is facilitated by earthworms and other soil dwellers.

VERMICOMPOST

Vermicompost is a stable fine granular organic matter, when added to clay soil loosens the soil and provides the passage for the entry of air. The mucus associated with the cast being hygroscopic absorbs water and prevents water logging and improves water holding capacity. Thus, in the sandy soils where there is the problem of water retention, the strong mucus coated aggregates of vermicompost hold water for longer time.

In the vermicompost, some of the secretions of worms and the associated microbes act as growth promoters along with other nutrients. It improves physical, chemical and biological properties of soil in the long run on repeated application. The organic carbon in vermicompost releases the nutrients slowly and steadily into the system and enables the plant to absorb these nutrients. The soil enriched with vermicompost provides additional substances that are not found in the chemicals. The multifarious effects of vermicompost influence the growth and yield of crops.

Recycling of Wastes through Verm-Composting

Vermitech is useful in recycling the agro-wastes through and the effective utilization of VEMP reduces. Nutrients presenting vermicompost are readily available and the increase in earthworm populations on application of vermicompost and mulching lead to the easy transfer of nutrients to the plants providing synchrony in interfered ecosystems. The availability of microbes in the compost and the nutrients in soils tilled by earthworms are definitely more due to the formation of drilospheres. Vermicompost, therefore, leaves no doubt on its potential for good yields, the yield however depends on the management practices followed by the farmer.

Minimizing Pollution Hazard

Earthworm can minimize the pollution hazards caused by organic waste by enhancing waste degradation. The experiences of the farming community have shown that the exclusive use of chemicals to keep away the famine has turned out be a mirage. It is like giving alms for a day, instead of providing an opportunity to a person to earn for his life time to escape from poverty.

The ideal earthworm species of earthworms help to convert waste into wealth. If this operation is made successful by the own efforts of the farming community, one can certainly by hopeful of converting the waste land into a fertile land. It is this hope of the possibility of producing the needed top soil or humus by earthworms that has encouraged this effort. This may help in reaching the goal of green revolution from a different angle.

Advantages of Vermicompost

1. It enable easy management.
2. Vermicomposting wastes at source or place where one wishes or has facilities, compost can be more profitably utilized and commercialization for economic gains.
3. Management of live population in field too first necessitates development of suitable conditions which are not easy due to modern agriculture practices. Even if these are made available, a large culture of suitable earthworm species, *viz.*, approximately on tonnage level = about 1 million live individuals per hectare land are required. All such requirements necessitate development of earthworm culture (vermiculture).
4. Huge quantities of domestic, agricultural and rural industrial organic wastes can be recycled for various usages. This also reduces pollution.

The other applications of vermicompost, in the case of paddy, at the time of tillering, panicle initiation, flowering and drain fillings essential. Unlike the chemical fertilizers, the loss of nutrients by leaching and other phenomena related to nutrient loss are minimal in case of VEMP interactions. In Azolla applied systems, the decomposition and release of nutrients maybe hastened by the application of vermicompost. The effectiveness of blue-green algae with vermicompost is not much understood compared to other biofertilizer. In sugarcane, paddy and other crops, vermicompost can be effectively utilized as a carrier medium for Azospirillum, Rhizobium and phosphate solubilizers, which are recommended with FYM (Farm Yard Manure). These also provide very good association for enhancing their interactions with plant roots. Experiments conducted on the effect of earthworms and/or vermicompost and vermiwash through vermitech on the cultivation of vegetables like tomato, brinjal and Okra have yielded very valuable results.

Adverse Effects on Crops

Earthworm activity can also to a less extent damage crops by either seizing fallen leaves of live plants and pulling them down or by uprooting delicate seedlings. Casts may show up as ugly structures in ornamental lawns or may form an obstruction on golf courses or football pitches. Earthworms have also been occasionally reported to transmit many parasites and diseases of plants and animals.

Economic Viability

Vermin-technology is popular because it is a simple methodology with low investment. It technology does not need sophisticated infrastructure. All that is required is some pits or containers to initially decompose organic waste and tanks made out of stone slabs or wooden planks, trenches or plastic troughs or for any convenient material just to protect earthworms from pests and predators. Therefore, the investment on this is directly related to the choice of the material used to build tanks. Stone slab is subjected to availability while plastic bins would be less durable and expensive. Thus, tanks built using cement and bricks could suit all situations. However, trenches without any cement are popular in low rainfall areas. The cost of bricks and amount spent on construction vary from place

VERMICULTURE PROCESS

It is very simple and easy to method. It needs careful step wise programme. However, like all live animal or culture of any biological material, first step is to start purely as a test experiment on a small scale. This enables acquire personal experience(s) on various aspects. For beginners, hobbyist and farmers, simple procedure should be first to start on experimental scale and record every detail of steps followed and responses on earthworms.

The care should be observed that how these fair in particular medium with increase or decrease in their numbers, waste matter conversion, cast production and composting etc. Following are the important levels of planning:

1. Site selection
2. Availability of decomposable organic waste, its daily quantity and quality, alternate organic material, transportation and stocking etc.,
3. Marketable outlet and requirement with future scope.
4. Collection and study of knowhow on earthworms to be cultured.
5. Collection, procurement of suitable species.
6. Testing of suitability of species on performance in available organic waste.
7. Maintenance of seed culture for eventual large scale culture.

Other important considerations should also be kept in mind, like funding, availability of labour, quality of waste and packaging etc. etc. are to be kept in mind. These aspects comprise essential steps in any project formulation In terms of wider utilization of vermiculture, first step is selection of suitable species for culture and eventual vermicomposting.

Selection of Suitable Species

The selection of suitable species for vermiculture is based on the requirement, *viz.*, for composting, poultry and animal feed, marketing for fish or other forms of culture and sale for fish baits. As yet demands in India for fish bait is meagre and lack for wanting development of fresh water sport fisheries, including for selected fish species (like Eel Fish) culture and also large scale culture. For all culture feeds, care has to be taken that earthworm feed does not harbour parasites and pathogens.

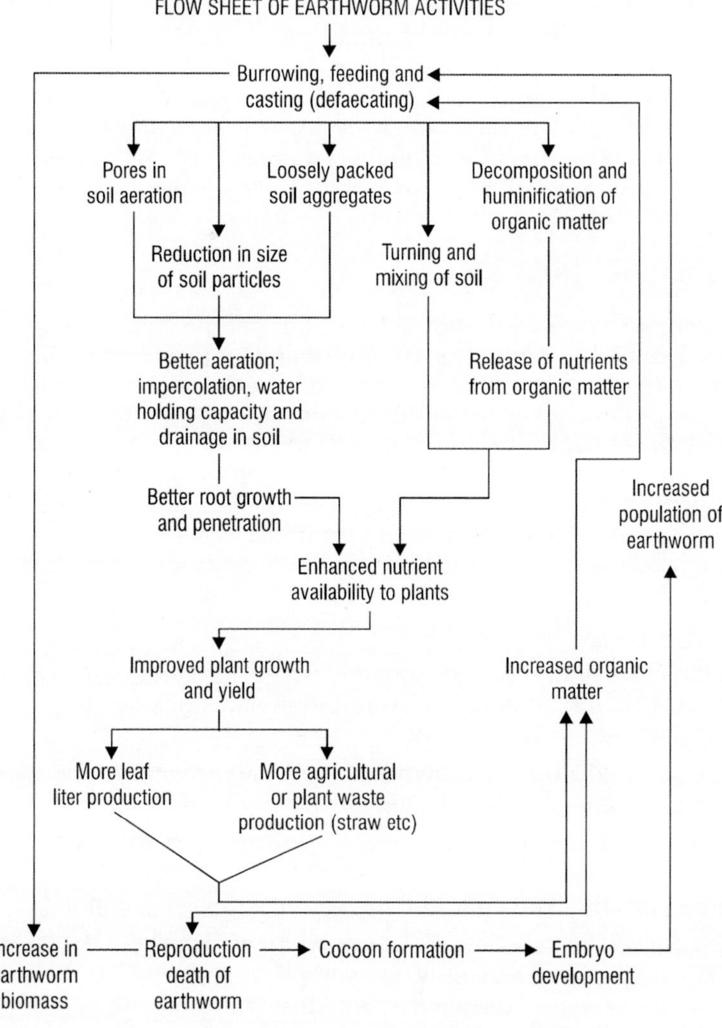

FIG. 8.1: Action of earthworm for production of vermicompost.

The selection for suitable species is based on biological and ecological parameters, like habitat characteristics, distribution in soil or feed media and trophic functions. Usually, following three types of earthworms which can be used in vermiculture:

1. Epiges
2. Endoges

3. Aneciques

Epiges

These are small in size. They have uniform body colouration and live on surface litter or dung. These have short life cycle with high reproduction and regeneration rate. They can tolerate disturbance and have active gizzard; but have limited period of activity. These are phytophagous. These species are good bio-depredators and can release good nutrients in the soil. They do not re-distribute nutrients. Thus have insignificant role in humus formation and are not good for use in field conditions for soil reclamation, unless careful manipulations are done

Endoges

These are small or large in size. They have weak pigmentation and found in top soil layer of organic and mineralized matter. Burrow branchings are horizontal and worms moderately tolerate disturbance. They may show some preference in selection of feed substrate (buffered and predictable) conditions. These are **geophagous**. Life cycle is of intermediate duration. They have potential for soil improvements because of high efficiency in energy utilization from poor soils, therefore, can be used in field with some manipulation.

Aneciques

They are large in size. They have pigmentation only at anterior and posterior ends. They are largely nocturnal, deep burrowing and pull leaves or litter matter into soil (burrows) and are **phytogeophagous** and tolerance to disturbance is poor. Their reproduction rate is low. They may have useful role in mixing nutrients as from deep burrow come up to surface for casting.

BASIC CHARACTERISTICS OF SUITABLE SPECIES

Farmers or gardener needs only those species which can quickly convert organic waste to vermicompost. Therefore, a worm should have following characteristics:

1. Worm species should be tolerant to disease.
2. The culturing techniques should be simple enough to adopt.
3. Worm should be efficient convertor of plant or animal bio mass to body proteins, so that its growth rates are high.
4. It should have high consumption, digestion and assimilation rate (composting qualities).
5. The worm should have wide adaptability (tolerance) to environmental factors (capability to live in varying temperature conditions).
6. The worms should have feeding preference and adaptability for wide range of organic material (high and rich organic matter).
7. The worm should produce large numbers of cocoons that should not have long hatching time, so that multiplication and organic matter

conversion is fast.

8. Growth rate, maturity from young one to adult stage should be fast.
9. Worm should have compatibility or tolerance with other worms (as with possibility of mixture of species by amateurs) as would add to productivity of biomass (worms) and conversion rate at different strata (layers) of organic matter, *i.e.*, faster composting. This feature, *i.e.*, composting with different layer feeders have so far received very little attention.
10. Worms should be disease resistant.
11. Worms on introduction in substrate, should have least inactivity period (= vermistabilization period).
12. The worms should feed near the surface of organic matter.

Fixing Earthworms for Identification

1. Paralyze the worms by placing them in water and gradually adding alcohol.
2. When they are narcotized transfer them to a petri plate having 10% formalin solution by holding the worms straight for a few seconds. Worms stiffen.
3. After 24h, transfer them into 70% alcohol.

Transport of Fixed Worms to Laboratory

1. Dip cotton pads in alcohol (70%) and place fixed worms on these pads.
2. If there are more than one species mark the numbers on small pieces of paper with pencil and stich them posterior end of the worms.
3. Roll the cotton pads and tie with thread.
4. Place the rolled pads in a polythene cover and seal it to prevent drying of cotton pad.
5. Place the polythene cover in paper covers and transport to laboratory.

Description of Suitable Species

In whole world, only 3-4 species are in wide use. In India, only two peregrine species are being extensively used for vermiculture namely *Eisenia foetida* and *Eudrilus eugeniae.*

Several species are also available but they have to be tested on various grounds like, their inbuilt adaptability to climatic conditions, per existing distributional range, local availability and in view of needs for better waste conversion rates etc. The important species for Indian sub-continent are as follows:

Family: Lumbricidae
Bimastos parvus (= Allolobophora (Bimastos) parvus Eisen)

They are distributed in Kashmir (Gorai, Srinagar); Himachal Pradesh (Kasauli, Barog, Simla, Chail, Narkanda, Kothai); Punjab (Ferozepur and other areas?);

Rajasthan (Pratapgarh); and Uttar Pradesh (Nainital and other areas?). They are brownish red body with saddle shaped clitellum, dorsally covering 6 or more segments, from XXIV or XXV to XXX; length 25-40 mm; diameter 1-2 mm; total segments 85-111 usually 90. Other characters morphological and are differentiable by experts only for confirmation. In many temperate countries, allied species, namely, (1) *Allolobophora subrubricunda* and (2) *Allolobophora chlorotica,* (3) *Dendrobaena rubida,* (4) *Eisenia hortensis* and *Lumbricus rubellus* are used in vermitechnology. These species, barring perhaps *E. hortensis* are more or less established in temperate regions of India. In such regions, vermitechnology with these species deserves attention, as live individuals would be more readily available in nature and would be better suited to local conditions. Only one has to search and know local distribution or occurrence in different areas.

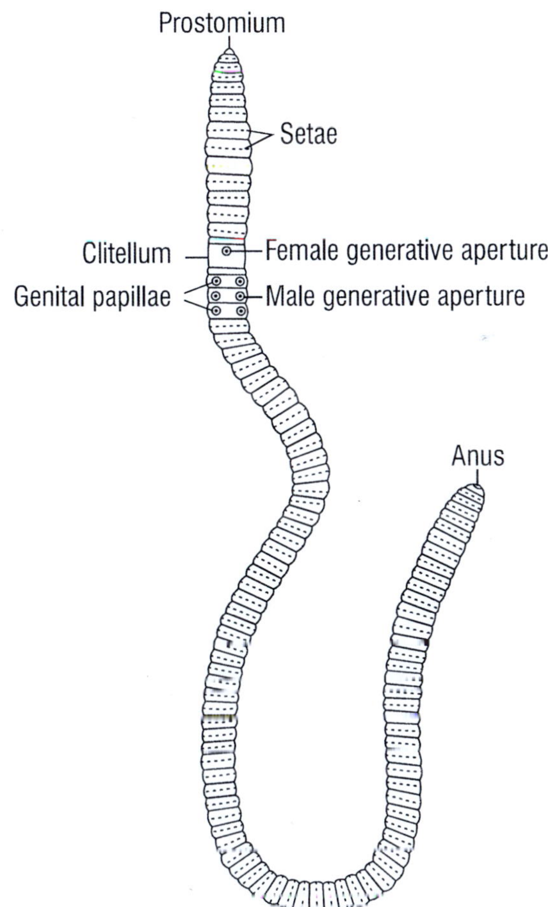

FIG. 8.2: Ventral view *of Metaphire posthuma.*

Eisenia foetida (Sav.)

Eisenia foetida is perhaps world's most widely used earthworm in vermitechnology for vermicomposting. They are distributed throughout India in nature. They found in sewage dumps (sewage filter mesh grills), sludge, tanks and filters. They are marketed under different names like Red worm, Pink worm or Purple worm, Tiger worm, Brandling worm or even vermiculture or vermicomposting worm. Living worms appears coloured (red, brown or purple or even darker). Careful observation indicates that they have dorsally coloured bands and are often two per segment. Ventrally body is pale. On maturity, clitellum spreads over 7-9 segments in length over XXIV, XXV or XXVI to XXXII body segments. On these, ridges wall develop on maturity. They are quite hardy and can tolerate wide fluctuations of temperature and humidity. The culturing process of species is quite easy due to feeding adaptability or adjustment to organic matter. They can live on wide variety of degradable/decomposable organic wastes. On these, an average conversion efficiency ranges upto 7 mg/worm/day. They have fast growth rate. Mature individuals can attain up to 1500 mg body weight. Mature adults attain body weight upto 1.5 gm and attain reproduction capability within 50-55 days of hatching from cocoon. A mature worm on average can produce one cocoon every third day and from each cocoon on hatching (within 23 days) emerge from 1 to 3 individuals. However, generally, survival of all emerged individuals is not seen. Still, on average, numbers can multiply or increase as high as 30 times. Duration of life is 70 days. They do not show movement outside feed substrate (decomposing organic waste).

Family: Eudrilidae

Eudrilus eugeniae (Kinb.)

This species is perhaps fastest growing and second most widely used earthworm in vermitechnology for composting and for use as vermi-protein. These species are originally distributed in Equatorial West Africa. They are presently distributed (may now be naturally occurring also) in most parts of the world and have been introduced for various usages in vermitechnology. It is common in parts of even America and Europe. In parts of America, species is common and is popularly known as *Night Crawler*. In India, common in many vermiculture establishments, particularly in Southern India this species has been recommended for vermiculture. It is distributed in Travancore, Pune and North Konkan. The living worms appear brown and red to dark violet. The general colouration is like animal flesh. Their length ranges from 32-140 mm, diameter is 5-8 mm and total body segments range 145-196. In mature adults, clitellum spreads over 5-6 segments numbering XIII or XIV to XVIII. Spermathecal and female pores are fused commencing after 13/14 body segments. Male pores commence before 17/18 body segments. It grows fast (assimilation and conversion to biomass) and is better than other species and ranges upto 12 mg/day. It maximum body weight ranges up to 4.3 mg/individual. They attain maturity within 40 days, and a week later individuals commence cocoon laying, almost 1 cocoon/day or upto 4 cocoons/3 days for 46 days, depending upon appropriate conditions. Total life span in laboratory is estimated to range 1 to 3 years. Cocoon incubation period that at times, number of individuals per cocoon may range 1-5. They are low temperature tolerant, however, they can tolerate high temperature as well (over 48°F in room conditions, *i.e.*, under indirect sun exposure). Harvesting of the species from culture heaps is little difficult. However, simple solution is to up turn the heap, leave it for 1-3 hours, all worms move to lower layer. From top layers, compost can be collected and worms can be replaced in culture media or harvested for appropriate utilization. Their reproductive activity gradually decreases and so composting activity decreases. In such state is advisable to utilize worm for animal feed or just for burying in fields (a nitrogenous waste).

Family: Megascolecidae

They are available throughout world. So far only six species of this family are in use for vermitechnology work. These are *Amynthas diffrigens, Lampito mauritii, Metaphire anomala, Metaphire birmanica, Perionyx excavatus* and *Perionyx sansbaricus* which is endemic to India. It is essential to know about their distribution in of India. For all taxonomic characteristics consultations with specialists is necessary

Lampito mauritii (Kinb.)

Species is also referred in old documents as *Megascolex mauritii* (Kinb.). They are distributed throughout India (excepting Uttar Pradesh, wherein present distribution is not known). They are reported occurring in Kapurthala (Punjab), many areas of Bengal, Gujarat, and Maharashtra. Gujarat, Madhya Pradesh, S. Rajasthan, Andhra Pradesh, Tamil Nadu, Goa, Kerala, Andaman Islands, Lakshdweep Island. They are also common in many parts of South and South East Asia. The body colouration is dark yellow except at anterior end which has purplish tinge. The total body length 80-210 mm with a diameter of 3½-5 mm. They have 166-190 total body segments. Clitellum spread over to four body segments, on XIV to XVII and is ring shaped. Male pores on papillae, female pores near each other in pairs on XIV segment and spermathecal pores, in three pairs.

Metaphire anomala Mich. (= Pheretima anomala)

They found in Sibpur (Calcutta). We still need their distributional status. They are excepted to be distributed in Orissa, Bihar, South-East Madhya Pradesh, Andhra Pradesh, Tamil Nadu, Karnataka and Kerala, also in parts of Maharashtra. The body colorations have not yet been studied. The total body length 80-90 mm, diameter 5 to 5.5 mm and total body segments 130 (variation range?). Clitellum, ring shaped covering three body segments, i.e. XIII-XVI with XIV ventrally having setae. Spermathecae pores absent but copulatory papillae in pairs present. The details on performance of species in vermiculture work are needed.

Metaphire posthuma (= Pheretima posthuma)

This is common species in India. Unfortunately other details on vermitechnological usages, if any, to-date yet to be studied.

Perionyx excavatus E. Perr.

In India, this genus is spread from eastern or western Himalayas downwards in regions of heavy rainfall. Now perhaps, species are widely distributed and two exotic species, *Perionyx excavatus* and *Perionyx sansbaricus,* believed to introduce from Australia or New Zealand. Only *P. excavator* is in use for vermitechnology in many parts of the world. In India, the species is common in Eastern Himalayas (Dibrugarh, Sadiya and Darjeeling), Western Himalayas (Kumaon Dist., Sahastradhara, Dehradun, Simla) Pilibhit, Bengal (Calcutta, Rajasthan, Sibpur), Maharashtra (Bombay) and Little Andaman Islands. The body colouration on dorsal surface (upper part) deep purple to reddish brown and lower side pale. The total body length ranges from 23 to 120 mm with a diameter of 2.5 mm. They are chitellum ring shaped, covering five or less segments on XIII or part, XIII to XVII. Male pores in a depression (line or area). Spermathecal pores in 7/8 and 8/9, nearly at same distance as male pores. The life cycle duration is of ± 46 days, growth rate is 3.5 mg/day, body weight (max), 600 mg. Reproduction rate is fast. They attain maturity within 21-22 days and reproduction (cocoon laying) commences by 24th day. The average cocoon production is 1.1/worm/day. This is almost double reproduction rate of most commonly used worm in vermiculture. The incubation period reportedly ranges to ± 18.7 days and mean number of hatchings is 1 to 3 per cocoon.

Perionyx sansbaricus Michaelson

They are widely distributed in many parts of India. The species is found in natural environments of Baroda (Gujarat), Igatpuri, Manmad, Wathur near Mahabaleshwar and Londa near Castle Rock (Maharashtra) Khandwa and Kala Khund near Indore (Madhya Pradesh), Coonoor and Kotagiri in Nilgiris, Kodaikanal (South India) and New Delhi and vicinity area (North India). The dorsally body colour is purple which is darker at anterior portion. On lower side (ventrally) body is pale. The body length is 32-63 mm with a diameter of 2.5 to 3.5 mm. The total body segments range 84-108. Clitellum is ring shaped, extends on 5 segments, over XIII-XVII body segments. Area of male pores variable (XVIII). Spermathecal pores near the middle line-6/7,6/8 and 8/9.

The other species which are used for vermiculture are, *Octochaetus (Octochaetoides) surnensis* Mich., *Ramiella bishambari* (Steph.), *Dichogaster bolaui* (Mich.), *Dichogaster affinis* (Mich.), *Dichogaster curgensis* (Micha.), *Dichogaster saliens* (Bedd.), *Ramiella bishambari* (Steph.), *Erythrodraeodrilus suctorius* (Steph.), *Ocnerodrilus (Ocnerodrilus) occidentalis* (Eisen.), *Moniligaster perrieri* (Mich.), *Drawida willsi* (Mich.), etc.

MAINTENANCE OF BASE CULTURE

Base or mother or seed culture is used for large scale culture. These should be carefully maintained and should be checked for health. The selection of feed of an individual is necessary. It should have any chemical that is lethal to worms. The procurement of organic waste (feed) should also be checked for seepage and contamination. For initial multiplication best feed substrate is cow dung manure. It should be at advance stage of decomposition and should bc devoid of parasites.

VERMICOMPOSTING MATERIALS

The biologically degradable and decomposable organic wastes are used in vermiculture and vermicomposting. Commonly used composting materials are as follows:

Animal Dung

Cattle dung, sheep dung, horse dung, goat dung and poultry dropping etc may be used for this purpose. In use of animal dungs other than cattle dung, various preliminary testing and precautions for pathogens and responses to earthworms are necessary. The uses of horse dung should be done carefully because tetanus virus is common in horse dung and is lethal to human beings.

Agricultural Waste

Agricultural waste obtained after harvesting and threshing may be used. They include stem, leaves, husk (excepting paddy husk), peels, vegetable waste, orchard leaf litter, processed food wastes, sugarcane trash and bagasse; and processing wastes.

Forestry Wastes

These are plant products such as wood shavings, peels, saw dust and pulp. All these besides various types of forest leaf litter can be used. The unutilized forest waste such as leaf litter may also be used for vermicomposting.

City Leaf Litter

The burnt leaf litter from avenue or residential areas may be used, however, reports are not available. If it is used, this would keep cities clean and would provide useful product. The leaf litter of mango, guava, grasses and certain weeds may be used, but we need more information on this aspect.

Waste Paper and Cotton Cloth etc.

These are decomposable organic waste. These if are not being recycled for other useful products, can be recycled with vermicomposting.

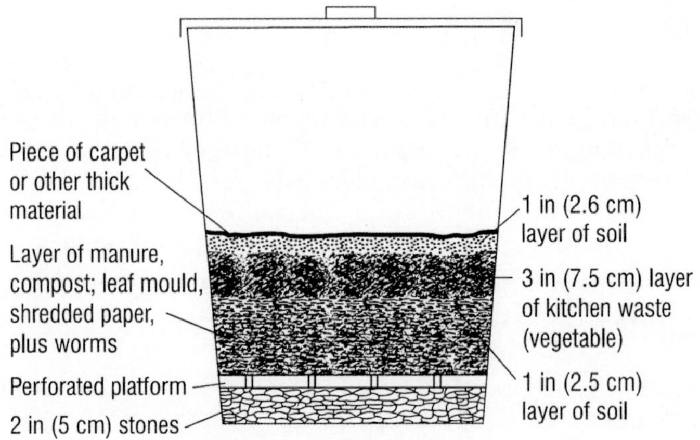

FIG. 8.3: An earthworm compost bin.

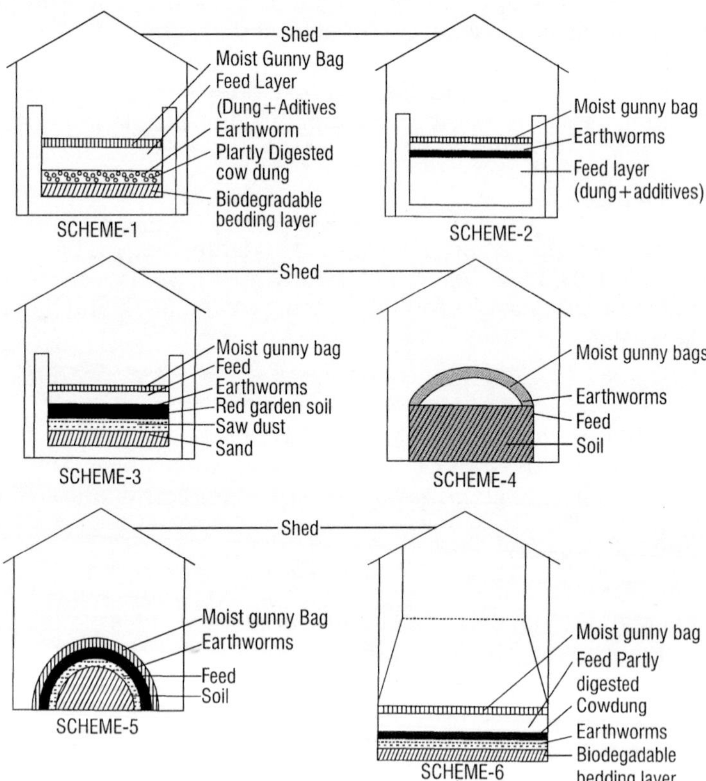

FIG 8.4: Various schemes for vermicomposting.

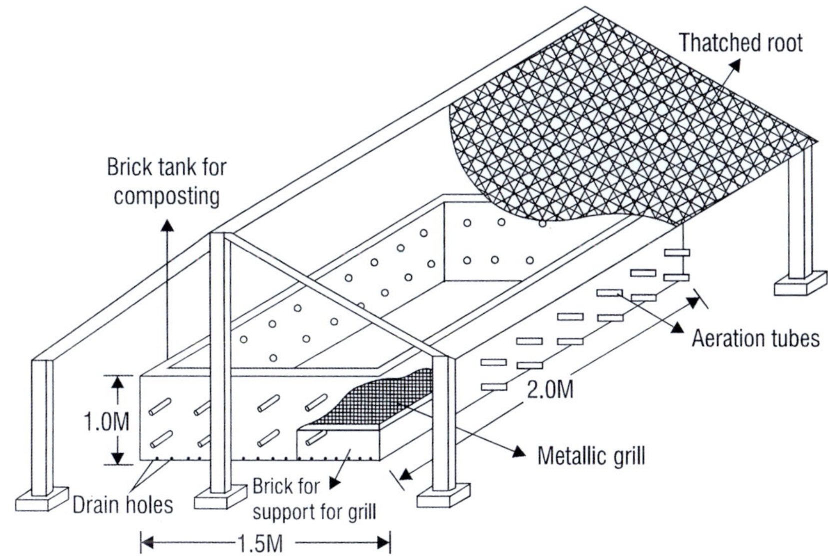

FIG 8.5: Brick Composting Unit.

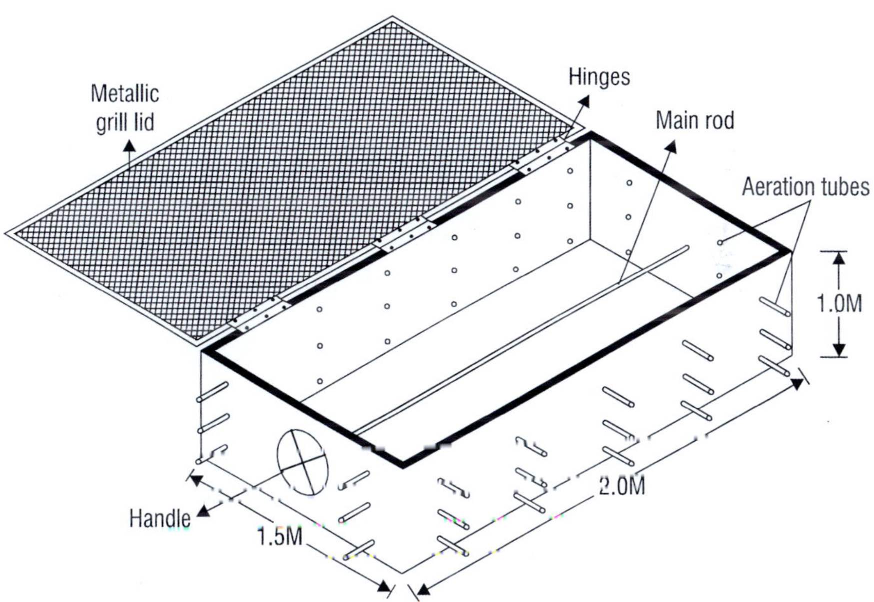

FIG 8.6: Composting mixture unit.

City Refuge

City refuge or garbage on daily production basis comprise important items of city factors and a considerable portion of city refuge can be sorted and recycled or composted. Most of household as kitchen waste with little manipulation can be used for vermicompose.

Biogas Slurry

After recovery of biogas, if not required for agricultural use, viz., in conventional composting can be used for vermicomposting.

Industrial Wastes

The industrial wastes like waste from food processing, distillery etc. can also be used in vermiculture with some manipulations.

PRELIMINARY TREATMENT OF COMPOSTING MATERIAL

Initially, proper collection, sorting or separation of compostable, non-compostable and non-biodegradables like plastics, stone, glass, ceramics and metals should be done.

Heavily contaminated wastes (even in kitchen wastes, heavily spicy wastes) with chemicals should be separated. The cleaned matter selected for composting should be heaped and large lumps should be broken. The separated matter should be spreaded in a layer upto 1 foot and to be exposed to sun for a day. This helps in killing several unwanted organisms and removes foul smell. The mixing of daily organic waste produce may be done with somewhat pretreated leaf litter in approximate ratio ranging from 10 to 40% of the waste to be vermicomposted.

Pre-Treatment of Leaf Litter and Agricultural Waste

They are heaped on ground and exposed to sun. Heaps are then be beaten with some stick or bamboo to break into smaller pieces. Materials like crop plant stems and bagasse etc. are required to be cut into smaller pieces for enhancing decomposition and vermicomposting processes. The non-marketable vegetable produce which are sometimes infected with insect pests should be chemically treated. The common example to these is insect pest popularly known as brinjal borers which effect several vegetables.

These insects can be controlled by chopping waste into smaller pieces up to 4" size and spread over ground. This layer be then sprayed with 4% aqueous solution of some Neem insecticidal formulation. This would gradually kill insect pests stages without making waste unfit for vermicomposting. Neem formulation can also be prepared by collecting Neem fruits, drying and taking out seed kernel as we take out almonds by removing hard outer covering. 20-25 gm Neem seed kernel be wrapped in a cloth over night and squeezed in half liter of water with repeated dipping and squeezing with turning movements. Solution so obtained is insufficient concentration to kill most insect pests stages.

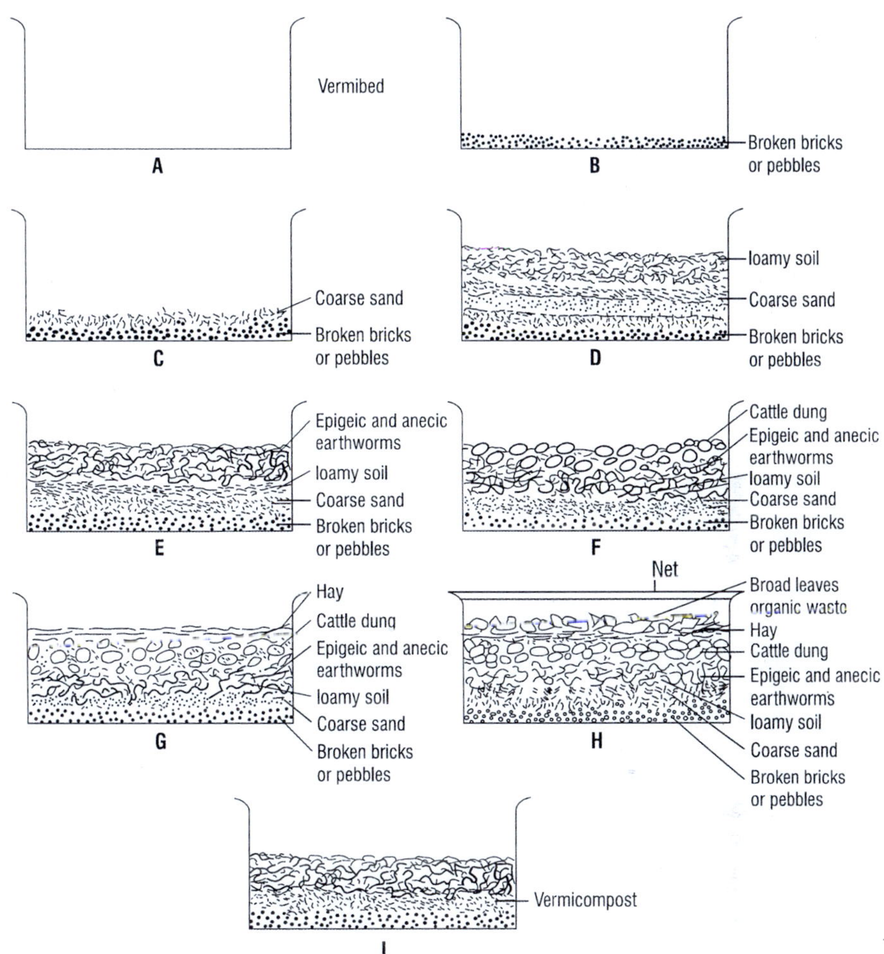

FIG 8.7: Steps of vermicomposting.

SMALL SCALE OR INDOOR VERMICOMPOSTING

It is done under covered areas (with a shade). It is should be protected from climatic adversaries like high rains, prolonged spells of high or low temperatures (from less than 10° F to more than 45° F) and predators like ants, rodents and large insectivorous birds are abundant. Small scale indoor vermiculture and composting is easy and can be adopted.

LARGE SCALE OR OUTDOOR VERMICOMPOSTING

Simple promotion of vermic activity in fallow fields after harvesting crop(s).

Large or small quantities of plant wastes to be vermicomposted, after pretreatment be heaped and mixed with small quantities of mature cow dung manure (FYM). The add bacterial inoculum for enhancing decomposition process and eventual faster vermicomposting. These heaps be sprinkled with little water (roughly 5 ltrs per 20 kg wastes) and covered with a hesian cloth or left on ground. Whole of the waste heap should be kept moist for 3-7 days and periodically it is upturned and mixed. After 2-7 weeks, heap comes in a stage that vermicomposting is fast. This waste is now ready for vermicomposting.

REQUIREMENTS FOR VERMICOMPOSTING

Container

The shape and size of the vermicomposting container, depend on the requirement that is quantity of waste to be composted and number of live earthworms we want to culture. On an average, 2000 adult earthworms can be maintained in containers of 1 m^2 dimension. These with appropriate conditioning of composting material would convert approximately 200 kg wastes every month. Interestingly, roughly in a container of 2.23 × 2.23 mt. About 10 kgs of earthworms can convert approximately 1 ton per month. However, to have optimal conversion normally only upper 9-12" layer is composted. This should be softly scrapped off.

Bedding Material

This is the lower most layer of earthworm feed substrate that is required to be vermicomposted. For this any biodegradable matter is used like banana stem peels, coir pith, coconut leaves, sugarcane trash, stems of crops, grasses or husk. Waste or discarded cattle feed can also be used for bedding.

Moisture Content

Moisture content during vermicomposting should be maintained between 30-40 per cent. If moisture is high, dry cow dung manure or leaf litter should be mixed with substrate.

Temperature

Requirement for optimal results is 20-30° C. However, survival of earthworms is even at lower temperatures and upto 48° C air temperature. Obviously with little provision of shade, temperature within worm feed substrate (material to be vermicomposted) can be reduced. For this it is desirable that substrate should not be tightly packed in containers.

INITIATION OF VERMICULTURE IN INDIA

The use of *Perionyx excavatus* was the beginning of pioneer attempt in India to employ earthworms for the degradation of solid waste. Later *Eudrilus eugeniae* was brought to the laboratory in University of Agricultural Sciences, Bangalore, India to test the possibility of their establishment in this subcontinent which has similar climatic conditions as that of its place of origin (Nigeria). Since 1982 *Eudrilus eugeniae* has been promoted for waste degradation. When compared to *Perionyx excavatus*, this species has surpassed both in feeding and reproductive rate. In the beginning of the experiments, it was thought that the response may be short lived, as many organisms show a tendency to perform extremely well in a new environment till they get established. Such a change in their efficiency was not observed in these last 15 years. A series of experiments has shown that it is an ideal species to work on agricultural and agro based industry waste. Their sensitivity to density pressure and meandering nature demand larger area of space for their establishment. *Eisenia fetida* also entered the laboratory from Germany and is preferred species for composting activity in the urban areas (urban waste).

The University of Agricultural Sciences, Bangalore, India initiated vermiculture in India and they propagated the knowledge of vermiculture to farming community. The university released Press Note by the Technology Release Committee in the year 1984. Till 1990, the technology did not induce much interest in many people. At initial stage, several controversies were developed about the composting and selection of species of earthworms.

Most of the researchers were of opinion that indigenous species of earthworms should be used for composting and not the exotic species like *Eisenia fetida* and *Eudrilus eugeniae*. The group that opposed the use of exotic species in this country lacked the knowledge of their feeding habits, habitat preference, their reproductive potentials and ability to survive in captivity to accept them for vermicomposting. But a detailed and in-depth study of the biology of 15 out of the 60 species of earthworms collected during the survey of Karnataka State helped me to put forward my views that a very few species of earthworms can be used for composting purpose and a few more as co-inhabitants to hasten up the process of decomposting along with other soil organisms. In India, which was under colonization for several centuries, exotic species of plants and animals have been transported either intentionally or unintentionally. The northern part of the country has more exotic species of earthworms than the southern peninsula.

CHAPTER 9

Bio-Intensive Nutrient Management

The practice of biological management of soil fertility is the old concept. It is the use of biological resources of the ecosystem, particularly those of the soil itself, for the manipulation of soil fertility. Emergence from the circumstances of nutritional deficiency and poverty that characterize the lives of many of the small-scale farmers of the world can only be achieved where there are sufficient resources to raise food productivity. Biological and physicochemical management are essentially based on integrated approach to soil fertility management.

According to Sanchez (1994) we should rely on biological processes by adapting germplasm to adverse soil conditions, enhancing soil biological activity and optimizing nutrient cycling to minimize external inputs and maximize the efficiency of their use

It is an ecological approach to soil fertility management, which has favored balanced farming systems. Biological approach to soil fertility management will help to restore soil fertility and will solve many problems related to soil management.

It is also established that many of the great gains in production made in the green revolution by use of high-yielding varieties with high inputs of inorganic fertilizer cannot be maintained indefinitely. Among the causes attributed to yields declines under long-term cultivation are changes in soil fertility associated with loss of organic matter and the accompanying decline in soil physical and chemical properties.

The biological soil fertility management is an ecological approach for sustainable development and is mainly concerned with the maintenance of yield, closely associated with desires to conserve natural resources, including a greater value accorded to maintenance of biodiversity. An increased efficient use of resource, including the need to utilize all available resources within economic limits will be realized in the long term as well as profitable in the short term too

SOIL POPULATIONS AND PROCESSES

Biological populations and processes influence soil fertility in a variety of ways each of which can have an ameliorating effect on the main soil-based constraints to productivity:

1. Symbionts such as rhizobia and mycorrhiza increase the efficiency of nutrient acquisition by plants.
2. A wide range of fungi, bacteria, and animals participate in the process of decomposition, mineralization, and nutrient immobilization and

therefore influence the efficiency of nutrient cycles.

3. Soil organisms mediate both the synthesis and decomposition of soil organic matter (SOM) and therefore influence cation exchange capacity; the soil N, S, and P reserve; soil acidity and toxicity; and soil water-holding capacity.
4. The burrowing and particle transport activities of soil fauna, and soil particle aggregation by fungi and bacteria, influence soil structure and soil water regimes.

The above practices are not unique to biological management but common to farming practice the world over. The use of biological inputs such as N-fixing bacteria, mycorrhiza, or soil fauna as a means of enhancing the endemic biological activities are the means of biological soil management. Direct management is also achieved by the use of organic matter inputs – in effect a means of selectively feeding the heterotrophic biological populations of soil – a practice of very ancient origin but at times eschewed in modern agriculture. Equally direct but usually unintentional effects are also achieved by the use of pesticides, which may kill particular groups of soil organisms that are involved in processes of significance to soil fertility. Management techniques such as tillage and fertilization also influence the activity of the biota indirectly by altering the physical and chemical environment of the soil.

The success of biological management practices depends on following two preconditions, which must be satisfied:

1. The availability of a management practice that is practically and economically acceptable to the farmer.
2. The demonstration by the scientist that the practice leads to enhanced soil fertility.

Some soil microorganisms play an important role in improving soil fertility and crop productivity due to their capability to fix atmospheric nitrogen, solubilize insoluble phosphate and decompose farm wastes resulting in the release of plant nutrient. The extent of benefit from these microorganisms depends upon their number and efficiency, which however, is governed by a large number of soil and environmental factors. When the number and activity of specific microorganism called microbial inoculant or biofertilizer is used to hasten biological activity to improve availability of plant nutrient. A number of products are now available that are generally referred to as soil and plant additives, of non traditional nature. These products include:

1. Microbial fertilizers and soil inoculants contain unique and beneficial strains of soil microorganisms.
2. Microbial activators that supposedly contain special chemical formulations for increasing the numbers and activity of beneficial microorganisms in soil.
3. Soil conditions that claim to create favourable soil physical and chemical conditions, which result in increased growth and yield of crops, and (iv) vermin-compost, which helps in improving soil health and fertility.

Nitrogen fixing organisms can be provided to the farmers in the name of microbial inoculants otherwise termed as biofertilizers. The biofertilizers containing biological nitrogen fixing organisms are of utmost important in agriculture in view of the following advantage:

1. They help in the establishment and growth of crop plants and trees.
2. They enhance biomass production and grain yields by 10-20%.
3. They are useful in sustainable agriculture.
4. They are suitable in organic farming.
5. They play an important role in agroforestry / silvipastoral systems.

After the advent of the acetylene test for assessing nitrogen fixing potential, rapid progress has been made in identifying nitrogen-fixing organisms. Out of a large number of microorganisms possessing the property of nitrogen fixation, only a few such as *Rhizobium, Azotobacter, Azospirillum,* BGA, Azolla, etc. have been commercially exploited. The reaction in biological nitrogen fixation is essential the same as in production of chemical fertilizers (Haber-Bosch process) i.e. the catalytic reduction of dinitrogen (N_2) to ammonia (NH_3).

USE OF BIOFERTILIZERS

The use of biofertilizers is currently gaining interest as a cheap, safe alternate to conventional chemical fertilizers. Air has 78% Nitrogen and 88,000 t of N in every ha. of land on the world. Earth can port billions of bacteria that can trap the atmospheric nitrogen from the air and put into observable form of the plant. Azospirillum forms an associative symbiosis with many plants, they grow and fix nitrogen salts of organic acids such as malate, succinate, pyruvate an dlactate. Azospirillum is applied as seed treatment and soil application in crops like rice, sugar cane, pulses, soybean and vegetables. Significant increase in root length, top dry weight, root dry weight, total leaf area and yield were reported. The inoculants, like nitro plus (Legume inoculant) and VAM (vesicular Arbuscular mycorrhizae) are also effective for crop yield improvement. The *Bacillus sp.* and *Pseudomonas sp.* are helpful in synthesizing the insoluble form of P. The combined applications of phosphobacteria, rock phosphate and FYM to commercial crops have greatly enhanced biomass production, uptake of nutrients and yield.

ENRICHMENT OF COMPOST WITH MICROBIAL INOCULANTS

Compost prepared by traditional method is usually low in nutrients and there is need to improve its quality. Enrichment of compost using low cost N fixing and phosphate solubilising microbes is one of the possible ways of improving nutrient status of the product. It could be achieved by introducing microbial inoculants, which are more efficient than the native strains associated with substrate materials. Both the nitrogen fixing and phosphate solubilising microbes are more exacting in their physiological and ecological requirements and it is difficult to meet these requirements under natural conditions. The only alternative is to enhance their inoculum potential in the composting mass. Studies conducted at IARI, New Delhi, showed that inoculation with *Azotobacter/Azospirillum* and phosphate solubilising culture in the presence of 1 % rock phosphate is a beneficial input to obtain good quality compost rich in nitrogen (1.8 %). The humus content was also higher in materials treated with microbial inoculants.

NITORGEN FIXING MICROBS

Rhizobium

Rhizobium, in association with legumes, fixes atmospheric N. The legumes and their symbiotic association with the rhizobium bacterium result in the formation of root nodules that fix atmospheric N. Successful nodulation of leguminous crop by rhizobium largely depends on the availability of a compatible strain for a particular legume. *Rhizobium* population in the soil depends on the presence of legume crops in the field.

The presence of *Rhizobium* bacteria in the root nodule can be seen under high power microscope by crushing a washed nodule between two glass sliders. This kind of bacteria, preferentially infect legumes as compared to non-legumes. Structurally, *Rhizobium* is rod shaped. It shows a great variations, which can be observed during their life cycle. These are coccoid, very small, highly motile and ellipsoidal forms. The bacteroids are usually irregular with X, Y, star and club-shaped forms. They are 0.5 to 0.9 micron by 1.2 to 3.0 micron in size. All the species are non-sporing and non-acid fast. These symbiotic bacteria (*Rhizobium*) are difficult to cultivate in ordinary culture media but they grow on manitol agar. *Rhizobia* are susceptible to antibiotics (produced by other microorganisms of soil) and also to the action of bacteriophages. Fungicides, herbicides and other plant protectants are also toxic to *Rhizobia*.

The most widely used biofertilizer is *Rhizobium,* which colonizes the roots of specific legumes to form tumour like growths called root nodules. It is these nodules that act as factories of ammonia production. Rhizobia have the ability to fix atmospheric nitrogen in symbiotic association with legumes and certain non-legumes like *Parasponia*. A mature nodule consists of a central 'bacterial zone' surrounded by several layers of cortical cells. The process of N-fixation is wholly dependent on the activity of the enzyme nitrogenase, which is located within the bacteroids. The volume and number of bacteroids have a direct positive relationship with the amount of nitrogen fixed.

In the absence of legumes, the population of rhizobium in the soil diminishes. Most strains of rhizobia native to the soil are effective in N fixation. Some of them may be ineffective and produce ineffective nodules. These nodules are smaller in size, soft and whitish in color. Their N fixing capability is much low. The effective strains produce large, pink and firm nodules. Their N fixing capacity is far greater. Artificial seed inoculation is often needed to restore population of effective strains of rhizobium near the root zone to ensure high yields. Each legume requires a specific species of rhizobium to form effective nodules.

TABLE 9.1: Cross inoculation group of *Rhizobium*

Rhizobium species	Cross inoculation group	Legume types
R. leguminosarum	Peas	Pisum, Vicia
R. phaseoli	Beans	Phaseolus
R. trifolii	Clover	Trifolium
R. meliloti	Alfalfa	Melilotus, Medicago, Trigonella
R. lupini	Lupini	Lupinus, Ornithopus
R. japonicum	Soybean	Glycine
R. sp.	Cowpea	Vigna, Arachis

There is specific *Rhizobium* for every legume and the inoculation with efficient strains of *Rhizobia* are essential for the nitrogen gains and better crop yields. Inoculation of pulse crop with appropriate cultures can give yield increases of the order of 15-30 per cent. Thus *Rhizobium*-legume symbiosis can generally meet 80 per cent of the N requirement of the legume crop. A starter does of N, as chemical fertilizer, is required (equivalent to about 20 per cent requirement of legume crop) for good crop yields.

The *Rhizobium*-legume association can fix upto 100-300 kg N/ha in one crop season and in certain situations can leave behind substantial nitrogen for the following crop. The range of nitrogen fixed per hectare per year by different legumes is 100-150 kg for clover, 80-85 kg for cowpea, 100-300 kg for alfalfa, 240-320 kg for faba bean, 90-100 kg for lentil, 150-200 kg for lupins, 50-60 kg for groundnut, 60-80 kg for soybean, 50-55 kg for mungbean and 100-400 kg for pasture legumes.

Rhizobium Culture

Rhizobium culture has been used for leguminous vegetable crops is in practice among farmers. *Rhizobium* forms symbiotic association with roots of leguminous crops forming nitrogen-fixing nodules on the root system. Rhizobia enter the roots of legumes either through root hairs or directly at the emergence of lateral roots. They form infection threads through which the bacteria enter the root system. The entry of *Rhizobium* into the roots of plans is a perfect symbiosis. Initial stages is symbiosis is the reciprocal molecular conversation, transmission of a gene inducer from legume host to bacterium and the bacteria in response synthesizing a morphogen that is transmitted to the plant, switching the developmental fate of the legume root, the bacteria and formation of bacteroids in the nodules. It is enzyme nitrogenase that reduces atmospheric nitrogen to ammonia. This enzyme has been studied in a very great detail. (Das, 1991)

There are 2 types of rhizobia:

1. Slow growing *Bradyrhizobium* sp.
2. Fast-growing *Rhizobium* sp.

Soil contains native species of *Rhizobium* but all are not capable of forming nodules. Some of the strains are highly specific to certain species. These are called cross inoculation groups. For example, *Rhizobium leguminosarum* could nodulate peas only, while *R. leguminosarum bv. phaseoli* can nodulate French bean and other beans and *Rhizobium* sps. Nodulate cowpea. Artificial inoculations help increase the native population of a particular species in the soil when that crop is grown.

Vegetable legumes – French bean, pea and cowpea – have been extensively studied. Effective nodulating strains have been isolated and developed after field-testing. French been strains have been isolated and developed after field-testing. French been strains have been modified from time to time to suit salinity, alkalinity and acidity of soil. In iron mine waste soil also rhizobium inoculation could improve nodulation in cowpea. Mutant strains of *Rhizobium leguminosarum* bv phaseoli superior to wild strains in nodulation have been induced in French bean through UV induction, antibiotic resistance, acid and azide resistance at the CCS Haryana Agricultural University, Hissar. For cowpea effective strains of *Rhizobium* sps. have been identified at University of Agricultural Sciences, Bangalore, Tamil Nadu Agricultural University, Coimbatore and CCS Haryana Agricultural University, Hissar.

20 kg/ha of *Rhizobium* inoculum increased the number of nodules 38%. The *Rhizobium* sp. also increased nodulation, nitrogen fixation and biomass yield by slow and fast growing cowpea rhizobia in gaur under different environmental conditions

It is estimated that about one-third off the electrons made available for reduction of nitrogen in legume nodules are used instead to reduce hydrogen and this gas just escapes into the soil atmosphere carrying away wasted energy. Certain species of *Rhizobium* contain uptake hydrogenase enzyme (Hup+ strains) that oxidizes much of the hydrogen to water before it escapes. During this oxidation ATP is produced from ADP and Pi. This process therefore, conserves energy and is more efficient N fixers. The *Rhizobium* strains possessing this enzyme yield more than the strains that do not possess them. The Hup+ strains have been identified by Dadarwal (1986) to be more competitive than Hup- strains.

Genetic Modifications in Rhizobium Strains

Efforts have been made to develop genetically modified strains incorporating traits for better competitiveness; Khanuja and Suman (1993) obtained heat shock tolerant mutants in the process of improving rhizobial strains. Sharma *et al* (1997) introduced gene markers (Gus) in *Rhizobium* strains to monitor the release of genetically engineered strains in the environment. However, genetically modified strains have not made any impact on legume production under field conditions.

Rhizobium in Root Nodule

The presence of these bacteria in the root nodule can be seen under high power microscope by crushing a washed nodule between two glass slider. Frank (1877) and Beijernick (1888) discovered *Rhizobium* a free living or symbiotic bacteria of leguminous as well as non-leguminous plants. This kind of bacteria, preferentially, infect legumes as compared to non-legumes. Structurally, *Rhizobium* are rod shaped but great variation can be observed during their life cycle. These are coccoid, very small, highly motile and ellipsoidal forms. The bacteroids are usually irregular with X, Y, star and club-shaped forms. They are 0.5 to 0.9 micron by 1.2 to 3.0 micron in size. All the species are non-sporing and non-acid fast. These symbiotic bacteria (*Rhizobium*) are difficult to cultivate in ordinary culture media but they grow on manitol agar.

Rhizobia are susceptible to antibiotics (produced by other microorganisms of soil) and also to the action of bacteriophages. Fungicides, herbicides and other plant protectants are also toxic to *Rhizobia*. Occurrence of different species of *Rhizobium* in leguminous plants can be enumerated as follows:

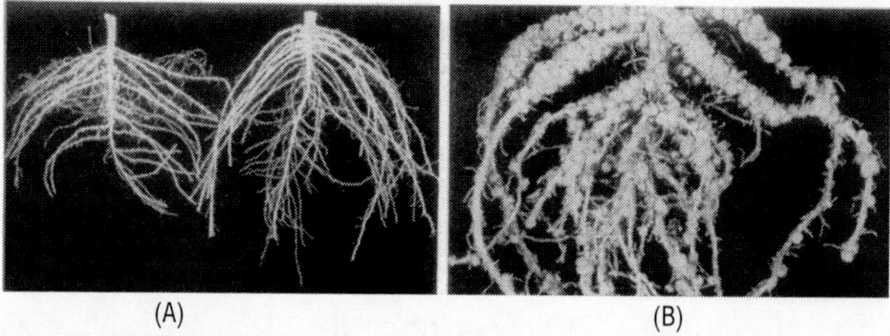

(A) (B)

FIG. 9.1: Bare uninoculated roots of soybean plant (A) and nodulated roots of soybean (B) inoculated with *Rhizobium.*

Agrobacterium (A. radiobacter, A. tumefaciens)

It has also been included in the species: *Rhizobium radiobacter,* but this splitting of *Rhizobium* at the generic level into *Agrobacterium* is not accepted by most of the microbiologists.

One more kind of slow growing asymbiotic bacteria of lupin, soybean, cowpea have been identified and placed under a separate genus *Phytomyxa* (*P. japonicum*). In short, the root nodule genera can be delimited into two genera: *Rhizobium* and *Phytomyxa*: the former comprises of three species: *R. leguminosarum, R. meliloti* and *R. radiobacter*, while the latter bears only one species *P. japonicum*.

Rhizobium can also be classified on the basis of DNA composition, for example guanine and cytosine composition of *R. leguminosarum* and *R. meliloti* is 56.6 to 63.1 per cent while G+C composition of *P. japonicum* is 62.8 to 65.5 per cent.

Structure and Function of Nodule

The outermost layer of the nodule constitutes the bacteroid zone which is enclosed by several layers of cortical cells. The rate of nitrogen fixation of nodule is directly proportional to the volume of the effective nodule.

Sometimes ineffective strains produce ineffective nodules. These nodules are relatively small and contain poorly developed tissue associated with morphological abnormalities.

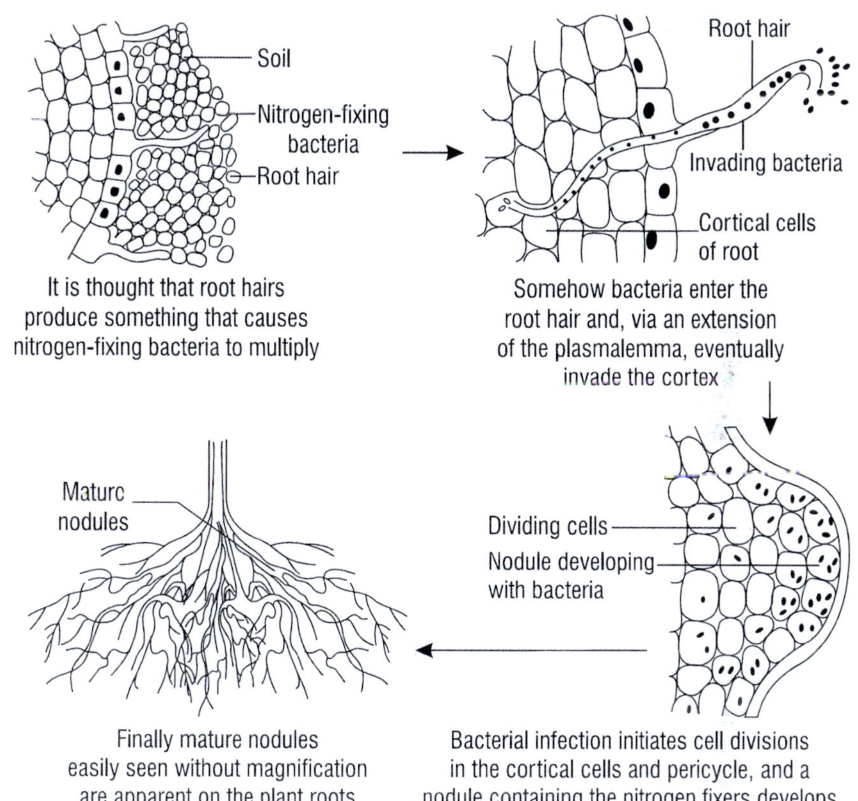

FIG. 9.2: Nodule formation by *Rhizobium* on legume plants. The root hairs are the site of infection. Note the change in the root hair. A characteristic deformation is the curling of the root hair tip to produce a Shepherd's crook. The nodule can be seen on the root.

Formation of Nodule in Leguminous Plants

Biological nitrogen fixation is the reduction of atmospheric nitrogen to ammonia by microorganisms in the soil. It involves a highly specialized and intricately evolved interactions between soil microorganisms and higher plants for harnessing the atmospheric nitrogen. It is a fascinating biological phenomenon studied extensively to provide low-cost nitrogen and improve crop productivity. The nitrogen fixing organisms associated with horticultural crops are the Rhizobium species, living in symbiotic relationship with the leguminous plants and free living fixers belonging to the *Azotobacter* family and the *Azospirillum* living in association with the root system of crop plants. Roots of most of the leguminous plants possess circular out-growth or swelling called nodules or tubérculos, which are formed after infection of certain types of nitrogen fixing bacteria particularly the different species of *Rhizobium.*

These bacteria have the capacity to fix free atmospheric nitrogen of soil in nodule. They infect or get their entrance in root through the soft-hair or other epidermal cells by damaging them. In the first stage of the infection, bacteria grow very profusely at the tip of the root hair and form a long filament in the root hair- called *infection thread*. This thread reaches the endodermis and pericycle area through cortex tissue. Cells of this area (cortex) go on dividing and form a young nodule. The newly formed young nodule pushes the overlying parenchyma and the epidermal tissue towards outside and produces a small swelling on the surface of the root. Anatomically, a nodule is made of thin walled parenchymatous cells which is filled up with the nitrogen fixing organisms. The size and shape of the root nodule varies according to plants in which occur. It is not necessary that all the bacteria, which infect the root, produce nodule.

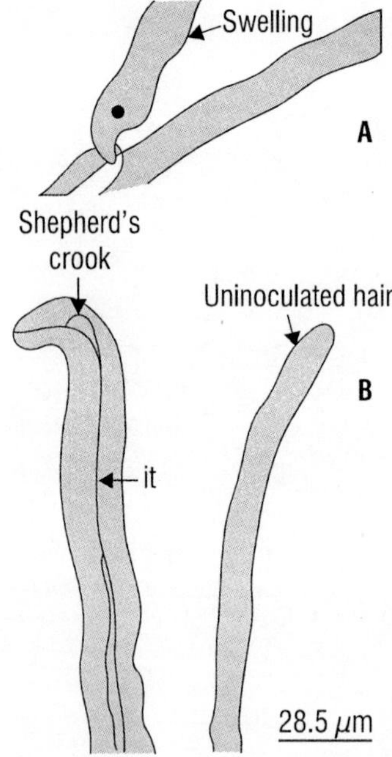

FIG. 9.3: The early host responses to infection leading to the formation of N_2-fixing root nodules. (A) A root hair two hours after inoculation, showing curling and swelling. (B) A root hair (left) two days after inoculation, showing a shepherd's crook and an infection thread (it). A thin uninoculated root hair is shown on the right.

The root nodule always contains double number of chromosomes as against of normal somatic tissue. If the root lacks such cells (cells with double chromosome number), there will be no formation of root nodule, even after the formation of infection thread. Those plants which do not bear root-nodules will never be able to fix atmospheric nitrogen in the plants. It has been observed that out of 429 genera of family Leguminosae about 179 genera containing 949 species form nodules.

Leghaemoglobin

The effective nodules are usually larger and pink in colour. This is due to the presence of red coloured leghaemoglobin (leg indicating its presence in legume root nodule). This pigment is similar to haemoglobin of blood and is found in nodules between bacteroids and membrane envelops, enclosing them. Chemically, it is haem-protein and it contains a haem meoiety attached to a peptide chain and represents the globin part of the molecule. The molecular weight of leghaemoglobin is 16,000 - 17,000 Daltons. The amount of leghaemoglobin in nodules has the direct relationship between amounts of atmospheric nitrogen fixed by legumes. The functions of leghaemoglobin are as follows:

1. It represents an active site of nitrogen absorption and reduction.
2. It acts as a specific electron carrier in nitrogen fixation.
3. It regulates the oxygen-supply in the nodule.
4. It acts as an oxygen-carrier.

Leghaemoglobin acts as a biological valve in regulating the supply of oxygen to bacteroids in nodules.

Site of Nitrogen Fixation in Nodules

The membrane envelop surrounding the groups of bacteroids are the site of nitrogen fixation, but in recent years it has been proved that bacteroids are the seat of nitrogen fixation.

Leguminous Plants / Rhizobiaceae Symbiosis

Symbiotic N_2 fixation in agriculture can be attributed mainly to legumes – the plants in the *Leguminosae*. It is estimated that *Leguminosae* contains more than 200 genera, and 20,000 species, which ranges from small plants such as the clover to the large trees such as *Acacia* species. Approximately 90% of them can fix nitrogen from the atmosphere with *Rhizobiaceae*, either with *Azorhizobium*, *Bradyrhizobium*, *Rhizobium* or with *Sinorhizobium* in root nodules. The important agricultural legumes can be divided into three groups: crops that grown for their commodities (e.g. grain legumes); forage legumes that may be grazed or harvested for animal fodder; and trees or shrubs in agroforestry systems. Worldwide approximately 1.5 million km^2 of land are cultivated with grain legume plants, mainly in *Glycine max*, *Phaseolus vulgaris*, *Pisum sativum*, *Arachis hypogaea*, *Cajanus cajan* etc. The annual harvest of grain legumes are about 200 million tonnes, which provides the plant protein source for human and animal consumption, or vegetable oil and other raw materials. The areas partially covered by the forage legumes mainly in *Trifolium*, *Lotus*, *Medicago*, *Stylosanthes*, *Macroptilium* and *Mimosa* etc., is even much larger, about 30 million km^2 of grassland in the five continents. The third group consists of the genera *Acacia*, *Albizia*, *Alnus*, *Leucaena*, *Robinia* etc., which are mostly used as timber, fuelwood or craft wood, and in the pharmaceutical industry as antibacterial and antifungal agents, or food additives and any other functions.

There are five genera of the *Rhizobiaceae*: *Azorhizobium*, *Bradyrhizobium*, *Rhizobium*, *Sinorhizobium* and *Photorhizobium*. All members of *Rhizobiaceae* are characterised by a gram-negative cell wall structure. Cells are generally rod-shaped, non-spore-forming and motile with differently arranged flagella. All rhizobia are aerobic bacteria that persist saprophytic ally in the soil until they infect a root hair cell. The enzymes from the bacteria degrade part of the cell wall and allow bacteria entry into the root-hair cell itself, which lead the root hair to produce a threadlike structure called the infection thread. The bacteria multiply extensively inside the thread, which extends inwardly and penetrates through and between the cortex cells. In the inner cortex cells the bacteria are released into the cytoplasm and stimulate some cells (especially tetraploid cells) to divide. Each enlarged, non-motile bacterium is referred to as a bacteroid. These result in a proliferation of tissues, eventually forming a mature root nodule. A typical root nodule cell contains several thousand bacteroids. The nodule contains a protein called leghemoglobin, which gives legume nodules a pink color due to its prosthetic hem group. Leghemoglobin is thought to help transport O_2 into the bacteroids, which is essential for bacteroid respiration. Nitrogen fixation in root nodules occurs directly within the bacteroids. The host plant provides bacteroids with carbohydrates, which they oxidase and from which they obtain energy. These carbohydrates are first formed in leaves during photosynthesis and then are translocated through phloem to the root nodules. Sucrose is the most abundant carbohydrate translocated, at least in legumes. Some of electrons and ATP obtained during oxidation by the bacteroids are used to reduce N_2 to NH_4^+, which is catalyzed by nitrogenase.

Azotobacter Inoculant

It is a common soil bacterium and the species *A. chroococcum* is present widely in Indian soils. Soil organic matter is the important factor that decides the growth of this bacteria. Poor organic matter content in the Indian soil is a limiting factor that not only affects the proliferation of Azotobacter in soil but also limits its N fixing capacity.

Azotobacter is a free living (non-symbiotic) aerobic nitrogen fixing bacteria commonly found in a close relationship with vegetable crops. Azotobacter was first isolated by Beijerinck. Five species of *Azotobacter* are poplar. They are: *a. chroococcum*, *A. chroococcum*, *A. vinelandii*, *A. beijerinckii* and A. paspali. *Azotobacter chroococcum* appeared in acid soils and arable soils while *A. beijerinckii* in neutral and alkali soils. They form cysts to withstand adverse condition. Each cyst has a living cell with two coats. The cysts accumulate poly hydroxybutyric acid. With the onset of favourable conditions, the cysts give rise to vegetative cells. Pigment production is one of the important characteristics of *Azotobacter* sp.

Azotobacter chroococcum, A beijerinckii species are deemed to be the most commonly occurring species in India. These strains also produce growth substances and antifungal substances. The beneficial effects of *Azotobacter* biofertilizer on cereals, millets, vegetables, cotton and sugarcane under both irrigated and rainfed field conditions have been substantiated and documented. Application of Azotobacter and sorghum by 5-30% over control. Apart from nitrogen, this organism is also capable of producing antibacterial and antifungal compounds, hormones and siderophores.

Field experiments have been done extensively in India on *Azotobacter* inoculation of seed or seedling of onion, brinjal, tomato and cabbage under different agroclimatic conditions. *Azotobacter* inoculation saves addition of nitrogenous fertilizers by 10 to 20%.

Azospirillum Inoculant

Azospirillum is known to have a close associative symbiosis with the higher plant system. These bacteria have association with cereals like sorghum, maize, pearl millet, finger millet, foxtail millet and other minor millets and also fodder grasses. Bacteria colonizing the roots not only remain on the root surface but also a sizable proportion of them, penetrates into the root tissues and live in harmony with the plant. They do not, however, produce any visible nodule or out growth on the root tissue.

The ability to fix nitrogen by certain spirilla (Spirillum lipoferum) was first recorded by Beijerinck in 1925 and later on confirmed by several workers using not only conventional micro-Kjeldahl assay but also by the methods such as acetylene reduction technique and the isotopic enrichment method involving N. Tarrand and his associates (1978) renamed this organism as *Azospirillum* (N_2 fixing *Spirillum*).

Azospirilla are a group of bacteria found in loose association with the root system of many crop plants. It is a type of symbiosis where the bacterial cells are found colonizing the root cortical cells or the intercellular spaces in the cortex. These bacterial grow better under reduced oxygen levels. They fix nitrogen from 10 to 40 kg/ha. The *Azospirillum* inoculation helps the plans in better vegetative growth, saving nitrogenous fertilizers by 25-30%. *Azospirilla* are microaerophilic in nature. They are found in association with vegetables crops either freely in soil or inside the root system. Four species of *Azospirillum* have been identified. They are *A. lipoferum, A. brasilense, A. amazonense, A iraqiensis*. The *A. brasilense and A. lipoferum* are very popular in India soils. The *A. bangalorense* a new species is found in association with tomato Sukhada (1988). One of the characteristics of *Azospirillum* is its ability to reduce nitrate and denitrify. Hence it is necessary to select strains which do not possess these characters. Sukhada and Shaila (1996) detected the *nif HDK* genes on the chromosomal DNA rather than on the plasmids of *Azospirillum bangalorense* isolated from tomato.

The bacterium, *Azospirillum* colonizes root mass and fixes N in loose association with plants. Their inoculation has shown positive interaction with applied N in several field crops with an average response equivalent to 25-30 kg ha^{-1} of applied N. The crops, which respond to *Azospirillum* inoculation are maize, barley, oats, sorghum, pearl millet and forage crops. *Azospirillum* applications increase grain yield of cereals by 5-20% of millets by 30% and of fodder by over 50%.

Inoculations with *Azospirillum* have registered increased yield in different vegetable crops. Subbiah (1994) reported increase in nutrient content and yield in chilli and Bellary onion when *Azospirillum* is applied @ 2 kg/ha. Similarly in okra, cabbage, brinjal, tomato, capsicum and chillies *Azospirillum* inoculations improve yield. Its application is common practice in vegetables in Tamil Nadu.

Blue-Green Algae Inoculant

Blue green algae are referred to as paddy organisms because of their abundance in the paddy fields. Many species belonging to the genera, *Tolypothrix, Nostoc, Schizothrix, Calothrix, Anabaenopsis* and *Plectonema* are abundant in tropical conditions. Most of N fixing BGA are filamenters, consisting of chain of vegetative cells including specialized cells called heterocyst which function as a micronodules for synthesis and N fixing machinery. BGA, *Anabaena* Azoll*ae*, also occurs in symbiotic relation with the aquatic fern, Azolla.

In addition to N fixation, BGA also synthesis and liberate growth promoting substances such as auxin and amino compounds which stimulate the growth of rice.

As early as 1939, Dr. P. K. De suggested that fertility of tropical rice fields was largely due to activity of the N-fixing BGA. Blue-green algae are photosynthetic nitrogen fixers i.e. they use energy derived from photosynthesis to fix atmospheric nitrogen. They are also called cyanobacteria and are free-living organisms. In general BGA show dominance over other classes of algae in India. The utilization of blue-green algae as a biofertilizer for rice is very promising. A judicious use of these algae could provide to the country's entire rice acreage as much nitrogen as obtained from 15-17 lakh tonnes of urea. Methods have been developed for mass production of algal biofertilizer and rice growers in many parts of the world have already started using these algae. Recent researches have shown that algae also help to reduce soil alkalinity and this opens up possibilities for bio reclamation of such inhospitable environments.

Blue green algae (BGA) when cultured in small pit and used as an inoculum in paddy fields at the rate of 12-15 kg ha^{-1} resulted in nitrogen economy to rice crop of the order of 20-25 kg N ha^{-1} with yield increase of 5-14 per cent.

Multiplication of BGA

Algae may be multiplied in troughs or cement tanks for small quantity and in the field for large scale production.

Trough Method

Shallow trays of 2 – 0.2 m of galvanized iron sheet or permanent tanks are constructed. They are kept in the open air and exposed to the sun. The size may vary according to the quantity of algae material to be produced. About 10 kg of soil is taken and mixed well with 200g of super phosphate. Water is stagnated to a depth of 5-15 cm in the trays depending on the local conditions to meet the evaporative demand. The soil should be neutral in pH and if it is acidic, lime should be added. To prevent predators, carbofuran 3 G is to be applied @ 25 g/tray. About 250 g of algal culture is sprinkled on the surface of the standing water.

Growth of algae is rapid during hot summer months. In about 7-10 days, a thick mat of algae is formed. If the daily evaporation loss is high, water is added intermittently. When the algal growth becomes sufficiently thick, watering is to be stopped. Water is allowed to evaporate completely. The dry algae crack into flakes, which are collected and stored for further use. About 1.5 to 2.0 kg of algae material can be collected from each trough of 0.4 m^2 area.

The troughs are again filled with water and a small amount of dry algal flakes are added as inoculum. Multiplication can be continued in this manner. Once the soil in the trough is exhausted (usually after three or four harvests) it can be replaced with fresh soil mixed with super phosphate and continued as above.

Mass Multiplication of BGA in the Field

Field production of algae is adopted for multiplication of the material on a commercial scale. In the wetland, about 40 m^2 area is demarked and puddled to facilitate water stagnation. The area is encircled with strong earthen bunds of about 15 cm height. Super phosphate is applied at 2 kg for 40 m^2 area. About 5 kg of composite algal culture is required for 40 m^2 area. To control predators like daphnids, snails, and mosquitoes, carbofuran 3 G is applied at 200g. In about 7-10 days period a thick mat of algae develops in clay soils. In loamy soils, about 3-4 weeks are required for full growth of algae. Once the algae have attained full growth with complete coverage, algal flakes are collected and stored in sacks for further use.

During summer months (April - June), the average yield of algae ranges from 20 to 30 kg/40m^2. Continuous harvest of algae may be made from the same area by reflooding the plot and applying super phosphate and pesticides. Addition of further inoculum is not necessary for subsequent production.

Azolla-Anabaena Symbiosis

A small floating water fern, *Azolla* is commonly seen in low land fields and in shallow fresh water bodies. This fern harbours a blue-green algae, *Anabaena azollae.* The *Azolla-Anabaena* association is a live floating nitrogen factory using energy from photosynthesis to fix atmospheric nitrogen amounting to 100-150 kg N/ha/year from about 40-60 tonnes of biomass. The studies carried out under AICARP showed that incorporation of 5 t ha-1 Azolla gave rice yields equivalent to about 30 kg N ha-1 at several locations. Reports of its current use as a biofertilizer for rice China, Vietnam, India, Indonesia,

Thailand and other East and South Asian countries are available. An integrated system of rice Azolla fish has been developed in China.

Azolla is widely distributed in the rice growing tracts of tropics and temperate zones and grows on the irrigated rice fields. There are six species of *Azolla*: *A. caroliana, A. nilotica, A. filiculoides, A mexicana, A. microphylla and A, pinnata*. The common species occurring in India is *A. pinnata*. A heat tolerant species *A. microphylla* is introduced to India from the Latin America. *A. filiculoides* is a temperate species got from Europe. New crosses between *A. microphylla* and *A. filiculoides* are also available. Several of hybrids have high nutritive value (4.0 – 4.5 % N on dry weight basis) and grow fast. The hybrids can also withstand both heat and cold.

Azolla mutants with a high degree of tolerance to salinity have been developed at International Energy Agencies, Vienna. Scientists in China and U.S. are now working on introducing a *Bacillus thuringiensis* (Bt) toxin in Azolla to fight against insect pests.

Multiplication of Azolla

Propagation

The water fern is mainly propagated by vegetative means. It will yield on an average about 1.5 kg per square meter in a week. Under normal conditions, the fern doubles itself in three days. Multiplying Azolla through sporocarp technique have also been developed in many research institutions in the world.

Azolla can be multiplied in small cement cisterns or galvanized iron trays of convenient size. Azolla can also be multiplied in the field nursery.

Cement Cisterns

Azolla can be multiplied in cement cisterns of 2– 1 x 0.3 m. A layer of 7.5-10.0 cm soil is provided at the bottom of the cisterns. About 5.0 – 7.5 cm depth of water is maintained in the cisterns. Super phosphate @ 8-12 g and carbofuran at 2.0 g are applied. About 200 g of fresh Azolla is inoculated to each cistern. In about 12-15 days, about 2.0 –2.2 kg of Azolla can be harvested. After each harvest, super phosphate is applied to facilitate fast multiplication. Every time, 200g of Azolla is inoculated for further multiplication.

Mass multiplication in Nursery Plots

The field selected for Azolla nursery should be nearer to shade, thoroughly ploughed and inoculated. The field is divided into plots of 40 m^2 area (20 □ 2 m) by providing suitable bunds and irrigation channels. Water is maintained to a depth of 10 cm. About 10 kg of fresh cattle dung is mixed with 20 litres of water and sprinkled uniformly throughout the plots. Fresh Azolla is inoculated at 8 kg/plot and super phosphate (100g) is applied twice at an interval of 4 days. Carbofuran is added at 100g per plot, 7days after inoculation to control insect pests. About 50-75 kg of fresh Azolla can be harvested 15 days after inoculation. Azolla can be again inoculated with the addition of cattle dung, super phosphate and carbofuran granules.

Frankiaceae Symbiosis

Some of the actinorhizal plants like Casuarina and Alnus are found to have nodules infected by Frankia. In has been reported that N fixed by *Frankia* in Alnus ranges from 50 to 150 kg/ha/year. The potentiality of these microorganisms has been demonstrated in increasing the biomass yield of *Casuarina* and *Alnus*.

About 200 plant species from 25 genera, 8 families and 7 orders, all non-leguminous angiosperms, have been found to form nodule symbioses with N_2-fixing *actinomycetes* belonging to the genus of *Frankiae*. All of these species are perennial dicots and, with few exceptions, woody shrubs or trees. They are named actinorhizal plants from *actino* in *actinomycete*, and from *rhiza* in the Greek word for root. Actinorhizal plants can be found both on every continent except Antarctica and in most climatic zones. They typically grow on disturbed marginal soils and are pioneer species early in successional plant community development, such as *Dryas* species in arctic tundra; *Casuarina*, *Hippophae*, *Myrica* and *Elaeagnus* species in coastal dunes; *Alnus* and *Myrica* species in riparian; *Alnus* and *Dryas* species in glacial till; *Casuarina*, *Purshia*, *Ceanothus Cercocarpus*, *Comptonia* and *Cowania* species in chaparral and xeric; *Alnus* species in alpine; *Alnus*, *Casuarina*, *Coriaria* and *Shepherdia* species in forest. Globally, especially in wherever indigenous legumes are absent or rare, actinorhizal plants have potential applications in soil amelioration and reforestation, as fuelwood, timber and pulp, and as windbreak or even for addressing **pyrodenitrification**.

The symbiont *Frankia* is a gram-positive, filamentous bacterium belonging to the family *Frankiaceae* within the order *Actinomycetes*. Speciation in *Frankia* is not yet clear and within the genus different isolates are classified. Almost all of *Frankia* are characterised by three structural forms of hyphae, sporangia and vesicles. The hyphae are branched with a diameter of 0.5 to 1.5 μm and the mature vesicle is spherical with a diameter of 2 to 4 μm. Both the hyphae and the mature vesicles are septate. When actinorhizal plants, such as *Alnus* and *Casuarina* seedlings are excavated from soil containing *Frankia*, numerous small, multi lobed, coralloid-type, amber or whitish nodules are found on their root systems. The oldest and biggest nodules are close to the stem base and the youngest are on the distal parts of the root system. Some *Casuarina* nodules may reach a size of around 5 – 10 mm in diameter and a weight of about 1 g in dry matter. It has been reported that *Frankia* produced sporangia and vesicles as soon as the microorganism escaped from the mother nodule. This indicates that the formation of *Frankia* structures inside the nodule might be under host control. In general, the nitrogen fixation rates of *Frankia*–actinorhizal plant symbioses are comparable to that of *Rhizobia*–legume symbioses.

P-solubilising Microorganisms

Phosphorus is a very important nutrient of plants required in large quantities. Inorganic forms of this nutrient are compounds of Ca, Fe, and Al, while organic forms are phytin, phospholipids and nucleic acids, which are added to the soil by decaying vegetation. Rock phosphate is the raw material for production of phosphatic fertilizers. In India more than 75 per cent of phosphatic fertilizers are imported. Large amount of phosphorus applied to various soils get fixed and become unavailable to the plants. Several soil bacteria, particularly those belonging to the genera *Pseudomonas* and *Bacillus* and fungi *Penicillium* and *Aspergillus* possess the ability to bring insoluble phosphates in soil into soluble forms by secreting organic acids such as acetic, formic, propionic, lactic, glycolic, fumaric and succinic acids. These acids lower the pH and bring about dissolution of bound form of phosphate. Some of the hydroxy acids may chelate with Ca and Fe resulting in effective solubilization and utilization of phosphates by crops.

A group of heterotropic microorganisms are known to have the ability to solubilize P from insoluble sources. The organisms include bacteria: *Bacillus megatherium, B. circulans* and *B. subtilis;* and, fungi: *Aspergillus awamori, Penicillium digitum* and *Trichoderma* spp. These organisms secrete various organic and inorganic acids. They act on insoluble phosphates and convert them into soluble phosphates in the rhizosphere. Among the microorganisms, bacteria are found to be more efficient in the secretion of organic acids. They are generally called *phosphobacteria*. Addition of organic manures helps in increasing the solubilising power of the microorganisms.

Phosphate solubilizing bacteria belonging to the genera *Pseudomonas* and *Bacillus,* and fungi to the genera *Penicillium* and *Aspergillus* possess the ability to solubilize the bound phosphates in the soil by secreting organic acids like formic, acetic, propionic, lactic, glycolic, fumaric and succinic. Addition of these bacteria to soil increases the availability of phosphorus in the soil thereby bringing about a saving the input of phosphorus.

The efficiency of phosphatic fertilizers is very low (15-30%) due to its fixation in soil. Besides, native soil phosphorus is mostly unavailable to crops because of its low solubility. The introduction of efficient P solubilizers in the rhizosphere has been found to increase the availability of phosphorus from both applied and native soil P. The phosphorus, which gets fixed in soil is made available to crops by the action of microorganisms which solubilize insoluble/fixed forms of phosphorus into forms which are readily taken up by the plants. The common phosphate solubilizing bacteria and fungi are *Pseudomonas striata, Bacillus polymyxa, Aspergillus awamori, Penicillium digitatum.* These microorganisms can grow on insoluble phosphate sources such as tricalcium phosphate, ferric, aluminium and magnesium phosphate, rock phosphate and bone meal and convert them into soluble forms. Inoculation of seeds or seedlings with microphos biofertilisers can provide 30 kg P_2O_5 ha^{-1} equivalent P applied from superphosphate by solubilizing the soil phosphorus. Another factor which affects the effective utilization of P is its low mobility in soil. Fortunately some filamentous organisms like vesicular arbuscular mycorrhizae (VAM) which form symbiotic association with plant roots have the ability to mobilize phosphorus from soil and thereby helping in absorption of P by plant roots. Commercial exploitation of phosphatic biofertilizers can play an important role particularly in making the direct use of abundantly available low grade phosphate possible.

Azotobacter and phospho-bacteria also produce hormones viz., indoleacetic acid and gibberellin. These hormones stimulate root growth and development. The use of growth stimulating seed inoculants help in greater uptake of plant nutrients from applied chemical fertilizers by increasing the root growth.

In India, Indian Agricultural Research institute, New Delhi conducting researches on phosphate solubilising bacteria and the bacterial inocula. From chilli Arka Jwala nitrogen-fixing and phosphate solubilising bacteria have been isolated and identified as *Bacillus* species. These bacteria improved the growth of chilli plants on inoculation. Brinjal cultivars also harbour nitrogen fixing and phosphate solubilising bacterial association. These have also been isolated and identified as *Bacillus* species. These bacteria fix nitrogen and solubilize phosphorus thereby increasing the availability of these elements in the soil. These bacteria fix nitrogen and solubilize phosphorus there by increasing the availability of these elements in the soil. Besides, these plans are also found to release plant growth hormones, which increase the growth and yield of plans on inoculation In India, inoculum of phosphate solubilising microorganisms (PSM) is in market in a large quantity as *Microphos*, *Phosphobacteria*, and *Superphos*. Inoculation is done to improve the population of microorganisms near the root zone of crop plants, so that more phosphorus is available to the crop plants. Even in orchards, phosphobacteria is applied by many growers to save inorganic phosphorus.

Large Scale Inoculum Production

The *Rhizobium, Azotobacter, Azospirillum* and phosphate solubilising bacterial inocula are commercially produced in several Universities, Regional Biofertilizer Development corporations and by private entrepreneurs. Presently around 2 million tonnes of microbial inocula are produced in India, which is insufficient, because India needs around 35 million tonnes. The major problem in inoculum production is inadequate quality control, scientific selection of strains for production using fermentors under controlled conditions, survival of the inoculum during storage in distribution centres, and lack of involvement of organized industry.

VA mycorrhiza are obligate symbionts, they require a plant host for its production, therefore, they cannot be produced artificially like other microbial cultures. At IIHR, Bangalore an on-farm inoculum production technique has been standardized. From an area of 5m □ 5m more than 3 tonnes of inoculum could be produced. Attempts to culture VAM fungi in artificial medium have met with little or no success. However attempts have been made to culture VAM fungi using Ri DNA transformed roots. Plant roots are transformed with *Agrobacterium rhizogenes* to develop *in vitro* hairy root cultures on which mycorrhiza is cultured.

Key Factors in Organic Farming

THE world population of 5.3 billion in 1980 will increase to 6 billion in 2000 and to over 10 billion by continue to create unprecedented pressures on the limited natural resource base to produce additional food, fibre, fuel and raw materials. In the past, these increased requirements were met mainly through technological innovation, institutional and infrastructural development and policy initiatives that promoted growth in agricultural productivity. As a result, cereal production nearly doubled between 1960 and 1990. Increased fertilizer use played a lead role in this doubling of cereal production. Although consumption of all three primary nutrients, namely, N P and K, increased in tropical soils, N use increased more rapidly.

India and China are the two largest countries, which depend mostly on rice for their population of about 900 and 1200 million respectively. India had produced 112.5 million tonnes of paddy in 41.8 million hectares, while China produced 188.4 million tonnes in 32.9 million hectares, with productivity of 2691 and 52278 kg/ha respectively. China could obtain this spectacular yield advantage mainly due to the cultivation of hybrid rice varieties and organic farming including the *Azolla* biofertilizer technology. In India, the total supply of nitrogenous fertilizer in the recent past was 9 million tonnes against the demand of 11 million tonnes resulting in a gap of 16 per cent. This gap is likely to widen further and could be narrowed down through the nitrogen that is naturally available in soil through continuous biological nitrogen fixation (BNF). Studies show that biological nitrogen fixation in flooded rice paddies can contribute around 60 kg per hectare per crop.

The main agents of BNF in rice field are fee-living BGA, symbiotic BGA with *Azolla* and heterotrophic bacteria in the rhizosphere and soil. Free-living bacteria contribute 15 kg N per year. Cyanobacteria 30-40 kg, associative bacteria 36 kg N. Azolla-Anabaena symbiosis 40-60 kg N per hectare per season and the highest by the nitrogen fixing leguminous green manures around 60-80 kg N per hectare per season. Thus, these potential groups of nitrogen fixing system may be exploited for harnessing atmospheric nitrogen so as to increase crop productivity.

The use of these potential biological systems would provide 15-20 lakh tonnes of nitrogen for crop production in India, while the equivalent urea fertilizer needed is around 33-44 lakh tonnes. Moreover, they also help maintain the long term soil fertility and ecological sustainability required for increasing crop productivity. The use of biofertilizer also enriches the soil organic matter, soil enzymes and soil microbial population. The long tern soil fertility largely depends on the above parameters, which is essential for achieving higher productivity of crops and sustainability of yields.

Significance of BNF

Nitrogen is the most critical nutrient for rice productivity. Rice take s1 kg of nitrogen to produce 15-20 kg of grain. Low-land rice in the tropics can use the nitrogen that is available either naturally through BNF or from mineralization in soil to produce 2-3 tonnes of grain per hectare. However, additional nitrogen must be supplied for higher yields. Innovative research is needed to help keep rice supply ahead of demand while sustaining the world's natural resource base. To feed the ever-increasing population, the world annual rice production must increase from the present 460 million to 560 million tonnes by the year 2000 and to 760 million tonnes by 2020. At present nitrogen use efficiency level, at least double the 10 million tonnes of nitrogen fertilizer currently used each year for rice production will be required. Manufacturing the fertilizer for today's need requires 544×10^7 MJ of fossil fuel energy which is equivalent to 13 million tonnes of oil. A non-renewable source, the oxidized products of which pose hazards to human health and to the environment. With the gradual depletion of petroleum resources, increased production costs and low efficiency of chemical N fertilizer, it is time to think of supplementary and or alternative sources of N besides exploring ways for efficient use of soil and fertilizer N, in order to meet increasing demands in the coming years.

It is in this context that the BNF derived N assume significance in the low land system that provide about 80 per cent of the world's rice. Conventional BNF systems have only a limited capacity to render rice independent of external sources of N. Among the conventional BNF system of rice, the free living and associative diazotrophs in rice field are low and inefficiently utilized as compared with those of legumes under suitable conditions. If a BNF system could be assembled in non-legumes such as rice, wheat, maize and sorghum and in sugarcane, we could enhance the N supply potential as fixed N would be available directly to the plant with little or no less. Thus, it has become imperative to consider unconventional BNF systems to supply crops with N.

MYCORRHIZA

The term *mycorrhizae* comes from the Greek *myco* or fungus and *rhiza* or root and literally means *fungus root*. Mycorrhizae have been described as perhaps the ultimate in reciprocal parasitism (symbiosis) wherein the fungus supplies the higher plant partner with mineral nutrients and plant growth hormones and protects the roots against pathogens, and the higher plant supplies the fungus with energy substrates. Almost all higher plants are dependent upon one of the mycorrhizal associations.

Mycorrhizae are highly evolved, mutualistic symbioses or associations between soil fungi and plant roots. The partners in this association are members of the fungus kingdom (*Basidiomycetes*, *Ascomycetes* and *Zygomycetes*) and most vascular plants. Mycorrhizal plants have been found in every continent and in every major vegetation type. Depending on the plant and fungal species involved as well as distinct morphological patterns, at least seven different types of mycorrhizal associations have been recognised. They are: vesicular arbuscular mycorrhiza (VAM); ectomycorrhiza (ECM); orchid mycorrhiza; ericoid mycorrhiza; ectendomycorrhiza; arbutoid mycorrhiza and monotropoid mycorrhiza. In addition, dual associations of both ECM and VAM have been found in some trees and shrubs such as *Acacia*, *Casuarina*, *Eucalyptus* etc.

Roots as Sinks and Sources of Nutrients and Carbon in Agricultural Systems

Roots are a special category of *soil biota*, being part of organisms that live in two very different types of environment, aboveground and belowground. As they belong to the organisms that form the main target of all agricultural interventions and are directly involved in the efficiency with which nutrients and water are used for crop production, root knowledge should be "at the root" of any discussion on sustainable agriculture. Yet, little direct attention is given to roots, possible because of a lack of:

1. Clear concepts on what to look for
2. Easy (be it quick and dirty) methods to observe the relevant properties in actual root systems.

The need for maximizing resource use efficiency and minimizing environmental pollution has given a new drive for root ecological research. Within the soil, roots are the main sinks for nutrients during the life of the crop, but may also play a role in immobilizing nutrients during the initial stages of their decomposition; subsequently, they will form a source of nutrients for soil organisms and future plants. The efficiency of roots as sinks for nutrients, both during and after their life, partly determines the conflict between environmental and production aims of agriculture.

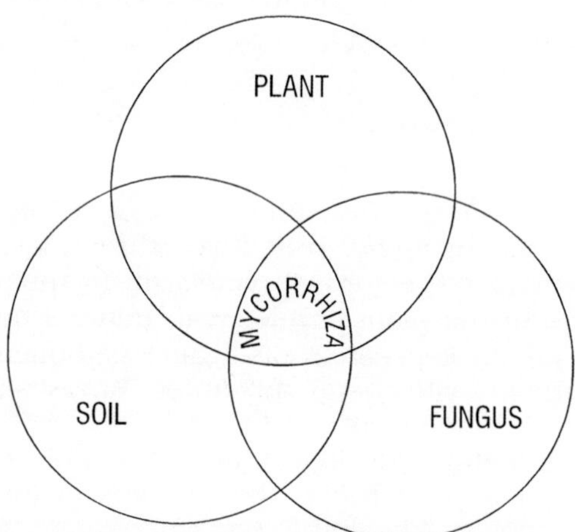

FIG. 9.4: Interrelationship between mycorrhiza, soil and plant.

For a long time the prevailing concept was *the more extensive the root system, the higher crop production*. We know that this is not true across the full range of agricultural production conditions; maximum plant production can be obtained with relatively small root systems if the daily water and nutrient requirements are met by technical means as in intensive horticultural production systems. Roots are, however, directly involved in the efficiency of plants to use available water and nutrient reserves in the soil, and therefore in reducing negative side effects of agricultural production by leaching and losses to the atmosphere. As a first estimate, we may still expect that *the more extensive the root system, the higher may be the nutrient and water use efficiency*. The possibility of obtaining greater resource use efficiency can only be realized if total supply of nutrients and water is regulated according to the crop demands and the attainable resource use efficiency. On a field scale, both resource supply and possible crop production show spatial variability. Inadequate techniques for dealing with this variation may reduce the resource use efficiency well below what is possible in the normally small experimental units considered for research.

If the long-term organic matter balance is secured, system outputs (harvested products, leaching, and gaseous losses) must be balanced by inputs. In the northern temperate zone the main crop growing season normally has a rainfall deficit, and leaching (beyond the crop root zone) is mostly confined to autumn and winter. Denitrification may occur in summer, but in early autumn conditions winter. Denitrification may occur in summer, but in early autumn conditions are especially conducive for this process as relatively high temperatures are combined with wet soil conditions and ample availability of organic substrates.

Rates of net N mineralization are less than required by crop demand during the growing seasons, but continue after the uptake period. Fertilizer requirements for avoidance of any N shortage in the crop have to allow for the weather-induced variability in crop demand and in rates of net N mineralization and also for the spatial variability within a field treated as a single management unit. Possibilities for *response farming* are limited, however, as late fertilizer applications may not reach the roots under dry conditions.

The relevance of various root parameters for predicting uptake efficiency depends not only on the resource studied but also on the complexity of the agricultural system. In intensive horticulture, with nearly complete technical control over nutrient and water supply, fairly small root systems may allow very high crop production in a situation where resource use efficiency ranges from very low to very high. It depends on the technical control over nutrient and water supply. Fairly small root systems may allow very high crop productions in a situation where resource use efficiency ranges from very low to very high, depending on the technical perfection of the often soilless (hydroponic) production system. In field crops grown as a monoculture, the technical possibilities are far lower for ensuring a supply of water and nutrients where and when needed by the crop, and it is difficult to achieve maximum crop production rates at high efficiency. The soil has to act as a buffer, temporarily storing these resources. Root systems are important in obtaining these resources as and when available and needed. Adjustment of supply and demand in both time and space (synchrony and synlocation) become critical factors. Critical values of root length density (root length per unit volume of soil) needed to obtain a specified water or nutrient use efficiency can be estimated from existing models. Normally, the rate-limiting step is formed by the transport from the soil matrix to the root surface and therefore by the geometry of the root-soil system and the required transport distances. In mixed cropping systems (including grasslands), the belowground interactions between the various plant species add a level of complexity to the system. On the one hand, mixed cropping system allow complementarity in use of the space and thus of the stored resources, hence, improving overall resource efficiency. On the other hand, it means that root length densities that would be sufficient for efficient resource use in a monoculture may not be sufficient in a competitive situation. The plant mixtures found in agroforestry system increase the complexity by another dimension, as the perennial and annual components have separate time frames on which to evaluate the interactions.

The soil water balance, as affected by climate, irrigation, and drainage, has a major influence on the required root functions in uptake. As the main crop-growing season normally has a rainfall deficit in the northern temperate zone, dry soil conditions hamper diffusive transport and thus increase the root length density required for uptake. A lack of synchrony between N mineralization and N demand, which would lead to a build up to mineral N in the topsoil, is not a real problem under these conditions, as the main leaching risk occurs after the growing season. The main problem for effective use of organic N pools in the temperate zone is that mineralization is too slow in spring. In the humid tropics, by contrast, with a net rainfall surplus during most of the growing season, any accumulation of mineral N will be leached rapidly from the topsoil to deeper layers. Under such conditions synchrony of N mineralization and N demand is essential for obtaining high N use efficiencies and reducing leaching.

Roots are a poorly quantified source of carbon in the soil ecosystem, and widely different estimates are available in the literature, at least partly due to uncertainties and biases caused by the methods used. Fine root residues of annual crops at harvest time tend to be low in nutrient content, as annual crops generally remobilize nutrients toward the end of their growth cycle, redistributing them to the seeds or vegetative storage tissue. For perennial crops with a less pronounced phenology, root decay may be a continuous process during the growing season. During their decomposition by microorganisms, roots may form a temporary sink of N. Little is known on the timing of net N mineralization from root residues and its role in the synchrony between net N mineralization and demand by the following crop.

Importance of Mycorrhiza

In the sand soil ecosystems of India, pioneer plants are subject to deficiency in major nutrients (N, P, and K) and organic matter, wide fluctuation in soil moisture and temperature. It is widely accepted that arbuscular mycorrhizae are important for the establishment, growth and survival of seedlings, particularly in marginal habitats, where the symbiosis improves stress tolerance as well as conserves soil structures. A lot of work done in the past few decades has enabled these fungi to emerge as a potential biofertilizer; a cheap and environment friendly alternative to expensive, petroleum based chemical fertilizers. This aspect especially gains significance for a developing country like India where judicious and large-scale utilization of the technology can prove very useful for getting maximum and long-term gains in various wasteland reclamation, reforestation and afforestation programmes. In addition it gives a much needed thrust in the production of important agricultural crops, on which the economy of the country is dependent.

Mycorrhizal associations are believed to help plants grow. The potential for manipulating mycorrhizal associations to increase productivity in plantation forestry, or plant establishment during ecosystem recovery after severe disturbance, are the focus of major research initiatives. There is also much interest in their potential utilisation in agriculture and horticulture. However, it could be argued that we do not know enough about the role of mycorrhizal associations in natural, disturbed, or managed ecosystems to evaluate their potential for applied use. Suggested roles of mycorrhizal fungi in natural and managed ecosystems

Benefits to Plants

1. Increased plant nutrient supply by extending the volume of soil accessible to plants.
2. Increased plant nutrient supply by acquiring nutrient forms that would not normally be available to plants.
3. Some ECM and ericoid fungi have the capacity to breakdown phenolic compounds in soils which can interfere with nutrient uptake.
4. Root colonization by ECM and VAM fungi can provide protection from parasitic fungi and nematodes
5. Non-nutritional benefits to plants due to changes in water relations, phytohormone levels, carbon assimilation, etc. have been reported, but are difficult to interpret.
6. Mycorrhizal benefits can include grater yield, nutrient accumulation, and/or reproductive success
7. Mycorrhizas can cause growth form changes to root architecture, vascular tissue, etc.
8. Suppression of competing non-host plans, by mycorrhizal fungi has been observed.
9. Significant amounts of carbon transfer through ECM fungus mycelia connecting different plant species has been measured. This could reduce competition between plants and contribute to the stability and diversity of ecosystems.
10. Networks of hyphae supported by dominant trees may help seedlings become established or contribute to the growth of shaded understorey plants.
11. Nutrient transfer from dead to living plants can occur.

Other Roles in Ecosystems

1. Soil hyphae are likely to have an important role in nutrient cycling by helping to prevent losses from the system, especially at times when roots are inactive.
2. Hyphae are conduits that may transport carbon from plant roots to other soil organisms involved in nutrient cycling processes – cooperating with other members of the decomposition soil food-web

3. Soil hyphae may have an important role in nutrient cycling by acquiring nutrients from saprophytic fungi
4. Epigeous and hypogeous sporocarp of ECM and VAM fungi are important food sources for placental and marsupial mammals
5. Mycorrhizal roots and fungus fruit bodies are important as food sources and habitats for invertebrates
6. Mycorrhizal fungus hyphae are an important food source for soil invertebrates
7. Mycorrhizas influence soil microbial populations and exudates in the mycorrhizosphere and hyphosphere
8. Hyphae of VAM fungi are considered to contribute to soil structure. Their role in mechanical aggregation has been questioned, but secretions such as glomalin may be more important. Hyphal mats produced by ECM fungi considerably alter sol structure
9. Mycorrhizal fungi contribute to carbon storage in soil by altering the quality and quantity of soil organic matter.

Values of People

1. ECM fungi are economically and nutritionally important s human food resources.
2. These mushrooms have also have been used as medicines and natural dyes.
3. Fungi have aesthetic values and are an important part of the as culture, folklore and appreciation of nature by many people.
4. Fungal diversity is a bio-indicator of environmental quality.
5. Fungi, which have adapted to local soil conditions are required for agriculture, horticulture and forestry.

Mycorrhizal Interactions with Plants and Soil Organisms in Sustainable Agroecosystem

Although farming has been affected by technological development, it remains basically an ecological enterprise. It is an activity in which natural ecosystems, open to the influence of climate, substrate, and wild biota, are modified to increase yields of desired food and fiber products. The greater the changes in the basic patterns of structure and function that prevail in the natural system, the greater is the human effort necessary to maintain the agricultural system. Therefore, at the present time, it has been proved that conventional agriculture products ecological disturbance and lack of sustainability, resulting in a reduction of soil fertility and increased damage by pathogens to cultivated plants. In addition, it has been observed that some traditional agricultural systems have higher sustainability and produce less ecological damage. These systems have been called **low external-input agricultural** (LEIA) systems. Studies related with mycorrhizal research in LEIA agroecosystems have been scarce. In sustainable agroecosystems, mycorrhizal incidence is low and other kinds of microorganisms are abundant and play important role in the sustainability of the systems. So far, there has been insufficient investigation of interactions between AM and soil organisms in LEIA systems.

As opposed to the conventional systems, LEIA agro-ecosystems have high genetic and cultural diversity, multiple uses of resources, and efficient nutrient and material recycling. The search for strategies to improve yields and to maintain these increases is a great challenge to human population at the present time. The role of biological alternatives, because of the intrinsic nature of farming, is of key importance to the search for reduced use of fertilizers, pesticides, and other chemicals. Among these alternatives, mycorrhiza management is particularly important because it strongly influences the plant nutrition processes and the soil stabilization.

It the physiological and ecological importance of mycorrhiza in natural ecosystems and its beneficial effects on cultivated plants have been widely documented. It is well known that the characteristic dominant plants of each major terrestrial community associate mutualistically with soil fungi to form typical kinds of mycorrhiza. Therefore, cricaceous plants, which form the major component of the heathland biome, form with ascomycetous fungi distinctive *ericoid* mycorrhiza. In a similar way, the dominant trees of the boreal and temperate forest biomes associate primarily with basidiomycetous fungi to form ectomycorrhiza, and the natural grasslands and most of the tropical rain forest species of the world form arbuscular mycorrhiza in association with fungi of zygomycetous affinity. As natural ecosystem plants, most of the cultivated plants tend to form mycorrhizal associations. AM is the most widely distributed and colonizes most species of agricultural crops. The objective of this contribution is to discuss the importance of the mycorrhizal fungi and their interactions with plant management and other organisms in in LEIA agroecosystems.

Symbiosis

The term *symbiosis* was first used by Frank in 1877 to describe the regular coexistence of different organisms such as fungi and algae in lichens. It was used as a neutral term that did not imply parasitism. A decade later De Bary (1887) used symbiosis to include parasite as well but, the meaning of the terms *symbiosis* and *parasite* has been changed later on. Symbiosis was used more and more for mutually beneficial associations between dissimilar organisms, and parasite and parasitism came to be almost synonymous with pathogen and pathogenesis.

In recent years and symbiosis is defined as *the living together of differently named organisms*. This definition includes all associations ranging from *mutualistic*, in which all organisms involved are believed to derive benefit; to *parasitic*, in which one organism benefits to the disadvantage of other member of the association. A more precise definition of mutualism is that associations are *mutualistic* if the fitness (offspring produced) of the associating organisms is greater than they are living apart. In a symbiosis, the organism with the large size is the host and the smaller is the symbiont. The symbiont can either be external to the host (ectosymbiotic) or within it (endosymbiotic). The symbiosis is *obligate* for an organism which is unable to survive and reproduce in the absence of its living partner, and *facultative* if it is able to do so even if the partner does during the association.

In a dual association, symbiosis is wide-spread especially between higher plants and bacteria in N-fixing symbiosis; between higher plants and fungi in most mycorrhizae; between algae and fungi in lichens; between algae and coelenterate in corals and so on. However, it is not currently known whether symbiosis occurs between higher plants. In addition, the tripartite symbiosis of plants, N-fixing bacteria and mycorrhizae (either vesicular-arbuscular mycorrhiza or ectomycorrhiza) has also been found in some *Acacia*, *Albizia*, *Casuarina* and *Leucaena* species. Symbioses are often found in nutrient limiting conditions and nutritional interactions play a key role in most symbioses. There are two important types of symbioses: mycorrhizal symbiosis and N-fixing symbiosis.

Mycorrhizae is the symbiotic association of fungi with roots of vascular plants. The main advantage of mycorrhizae to the host plants lies in the extension of the penetration zone of the root fungus system in the soil, facilitating an increased phosphorus uptake. The interconnected net-work of external hyphae acts as an additional catchment and absorbing surface in the soil beyond the depletion zone that would otherwise be inaccessible to the plant roots. Endotrophic mycorrhizae have been shown to be present in a wide range of horticultural species including apple, walnut, almonds, citrus, avocado, strawberry and grape. In many cases, the mycorrhizae have been shown to markedly improve the growth of plants.

Some soils are deficit in mineral salts and plants living in them have special means of obtaining nitrogen and other essential elements, for example, the roots of many moorland and woodland plants, particularly coniferous trees, which live in rich humus but deficient in mineral salts possesses a mycorrhizae. Mycorrhizae (**fungus-roots**) are symbiotic associations between fungi and the roots of vascular plants. Two main groups of mycorrhizae are recognized: the **ectomycorrhizae** and **endomycorrhizae,** although the rarer group with intermediate properties, the ectendotrophic mycorrhizae. In the ectomycorrhizae, the fungal hyphae form a mantle both outside the root within the root in the intercellular spaces of the epidermis and cortex. No intracellular penetration into epidermal or cortical cells occurs, but an extensive network called the Harting net is formed between these cells. Ectomycorrhizae are common on trees, including members of the families Pinaceae (pine, fir, spruce, larch, hemlock), Fagaceae (willow, poplar, chestnut), Betulaceae (birch alder), Salicaceae (willow, poplar, and few other show scanning electron micrographs of two *Pinus contorta* roots, one uninfected and having root hairs and the other infected with an ectomycorrhizae fungus.

Endomycorrhizae consist of three subgroups, but by far the most common are the vesicular arbuscular mycorrhizae (VAM). The fungi present in VAM are members of the Endogonaceae, and they produce an internal network of hyphae between cortical cells that extends out into the soil, where the hyphae absorb mineral salts and water. Although VAM fungi seem to penetrate directly into the cytosol of cortical cells (where they form structures called vesicles and arbuscules, giving them their name), the hyphae are surrounded by an invaginate plasma membrane of the cortex cell. The VAM are present on most species of herbaceous angiosperms, whether monocot or dicot, annual or perennial crops, or native or introduced species; they also occur in the gymnosperm genera *Cupressus, Thuja, Taxodium, Juniperus,* and *Sequoia* and in ferns, Lycopods, and bryophytes.

Root System Form

The main role of mycorrhizal associations is to acquire nutrients by exploring the soil volume with hyphae that are both more responsive and more extensive than the roots themselves. However, roots of some species of plants are also capable of effectively exploring large soil volumes and responding to temporary soil resources, and these species are likely to have facultative mycorrhizal associations, or be non mycorrhizal. The production of fine roots or associations with mycorrhizal fungi are the two alternative strategies that both require plants to expend metabolic energy on their root systems. It is considered likely that mycorrhizal fungus hyphae should be a more cost effective means of exploring large soil volumes, because of their narrow diameter. Plants generally do not support both high levels of mycorrhizal colonization and root systems with fine/active roots, because of the high metabolic cost that would result.

Considerable variations in root system extensiveness, geometry, depth distribution, and plasticity occur between plant species. In experiments, total biomass is the parameter most often used to quantify roots, but provides much less information than root length or specific root length (length/unit root weight) data. Parameters that could be used to predict nutrient absorption are: total root biomass <fine root biomass <root length <root surface area <rhizosphere volume, arranged in increasing order of predictive ability. When root parameters of plant species are compared, root hairs have been found to be most highly correlated with their ability to absorb P from soils. Plants with extensive (highly branched, fine, long roots with numerous root hairs) have often been observed to derive little benefit from mycorrhizas in experiments.

In addition to have fine roots with long root hairs, there is a tendency for mycorrhizal plans to have roots which are less active (grow slowly, live longer, etc.) than species with little or not mycorrhizal colonization of their roots. The responsiveness, or capacity of plants to exploit small scale or short duration changes in water or nutrient availability by rapidly producing new roots, is thought to be an important determinant of their success during competition for soil resources. Some plants can also change rhizosphere conditions, such as pH, which may influence nutrient availability.

Australian non mycorrhizal plants in the family Proteaceae have cluster roots that likely promote nutrient uptake by their large surface area and production of leachates, which contain organic acids and other substances that can increase nutrient solubility. The capacity of plants to influence nutrient availability in soils will also depend on the extensiveness and activity of their root system, since young roots are the primary source of exudates.

TABLE 9.2: Generalized relationship between features of root systems and the mycorrhizal dependency of plants

Root feature	Mycorrhizal dependency of plant High	Low
ROOT SURFACE AREA	low	high
Root system surface area	low	High
Branching orders of lateral roots	few	many
Branching frequency	sparse	frequent
Root hair abundance and length	few/short	many/long
ROOT ACTIVITY	slow	fast
Root growth rate	slow	fast
Response to soil conditions	slow	fast
Primary root lifespan	long	short
Protective structural features	strong	weak
Root exudation	? less	? more
MYCORRHIZAL FORMATION	efficient	inefficient or inhibited

Notes: This table is based on differences in root system properties between plants of different mycorrhizal status in Canadian forests.

Host and soil factors which can be used to help predict the potential benefit of mycorrhizas are outlined in the Table 9.2 (*Assuming that soil nutrient levels are not unusually high and inoculum of appropriate mycorrhizal fungi is available, root strategy differences will primarily determine the magnitude of benefits from mycorrhizal associations).*

Soil and Site Factors Influencing Mycorrhizas

Soil factors that can influence the benefits provided by mycorrhizal fungi to plants are considered below.

Mycorrhizal Inoculum

Plants with mycorrhizal associations predominate in most natural ecosystems, so inoculum of mycorrhizal fungi is present in most soils. The quantity inoculum of mycorrhizal fungi which are compatible with a host plant in soils can be measured by bioassay experiments. In these experiments, seedlings are grown in intact soil cores or mixed soil samples for sufficient time to allow mycorrhizas to form, then roots are sampled, processed and assessed to measure mycorrhiza formation.

In Australia, there are fewer species of ECM fungi in sites where eucalypt plantations are established than in undisturbed forests. There are also cases where plantations of trees have been introduced to habitats where they do not occur naturally. For example, Australian eucalypts are now grown in exotic habitats in Africa, Europe, Asia and South America, habitats which are few compatible ECM fungi, because local fungi are specific to other hosts such as conifers.

Soil Disturbance

Propagules of mycorrhizal fungi maybe absent from soils where severe soil disturbance has resulted in topsoil loss, or where host plants are limited by adverse soil or site factors such as salinity, aridity, waterlogging, or climatic extremes (Brundrett 1991). Most studies of mycorrhizal associations in highly disturbed habitats such as mine sites have found reduced levels of mycorrhizal propagules. Less severe forms of soil disturbance, including agricultural tillage, soil animal activities, fire and erosion can also reduce levels of mycorrhizal fungus propagules.

Spores and other propagules of mycorrhizal fungi can introduced to new sites by wind or water erosion or by the activity of animals, which feed on fungi. In a disturbed habitat, the effectiveness of natural vectors will depend on the proximity of undisturbed habitats containing suitable fungi (and their associated animals) as well as the phenology of fruiting of fungi. Effective colonization of disturbed habitats, such as minesites in Australia, by mycorrhizal fungi has been observed, but there is as yet insufficient information about the time required for this process to occur.

Soil Fertility

A key factor, which effects the potential for mycorrhizas to benefit plants in particular sites is the supply of phosphate and nitrogen in soil. Phosphorus is generally considered to be the most important plant-growth limiting factor which can be supplied by mycorrhizal associations, because of the many abiotic and biotic factors which can restrict its mobility in soils. Reductions in the benefit provided by mycorrhizal associations to plants are caused by increasing soil phosphorus levels. High rate of P and N fertilizers suppress ectomycorrhiza development in the field. High concentrations of soil N can influence the relative abundance of different ECM types.

Adverse Soil Conditions

Land degradation due to salinity, waterlogging, erosion, etc., are serious and growing, problems in Australia and other countries. Excessive NaCl levels in soil inhibit mycorrhizal formation and restrict the activity of most mycorrhizal fungi, but some can tolerant these conditions. Observations in natural ecosystems have shown that plants with mycorrhizal associations are often less common than non-mycorrhizal species in soils which are waterlogged or saline, but that some mycorrhizal plants are normally present in even the worst sols. ECM fungi can be highly sensitive to water logging of sols, while VAM fungi may be less sensitive.

Characteristics of Fungal Isolates

Populations of mycorrhizal fungi are thought to have occupied the same soil habitats for millions of years, slowly adapting to changes in site conditions. Some mycorrhizal fungi appear to have worldwide distribution patterns and have apparently adapted to a wide range of habitats, but it is known that soil factors such as pH restrict the distribution of many taxa. Unfortunately, information about the soil and climatic conditions where isolates used in experiments were originally obtained is often not provided. These data are urgently needed to allow isolates to be selected for use in inoculation programs and to establish correlations between their taxonomy and physiology.

The fact that most mycorrhizal research has been concerned with plant growth responses with little consideration of the fungi involved has also helped to create the false impression that most of these fungi are functionally equivalent. Properties of individual mycorrhizal fungi result from their adaptations to soil/environmental/host factors. Some Properties of fungi considered to be important are:

1. The amount of soil hyphae produced relative to root colonization.
2. The rates of hyphal growth and root colonization.
3. Physiological characteristics, which regulate nutrient absorption or nutrient translocation by hyphae and exchange with the host.

Manipulating Mycorrhizal Fungi

Methodologies have been developed for the inoculation of plants with mycorrhizal fungi. However, there has been less consideration about how we should determine where and when, if ever, it is necessary to do this. The diagram provided below, outlines the steps in a decision making processes that can be used to determine if mycorrhizal fungi that are compatible with host plants are present in soils, or are likely to be introduced by natural mechanisms. This information is required to evaluate the potential benefits of mycorrhizal inoculation in a specific situation. There would only be a need to manipulate mycorrhizal fungi by inoculating plant seedlings of changing site management practices it substantial benefits from these manipulations are expected. Methods for the production of mycorrhizal plants in the glasshouse and nursery to allow experimentation or for use in the field are described in the accompanying manual.

Types of Associations

At least seven different types of mycorrhizal associations have been recognised, involving different groups of fungi and host plants and distinct morphology patterns.

TABLE 9.3: Kinds of mycorrhizal associations

Endomycorrhizae	1. Vesicular-Arbuscular (Va)	Higher Plants-Great Majority Of The Vesicular Plants Including Ferns, Monocots, Dicots, Fungi-Endogonaceae (Phycomycetes)
	2. Ericaceous	Higher Plants: Ericales. Fungi-Generally Unknown Although A Basidiomycete Has Been Found In Association With Azalea.
	3. Orchidaceous	Higher Plants: Orchids. Fungi: Agaricales and Aphyllophorales (Basidiomycetes).

Ectomycorrhizae	Higher Plants: Great Majority Of The Vesicular Plants Including Ferns, Monocots, Dicots, Fungi-Endogonaceae (Phycomycetes) Fungi: Many thousands Of Species In Basidiomycetes, Ascomycetes, And Endogonaceae (Phycomycetes).
Ectendomycorrhiza	

1. **Vesicular arbuscular mycorrhizas** (= arbuscular mycorrhizas. VAM or AM) are associations where Zygomycete fungi in the Glomales produce arbuscules, hyphae and vesicles within roots. Spores are formed in soil or roots. These associations are defined by the presence of arbuscules. Fungi within roots spread by linear hyphae or coiled hyphae.
2. **Ectomycorrhizas (ECM)** are associations where Basidiomycetes and other fungi form short swollen lateral roots covered by mantle hyphae. These roots have Harting net hyphae around cells in the epidermis or cortex.
3. **Ectendo-, arbutoid- and monotropoid mycorrhizal associations** are similar to ectomycorrhizas. These associations have not been found in Australia.
4. **Orchid mycorrhizas** consist of coils of hyphae within roots or stems of plants in the family Orchidaceae. Young orchid seedlings and some adult plants which lack chlorophyll are entirely dependant on mycorrhizal fungi for their survival.
5. **Ericoid mycorrhizas** have hyphal coils in outer cells of the narrow "hair roots" of plants in the plant order Ericales. These associations also occur in thicker roots of Australian members of the Epacridaceae.
6. The Australian lily ***Thysanotus*** has a unique mycorrhizal association where fungus hyphae only grow under epidermal cells.

The distinction between ecto-and endomycorrhizae is based upon the morphology of the structure formed by fungus and higher plants. Ectomycorrhizal forming fungi are able to penetrate between, but not into, the cortical cell of the host root, thus forming a network of fungal strands within the cortex commonly called a "Hartig net". The fungus also normally makes a mantle which completely encloses the infected root tip and from which fungal hyphae extend into the soil. This network of hyphae greatly increase the ability of the root to take up nutrients in three ways:

1. The surface area in contact with soil particles is much greater, and the much smaller hyphae can penetrate fissures in soil particles too small for roots;
2. The fungus excretes enzymes which allow it to dissolve soil nutrients otherwise unavailable to the roots; and
3. The fungus extends the life of an absorbing root from a few days to a year or more.

Inoculation of tree seedlings with appropriate fungus species to stimulate development of mycorrhiza has been shown to be critical to reforestation, especially, in a diverse range of sites and climates from timberline in the Alps to subtropical islands like Puerto Rico. The last fifteen years have seen great interest in mycorrhiza in the southern states where fungi such as Pisolithus tinctorius are believed to be responsible for increased growth and survival of loblolly pine, especially on harsh sites such as the spoils of mining operations.

Endomycorrhizae differ from the above primarily in that the fungus hyphae penetrate directly into the cortical cells of the root. Neither endo-nor ectomycorrhizal forming fungi are, however, able to penetrate beyond the cortical region – the vascular tissues of the root are not infected.

Types of Mycorrhiza

Mycorrhizae are classified on the basis of the extent of root penetration, production of external mantle or sheath and the inter- and intracellular structures they form once inside the plant root. Since enormous amount of literature exists on classification and types of mycorrhizas only salient features of different types are mentioned here. So far seven types of mycorrhizae are recognized.

Ectomycorrhiza

Here the fungal members belong to 25 families of Basidiomycotina 7 families of Ascomycotina or 1 genus of Zygomycotina i.e. *Endogone.* The plant symbionts include both Gymnosperms and Angiosperms. A fungal sheath or mantle is present. There is no hyphal penetration of cells. Fungal hypha is generally septate. A distinct Hartig's net is present between the cells. Hartig's net is plexus of fungal hyphae between epidermal and cortical cells. It provides a large surface area for the interchange of nutrients between host and the fungi.

Ectendomycorrhiza

The fungi belong to Basidiomycotina, which covers both Gymnosperms and Angiosperms plants. The fungal sheath if present is very thin. Septate fungal hyphae penetrate the root cells and coil inside it. The fine network of fungal hyphae (Hartig's net) is present in the intercellular spaces and there is intracellular penetration of living hyphae in cells.

Endomycorrhiza/Arbuscular Mycorrhizal Fungi (AMF)

Fungi are confined to the order Glomales of Zygomycotina. The plant symbionts range from Bryophytes to Angiosperms. There is no fungal sheath. Aseptate hyphae enter the root cortical cells and form characteristic arbuscules and/or vesicles hence the name. The plasmalemma of the host cell invaginates and encloses the arbuscules.

Arbutoid Mycorrhiza

Here the fungal partner is Basidiomycetous and colonizes members of order Ericales (Arbutoidae). Fungal sheath present. Septate hyphae penetrate only the epidermal cells and coil inside it. Hartig's net is present in the intercellular spaces.

Monotropoid Mycorrhiza

The fungi belong to Basidiomycotina colonizing achlorophyllous member of Angiosperms belonging to family Monotropaceae. Fungal sheath present. Branches from septate hyphae penetrate the epidermal cells forming haustorial pegs inside. Hartig's net not seen in mature roots.

Ericoid Mycorrhiza

Fungal members are usually Basidiomycetous and one, *Hymenoscyphus,* belongs to Ascomycotina. This is found in roots of plants belonging to order Ericales. Rootlets are covered by a loosely woven mesh of dark brown septate hyphae from which branches penetrate the cortical cells, forming intracellular coils invagination and enclosed by the host's plasmalemma. Hartig's net formed in the intercellular spaces.

Orchidoid Mycorrhiza

Fungi belong to Basidiomycotina and mycelia sterilia. These colonies only members of family Orchidaceae (Angiosperm). Septate hyphae enter the cells and develop into coils often called "peltrons". This association is probably pseudo mycorrhizal but plays an important role in the establishment of orchid seedlings.

Host Plants

The relative importance of different mycorrhizal associations for plant species is graphically summarized below using data for Angiosperms from the mycorrhizal literature. Many Pteridophytes are also known to have VAM associations, while Gymnosperms may have ECM or VAM associations. Most information about the mycorrhizal status of plants if from the northern hemisphere in temperate regions, especially for crop plants and weeds.

Place of Availability of Mycorrhizal Plants

Field surveys have found that plans with mycorrhizal associations predominate in most natural ecosystem sin Australia and throughout the world.

TABLE 9.4: Briefly summarizes the importance of mycorrhizal associations throughout the world

Association	Occurrence
Vesicular Arbuscular Mycorrhizal (VAM) Plants	Plants with VAM are common In most habitats It is easier to say where they are not found
Ectomycorrhizal (ECM) Plants	Trees with Ecm are dominent in coniferous forests, especially in cold boreal or alpine regions ECM trees and shrubs common in many broad-leaved forests in temperate or Mediterranean regions ECM trees also occur in some tropical or subtropical savanna or rain forests habitats

Association	Occurrence
Non mycorrhizal (NM) Plants	NM plants are most common in disturbed habitats, or sites with extreme environmental or soil conditions NM plants appear to be more common in Australia than in other continents.

Mycorrhizal Fungi

Members of the fungus kingdom obtain nutrition from many sources, including decomposition of organic substrates, predation and parasitism, and involvement in mutualistic associations (Christensen 1989, Kendrick 1992). Many soil fungi are saprobes with the enzymatic ability to digest organic substrates of varying degrees of complexity, but some subsist on very low levels of organic or inorganic substrates (Wainwright 1988). Mycorrhizal fungi are a major component of the soil microflora in many ecosystems, but usually have limited saprophytic abilities (Harley & Smith 1983). Mycorrhizal fungi are considered to have many important roles in natural and managed ecosystems.

Very little is known about the taxonomic or functional diversity of mycorrhizal fungi. Indeed, there is a paucity of knowledge about all Australian fungi, with only about 10% of these organisms having been named, while an estimated 80% of them remain undiscovered. Laboratory and glasshouse experiments have shown that different fungal taxa vary in their capacity to utilize resources, withstand adverse conditions etc. For this reason, mycorrhizal fungus diversity is thought to contribute to the resilience of ecosystems, but there is very little comparative data that can be used to test this hypothesis. More information about fungi forming ECM and VAM associations is provided in the relevant sections.

Arbuscular Mycorrhizal Fungi (AMF)

AMF are the most widespread in their distribution both among plant species and geographically. They are believed to be disseminated inter-continentally prior to continental drift as supported by fossil record of earlier plants. They occur more frequently in cultivated soils usually in top 15-30 cms. Because most economically important plants form this type of mycorrhiza, this area of research has attracted much attention. Commercial use of AMF may be an alternative to the rising agricultural and fertiliser costs, because of their potential in increasing the crop yields.

Establishment of Symbiosis

These associations do not usually cause observable external changes in the morphology of plant roots, and can not be detected without appropriate staining. They colonies a wide range of host plants that include most agricultural and horticultural plants both annual and perennial, fruit trees and a large number of forest tree species. AMF do generally not colonies plants that enter into other types of mycorrhiza. They lack host specificity from the host point of view as shown by their wide host range. However from fungal view point the specificity does exist. Of the 150 species that have been described in *Glomales* only a small proportions are presumed to be mycorrhizal. Since these fungi lack sexual reproduction recognition, characterization and classification is based upon their chlamydospores which show variation in size, shape, cooler and wall characteristics depending upon the species.

These fungi are able to colonies the epidermal and cortical cells of the root, but do not enter the vascular system or the root meristems. The morphological stages of development are variable, depending mainly on the plant species involved.

In general, fungal hyphae growing from spores in the soil or from adjacent plants roots contact the root surface, where they differentiate to form an appressorium. This is the first indication of recognition between the fungus and the plant, and appressoria are not formed in the roots of non-host plants or on synthetic media or membranes. Penetration of the root occurs via the appressoria and the fungus may produce intracellular coils in the subepidermal cell layers, followed by intercellular growth into the inner cortex in the subepidermal cell layers, followed by intercellular growth into the inner cortex of the root (although sometimes only intercellular hyphae are observed). On reaching the inner cortex, hyphal branches penetrate the cortical cell walls and differentiate within the cells to form highly branched, terminal structures known as arbuscules. Here the symbionts are in intimate contact, separated only by the their membranes. This interfere is assumed to be the site of carbon and phosphate exchange between the symbionts, and the arbuscule generally regarded as a key structure in the symbiosis. Following the formation of arbuscules, some species of fungi also form vesicles in the root. Vesicles are small darkly staining globular or spherical structures formed later. They store oil and Polyphosphate granules, which are utilized by the plant under phosphorus deficiency conditions.

Classification of AMF

There are only six genera of fungi that contain species, which are known to produce AMF with plants. Two of these genera, *Glomus* and *Sclerocystis*, produce chlamydospores only. Four genera form spores that are similar to azygospores: *Gigaspora, Scutellospora, Acaulospora* and *Entrophospora*. Their spore characteristics and the relationship of the spore to the associated hyphal attachments distinguish the genera. The existent genera were establishment to reflect the manner in which the spores were produced, except for the genus *Scutellospora*. Species in this genus are distinguished from species in the closely allied genus *Gigaspora* by wall characteristics, method of germ tube formation, and auxiliary cell characteristics.

The fungi forming AM are placed in the class Zygomycetes, order Endogonales, family Endogonaceae. The type genus for Endogonaceae is *Endogone*, a genus in which species form zygospores and azygospores but do not produce arbuscular mycorrhizae.

Over the years, numerous attempts have been made to monoculture AMF on artificial media. These attempts, however, have met with little success. Thus, AMF must still be collected from field soils, increased in the greenhouse on living host plants (nurse crops) and collected once again from these "pot culture" soils, where the TERI is specialized. Clearly, the production and collection of propagules from soil remains an important, time-consuming and fundamental part of AMF research, both for inoculum preparation and population estimation. On the other hand, all AMF are not morphologically and physiologically the same, so it is important to know the identity of the fungus with which one is working. Although these fungi have extremely wide host ranges, they are very specific in a single plant species and they can differ also in their effect on plants in different ecosystems, soil types, and in similar soils with differing biological, chemical, and physical components.

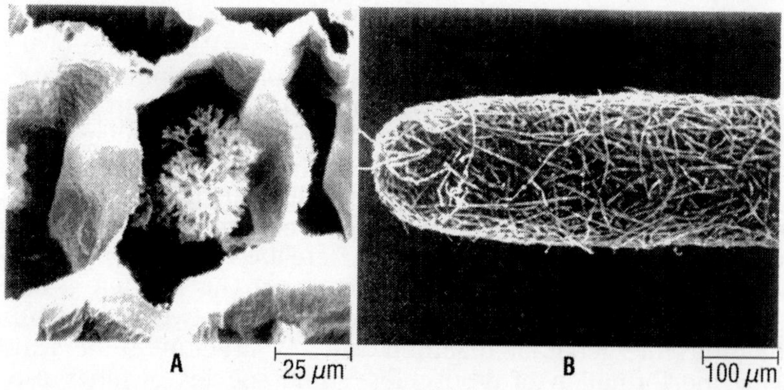

FIG. 9.5: Scanning electron micrographs of mycorrhiza. (A) A fully developed arbuscule of an endomycorrhiza in a plant cell. (The term arbuscule means little bush). As the arbuscule decomposes, it releases accumulated nutrients from plant. (B) The mycelial mantel of a typical ectomycorrhizal fungus surrounding a eucalyptus tree root.

Ectomycorrhizae (ECM)

They are characterised by dense mycelial sheaths around the roots and a Hartig net between root cells, are limited to mostly temperate forest trees and *Basidiomycetes* and some other fungi. Mycorrhizae involve an intimate association of the host plant's root tissue and the fungus, but there is also fungal tissue that extends into the soil as individual hyphae and in some species as more complex strands. The external hyphae can take up mineral nutrients from the soil and transport them into the host root. Mycorrhizal associations can therefore be of great potential benefit to the host plant in nutrient limited systems.

Vesicular-Arbuscular Mycorrhiza (VAM)

Zygomycete fungi form arbuscules and/or vesicles and external hyphal networks in the soil and grow extensively within the cells of the cortex, are formed by nearly all vascular plants. The endomycorrhizae, a fungus, (arbuscular mycorrhizae) found in close association among angiosperms. The VA mycorrhizae are formed by the non-septate phycomycetes fungi belonging to the genera *Glomus, Gigaspora, Acaulospora* and *Sclerocystis* in the family Endogonaceae of the order Mucorales. They produce vesicles and arbuscules inside the root system. Arbuscules are highly branched fungal hyphae while vesicles are the bullous swellings of these hyphae. These VAM fungi are beneficial to the plant which they colonies. They make more nutrients available to the plant, improve soil texture, disease resistance and help in better plant growth. Besides, mycorrhizae also help in the biological control of root pathogen.

An extensive research on VA mycorrhizal association in horticultural crops have been done at the University of Agricultural Sciences, Bangalore and Indian Institute of Horticultural Research (IIHR), Bangalore. Tomato, brinjal, okra, bell pepper, chilli and onions have been studied for their response to VAM inoculation (Sukhada, 1993). He also reported beneficial response of VAM fungi in cowpea has been reported (Sukhada, 1993).

In India the beneficial effects of vesicular-arbuscular mycorrhizae (VAM) have been observed in fruit crops like citrus, papaya and *litchi*. Inoculation with *Glomus fasciculatum* fungus was found to increase the dry matter content and the uptake of Fe, Zn, Mn by different rootstocks of citrus (Manjunath *et al.*, 1983).

In litchi, positive correlations of VAM population in the rhizosphere of soil and mycorrhizal infection in the roots with leaf area and fruit have been noticed (Rana and Srivastava, 1984). In papaya, inoculation with *Glomus fasciculatum* and *Glomus mosseae* was found to increase the plant height, dry matter content and nutrient uptake with lower doses of applied phosphorus.

Recent studies have shown the possibility of domesticating *Mycorrhizae* in agricultural systems is not in doubt. Lack of a suitable inoculum production technology is the major limitation for commercial exploitation of this system.

TABLE 9.5: Effect of mycorrhizal (VAM) inoculation on crop yields

Crop	P-fertiliser supplementation (kg P_2O_5/ha)	% yield increase over control
Finger millet	19	18
Soybean	25-50	10
Chillies	37.5	55
Chickpea	40	25
Groundnut	-	10-20

Source: Venkatraman and Tilak (1990).

How Mycorrhizas Work

Mycorrhizal associations are defined in the introduction section. Mineral nutrient acquisition from soil is considered to be the primary function of mycorrhizas, but other roles for these fungi have been suggested. These roles are listed below, but many have not been quantified in natural ecosystems. Only the uptake of nutrients is discussed further here.

Nutrient Depletion Zones

The demand for a particular mineral nutrient depends on plant internal requirements, while the supply of that nutrient primarily depends on its availability and mobility in soils. Mineral nutrients such as phosphorus have very limited mobility in soils so that depletion zones – where all the available nutrient has been utilized, quickly from around roots. Thus to obtain more phosphorus, plants must bypass these depletion zones by further root activity elsewhere in the soil. The outcome of this quest for phosphorus (and other relatively immobile soil resources) should largely be determined by the surface area of a plant's root system. The most important role of mycorrhizal fungus hyphae is to extend the surface area of roots.

Hyphal Activity

Mycorrhizal fungus hyphae are considered to function primarily by increasing the soil volume from which available forms of phosphorus are absorbed and provided to roots. Hyphae of VAM fungi can respond to localized sources of soil nutrients more rapidly than roots and to produce tine highly-branched *absorptive* hyphae in decomposting organic substrates. ECM fungus hyphae also proliferation in decomposing soil organic matter.

It was considered that hyphae of VAM fungi utilized the same forms of nutrients as roots. However, there now is evidence that these associations have a greater benefit when phosphorus is present in less-soluble forms. Hyphae of ECM fungi can utilize both inorganic and simple organic sources of nitrogen and phosphorus. There are some ECM fungi which specialize in growth is soils enrichment with ammonium. Soil mineralisation by organic acids produced by hyphae has been measured within Ecm fungus mats. ECM fungus hyphae can also grow within rocks where they contribute to mineral weathering. Thus, mycorrhizal fungi often have access to forms of nutrients which are not directly available to plants, as well as acquiring nutrients which are spatially or chronologically separated from roots.

Mycorrhizal Dependency

Mycorrhizal dependency is a measure of the benefit provided by mycorrhizas and will depend no relative contribution of root and mycorrhizal mediated nutrient uptake to plants (Janos 1980). Mycorrhizal dependency has often been quantified by calculating the yield ratio between mycorrhizal plants and uninoculated control plants grown in a particular soil at a single soil P level. However, it is better to analyse mycorrhizal benefits across a range of soil P levels, by producing nutrient response curves.

The concept that plants have varying degrees of dependence on mycorrhizal associations is gaining acceptance. Detailed examinations of plants in natural ecosystems often show consistent differences between host plant species in the intensity and consistency of mycorrhiza formation (proportion of root system involved). These observations have shown that species generally either have:

1. Consistently high levels of mycorrhizas

2. Intermediate, or variable levels of mycorrhizas.
3. Or are not mycorrhizal.

Plants belonging to these categories can be called obligatorily mycorrhizal, facultatively mycorrhizal, or non mycorrhizal, as are defined below.

In natural ecosystems, plant with facultatively mycorrhizal associations or non mycorrhizal roots are more common in very dry, wet or cold habitats where plant productivity is limited by soil/environmental conditions, or in disturbed habitats where mycorrhizal fungus inoculum is limited. Non mycorrhizal trees are rare (members of the Australian family Proteaceae are one exception). There are a number of non mycorrhizal genera which are important in agriculture and horticulture including members of the families Chenopodiaceae, Amaranthaceae, Caryophyllaceae, Polygonaceae, Brassicaceae, Scrophulariaceae, Commelinaceae, Juncaceae and Cyperaceae.

Obligatorily Mycorrhizal Plants

Obligatorily mycorrhizal plants are those plants which will not survive to reproductive maturity without being associated with mycorrhizal fungi in the soils (or at the fertility levels) of their natural habitats. These species consistently support mycorrhizal colonization throughout most of their young roots.

Facultatively Mycorrhizal Plants

Plants that benefit from mycorrhizal associations only in some of the least fertile sols in which they naturally occur. In ecosystem surveys, inconsistent mycorrhizas or lo w levels of mycorrhizal colonization have been used to designate facultatively mycorrhizal species when soil fertility levels could not be manipulated.

Non mycorrhizal Plants

Plants with roots that consistently resist colonization by mycorrhizal fungi, at least when they are young and healthy. Nutrient levels and other soil properties and mycorrhizal propagules dynamics can also reduce mycorrhizal formation, but usually do not prevent it completely.

Mycorrhizal Fungi and Pathogen Interaction

Tomato plants inoculated with VAM fungi were highly resistant to *Pythium aphanidermatum* causing damping off. Control of nematodes was also reported due to VAM inoculations. Accumulation of phytoalexin in cowpea root during the infection of *G. fasciculatum* and *G. aggregatum* which could be correlated with the degree of infection. Nematode infection had reduced significantly in mycorrhizal plants. The root extract of mycorrhizal plants caused mortality of nematode larval compared to non-mycorrhizal root extracts. Tomato seedlings inoculated with VAM fungi were more resistant to *Meloidogyne javanica*. A similar control of nematode infestation in tomato and brinjal by inoculation with *G. mosseae* and neem cake.

TABLE 9.6: Crop types and responses of inoculation

Crop	Inoculation and Response	Reference
Banana	Responds excellently to VAM inoculation. The *G. mosseae* and *G. fasciculatum* inoculation in combination with phosphate solubilizers and *Azospirillum* improved bunch weight in bale Elakki	Sukhada (1997)
Banana Dwarf Cavendish	Treated with *G. fasciculatum* 7 days prior to nematode treatment was able to offset the ill effects of the latter and had more vigorous root nematode population reduced.	Umesh *et al* (1988)
papaya, Citrus, Mango and Pomegranate	Plans could be inoculated effectively with VAM fungi in the nursery and transplanted to the field The inoculation improves the plant establishment, improves of phosphorus in soil and their uptake (Sukhada, 1992). Due to the increased availability of phosphate in soil a saving in the addition of inorganic phosphate in the soil can be brought about.	(Sukhada, 1997). work conducted at IIHR, Bangalore
Litchi *(Litchi chinensis* Son)	Distribution of endomycorrhizal spores in the rhizosphere	Rana and Srivastava (1984)
Papaya (Coorg Honey Dew)	Studied the uptake of ^{32}P labeled super phosphate by endomycorrhizal and derived that mycorrhizal fungi besides acting as scavengers of phosphate also increased the availability of phosphates in the soil	Sukhada *et al* (1995)
	Standardized PCR amplification technique to identify different genera of VAM fungi	Sukhada *et al* (1997)

Nitrogen Transfer In Mycorrhizal Plants

The idea of nitrogen fixed by a legume component may be available to its associated non-legume plant originated from earlier studies in the late of. In several of their experiments they found that as much as 10 to 30% of the total N fixed in field pea was deposited in the growth media of sand, predominantly as amino acids, and a major part was taken up by an associated cereal. In general, the process of N-deposition and uptake in a neighbouring plant is often termed **N-transfer**. Since then substantial N-transfer have been observed in several legume/crop intercropping systems such as soybean or cowpea/maize or sorghum, cowpea/rice, gram clover/wheat, barley or oat/vetch or lupin etc., although some experiments showed little or no N-transfer between them

Compared to pure stand cultivation, possible benefits obtained by non-legumes or trees may partly be due to transfer of the symbiotically fixed N by legumes, either through directly release from nodulated roots or through decomposition of dead nodule and root tissue in the soil or through a common mycorrhizal network or even enhanced by mycorrhizal hyphae. However, the mechanism of N transfer and the role of mycorrhizal hyphae in the direct transfer of nitrogen are not well established. Therefore, studies in identifying the pathways of N transfer and whether mycorrhizal hyphae play a direct role in facilitating N transfer are needed.

Nitrogen Nutrition in Mycorrhizal Plants

Forms of Nitrogen Used by Mycorrhizal Associations

The two most available forms of inorganic N for plants and both VA mycorrhizal and ectomycorrhizal fungi, are NO_3^- and NH_4^+. Originally, NH_4^+ is mainly derived from the reduction of atmospheric dinitrogen symbiotically fixed by nitrogen fixing plants, either by *Rhizobia* in legumes or by *Frankia* in actinorhizal plants. In general, NO_3^- is the dominant form of nitrogen available to plants and fungi in almost all agricultural soils due to the rapid nitrification of NH_4^+. In contrast, NH_4^+, which is released from soil humus and other organic N sources by ammonifying organisms, predominates and NO_3^- may be almost entirely absent in many undisturbed or very acidic soils. NO_3^- is highly mobile and is readily transported towards the plant roots by mass flow and diffusion. NH_4^+ is absorbed to negatively charged soil particles and transported towards the plant roots mainly by diffusion.

Mycorrhizal Effects on Nodulation and Nitrogen Fixation

The possibility of a direct involvement of ectomycorrhizal fungi in N acquisition by plants was first suggested by Frank in 1894. Then after 50 years of Frank's work, the probably first observation of growth, nodulation and mycorrhizal status of a large number of legumes was made by Asai in 1944. In most cases, improved nodulation and N fixation in mycorrhizal plants appears to be the result of relief from P stress and possibly uptake of some other essential micronutrients, which result in both a general improvement in growth and yield and indirect effects upon the N-fixing system. The differences between mycorrhizal and non-mycorrhizal plants usually disappear if the latter are supplied with a readily available P source.

It was known that cereals intercropped with grain legumes generally benefit from the association in terms of increased grain and nitrogen yields per unit area compared with mono cropped cereals. This indicates that N-transfer from the nitrogen fixing plants (donor) to an associated non-nitrogen fixing plant (receiver) is most likely to happen in nature either via the interception and uptake of released fixed nitrogen from donor plant by the roots of receiver plant or through the root mycorrhizal between donor and receiver plants, or even enhanced by mycorrhizal hyphae. Therefore, in agricultural and/or natural ecological communities, mycorrhizal associations may be important factors influencing the performance of both donor and receiver plants through the acquisition and translocation of nitrogen by the mycorrhizal fungus, particularly when the relatively immobile NH_4^+-N rather than the mobile NO_3^--N is the major source of plant available N.

It is therefore, concluded that Nitrogen fixing and non-nitrogen fixing plants can provide a good example for investigating the N-transfer between the plants where neighbouring plants may have differing nitrogen status. In general, with or without a split root system, combined with the fine nylon mesh to allow the direct mycorrhizal link but not root contact, together with ^{15}N-isotope labelling technique to enrich the N in the donor plant, some experiments have demonstrated a below-ground N-transfer whereas others found no evidence for transfer of nitrogen between donor and receiver plants via mycorrhizal hyphae (see Smith and Read, 1997). Furthermore, there is controversy about the extent to which direct N-transfer is actually being facilitated by mycorrhizal hyphae or by some other indirect means, and whether it is of agricultural or ecological significance if N-transfer do occur between these plants through mycorrhizal hypha.

The following questions still remain to be answered:

1. Does N transfer occurs between mycorrhizal plants at all? If so, how much N is transferred and how much N-transferred is enhanced by mycorrhizal hyphae?
2. How much net N transfer is from nitrogen fixing to non-nitrogen fixing mycorrhizal plants and *vice versa*?

Enhanced Crop Production

Nitrogen fixing microorganisms convert about 139-170 million tonnes of nitrogen every year into fertilizer nitrogen for which the energy bill is paid by Nature. It is at least several folds greater than the inputs of nitrogen from chemical fertilizer factories to the tune of 65 million tonnes per year. The fact that the total world biological nitrogen fixation is three times that of industrially produced nitrogen, demonstrates the significance of biological nitrogen fixation in agricultural and natural nitrogen cycles.

Nitrogen fixing systems offer an economically attractive and ecologically sound means of reducing external inputs and improving internal resources. Symbiotic systems such as that of *Azolla* and *Anabaena* complex and that of leguminous green manures with *Rhizobium* and *Azorhizobium* associations can be of particular value to wet land rice crop and renewable biological source supplementing inorganic N for cereals.

In TNAU significant contributions have been made on Azolla biofertilizer particularly on the selection of new strains of Azolla microphylla which is highly tolerant to high temperature and also tolerant to salinity with higher biomass producing capacity (25t/ha). *Azolla* hybrid has been developed by sexual hybridization. This hybrid is also highly adaptive and has higher N_2 fixing capacity in soil ecosystem. The nitrogen fixation by free-living blue green algae (cyanobacteria) also contributes significantly to the replenishment of soil N. Recently, several attempts have been made to improve the associations between rice and nitrogen fixing soil bacteria and possible nitrogen fixing rice endophytes to achieve a higher contribution of biological nitrogen fixation to rice crop. In these soils, sustainability depends on symbiotic and non-symbiotic BNF system.

Nitrogen fixing free-living cyanobacteria play a vital role in the maintenance of soil fertility and sustainability. The large scale production of soil based BGA inoculum developed at Tirur by TNAU during 1977 is a crude technology, which is two decades old, and hence has to be updated. Immobilized cyanobacteria inoculum is a recent development in the biofertilizer technology, which was developed in the biofertilizer technology, which was developed by TNAU, Coimbatore with the active collaboration of the Division of Life Sciences, King's College, London, and the Department of Biochemistry and Molecular Biology, University of Seville, Spain, under an EEC project.

Immobilization of nitrogen fixing cyanobacteria under rice field conditions can be done with solid matrices such as polyurethane foam, polyvinyl foam, sugar cane waste, paper waste, and rice husk. The immobilized cyanobacteria could continuously excrete the ammonia extracellularly which could be used for continuous supply of nitrogen to rice crop. The immobilized cyanobacterial system in solid matrix when inoculated in rice field is like a mini *in situ* fertilizer factory which produces ammonia continuously. The accumulation of ammoniacal N in the flood water as a result of immobilized cyanobacterial inoculation will reflect on the growth and yield of rice.

Rhizobium

This is one of the important nitrogen-fixing bacteria which helps the legume to maintain soil fertility by way of their symbiotic association and nitrogen fixation. A legume plant having effective root nodules can not only meet its own nitrogen requirement but can also enrich the soil nitrogen content thus improving soil fertility and sustainability. Legumes are cultivated widely in different parts of India. The practice of seed inoculation with Rhizobium is popular among farmers.

Inoculation of appropriate strains of rhizobia to legumes is known to enhance the yield but both successes and failures at field experiments have been documented. Failure to obtain the desired response is due to:

1. The presence of native ineffective strains which could not be displaced by introduced effective strains.
2. The presence of antagonists of rhizobia which minimize the numbers of rhizobia in the rhizosphere,
3. The soil conditions such as acidity, alkalinity application of pesticides and herbicides and
4. The presence of effective native rhizobial strains in large numbers.

Root Nodule Structure

Non-legumes such as cereal crops and grasses do not form the nodules in their root system. Recently, it has been investigated at TNAU and found that the symbiotic nitrogen fixing bacterium, Azorhizobium caulinodans isolated from the stem nodules of Sesbania rostrata, could colonize and produce nodular structures in rice root. Among the cereals tested, maize crop responded well for the inoculation of A. caulinodans and prominent nodular structures have been observed.

The bacterium, *A. caulinodans* is a unique one which fixes nitrogen under free-living condition and is also tolerant to combined nitrogen at low concentration. The bacterium could also tolerate flooded condition and A. caulinodans also produces the pectic and cellulolytic enzymes with good penetration capacity and, in general, it fixes nitrogen in rice root system at micro-agrophilic conditions (2-3% O_2 level). The nitrogenous enzyme which is responsible for nitrogen fixation was detected from the root nodule structures of rice, maize, sorghum, cumbu and ragi. The nitrogen fixing rate in the nodular structures of cereal crops is low but the colonization of the bacterium could be induced by biotechnological manipulation of rice plant.

Azospirillum

Field experiments conducted in different places in India have indicated the positive response of rice cultivars to Azospirillum inoculation. Experiments conducted at IRRI, Philippines, showed that inoculation with *Azospirillum lipoferum* promoted early tillering and also the growth of rice and significantly increased filling rate of grain and the grain weight per plant at harvest. Response of rice to inoculation of *Azospirillum* Lipoferum has been reported by several *Azospirillum* inoculation under low land rice field condition was variable, depending on the initial nitrogen status of the soil and the varieties of the rice. Beneficial effect of *Azospirillum* inoculation has been reported in maize, sorghum, sunflower, bhendi, cotton and pearl millet.

Acetobacter

More recently, a new species of nitrogen fixing bacterium, Acetobacter diazotrophicus, was isolated from sugarcane in different parts of the world. A diazotrophicus has been isolated from the root, leaf, bud and stem samples of sugarcane cultivars planted in 10 different locations of Tamil Nadu. Inoculation of the acid tolerant and sucrose loving bacterium to sugarcane enhances the cane yield by 17 to 20 tonnes per hectare with a 50 per cent reduction in the recommended chemical nitrogen. The results of the field experiments showed that inoculation of *A. diazotrophicus* to sugar cane enhances the cane yield and sugar yield than *Azospirillum* or *Azotobacter* inoculation.

Phosphorus Fertility

Phosphobacteria

Several soil bacteria and fungi, notably species of *Pseudomonas, Bacillus, Penicillum and Aspergillus* secrete organic acids and lower the pH in their vicinity to bring about solubilization of bound phosphates in soil Inoculation of composite culture of phosphate solubilizing microorganism comprising *Pseudomonas striata* and *Aspergillus awamori* recorded increase in the yield of cotton more than the single strain inoculation. Increases in the field of wheat and potato were demonstrated due to inoculation of peat based cultures of Bacillus Polymyxa and Pseudomonas Striata. Similar results were also obtained in wheat when the mixed cultures of phosphate Solubilising microorganisms were used with Azotobacter Chroococcum. Saving 50 per cent of recommended level of P_2O_5 is possible in sugarcane by inoculation with phosphobacteria as the cane yield and sugar yield of 50 per cent P_2O_5 and phosphobacteria treatments are on par with 100 per cent P_2O_5 application.

However, it is also observed that the positive influence of phosphobacteria was conspicuous even at 0 to 75 per cent of phosphorus levels in sugarcane. Combined inoculation of *Rhizobium* and *Phosphobacteria* have been tried in different crops and the increase in yield of soybean and groundnut was reported.

Vesicular Arbuscular Mycorrhiza

Mycorrhiza is the mutualistic association between plant roots and fungal mycelia. Vesicular Arbuscular Mycorrhiza (VAM) and endotrophic mycorrhizae formed by aseptate phycomycetous fungi are associated as an obligate symbiont with majority of agricultural crops growing under broad ecological range. Many graminaceous plants and legumes are highly susceptible to VAM colonization. VAM entophytes are not host specific although evidence is showing that certain endophytes may form preferential association with certain host plants. The transfer of nutrients, mainly phosphorus and also zinc and sulphur from the soil to the cells for the root cortex, is mediated by intracellular obligate fungal endosymbiont of the genera *Glomus, Gigaspora, Acaulospora, Sclerocystis* and *Endogone* which posses vesicles for storage of nutrients and arbuscules for funneling these nutrients into the root system. By far the commonest genus appears to be *Glomus*, which has several species distributed in soil.

Experiments with phosphate indicate that the hyphae of VAM fungi obtain their extra phosphate from the labile pool rather than dissolving in soluble phosphates. The better utilization of sparingly soluble rock phosphate is explained by the hyphae making closer physical contact with the particle surface where the ions are dissociating. These fungi produce phosphatases, which allow utilization of organic phosphorus especially under humid tropical conditions where hyphae can be in close with finely divided litter where acid soil reaction reduces the sorption of organic phosphates on soil surfaces. Vesicular arbuscular mycorrhiza (VAM) has also been reported to improve the uptake of the minor elements zinc, copper and also water.

Soil inoculum is produced using a traditional pot culture techniques containing all VAM fungal structures which are highly infective. The success of good soil inoculum production depends on the selection of host-plant and the ambient conditions under which a defined VAM fungus can be mass multiplied. The dual inoculation with VAM fungi and appropriate Rhizobium strains on the enhancement of nutrient uptake of legumes like soybean, chickpea, clover, cowpea and groundnut was reported. Several field, laboratory as well as green house experiments have demonstrated that VA-mycorrhizal inoculation can greatly improve growth and nutrition of host plants.

Scope of Biofertilizers in India

The total estimated demand for various biofertilizers for different crops grown in India is as follows:

Rhizobium....................35,730 tonnes
Azospirillum....................4,82,000 tonnes

Azotobacter...1,62,610 tonnes
Blue Green Algae..2,67,720 tonnes
Azolla..20,380 tonnes
Phosphate solubilizers ....2,75,500 tonnes

The demand is calculated based on the area under different crops and the dose of biofertilizer.

The present production capacities of different biofertilizers in India are as follows:

National Biofertilizer Development Centre (Central Sector)	375 tonnes
State Agricultural Universities	245 tonnes
State Agricultural Departments (Major Biofertilizer producing States include Tamil Nadu, Uttar Pradesh and Maharashtra)	665 tonnes
Agro Industries Corporations	1130 tonnes
Private Sector	195 tonnes

The annual requirement and production of different biofertilizers have clearly shown tremendous gap in this area. Thus a strategy for judicious combination of chemical fertilizers and biofertilizers will be economically viable and ecological useful. It should also be remembered that biofertilizers are not a substitute, but a supplement to chemical fertilizers for maximizing not only yield but also agro system stability.

Quality Control

Recently, the biofertilizer production technology has attracted much attention of the small and large producers because of its simple and cost effective nature. The current annual production of biofertilizers is in the range of 2000-2500 tonnes. Though there is lot of scope for scaling up the production level, the biofertilizer technology faces the problems of quality control, inconsistent field performances, poor transport and storage and lack of knowledge about these inputs. This has necessitated the implementation of better quality control mechanism to ensure the availability of quality inoculants to the farmers. The quality of biofertilizers depends on the use of specific and appropriate microbiological techniques for production and quality control of biofertilizers. The following points have to be taken into consideration while adopting quality control procedures for biofertilizers with the objectives of ensuring supply of quality inoculants to the farmers.

1. Biofertilizer production unit should have adequate microbiological laboratory facilities to produce quality inoculum. The unit should be maintained by a qualified microbiologist. The production unit should have one quality control laboratory to check their own biofertilizer packets.
2. Each biofertilizer production unit should maintain crop specific and pre-tested efficient N_2 fixing microorganisms or phosphobacteria either collected from a reputed scientific organization or of their own. The maintenance of microorganisms should be as per the standard microbiological procedures and repeated sub-culturing should be

avoided. If possible, lyophilized cultures should be used for each batch of biofertilizer production. The manufacturer should indicate the source of the culture in the biofertilizer packet produced.

3. A biofertilizer production unit should use fermentors so as to ensure high microbial counts at broth level itself.
4. The biofertilizer production unit should use good quality carrier material like peat or lignite or wood charcoal or any other suitable carrier and the biofertilizer products should indicate the name of the carrier.
5. The inoculum packet should contain 200 g carrier based culture packet in standard polythene or polypropylene bags. The BGA biofertilizer packets should be in 1 kg, 4 kg, 10 kg pack size.
6. Recently the cost of 200 g pack size biofertilizer packet is fixed at Rs. 5 and therefore there is no need to enhance the cost at present.
7. The microbial inoculum must contain the specified population count at the time of packing and expiry. The packet should be crop specific and should give effective yield increase.
8. The production centers and distribution centers should store the biofertilizer packets in cool temperature so as to ensure longer shelf life.
9. Sample packets of biofertilizer should be drawn randomly for quality testing by the Central quality control authority appointed by the state, and out of 5 samples in a batch, at least 3 should pass the specified quality standards.
10. To ensure microbial inoculum quality control testing and quality assurance of the biofertilizers produced by different production units of the state, an exclusive Central Quality Control Laboratory with growth chambers and immunological laboratory facilities should be established. A certification unit should be included in the Quality Control Laboratory for certification of the inoculants. An amount of Rs. 50 as certification charge should be collected for each batch of biofertilizer produced.
11. A Central biofertilizer germplasm collections should be established with subcentres at different agroclimatic zones so as to select efficient agriculturally useful microorganisms.
12. The government should make legislation for enforcement of quality standards for different biofertilizers produced in the state.

Promotional Strategies

1. The reason for inadequate popularity of biofertilizers is that biofertilizers cannot show instant and dramatic response like fertilizers.
2. In many cases, quality of culture supplied is poor.
3. Inadequate correct awareness among the farmers about inoculants.
4. Lack of strong promotion and extension work and insufficient publicity.
5. Lack of availability of the quality products in time in rural areas.

For successful launching of a biofertilizer programme it is necessary to equip the extension machinery at the grass-root level with the full technical back up of production as well as application technology. There is need to give emphasis to biofertilizers as supplementary, renewable and non-polluting sources of nutrients. Once the farmers are convinced that the biofertilizer are an inexpensive but effective way of keeping their soil fertility, they will accept them and this will lead to higher crop production and higher economic returns.

The nitrogen fixing biological systems such as blue green algae, Azolla, Azospirillum, Acetobacter, and Azorhizobium are compatible to different cropping systems and could fix considerable amounts of nitrogen.

The biomass of these biological systems decomposes rapidly in soil and supply the nitrogen to various crops grown in the field. Besides nitrogen addition, these biofertilizers also contribute significant amounts of phosphorus, potassium, sulfur, zinc, iron, molybdenum and other micronutrients. The biological fertilizers were known to influence the total soil microbial population and soil enzymes activity which directly reflect on soil fertility index.

The use of biological fertilizers would reduce the cost of chemical fertilizers involved in crop production. The effective utilization of these biological fertilizers for different cropping systems will not only provide economic benefits to rice farmers, but also improve and maintain the soil fertility and sustainability of the soil ecosystem.

Biological nitrogen fixation is an important process in rice farming systems because it is an inexpensive source of nitrogen for increasing the productivity of crops. The nitrogen fixing bio systems such as Azolla, cyanobacteria and legume green manures were able to adopt well under wetland rice field ecological conditions and fixing considerable amounts of nitrogen. The biomass of these bio systems decomposes rapidly in rice soil and supply the nitrogen to rice crop.

Besides nitrogen addition, these biofertilizers or bio-manures also contribute significant amounts of phosphorus, potassium, sulphur, zinc, iron, molybdenum and other micronutrients. The organic acids released during the mineralization process of the biomass of bio-manures would accelerate the phosphorus availability in rice soil.

These biological fertilizers or bio-manures were known to influence the total soil microbial population dynamics of nitrogen fixing free-living bacteria like, *Azotobacter*, *Azospirillum* and also the free-living symbiotic bacterium such as *Rhizobium*. It is evidently clear now that application of biological fertilizer enhanced the accumulation of soil enzymes, which directly reflected on soil fertility index. The use of biofertilizers would reduce the cost of chemical fertilizers involved in crop production. The effective utilization of biofertilizers for crops will not only provide economic benefits to the farmers but also improve and maintain the soil fertility and sustainability in natural soil ecosystem. Since these bio sources represent a great diversity in chemical, physical and biological characteristics, their efficient use depends upon the particular agro-ecological environment and local availability.

Therefore, it is very much essential to develop strong, workable and compatible package of nutrient management through organic sources for various crops based on scientific facts, local conditions and economic viability.

FUTURE THRUSTS

Developing strains of bacteria which can withstand high temperature and alkali, and are acid resistant and high pH tolerant; which produce antibiotics in the soil or any chemicals which impart the strain competitive edge over other strain is the need of the hour. Some of the genes responsible for production of plant hormones have been characterized and cloned. These genes could be introduced into *Rhizobium* strains. Strains that can fix nitrogen in the presence of added nitrogen and those that can increase nitrogen-fixing efficiency should be selected.

Efforts to mass multiply mycorrhizae in a large quantity using lighter carrier material should be intensified. The PCR techniques should be standardized to identify mycorrhizal colonization within plant roots using different species specific primers.

CHAPTER 10

Mechanism of Nitrogen Fixation

The universal presence of nitrogen in all living matter explains the close relationship that exists between environmental chemistry and biological systems. Although much of the work studying the biological transformation of nitrogen has been conducted on soil ecosystems, it is much the same for those conditions occurring in aquatic ecosystems.

The supply of nitrogen in the atmosphere is vast. However, it exists as an inert gas, a form unusable to life. In order for nature to utilize this immense reservoir of free nitrogen it has to be combined with other elements. The relationship that exists between the various forms of nitrogen compounds and the changes that can occur in nature are illustrated by a schematic diagram of the nitrogen cycle (Fig. 10.1).

This diagram shows how nitrogen in its free form is converted by electrical discharge to nitrates. These nitrates along with nitrogen fixing bacteria and algae form plant protein which in turn provides protein to animals and humans to carry on their life processes. Within the animal body, the protein is used for the growth and repair of tissue and also as a source of energy. Waste products produced by the life processes are formed and excreted as fecal matter and urine. The fecal matter, along with the protein from dead animals and plants, is converted by bacteria to ammonia; the urine is hydrolyzed rapidly to ammonium carbonate. The ammonia produced can be used to produce additional plant protein or be oxidized to form nitrites and nitrates, thus completing the nitrogen cycle.

The form in which nitrogen appears is indicative of the stage of stabilization of the organic matter. Illustrates the various nitrogen forms and how they vary with time. Initially most of the nitrogen will appear in the form of organic nitrogen and ammonia. As time goes on, the organic nitrogen is converted to ammonia. Under aerobic conditions, the ammonia will ultimately be converted to nitrites and nitrates.

Nitrogen in liquid waste can exist in 4 forms, all of which are of interest to the agricultural sanitary engineer. The 4 forms are organic nitrogen, ammonia nitrogen, nitrite nitrogen, and nitrate nitrogen. The total of these 4 forms constitute total nitrogen.

TABLE 10.1: Forms of nitrogen

Organic Nitrogen	All nitrogen present in organic compounds is considered to be organic nitrogen. The nitrogen-containing organic compounds are derivations of ammonia, the oxidation of which forms ammonia nitrogen.

Ammonia Nitrogen	The ammonia nitrogen is a result of bacterial decomposition of organic matter. Fresh sewage is generally high in organic nitrogen and low in ammonia nitrogen. The sum of organic and ammonia nitrogen should remain constant for the same liquid wastes, unless ammonia is allowed to escape to the atmosphere because of septic action. The total concentration of the two serves as a valuable index for evaluating the strength of liquid waste and for determining the type of treatment process to select.
Nitrite Nitrogen	Nitrite nitrogen is formed by bacterial oxidation of ammonia nitrogen. It is not present in fresh wastes but appears after bacterial activity has taken place. The presence of nitrite nitrogen indicate that the waste has undergone partial decomposition and is unstable. Nitrites can either be reduced back to ammonia or oxidized to nitrates.
Nitrate Nitrogen	Nitrate nitrogen is formed by the oxidation of nitrites and represents the most stable form of nitrogen. It is an indication of stability and is a determination of the completeness of the biological decomposition process.

Gaseous nitrogen is the most abundant element of atmosphere (78.08% v/v), and seems to have a highly complex nutrient cycle in the terrestrial and aquatic ecosystems. There is an interrelationship between different elements required for plant growth in environment. These all are affected by microbes. Atmospheric nitrogen is chemically inert. Therefore, it can be used, as such by most of the living organisms. The common source of nitrogen in the nature is the nitrate ions (NO_3). The nitrate ion is often absorbed by the plants as a mineral raw material from the environment. The absorbed nitrate ions are then converted into amino groups ($-NH_2$). The conversion of $-NO_3$ to NH_2 group is only possible by plants and not by animals. Minerals accumulation in soil, dissolution of rock by water, addition of fertilizers in soil are also the sources of NO_3 in the soil.

Apart from these sources, atmospheric nitrogen and oxygen combine together in presence of lightening which come down with the help of rain-water to the ground, and serve as NO_3 source on the earth. Absorbed nitrogen (in the form of NO_3) incorporates into organic structures of plants which remain intact until death. Animals also obtain nitrogen by eating plants. Finally, both plants and animals die or decay. Through gradual decay of all nitrogen of dead plants and animals, it gets concentrated into ammonia (NH_3). The so forming NH_3 is then utilized as a source of nutrition for nitrifying bacteria. Usually, these bacteria are of two kinds:

- Those bacteria which absorb ammonia and convert it into nitrite ions i.e.,

$$NH_3 \rightarrow NO_2$$

- Those bacteria which absorb nitrite ions and convert it into nitrate i.e.,

$$NO_2 \rightarrow NO_3$$

Thus, the environment again gets NO_3 by the activities of these nitrifying bacteria. The available NO_3 in the environment acts upon denitrifying bacteria which converts it into molecular, atmospheric nitrogen. The free molecular nitrogen is then, utilized by nitrogen fixing organisms in which few species of bacteria and blue green algae have been included. They are often present in soil and water. They absorb nitrogen and incorporate it into ammonia and protein. The process is called **Nitrogen fixation**.

Nitrogen is the mineral nutrient most needed by plants and it often limits plant growth. Some plant species have formed mutualistic symbioses with nitrogen fixing *prokaryotes* and some *eukaryotes*. Those organisms that can directly utilize atmospheric nitrogen as a nitrogen source, are called "*diazotrophs*", and belong to the kingdoms *eubacteria* and *archaebacteria*. They are able to live independently on soil N for growth. Inside the root nodules, the bacteria on the other hand are supplied with carbon from the host and are sheltered from competition with other organisms.

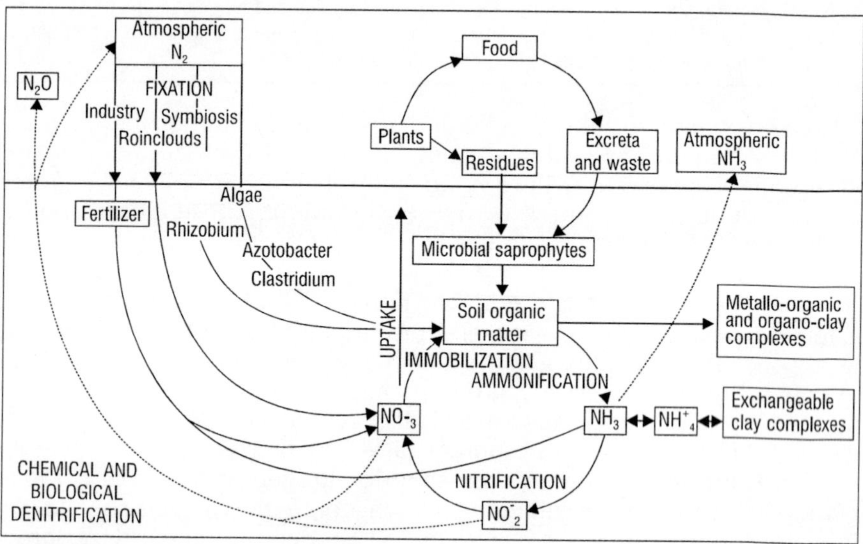

FIG. 10.1: The chemical conversion in the nitrogen cycle. The main activities are indicated together with some of the organisms concerned.

The most important mutualistic symbioses are root nodule symbioses, mainly formed by members of the genera *Rhizobium*, *Bradyrhizobium* and *Azorhizobium* on legumes, and by the genus *Frankia* on some non-legumes. The reduction of atmospheric dinitrogen (N_2) to ammonium (NH_4^+), which occurs in nitrogen fixing organisms is catalyzed by an enzyme complex called *nitrogenase*. A minimum 16 ATP and $8e^-$ are needed for a reduction of one molecule of N_2. It is estimated, that nitrogen fixation could range from one to a few kg ha^{-1} $year^{-1}$ in lichens and free-living bacteria, and up to a hundred or possibly a few hundred kg ha^{-1} $year^{-1}$ in legumes and actinorhizal plants.

Nitrogen fixation has always been a subject of great importance in biology, but since two decades the recent biotechnological development has lent its practical urgency. Nitrogenous fertilizers have become more and more expensive in terms of fossil energy, and hard cash; the population pressure has increased the need for high protein plant food. Biotechnologists are actively engaged in improving the efficiency of microorganisms to fix atmospheric nitrogen in fields. This will improve the functioning of the nitrogen cycle in biosphere, which will ultimately increase plant productivity. The programme of improvement also includes the creation of nitrogen-fixing symbiotic association other than those existing between leguminous plants and bacteria of the genus *Rhizobium.* This will help to improve agricultural production.

DIAZOTROPHIC MICROORGANISMS

The fixation of atmospheric nitrogen by prokaryotic organisms is known as diazotrophy. The phenomenon has also been observed in eukaryotic unicellular green alga, isolated from hot spring.

BIOLOGICAL NITROGEN FIXATION

This process is brought about by nitrogen-fixing microorganisms. These are soil inhabitants e.g., bacteria, blue green algae, molds, etc. Those microorganisms, which utilize gaseous nitrogen directly and independently in the soils are known as asymbiotic or non-symbiotic nitrogen fixing organisms and the process by which they brought about is referred as non-symbiotic nitrogen fixation. Those microorganisms, which utilize gaseous nitrogen indirectly through the mediation of other living organisms in the soil are called symbiotic nitrogen fixing bacteria and the process brought about by them is referred as symbiotic nitrogen fixation.

ASYMBIOTIC NITROGEN FIXATION

Microorganisms

Bacteria

As described earlier, soil contains free-living bacteria that are capable of fixing molecular nitrogen into its compounds. This process was first recognized in bacterium *Clostridium pastorianum* which is a Gram-positive, spore forming rod-shaped bacterium. After seven years of the isolation of *C. pastorianum,* two more important free-living nitrogen-fixing bacteria were isolated by Beijernick (1901). These newly identified bacteria were *Azotobacter chroococum* and *A. agile*.

Azotobacter chroococum is usually found in field-soils. They are short, thick, rod-shaped with rounded ends. Some larger ovoid forms are almost yeast-like in appearance. The cells are arranged singly or in end-to-end pairs. They measure 2 to 3 nm by 6 nm in size. They appear-brown in old cultures. They are Gram-negative and non acid fast. They are motile and the motility is due to the presence of polar flagellum. Physiologically, they are aerobic and remain active in between 25°C. They fix atmospheric nitrogen only after utilizing carbohydrate (dextrose, maltose, lactate, etc.) and give carbon dioxide as a byproduct. *A. glile* also actively fixes atmospheric nitrogen and produces carbon dioxide.

Apart from *Azotobacter*, other non-symbiotic bacteria like *Rhodospirillum pneumoniae, Rhodopseudomonas, Chlorobium, Diplococcus pneumoniae, Azotobacter aerogenes, Micrococcus sulfurens* have also been known to fix atmospheric nitrogen asymbiotically.

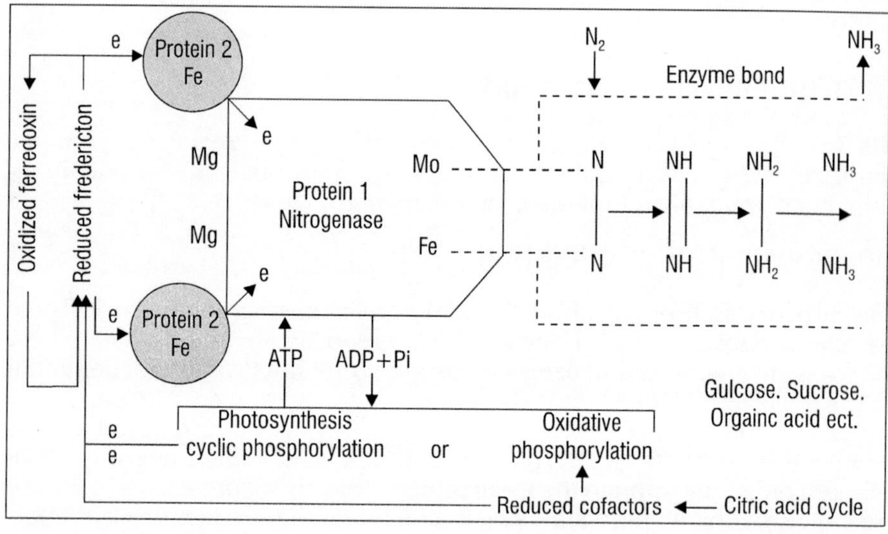

FIG. 10.2: A generalized scheme for the action of nitrogenase (incorporating results from studies on various organisms).

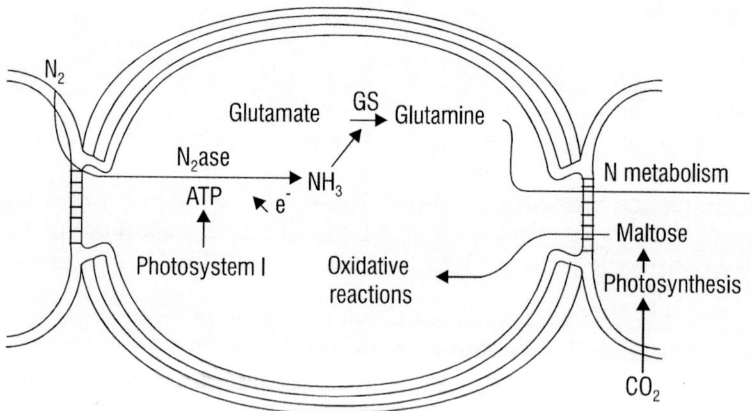

FIG. 10.3: Nitrogen and carbon flow associated with nitrogen fixation in the heterocyst of blue green alga. N_2ase = the nitrogenase enzyme as GS = glutamine synthetase, oxidative reactions including the oxidative pentose phosphate pathway. Polar nodule are not shown.

The ability to fix nitrogen by microorganisms was confirmed by means of the technique known as *acetylene reduction* to *ethylene* by diazotrophic microorganisms. The mechanism of this conversion is controlled by an enzymatic complex (or nitrogenous enzyme) which can reduce gaseous nitrogen to ammonia.

The enzyme nitrogenase is sensitive towards oxygen. There are reports which reveal that several groups of microorganisms fix nitrogen in presence of minute quantities of oxygen. Such microorganisms are known as **microaerobic fixers**. They oxidize methane. The examples are bacteria, Spirilla (*Aquaspirillum and Azospirillum*), *Xanthobacter autotrophicus* and *S. flavus, Thiobacillus ferrooxidans and* non-heterocystous microaerobic cyanobacteria (*Oscillatoria, Plectonema*). Certain coliform bacteria also fix nitrogen only in aerobic conditions (e.g., *Citrobacter, Enterobacter, and Erwinia).* A comprehensive lists of microorganisms have been presented in the book, but in near future they may need revision in light of findings of biochemical and physiological researches. It is established that oxygen not only inhibits the activity of nitrogenase enzyme but also regulates its biosynthesis, so that in conditions that are physiologically hardly favourable to nitrogen fixation, the enzyme is not synthesized.

Mechanism of Asymbiotic Nitrogen Fixation

The knowledge of asymbiotic nitrogen fixation is still insufficient. It was previously believed that *Azotobacter* fixes atmospheric nitrogen with the help of enzyme system called azatase and nitrogenase is only one of the compounds of this system. Beijerinck (1901) indicated that fixation of atmospheric nitrogen took place at the nitrous acid stage, but it was not proved. According to one more concept, nitrogen is fixed through the stage of hydrolysis to hydroxylamine or to ammonium nitrate which is produced from nitrogen and water as shown below:

$$N_2 + 2H_2O \rightarrow NH_4NO_2$$

But this concept has not been proved. The hydrolysis of nitrogen is very difficult under normal conditions.

It was further proposed that *Azotobacter* fixes nitrogen after its reduction by the active hydrogen. However, this hypothesis was also not proved.

At last the mechanism of nitrogen fixation was described on the physiological basis. It was proved that the protoplasm of *Azotobacter* produces special catalysts which take part in fixing free nitrogen. The catalysts contain carboxyl and amino group as an important parts which do not take any active part in fixing atmospheric nitrogen. Phenyl hydrogen and hydroxylamine are also present in the composition of the catalysts. It was presumed that molecular nitrogen reacts with catalysts of nitrogen fixation through oxygen of carboxyl group and produces hydrazine as the first product of the reaction. The hydrazine, is then, further reduced by active hydrogen and is transformed into amino acids. They finally enter the proteins of the cell protoplasm. This incorporation of amino acid into protoplasms helps into the growth of a bacterium.

Requirements of Nitrogen Fixation

1. Oxygen and carboxyl group of the catalysts play an important role in nitrogen fixation.
2. A very small amount of chemical energy (ATP) is used up during nitrogen fixation. About 1 kilo calories fix 2 mg of atmospheric nitrogen. It shows the relationship between respiration and nitrogen fixation.

Some strict anaerobic bacteria are more sensitive to oxygen. *Clostridium pasteurianum* is an obligate anaerobe, therefore, the mechanism of nitrogen fixation of such organism is similar to that *Azotobacter*, because they produce large amount of hydrogen during their metabolic reaction. The various possible pathways of nitrogen fixation have been shown in different illustrations.

Symbiotic Nitrogen Fixation

Agriculturists have noted that a relatively large group of plants- the legumes, are capable of fixing atmospheric nitrogen through a symbiotic association with soil bacteria called *Rhizobium*. These bacteria or nodule-bacteria, usually dwell in the roots of various beans, peas, lupines, cloves and other leguminous plants.

Hellriegal and Wilfarth (1888) first of all demonstrated that an increase in nitrogen content in soil is due to the presence of small tumor-like outgrowth on the roots of leguminous plants. It was found that in absence of root nodules, growth of succeeding plant-crop was retarded. The results of various experiments reveal that root nodule forming plants can feed upon atmospheric nitrogen and this capacity of plants can be observed only in those plants, which are growing in non-sterilized or bacteria containing soils. These symbiotic bacteria e.g., *Rhizobium* utilize the free atmospheric nitrogen and synthesize it into new nitrogenous compounds, which are utilized by plants for their growth; and bacteria get their food from these plants. Such a mutual beneficial association of bacteria and plants is referred to symbiosis.

Symbiosis was originally defined by Bary in 1879. Symbiosis is the living together of any two organisms. Mutualism is a relationship where both partners benefit, though it greads into those where the advantage or disadvantage to one of the partner may not always be obvious. Commensal relationships are usually rather loose association of the benefit of another without any effect on itself.

Mechanism of Nitrogen Fixation in Legumes

The mechanism of nitrogen fixation in legumes has not yet been understood. However, the basic prerequisite for nitrogen fixation in extracts of nodule-bacteroids are similar to those that have been described for asymbiotic bacteria such as *Azotobacter*. It has not been well established that how in both asymbiotic and symbiotic system ^{15}N is converted into NH_3 and then to amino acids, because the intermediate products between ^{15}N to NH_3 have not been determined.

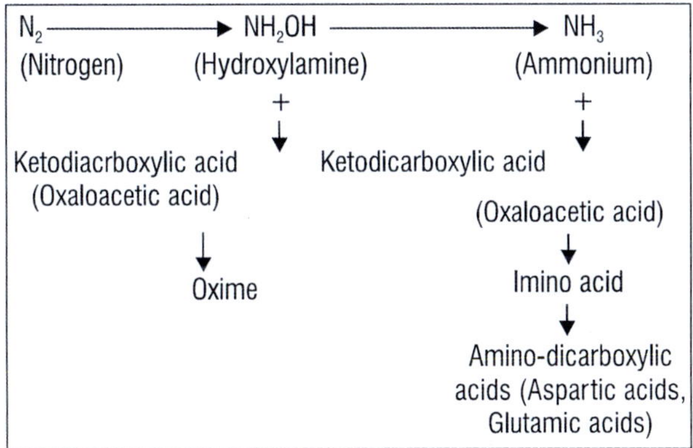

FIG. 10.4: Mechanism of nitrogen fixation (Virtanen, 1948).

A good number of theories have been proposed by various workers. Two of them are important, which are as follows:

Theory of Virtanen

Nitrogen fixation in root appears immediately after nodule formation. Virtanen (1948) proposed the following scheme of nitrogen fixation in root nodules:

According to him young plants fix more amount of nitrogen than the old plants. A great part of nitrogen is converted into L-aspartic acid and L-glutamic acid. Apart from these, α-alanine is also present in the nodule which is produced from L-aspartic acid by decarboxylation. In addition to this, small amount of Oxime-N and Nitrite-N are also present.

Theory of Burris and Wilson

According to this theory hydroxylamine is the central compound of nitrogen fixation, from which ammonia is formed through reduction or it can also directly react to the keto acid. The proposed scheme of Burris and Wilson (1945) is as follows:

Both these theories reveal the formation of amino dicarboxylic acid as the primary products in biological nitrogen fixation

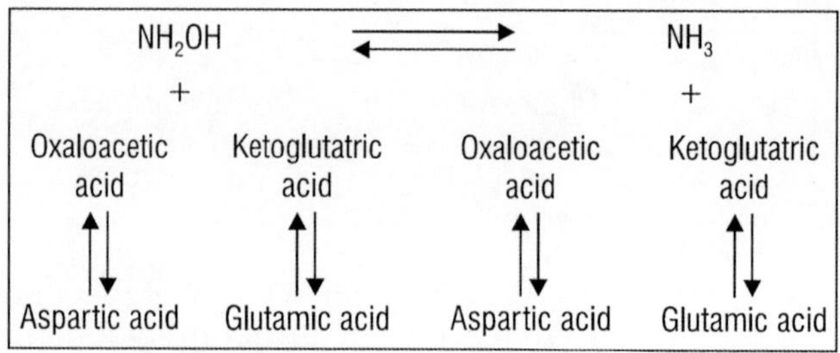

FIG. 10.5: Theory of Burris and Wilson

BENEFITS FROM NITROGEN FIXING PLANTS TO NON-NITROGEN FIXING PLANTS

The nitrogen fixation plant can directly fertilis the soil and indirectly neighbouring plants through above- and below-ground litter, through root exudates and leakage from leaves and roots, and through the excreta of grazing animals. When nitrogen fixing plants are grown in mixed plantations with non-nitrogen fixing plants, an increase in both growth and yield of the non-nitrogen fixing plants have often been found. Moreover, the legume with the non-legume intercrops, on average, yield more efficiently per unit land with higher area-□-time equivalence ratios than either intercrops with two legume species or two non-legume species.

The nitrogen fixing plants (donor) and non-nitrogen fixing plants (receiver) could provide a good system for investigating nitrogen transfer, which was often termed the process of **nitrogen decomposition** and uptake in a neighbouring plant, where neighbouring plants might have differing nitrogen status.

Nitrogen transfer could lead to increase productivity of intercrops and forests without increasing the use of nitrogenous fertilisers if properly managed and quantified. However, the roles of mycorrhizal hyphae in direct N transfer were somewhat inconclusive, and it is not yet clear if the transfer can be large enough to contribute significantly to the N status of the receiver in agricultural or natural ecosystems.

NITROGEN FIXING TREES AND FOREST MANAGEMENT

To maintain the sustained productivity of forest ecosystems, nitrogen input to the soil from both biological and non-biological sources is essential. Natural forest ecosystems contain climax vegetation in which the soil has been developed for a long period by the interaction of microorganisms and higher plants. Nitrogen fixation is one of the main processes in which nitrogen is brought into the soil.

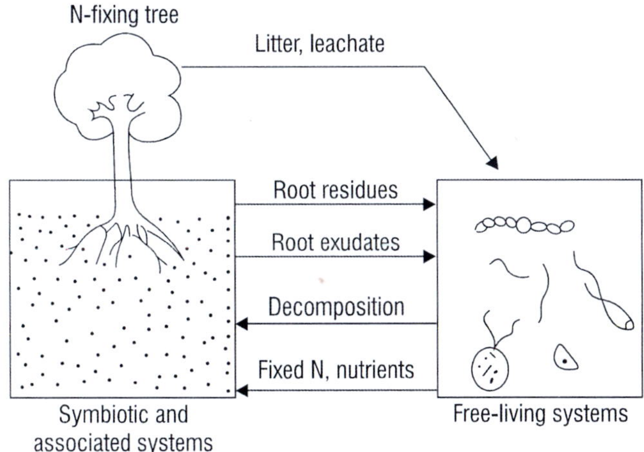

FIG. 10.6: Interaction of Nitrogen-fixing tree with symbolic or non-symbiotic soil organisms.

Future intensive forest cultivation, with consequent short rotations, will require replacement of nitrogen reserves at a greater rate than the present secondary succession leading to old-growth forests. With the increased use of trees genetically selected for increased yield, it will be necessary to maximize the potential for biological nitrogen fixation in future forest management, so as to maintain the long-term productivity of forest lands without significantly accelerating rates of nutrient depletion.

Nitrogen fixing microorganisms which grow symbiotically with C-autotrophic higher plants, can indirectly utilize much more energy than free-living N-fixing microorganisms, and contribute the majority of the N_2 input in forest ecosystems.

Woody nitrogen-fixing plants have a much greater influence on the forest ecosystem than herbaceous ones. A total of 997 nitrogen fixing woody species were included in a master list from a database maintained by the University of Hawaii Nif TAL Project; even this master list is not complete. About 44 species of 22 genera are recommended as economically important nitrogen fixing trees.

TABLE 10.2: Important nitrogen fixing trees, environment and uses

Species	Environment	Uses
Acacia auriculiformis	Humid	FL, OL, PP, TR
Acacia crassicarpa	Humid	TR
Acacia holosericea	Arid, Semi-arid	FL
Acacia leucophloea	Arid, Semi-arid	FL
Acacia mangium	Humid	FL, PP, TR
Acacia mearnsii	Highland tropics	FL, PP, ST, TN, TR
Acacia nilotica	Arid, Semi-arid	FD, FL, FR, GM, TN

Species	Environment	Uses
Acacia Senegal	Arid to Humid	FD, GM
Acacia tortillas	Arid to Humid	FL, FR
Albizia lebbeck	Arid to Humid	FL, FR, GR, OL, TR
Albizia procera	Humid	FL, OL, PL, TN, TR
Albizia saman	Humid	FL, FR, GR, OL, TR
Alnus nepalensis	Highland tropics	FL, PL, TR
Cajanus cajan	Arid to Humid	FD, FL, FR, GR
Calliandra calothyrsus	Humid	FL, FR, GR, HY
Casuarinas equisetifolia	Arid to Humid	FL, OL, PL, ST
Dalbergia sissoo	Arid, Semi-arid	FL, FR, SE, TR
Dalbergia latifolia	Humid	TR
Erythrina poeppigiana	Humid	GR, OL, SE, ST
Erythrina variegata	Humid	OL, SE, ST
Faidherbia albida	Arid, Semi-arid	FL, FR, GM, SE, TR
Flemingia macrophylla	Humid	DE, GM
Gliricidia sepium	Arid to Humid	FL, FR, GR, OL, SE
Hardwickia binata	Humid	FR, TR
Leucaena leucocephala	Arid to Humid	FL, FR, GR, PL, PP
Paraserianthes falcatoria	Humid	OL, PP, TR
Pithecellobium dulce	Arid to Humid	FL, FR, OL, PL, TR
Prosopis cineraria	Arid, Semi-arid	FL, FR, OL
Prosopis juliflora	Arid, Semi-arid	FL, FR
Prosopis pallida	Arid, Semi-arid	FL, FR, HY
Pterocarpus indicus	Humid	OL, TR
Robinia pseudoacacia	Temperate	FL, FR, HY, PE, SE
Sesbania grandiflora	Semi-arid, Humid	FD, FR, GR, OL, SE
Sesbania sesban	Arid to Humid	FR, GR
Tipuana tipu	Highland tropics	OL, TR

DE = Dye, FD = Food, FL = Fuelwood, FR = Fodder, GM = Gum, GR = Green manure, HY = Honey, OL = Ornamental, PL = Pole, PP = Pulp, TR = Timber, SE = Shade, ST = Shelterbelt/Windbreak, TN = Tannin.

TABLE 10.3: List of the leguminous and non- leguminous trees

Some leguminous trees that do not fix nitrogen	Some non-leguminous trees that fix nitrogen
Bauhinia species	*Alnus acuminata*
Cassia species	*Alnus nepalensis*
Delonix regia	*Casuarinas cunninghamiana*
Parkinsonia aculeate	*Casuarinas equisetifolia*

NITROGEN CONVERTERS IN THE SOIL

The oxidation of ammonia to nitrate in the soil is one of the important phases in nitrogen cycle. This is brought about by two groups of bacteria- *Nitrosomonas* and *Nitrobacter*. The required energy for the growth of these bacteria is obtained through the oxidation of ammonia or nitrate. Winogradsky (1891) first isolated these autotrophic bacteria and described the conversion of ammonia to nitrite, and nitrite to nitrate. This is called **nitrification**. The conversion of nitrate to nitrous oxide and nitrogen gas by various kinds of soil organisms is called **denitrification**. The process releases nitrogen gas to the atmosphere and completes the complex nitrogen cycle in nature.

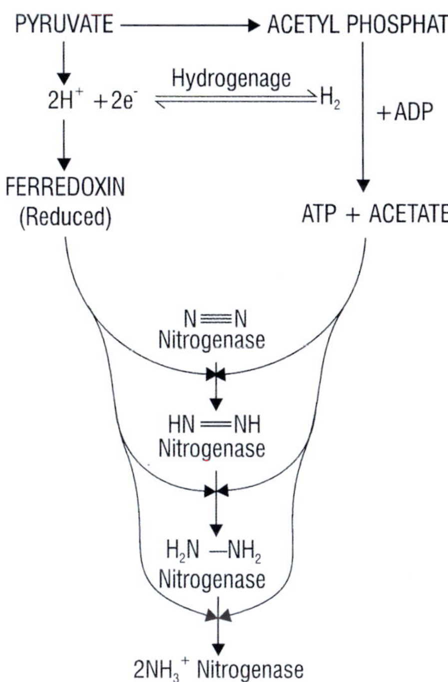

FIG. 10.7: Scheme proposed for mechanism of nitrogen fixation.

Cyanobacteria and Nitrogen Fixation

Frank (1889) first reported the ability of nitrogen fixation by blue green algae. *Nostoc punctiforme* and *Anabaena variabilis* have the ability to fix nitrogen in the soil. The principal genera of this group known to fix nitrogen are *Nostoc* spp. *(N. commune, N. punctiforme, N. muscorum); Anabaena* spp. (*A. ambigua, A. cylindrica, A. fertilissima. A. circularis*, etc.), *Cylindrospermum* spp. (*C. gorakhporense, C. licheniforme)*- all of them belong to the family Nostocaceae. A few of members of the family Microchaetaceae and Rivulariaceae like *Aulosira fertilissima* and *Calothrix brevissima* respectively are known to fix atmospheric nitrogen in rice fields. Significantly most of the free living nitrogen fixing members of this group produce **heterocyst** which is considered to be the site of nitrogen fixation.

The free living diazotrophic cyanobacteria are the largest contributors of the process of biological nitrogen fixation, the second most important biological process on this planet. Their nitrogen fixing ability was first related to the presence of specialized non-oxygen evolving cells called heterocysts in which nitrogenase; a highly oxygen-sensitive enzyme is produced from oxygen.

Nitrogen Fixing Fungi

The fungi also independently play an important role in fixing free nitrogen in soil. The fungi that are responsible for this function are of species of *Mactosporium commune, Cladosporium herbarum, Phoma, Alternaria tenuis, Rhodotorula* spp. etc. The occurrence of these fungi has been noted in forest-land.

BIOCHEMICAL ASPECTS OF DIAZOTROPHY

An American company (DuPont) in 1960 started the fundamental research on the biochemistry of nitrogen fixation, specially on nitrogenase activity in cell free extracts of *Clostridium pasteurianum.* The experiments were conducted in absence of air because of the very great oxygen sensitivity of nitrogenase. The techniques like chromatography, electrophoresis and fractionation under anaerobic conditions, helped the researchers to isolate nitrogenase from *Clostridium pasteurianum.* Thereafter, nitrogenase was easily isolated from other anaerobic nitrogen fixers while this extraction was more difficult with *Azotobacter vinelandii.* The so isolated enzyme has the property to accept electron from sodium dithionite and was oxygen-tolerant and evolved hydrogen in the absence of nitrogen. In 1981, nitrogenase in the form of crude preparations was isolated from about 30 microbes and out of which about 6 of them were purified.

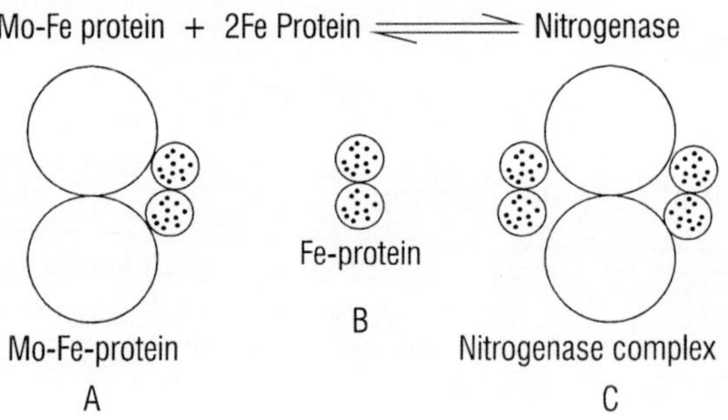

FIG. 10.8: Nitrogenase enzyme.

Biochemically, nitrogenase is a binary enzyme and consists of two brown metalloproteins, of which the joint activity is essential to reduce nitrogen to ammonia.

During nitrogen conversion, nitrogenase binds to molybdo-ferro protein (MoFe-protein) which was named **dinitrogenase**. When nitrogen binds at the molybdenum level in a separable fragment of protein, called **FeMo co**.

The two component proteins of the enzyme nitrogenase are quickly and irreversibly destroyed by exposure to oxygen.

Nitrogenase is also able to reduce other substrates in the place of nitrogen, including hydrogen ions and, in this case it evolves hydrogen.

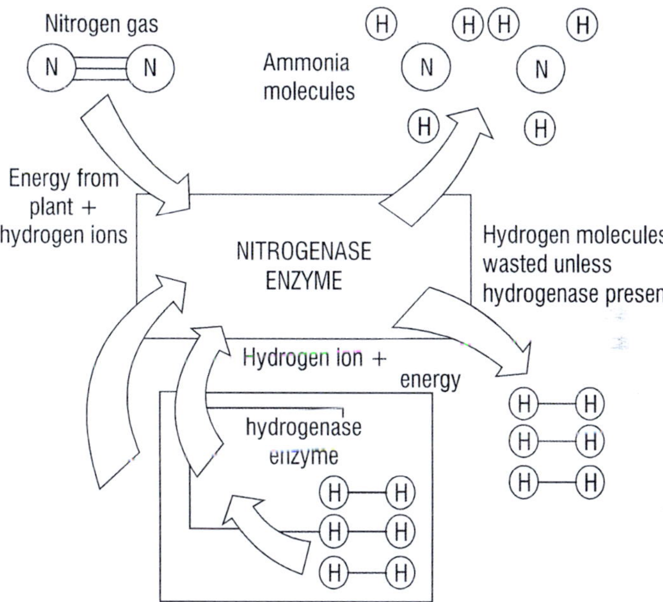

FIG. 10.9: Representation of the mechanism of nitrogen fixation and the action of hydrogenase.

Nitrogenase Producing *E. coli* Cells

Because of the close relationship between *E. coli* and *Klebsiella pneumoniae*, *E. coli* is specially qualified for the investigation of the nitrogenase system. With *E. coli* C-M74 cells, high nitrogenase activity can be attained in culture medium, supplemented with 50mg/l ammonium sulfate, 1% aspartic acid, and 0.1% yeast extract.

During reduction of nitrogen to ammonia, electrons are required which are transferred to nitrogenase by a specific transport proteins, the *Ferredoxins* or the *Flavodoxins*. ATP serves as energy source. The consumption of ATP molecules is 12 to 15 per nitrogen molecule. The reason for this high consumption of ATP has not been entirely elucidated.

As evident from the reactions that hydrogen is produced during nitrogen fixation. Certain nitrogen-fixing aerobic bacteria possess hydrogenase enzyme, which has the ability to recycle hydrogen produced by nitrogenase in vitro. Therefore, the re-utilization of this hydrogen generates more ATP (by oxidative phosphorylation) and consequently improves the efficiency of nitrogen fixation.

Rhizobium is also associated with soybean plants which contain hydrogen uptake (or *hup*) genes, which have the ability to recycle hydrogen and back into the nitrogenase system and fix nitrogen. Through this mechanism the energy is harvested by plants which is otherwise lost.

The nitrogenase enzyme of bacteria fix nitrogen by converting nitrogen gas into ammonia using hydrogen ions and energy by host plant. During the process there is a production of energy rich hydrogen gas. Some bacteria possess hydrogenase enzyme, which can capture hydrogen gas and converts it into hydrogen ions and release energy which can feed back into nitrogenase enzyme.

GENETICS OF FREE-LIVING AND SYMBIOTIC DIAZOTROPHS

Organization of Nitrogen-Fixation Genes

In molecular terms, very little is known of the means by which the prokaryote specifically recognizes and invades its respective host-plant. However, there has been a growing interest in approaching this problem by identifying and analyzing the relevant symbiotic genes and their gene products. Because of the agricultural importance of the *Rhizobium*-legume interaction, the most detailed studies have been conducted on this system. Also, because of bacteria are simpler than plants. Our knowledge of the symbiotic genes of *Rhizobium* is better than that of the plant genes. There have been several recent advances in the analysis of the *Rhizobium* genes that are involved in nodulation, host-range specificity and nitrogen fixation. By dissecting the structure, organization and regulation of these genes, a better understanding of their functions may be forthcoming.

In the fast-growing *Rhizobium* species, which nodulate temperate legumes such as clover, alfalfa, peas and field beans, the genes for nodulation and nitrogen fixation are clustered on large symbiotic plasmids. One such plasmid, pRL1JI, was isolated from a strain of *R. leguminosarum* (which nodulates peas) and in this case the genes that determined the ability to nodulate peas were located between two groups of genes required for nitrogen fixation. The study of the nitrogen-fixation (*nif*) genes in *Rhizobium* has been greatly facilitated by the fact that at least some of them are very similar to the corresponding *nif* genes in the free-living nitrogen-fixing bacterium - *Klebsiella pneumoniae.*

In *Klebsiella*, 17 *nif* genes have been identified which are variously involved in *nif* gene regulation, the synthesis of the iron-molybdenum cofactor (FeMoCo). Using DNA sequence comparisons, *Rhizobium* species have been shown to contain the *nifHD* and *K* genes (nitrogenase proteins) *nifA* (regulation) and *nifB* (synthesis of FeMoCo). Further studies may well demonstrate similarities between other *Rhizobium* and *Klebsiella nif* genes.

The study of nitrogen fixation genes was first carried out in *Klebsiella pneumoniae* (strain M5al); and spectacular progress achieved in 1970s. Although the first report came from the mutants of nitrogen fixing bacteria which was isolated and biochemically characterized as early in 1959. In 1971 the nitrogen fixation genes, called *nif*, successfully transferred to *Klebsiella pneumoniae.* Later on they could able to transfer these genes to

Escherichia coli. This work encouraged the researchers of this field specially for *Rhizobium* and the presence of *nif* genes in this microorganism. More than a dozen genes, termed the *nif* genes are involved in the assembly of nitrogen-fixation apparatus. The *nif* genes are not found in a free state but are clustered in one region (they are not scattered throughout the bacterial DNA). Therefore, it is much easier to stretch of DNA in a *Rhizobium* chromosome and insert the whole batch into another organism. This has also been done in *Agrobacterium tumefaciens* (Ti plasmids). In 1981, the number of *nif* genes was 17 and they were organized in eight transcription units (or operons), one of which was monocistronic. Later (1980) two other genes: *nif*-R between L and F, and *nif*-W included between F and M.

nod-Genes for Nodulation

Concerning the genes (called *nod* genes) for nodulation and host-range specificity, an important observation is that, despite the morphological complexity of the infection process, only a few genes on the symbiotic plasmid are required for nodulation. In the *R. leguminosarum* symbiotic plasmid pRL1JI, less than 10kb appears to be required for nodulation and for the determination of host-range specificity for peas. The reason has been analysed by DNA sequencing (and thus identifying genes as long open reading frames), isolation and characterization of mutants and by studying the regulation of the *nod* genes.

FIG. 10.10: Map of the *nif* gene cluster of the *Klebsiella pneumoniae* chromosome. The capital letters denote the genes (cistrons), the *nif* cluster being located between the genes for biosynthesis of histidine (his) and those for shikimic acid uptake (shi A). The arrow indicate the separate operations and the direction of their transcription. In 1981 the status of F and M as one or two distinct operons was not clear. The size of most of the genes, however, were known.

The organization of *nif*-genes of other free-living diazotrophs from which *nif* mutants were isolated (*Azotobacter vinelandii, Azospirillum lipoferum, Clostridium pasteurianum, Rhodopseudomonas capsulata* and a few cyanobacteria) was as like to that of *Klebsiella pneumoniae*.

NITROGEN FIXATION RESEARCH IN INDIA

The word on biochemistry of nitrogen fixation has mainly been done at Haryana Agriculture University, Indian Agriculture Research Institute, Punjab Agriculture University and BARC. At HAU, a pathway for ureides biogenesis in nodules of pigeonpea has been proposed. The further metabolism of ureides in aerial parts particularly leaves and developing pods has also been worked out. Detailed biochemical and physiological studies explaining nodule senescence have also been carried out at IARI. Attempts have been made to understand the regulation of uptake hydrogenase so as to reduce the energy cost of biological nitrogen fixation. In addition, studies are concentrated on increasing nitrogen fixing efficiency by insertion of uptake hydrogenase genes or by increasing nitrogenase activity in nodules. At BARC, early interactions in legume- *Rhizobium* symbiosis have been investigated, according to which, host-induced alterations of capsular polysaccharides and de novo synthesis of specific proteins by both the partners in response to each other, appear to be among the first steps in the series of signal response interactions. At PAU, the role of extracellular invertase of *Rhizobium* in free sugar metabolism in developing root nodules of legumes has been ascertained

The activity of nodule is also controlled by nitrogenase enzyme. This enzyme is about 75% efficient. This is because, when soil aeration is poor and nitrogen availability is limited, electrons instead of being used to reduce nitrogen, now combine with protons forming hydrogen. As a result, nodules evolve hydrogen instead of fixing nitrogen. This is a wasteful process. Sufficient progress is being made to identify strains of *Rhizobia* which evolve less hydrogen and fix more nitrogen.

The most ideal situation would be the one in which the higher plant is able to fix nitrogen without the help of the bacteria. Towards this end, attempts have been made to transfer nitrogen fixing genes (*nif* -genes) to higher plants from *Rhizobium*. However, it appears to be a long way before it is achieved.

In India, intensive forest cultivation of nitrogen fixing trees, with consequent short rotation is needed. This will require replacement of nitrogen reserves at a greater rate than the present secondary succession leading to old-growth forests. With the increased use of trees, genetically selected, it will be necessary to maximize the potential for biological nitrogen fixation in future forest management, so as to maintain the long term productivity of forest lands without significantly accelerating the rate of nutrient depletion. India is in urgent need of additional professional training and better transfer of technology in this context. Further, improved strains and new species of nitrogen-fixing microorganisms should be investigated/developed in order to reduce the use of chemical fertilizers in the fields.

CHAPER 11

Ecological Pest Management

Organic and Natural farming, which aim at 'cooperating rather than' confronting with the nature has been hailed as the only answer to bring sustainability to agriculture. In organic farming systems pest control strategies are largely preventative, rather than reactive. Organic farming is not a single method but rather a variety of techniques which are aimed at reducing costs, preserving the environment and protecting human health by eliminating the use of toxic farm chemicals. Masonobu Fukuoka of Japan, considered as the father of Organic farming, has been consistently recording yields of many crops which are comparable to the best yields obtained in Japan, by not using HYVs, chemical fertilizers or pesticides. Use of pesticides and chemical fertilizers does not necessarily lead to better farming than using natural agricultural methods

In the conventional agriculture pests are often controlled with man made chemicals which have following harmful effects:

1. Chemicals kill useful insects which eat pests
2. Chemicals can be very bad for the health of people who use them and people who eat food with small amounts of these chemicals still attached.
3. These chemical can stay in the environment and in the bodies of animals causing problems for many years (discussed in detail in chapter 1).
4. They are very simple chemicals and insect pests very quickly, over a few breeding cycles, become unaffected by them.
5. Many farmers in developing countries cannot afford to buy man made chemicals.

Pests-insects, disease causing pathogens, weeds and nematodes are estimated to cause an annual loss of about 33-35 per cent of the potential food production world-wide. In India, according to very conservative estimate, the annual crop losses due to these pests amount to a staggering Rs 25,000 crores. In order to increase agricultural production, many changes have been adopted in the recent past such as use of dwarf high yielding cultivars, higher fertilizer use, increased irrigation sources, intensive cropping, large areas under single crop, extension of new crops in non-conventional areas, large scale movement of seeds etc. All these changes have often led to serious pest problems. The new high yielding cultivars in use today are often more susceptible to pests than the old varieties were. Abundant irrigation and nutrient supply has enabled crops to be grown throughout the year-providing ideal conditions for the pests to multiply. Abandoning many time to planting etc. has further aggravated the problem. With the result, not only the infestation of known major pests has increased, but also hitherto unimportant or unknown pests have emerged as major pests.

The current pest control technology relies heavily on pesticides. The pesticide consumption world-wide which was 148,000 metric tonnes in 1966 has increased to over 302,000 metric tonnes in 1976, the most significant increase being for herbicides which increased from 51,000 to 181,000 metric tonnes. In India, it was nearly 200 metric tonnes in 1955 and about 75,000 tonnes in 1985 and may reach 100,000 tonnes in 1990 and 200,000 tonnes in 2000 AD. Pesticide residues in agricultural commodities are the issue of major concern besides their harmful effects on wild life, and other desirable flora and fauna. Equally worrying thing is about development of resistance in pests to pesticides. There are innumerable instances of insect pests, fungi and weed species showing resistance to pesticide for which they were previously susceptible. Outbreak of American bollworm *Heliothis armigera* in cotton crop in Andhra Pradesh in areas where synthetic pyrethroids were liberally used is indicative of things to come in the future. Such developments have negative impact as it makes room for using higher rate of encourages people to look for chemicals which are more toxic than the ones used previously.

Due to the dangers of using chemical pesticides, there has been a considerable resurgence of interest amongst the scientists to look for alternative methods of pest control. Slowly but steadily there has been growing appreciation about the role of pest resistant cultivars, biological control agents and cultural methods in pest control. These non-chemical or 'bio-environmental' methods are the major components in the integrated management of pests.

The non-chemical methods of pest management certainly form the major components of Organic farming. Management for integration of non-chemical methods of controlling insect pests, disease, weeds and nematodes, is essential for successful organic farming.

CULTURAL CONTROL

Cultural control is the deliberate manipulation of the environment to make it less favourable for pests by disrupting their reproductive cycles, eliminating their food or making it more favourable for their natural enemies. The following are just a few examples of the diverse ways in which the method can be used.

Sanitation

Sanitation includes the removal or destruction of breeding refuges and over-wintering sites of pests. Removing diseased plants (rouging) or pruning infested parts is important in disease management. To prevent introduction of inoculum, all means of carrying the pathogen should be looked into e.g., seeds or propagating material, water, plant debris, farm-yard manure, implements etc. Collection and destruction of egg masses could be very useful practice in insect management in orchards. Care need to be taken to use well composted manure, as high composting temperature, is expected to damage disease inoculum, weed seeds etc.

The government laws or quarantine play a great role in preventing the introduction of pests by propagating material. The seed certification programmes adopted by the various countries have proved immensely useful in containing the spread of pests.

While it is difficult of check the spread of diseases through irrigation water, systematic weed control along irrigation canals will help in stopping the spread of weed seeds. Similarly, machinery used in an infested field, unless cleaned properly, will aid in spread of disease pathogens and weed seeds. Disinfestation of smaller tools and movement of farm vehicles from less diseased to severely infected fields are general methods of preventing spread of diseases by man and machinery.

Pest population may often be effectively suppressed by destruction of their alternate plant hosts, many may in fact be weeds.

The parasitic weed *Striga* will continue to thrive on sorghum stubbles even after harvest. Unless pulled out, this may be a potential source of infestation in the coming years. Destruction of infested young shoots of sugarcane and the stubbles after harvest helps a great deal in keeping the incidence of sugarcane borer at a very low level.

Tillage

In many developing and under-developed countries, tillage still is the only known method of controlling weeds. The practice however is rapidly being replaced in much of the developed countries with herbicides.

Frequent shallow ploughings given before planting are very effective in controlling annual weeds. This techniques of sowing crop seed in the relatively weed-free soil is termed as 'state seed bed' techniques which is being very widely used in many crops. Being less expensive, this can form an important component of integrated weed management system in many places. However, it may bring back the buried weed seeds on to the surface. Alternatively, a very low dose of a contact herbicide may take care of weeds without disturbing the soil further. Deep ploughing particularly in summer months is very effective against deep rooted perennial weeds as it would expose the underground vegetative parts to hot weather.

With soil tillage, other pests are also killed mainly due to exposure or debris destruction. For example, the wheat sawfly in North Dakota has been reduced by as much as 75 per cent by cultivation. Summer tillage of wheat destroys wheat streak virus reservoirs and viruses' vector the wheat curl mite. Tillage, however, need to be timed to coincide with the susceptible stages of the insects.

Depth of ploughing is also an important factor as burial to atleast 8 cm is necessary to control *Cephalosporium* stripe of wheat, to atleast 15 cm for the bent grass nematode and to atleast 25-30 cm for sorghum downy-mildew. Although the long term effects of reduced tillage have not been studied adequately, some adverse effects are known For example, army worms and corn stalk borers often increase in crops grown under reduced tillage and surface accumulation of crop residues under reduced tillage conditions increases the infestations of termites, slugs, some nematodes and diseases.

Application of Manures and Soil Amendments

Application of soil amendments would change the rhizosphere environment by affecting porosity, aeration, temperature, water holding capacity etc. Healthy crop plants have a greater chance of fighting with pests than undernourished plants. Besides, soil amendments as such or their decomposition products may have a direct effect. More than thirty disease mainly root rots have been reported to be significantly controlled by application of organic amendments

Incorporation of various kinds of organic amendments to soil also holds promise in long-term control of nematodes. Use of chopped shoots, straw, saw dust, animal manure, cakes etc. have resulted in significant reduction in nematodes population. In India, oil cakes of linseed, mustard, peanut, castor bean, *Mahuva, (Madhuca indica)* and neem (*Azadirachta indica*) have proved very effective for nematode control in a variety of crops.

All plant feeding insects have specific nutrient requirements. The incidence of corn leaf hopper (*Dalbulus maidis*) which transmit corn stunt pathogen was reduced at low N levels and at high planting densities. Poor nutrition to crops is also known to affect the reproductive ability of some insects feeding on them.

Increased application of plant nutrients, although, increase the growth of both weeds and crop plants, it benefits crop more. However, increased application of nitrogen is known to reduce *Striga* incidence in cereals. However, other methods of weed control must accompany increased supply of nutrients as weeds are known to deplete large quantities of nutrients from soil, thereby depriving the crop plants of the precious and costly input.

Habitat Diversification

Cropping patterns have undergone drastic changes over the years. Provision of irrigation, development of short duration, photo-insensitive, high yielding varieties of crops have resulted in growing of more number of crops in a year. Consideration of potential pest problems while choosing crops/crop sequences would go a long way in pest management. Following are some of the examples.

Crop Rotation

Good crop rotation with non-host plants would result in reduction of pest population either directly by toxic action of the non-host crops or indirectly by depriving them of the food the host plant would have offered had it been grown continuously. Several soil borne diseases of cereal crops are successfully controlled by a crop rotation period of 2-3 years. Soil pathogens that can be controlled by a 3 to 4 year rotation with non-host crops include organisms causing cabbage black rot, bacterial leaf blight of bean and cabbage black leg. Garlic, onion, beet root and Lucerne are suggested as good rotational crops in controlling *Fusarium* wilt of melons and cucumber. Crop rotation is one of the most important measures for controlling plant parasitic nematodes and frequently it appears to be the only economical method for controlling this pest. Effective rotation is widely practised for control of the golden nematode of potatoes, the soybean cyst nematode and the root rust nematode.

Each crop has its own characteristic weeds and they thrive well when the same crop is grown successively. Monoculture of the cereals and use of hormone herbicides have led to serious problems of grass weeds in many parts of the world. In a 3-year study, high populations of *Xanthium pensylvanicum* were associated with soybean than with corn regardless of weed control measures. *Scirpus maritimus* a difficult-to-control weed in a continuously flooded low land field was effectively controlled when the rice was rotated with an upland crop.

Rotation is most effective against soil pests (e.g. white grubs, wire worms etc.), which take several years to reach maturity. Until synthetic organic insecticides became available, rotating corn with such crops as oats, clover and soybean was a standard procedure for controlling corn rootworms in mid western United States. In white ant infested areas crops like sugarcane, wheat and chillies should be avoided; instead tobacco or onions maybe grown.

Chief obstacles to success of crop rotations are the pronounced longevity of inoculum/weed seed in soil and frequently wide host ranges of pests. Often populations of a pest other than the target pest may increase on the alternative crop. Some crops used in rotation may often be of very low economic values.

Trap Cropping

The practice of attracting pests to small plantings of crops, which are then destroyed or sprayed with a toxicant has been quite successful against some plant nematodes, parasitic weeds and insect pests.

A susceptible crop such as sorghum or sudan grass may be grown before the main crop season to induce germination of *Striga* seeds before it is destroyed. Crops such as groundnut, linseed, cowpea, cotton, sunflower etc. form good rotational crops as they induce germination but are not parasitized by *Striga*.

Some nematodes may also be controlled by trap crops. Highly susceptible crops are allowed to grow in infested fields until the second stage larvae enter the roots and begin to develop. The crop will be destroyed before nematodes mature.

In all these, the plant destruction must properly be timed and implemented, lest the population of pest may increase many-fold.

Intercropping

Inter row space is a potential place for weeds which can be put to better use by intercropping. Intercropping in broad spaced crops such as pigeonpea, sugarcane, cotton, maize, sorghum with fast canopy forming plants, reduce the weed emergence and competition substantially

Lower incidence of insect pests were found on legumes intercropped with maize. Intercrops of clover, spinach, beans and tomato reduced the incidences of *Brevicoryne brassicae* and *Plutella xylostella* in cabbage substantially. Incidence of *Heliothis armigera* reduced in chickpea when grown in association with barley, mustard or whet.

Strip Farming

Intervening strip of non-suitable crop prevent movement of insect pests from one strip of a crop to another. Also adjacent strips share unspecialized natural enemies which would move when insect pests build up on the neighbouring strips. The abandonment of strip farming in Peru some years back has been given as the reason for boll worm outbreak on cotton there and certainly re-diversifying the cotton agro-ecosystem has now greatly reduced the incidence of the pest.

Time of Planting

Pest outbreaks occur at particular soil and climatic conditions and planting can be so adjusted that such outbreaks do not coincide with the susceptible stage of the crop.

With diseases, it is reported that the incidences of chickpea wilt and root rot of pea were considerably reduced when the planting was delayed. There are also reports of early sown wheat escaping rust damage and pearl millet sown before 15th July recording less incidence of ergot in North India.

Maize sown late suffers little from damage by maize borer as by then the egg parasite *Trichogramma* is able to keep down the population of the pest. Maize grown in winter is free from borer attack as this pest hibernates during winter. Rice is reported to suffer less from borer attack if transplanted early (before middle of June). Early Maturing cotton varieties (*Bikaneri norma*, H-777 and F-414) have become popular in Punjab and Haryana as they escape pink bollworm attack.

Late sown crop in winter is expected to have less problems with nematodes as they are inactive during this period.

Late planting of wheat reduces infestations of important weeds such as *Chenopodium album*, wild oats (*Avena spp*) and *Phalaris minor.* However, the limitation is, altered planting time may not always be the best one for obtained higher yields.

Water Management

The scab of potato is suppressed by irrigating potato at the time of tuber formation. Wet weather diseases such as halo-blight and anthracnose of beans, early blight and charcoal rot of potatoes can be checked by furrow rather than sprinkler irrigation. Diseases which are favoured by high moisture at the soil surface like damping off or collar rot can be minimized by planting crops in ridges or raised beds. Irrigation often makes it possible to avoid a pest by allowing the crop to be grown out-of-season.

It is a well-known fact that transplanted rice faces very less weed problem compared to other methods which is mainly due to pudding and impounding of water. Flooding upto 10 to 20 cm early in the season reduced the infestation of many weeds including *Echinochloa crus-galli* but may increase problems with other weeds such as blue green algae. Under irrigated conditions, planting crop-seed in the moisture zone in an otherwise dry seed bed and delaying the first irrigation reduces the weed infestations quite substantially.

Drip irrigation, by virtue of providing water at the base of the plant, results in far lesser weed problem than would be expected from other methods of irrigation. Similarly, overhead sprinkler irrigation in potato effectively controlled potato moth.

Crop Competition

Weeds and crop seeds germinate almost simultaneously and they start competing with each other for water, nutrients and space. Selection of vigorous growing crop cultivar, closer row spacing, higher crop seed rate, proper time and method of fertilizer application etc. are some of the husbandry practices which are likely to play a vital role in weed management. There is much meaning in the saying that *good crop is the best weed killer*.

Among winter cereals, barley is reported to be more competitive than wheat or oat. Maize is having the highest weed suppressing ability (92%) followed by pearl millet (88%), sorghum (81%) and *Setaria* (73%). Amongst non-cereals, cowpea (88%) is more competitive than groundnut (62%), castor (51%) and pigeonpea (54%). Similarly, *Cyperus rotundus* infests lower yield loss in maize and sorghum as compared to groundnut, soybean, and grain legumes.

The high yielding dwarf cultivars of wheat and rice are more vulnerable to weeds and more time is spent in weed removal than in traditional cultivars.

Higher plant population of the crop can put pressure on availability of space for weed growth from relatively early stages of crop growth. Narrow row spacing, higher seed rate and cross sowing all have suppressing effect on weeds.

Close planting tend to produce microenvironment ideal for some fungal diseases, while with others, the same environment may prevent diseases by discouraging an insect vector from feeding. Dense seeding of many crops reduces the losses in yield caused by pathogens and insect pests by providing more plants per hectare.

PHYSICAL AND MECHANICAL CONTROL

Manual Control

Despite major advances in chemical control, hand removal of weeds still remains to be the most practical method of weed control in many developing countries of the world for a variety of reasons. Although back-breaking and laborious hand weeding is quite effective if employed at the right time.

Such manual destruction of other pests is not practicable although, it could be followed to some extent in controlling of some insect pests in orchards, like collecting and destructing the egg masses, mechanical exclusion of insects by placing impenetrable barriers around the host plant etc.

Burning

Flaming with propane burners of infested plants or weeds was practiced widely in western countries before the introduction of chemical pesticides. Burning rice straw has been reported to result in reduced incidence of stem rot of rice caused by *Sclerotium oryzae*. Burning crop residues under direct seeding and reduced tillage conditions has shown to reduce infestation of several pests including weeds.

Solarisation

It involves covering soil surface with polyethylene sheets to increase the soil temperature which would be lethal to soil borne pathogens, insect-pests, nematodes, weed seeds etc. It was first practised successfully in Israel for the control of *Verticillium dahliae* and *Fusarium oxysporum*. A treatment period of 30-40 days appears to be adequate to control many diseases.

A mulching period of 2-6 weeks with clear polythene sheets has been reported to give effective control of many annual weeds including *Phalaris* spp. *Orobanche, Sinapis alba* etc. Irrigation prior to solarisation has a complementary effect as moisture-imbibed seeds are more sensitive to heat than dry seeds. The limitation however, is that normally weed seeds upto about 5 cm depth are only affected implying that deep preparatory cultivation would nullify the effect. Although the technique is limited by the cost of treatment, it may be made use of in controlling weeds in nursery areas, in high value crops etc.

With nematodes, effective control of *Pratylenchus thornei* in potatoes and of *Ditylenchus dipsaci* in garlic was obtained in Israel with solarisation for 4 and 8 weeks respectively.

Flooding

The classical example of use of flooding is for the control of banana wilt. Chlamydospores of *Fusarium* are killed by restricted supply of oxygen in the water stagnated fields. Other fungi showing significant control by flooding are *Verticillium dahliae* in cotton, *Sclerotinia sclerotiorum* in cauliflower and fungus causing black tobacco and bacterial blight of cotton Flooding also effectively controls many arable weeds and that is the reason for reduced weed infestation in transplanted rice as compared to direct seeded rice, as has been discussed earlier.

Many plant parasitic nematodes are reported to be susceptible to flooding. However, a flooding period of 4-8 months is required for effective control of root knot nematodes, which is naturally a limiting factor.

BIOLOGICAL CONTROL

Although still grossly underused, biological control is gaining world recognition as a primary and often essential component of successful integrated pest management.

Classical biological control involves deliberate introduction and establishment of natural enemies in areas where they did not previously occur. The approach is used largely against pests of foreign origin.

In addition to deliberate introduction of biocontrol agents, proper attention needs to be given for conservation and augmentation of natural enemies that already exist in an area. This need to be treated as an important element in pest management for many insect parasites are much more susceptible to insecticides than the pests they attack.

Biological control agents for insect pests are available in nature abundantly. Several pathogens including viruses such as nuclear polyhedrosis viruses (NPV) and granulosis viruses (GV), bacteria like *Bacillus thuringensis,* fungi like *Metarhizium, Beauveria* and *Verticillium* and protozoa like *Schizogregarine* cause diseases in insects and destroy them. Similarly, several insect-parasitoids (parasites thriving on insects) are also known in nature. *Trichogramma* is an egg parasitoid of several pests, particularly of sugarcane internode borer. *Goniozus, Elasmus, Eriborus, Bracon, Trichospilus, Tetrastichus* and *Chelonus* are some of the other parasitoids attacking insect pests. Insect predators such as *Chilocorus, Pharoscymnus, Cryptolaemus, Scymnus* and *Monochilus* feed on mealy bugs, coccids, scales and mites on citrus, grapevine and guava. The mirid predator *Cyrtorhinus* attack brown plant hopper of rice. The green lace wing *Chrysopa,* another most effective insect predator feeds on aphids, mealy bug and young caterpillars.

In biological control of weeds, there is always a fear of biological control agents attacking the crop plants and also a plant considered a weed in one place may not be so in another area. Because of this specificity, the attempts have been primarily against controlling weeds in range lands, forest areas, aquatic environment and parasitic weeds.

Some promising examples include: control of *Eupatorium odorata* by leaf eating caterpillar *Pareuchaetes pseudoinsulata;* lantana by *Teleonemia scrupulosa; Salvinia molesta* (aquatic weed) by *Paullinia acuminata* (grass hopper); water hyacinth by *Neochetina eichhorniae* and *N. bruchi; Parthenium* by Zogogramula bicolorata *and Smicronyx lutulentus.* Good control of a variety of aquatic weeds by Chinese grass carp (*Ctenopharyngodon idella*) has received world-wide attention. In Japan, there is limited use of tadpole shrimp (*Trilopus* spp) to control weeds in flooded rice fields. At the IRRI, the cover effect of *Azolla* spp. is being evaluated as a means of weed control in rice.

In aquatic weed control, biological control holds great potential, as use of chemicals need to be limited because of multifaceted uses of water, and mechanical or manual methods are cost prohibitive, and requires frequent treatments.

Increasing the Effectiveness of Bio-Control

***Artificial structures*:** Suitable nesting boxes are kept for predatory birds and wasps (*Polistes* spp.) and burlap traps are placed around citrus trees for attracting coccinellid predatory (*Cryptolaemus montrouzieri*) of mealybug.

***Food and shelter*:** Pollen-and nectar-bearing flowering plants are planted on the bunds to provide supplementary food to predatory like chrysopids, coccinellids, predatory mites, syrphid flies and parasitoids.

***Controlling ants*:** Ants interfere with the activities of predators and parasitoids. Control and physical exclusion of ants by putting a band of some botanical around the trunk in citrus orchards increases the efficiency of coccinellid predatory (*Cryptolaemus montrouzieri*) and several parasitoids.

***Cultural practices*:** Avoiding of trash burning in sugarcane protects the parasitoid *Tetrastichus pyrillae* of sugarcane pyrilla. Strip-cutting in alfalfa helps in shifting of the predators to nearby cotton corp. Regular removal of fallen fruits of apple and their destruction reduces the overall population of codling moth. The egg parasitoids released in such orchards are much more effective.

There are a number of examples wherein cultural lowering of the pest population gives more opportunity for the natural enemies to be effective. Thus, if the water shoots in citrus are removed, the populations of leaf-miner, scale-insects, citrus psylla and mealy-bug are reduced.

***Trap crops*:** This is a strategy by which biological control could be made more effective and economic. The target host is attracted to the border rows of a trap crop which is much more preferred than the main crop. The attracted pest on trap crop is more amenable to be economically tackled through biological control in a restricted area. Planting of castor around tobacco nurseries and marigold around main crop for attracting *Spodoptera* and *Helicoverpa* (*Heliothis*) respectively, and releasing egg parasitoids or spraying nuclear polyhedrosis virus could give effective suppression.

***Conditions for effectiveness*:** The time and period of release should be suitable for the natural enemy. It should synchronize with the availability of suitable stage of the pest the release of Trichogrammatid egg parasitoid would be more effective if the release is based on information collected through pheromone traps. Using suitable strain (through artificial selection) adopted to a particular crop ecosystem and pest-management practices increases the success of biological control.

The searching efficacy of parasitoids is increased by applying certain chemicals called fairomones, which are originally isolated from their host species (an example is Trichogrammatid egg parasitoids). Phago-stimulants are applied to the plants to increase the ingestion of leaves sprayed with microbial agents.

For discouraging the activity of hyperparasites, parasitoids should be released at a time when the activity of hyperparasitoids is the least.

***Legislation*:** The Government of India banned the export of frogs, because 90% of the food of frogs is agricultural pests. It has been estimated that catching frogs for export led to the survival of 200,000 tonnes of pests, thereby requiring farmers to spend more on pest control. Legislation is required to ban all such actions which directly or indirectly kill beneficial insects.

Conservation of Biodiversity

Biodiversity should be documented and properly inventoried, so that it can be utilized better. Otherwise, several species facing the threat of extinction will be wiped out before they are studied and named. It will also helps as plan conservation measures.

Conservation of Natural Enemies

Several ways of conservation of natural enemies have been attempted in the past. The use of relatively safer pesticides, and their selective use could conserve the biotic agents. Several other methods of enrichment and conservation of natural enemies include providing artificial structures (nesting boxes for wasps and predatory birds, burlap traps for C. montrouzieri), planting or retaining food and shelter plants of biotic agents on the bunds (retaining pollen and nectar bearing flowering plants Euphorbia, wild clover, etc. to provide supplementary food to parasitoids and predatory; rice gall midges and plant hoppers and their natural enemies survive on weeds like, *Cyperus* during off season and later shift to rice crop providing early suppression of the pests), placing bundles of rice straw for attracting spiders, retention of crop stubble, grass weed heaps for maintaining predator population, collecting egg masses of borers and putting them in a bamboo cage cum percher (which permits escape of egg parasitoids and allow the predatory birds to perch, and trap and kill the hatching larvae) are effective. In addition to nest boxes, nesting material, perching sites and water pans, retaining bushes (*Acalypha, Hibiscus*, crotons, etc.) helps retention or predatory birds in the desired area. Controlling ants (which interfere with the activity of biotic agents), avoiding complete trash burning (to conserve sugarcane pyrilla parasitoids), regular removal of fallen fruits (to increase the effectiveness of egg parasitoids due to overall lowering of population of the pests), strip cutting of alfalfa (to increase biotic agents in the adjoining crops), use of trap crops (planting castor around tobacco attracts *S. litura* to deposit the egg masses which are either collected or *T. remus* in released), use of semiochemicals or kairomones (to increase the searching ability and retention of the parasitoids, of gall midge, leafhopper and other pests from wild hosts to rice), use of organic manures instead of chemical fertilizers (to conserve predaceous arthropods such as carabid beetles and enhance the efficiency of plant disease antagonists), synchronizing the release of parasitoid with the availability of appropriate stage of the host and enacting suitable legislation (to prevent indiscriminate killing of beneficial organisms) have been suggested from time to time. Per cent parasitism of yellow stem borer (*Scirpophaga incertulas*) egg masses is significantly more in transplanted rice than direct seeding, which indicates the importance of the method of planting in attracting natural enemies.

Biopesticides

The natural occurrence of diseases caused by microorganism is common in both insect and weed populations and is a major natural mortality factor in many situations. For many practical reasons however, use of microorganisms for pest control involve their culture in artificial media and later introduction of comparatively large amounts of inoculum into the field at an appropriate time and place. The technique is possible only with those microorganisms that can be readily cultured in artificial media and moreover than can be induced to produce spores or other suitable resting stages which permit storage and application. Many fungi and bacteria can be handled in this way but insect viruses have the limitation that they have to be raised in living insects. This is unfortunate as insect viruses probably have greater potential for pest control than any other group of microorganisms because of their virulence and selectivity.

As these biocontrol agents (microbial pathogens) are applied on the targeted pests in much the same way as chemical pesticides, they are often termed as *biopesticides* or *natural pesticides*. Microbial phytotoxins, allelochemicals and other biocides of plant origin are also discussed here.

Bacillus thuringensis a bacterial pathogen infecting a wide range of insect pests is the most common microbial insecticide in use today. The disease agent is marketed by several companies and is registered for use against the insect caterpillar pests that attack a wide variety of vegetables, flower and ornamental crops. Unlike most chemical insecticides, it can be used on edible products upto the time of harvest. It is selective in action and does not harm parasites or predators of pests of any extent.

Another bacterium *B. popilale* is also commercially available for insect control which is effective against the white grubs *Popillae japonica* and *Holotrichia* spp.

Amongst insect pathogenic fungi, commercial preparations of *Verticillium lecanii* are now available for control of aphids, thrips and whitefly under glass house conditions Effectiveness out of doors is unlikely to be reliable because of the requirement for high humidity for infection to take place.

Several mycoherbicides have been registered in the USA for commercial use of weed control. One such is the formulation of soil-borne fungus *Phytophthora palmivora* for selective control of strangler (milk weed) vine *Morrenia odorata* in citrus groves. Marketed under the trade name 'De vine', it causes lethal root rots of its host plant and persists saprophytically in the soil for extended periods of time.

Extensive research during the last two decades has demonstrated that several plant secondary metabolities (allelochemicals) as well as fungal and microbial foxins posses good pesticidal activity. *Biolophos* is a natural phytotoxin isolated from the fermentation broth of *Streptomyces hygroscopicus* and *S. viridochromogenes* and exhibits strong herbicidal activity against a wide range of grass and broad leaf weeds following application to their foliage. It is currently being marketed in Japan under the trade name *Herbaceae*. Neem has long been known to have insecticidal properties and applications of neem extracts have been recommended for deterring insect attack, particularly locus. Vikwood Ltd., USA has formulated a neem seed extract 'Margosan O'. The National Chemical Laboratories, Pune in India has also been successful in developing neem formulations with antifeedant/insecticidal properties.

The allelochemicals from marigold-*terthicyl* was found to be highly nematicidal. Another compound *cucurbitacins* has also bee found to be equally effective in controlling nematode populations.

Botanicals Some useful weeds like Lantana, Notchi, Tulsi, Adhatoda etc; act as natural repellent to many pests. The indigenous trees like pungam, wood apple, anona and their byproducts have excellent insecticidal value in controlling diamond black moth, Heliothis, white files, leafhopper and aphid infestation. The fish oil resin soap (FOS) is a safe, non-poisonous natural product used for the control of developed whiteflies, mealybug, wooly aphid etc. A wider range of bio extracts are proved to be more effective in controlling many insect pests and diseases.

Farmers in their traditional wisdom have identified and used a variety of plant products and extracts for pest control, especially in storage. As many as 2121 plant species are reported to possess pest management properties, 1005 species of plants exhibiting insecticidal properties, 384 with antifeedant properties, 297 with repellent properties, 27 with attractant properties and 31 with growth inhibiting properties have been identified. The most-commonly used botanicals are neem (Azadirachta indica), Pongamia (Pongamia glabra) and manhua (Madhuca indica). Neem seed kernel extract (2 to 5%) has been found effective against several pests including rice cutworm, diamond back moth, rice brown plant hopper, rice green leafhoppers, tobacco caterpillar, several species of aphids and mites. Mahua seed kernel extract (5%) is effective against sawfly (Athalia lugens proxima) and others.

The efficacy of vegetable oils in preventing infestation of stored product pests such as bruchid (*Callosobruchus* spp.), rice and maize weevils (Sitophitus spp.) has been well documented. Root extracts of Tagetes or Asparagus work as a nematicide for plant parasitic nematodes. Similarly, leaf extracts of many higher plants can inhibit a number of fungal pathogens. A number of plants like *Chenopodium*, *Bougainvillea*, etc. have also been reported to be sources of antiviral principle.

Host Resistance

The use of plant varieties tolerant of or resistant to pest attack is a proven, effective, economical and safe method of pest control. Plants have diverse means to avoid, tolerate or recover from attack of other organisms. This trait is particularly important in low value crops.

In weed control, the phenomenon of allelopathy- the inhibition of growth of weedy plants by chemicals released by the crop plants-has not been extensively explored but may have a great potential. Superior weed suppressing types have been reported from searches of cucumber (*Cucumis sativus*), oat, sunflower and soybean collections. Some crops have superior ability to complete with weeds for critical growth factors such as moisture, nutrients and light. Cultivars specially bred for more vigorous canopies or root system should provide atleast partial degree of weed control. This along with allelopathy may provide a new valuable approach to weed management.

Occasionally, a variety developed to resist one pest may be more susceptible to another that had not caused much economic damage. Further, insects and diseases are continuously evolving strains that may adopt to formerly resistant plant varieties. The breakdown of plant resistance has become a greater problem with diseases as compared to insect pests, where it is fortunately much less common. Thus every effort should be made to use genes for general resistance and tolerance obtained from diverse sources and to add them to the genes for specific resistance.

Autocidal Control

Autocidal control involves rearing and release of insects that are sterile or are altered genetically in order to suppress members of their own species that are causing pest problems. Although a novel method, not much headway has been made in this direction because of difficulties with the high costs of rearing, sterilizing and liberating the insects

Behavioural Control

Pheromones

Pheromones are ecto hormones secreted by an organism, which elicit behavioral responses from the other members/sex of its own species. These are of different types – sex pheromones (attracting individuals of opposite sex), trail pheromones (helping the other individuals to follow the trial left by others), aggression pheromones (invoking aggression behaviour), etc. Pheromones are extremely selective, non-toxic, highly biodegradable and effective at very low application rates. Synthetic sex pheromones are commercially available and are used for surveillance, monitoring, and control of many lepidopterous pests such as old world bollworm, tobacco caterpillar, spotted bollworm, coding moth, potato tuber moth, diamond back moth, leaf folder, etc. Mass trapping has been attempted against rice stem borer, sweet potato weevil (*Cylas formicarius*) and red palm weevil (*Rhynchophorus ferrugineus*). Mating disruption or disruption of the insect communication system to prevent mating has been attempted against pink bollworm, codling moth, shoot borer and gypsy moth, etc. Pheromone components of *S. incertulas* impregnated in PVC resins and applied by hand at 40 g/ha are reported to be effective in causing up to 94 per cent reduction in the pheromone mediated communication compared to conventional methods.

Fairomones

Fairomones are volatile compounds that evoke behavioural responses adaptively favourable to the receiver. Fairomones are released either by the host plant or by the host insects. While the former issued by the pest and natural enemies to locate their habitats, the latter is used for prey finding and parasitisation/preying. Fairomones from host plant can be effectively used to mass trap the pest species as well as for monitoring. Tremendous strides have been made in identification, formulation and practical application. Use of Fairomonal compounds to increase the efficacy of the predator C. carnea and the egg parasitoid *T. chilonis* has been successfully demonstrated.

SUCCESS RATE OF ECOLOGICAL MANAGEMENT

In developing countries,

1. The rate of adoption is slow.
2. The majority of those programmes focus on pests within categories (insects, pathogens, nematodes or weeds) with little consideration of multiple pest interaction.

OTHER RELATED APPROACH

There are several other approaches for pest control yet two most popular approach i.e. IPM and BIPM is described here in brief. Both are having all above component and use of pesticide at minimum required level but not complete restriction on use of pesticide. Thus these are not acceptable to organic farming but need to be practised during transitional phase i.e. shift form conventional to organic farming.

Integrated Pest Management

Integrated Pest Management (IPM) is a component of sustainable agriculture with a sound ecological foundation. Although the concept of IPM had its origins nearly three decades back, the use of pesticides has not declined appreciably. The essence of the IPM concept, lies in the harmonious integration of compatible multiple methods, used singly or in combination against pests including insects, pathogens and weeds. It ensures economic benefits to the farmers and the society at large. The use of economic injury levels (EILs) and scouting for the assessment of pest populations have been the main criteria of IPM.

Biologically Intensive Pest Control (BIPM)

Relies on host-plant resistance, biological control and cultural control and use of biorational pesticides can be integrated. The Previous term IPM has historically aimed at developing systems based as much as possible on biological and other non-chemical methods. The BIPM also allows the selective use of pesticides after maximizing effectiveness of natural controls. BIPM, a more feasible and economical method of pest management has become popular, particularly among small and marginal farmers.

SUCCESS WITH BIOLOGICAL CONTROL

Rice

T. japonicum at 50000/ha/week starting from 30 DAT (days after transplanting) during rabi effectively suppress yellow stem borer.

Sugarcane

T. chilonis has been used six to eight times during the egg laying period of tissue borers. Sugarcane pyrilla has been effectively controlled by redistribution and periodic release of 8000-10,000 cocoons or 8,00,000-10,00,000 eggs per hectare of parasitoid *E. melanoleuca*. In 1999 there was an outbreak of pyrilla in 6 lakh hectares in north India, perhaps due to indiscriminate burning of trash, however, the parasitoid overpowered the pest within 20 days and the spraying cost of 6 lakh ha which comes to Rs. 48 crores was saved by this natural control.

Tomato

For the control of tomato fruit borer, *T. brasiliense* or *T. pretiosum* or *T. chilonis* are released 6 times at 50,000 per hectare starting the first release 30 days after transplanting.

Tobacco

For protecting tobacco nurseries from tobacco caterpillar, 3 fortnightly releases of egg parasitoid (*Telenomus remus*) at 40,000/ha and NPV at 250 le/ha along with 0.25% tannic or boric acid as synergist applied at 10 days interval, 4 weeks after transplanting have been found highly effective. NPV could be sequenced with bacterium Bt. var. *kurstaki*.

Cotton

Superior strains to *T. chilonis* have been selected for cotton ecosystem for north and peninsular Indian conditions. In cotton ecosystem, higher dosage of *T. chilonis* at 1,50,000 is applied at 7-10 days internal during the egg laying period of bollworms. The green lacewing, Chrysoperla carnea can be released at 2 larvae per plant during the peak egg hatching of cotton bollworms.

Horticultural and Plantation Crops

On apple release of parasitoid, *E. perniciosi* or *A. procila* at 2000/infested tree and *Chilocorus bijugus* at 20/infested tree give effective control of San Jose scale, *Quadraspidiotus perniciosus*. Release of parasitoid, *Aphelinus mali* for suppression of the population of wooly aphid. *Eriosoma lanigerum* is effective. Releasing egg parasitoid, *T. embryophagum* at 2000 per tree, during the active egg laying period of coding moth. *Cydia pomonella* and application of four sprays of granulosis virus 8 x 10^7 virus capsules/ml at fortnightly interval during active larval period have been recommended.

FUTURE THRUST

It is heartening to note the growing awareness among the farmers and policy makers about ecologically sustainable methods of pest management. More and more farmers are coming to realize the short-term benefits and long term positive effects of the use of bioagents and other ecologically safe methods to tackle pests that have increasingly driven them to desperation and sometime suicide.

Remote sensing techniques will provide information on the detection and quantifications of damage to plants, assessment of distribution of the principal host plants(s) or habitat of insect pests and surveillance of environmental factors favorable for the development and spread of insect and diseases. Computerized insect acoustic detection technique based on the closed organ pipe principle will be used to detect the presence of larvae of tissue borers, fruit flies and stored grain pests, etc. in oranges, mangoes, papaya and grains.

Exploitation of natural enemies until now underused (such as entomophilic nematodes and anthocorids), development of novel biorational pesticides and transgenic plants, resistance management to extend the life of essential pesticides and transgenic plants, and utilization of semi chemicals are essential to give us more options or tactics to enrich existing ecological management strategies.

The explosion of information technology is likely to make a certain impact and systems analysis approach will be applied effectively. Reliable predictive models of insect populations based on real time weather data can be achieved as programmes and weather information's are accessible through internet. Use of Geographical Information System (GIS) can also be attempted.

The ease and speed of dissemination of information through the internet will also help the researchers and the extension agencies. Successful ecological management has been a result of better fundamental understanding of the ecology of crop/pest interactions and rarely due to a revolutionarily new pest introductions. Recent introductions such as serpentine leafminer (*Liriomyza trifoli*), spiraling whitefly (Aleurodicus disperses), coffee berry borer (*Hypothenemus hampei*), and the coconut eriophyid mite (*Aceria guerreronis*) have already caused great damage.

Plant and animal quarantine has assumed considerable importance in this scenario. Hence, the governmental agencies have to be very vigilant. Efforts are on to formulate policy guidelines at the national level regarding export and import of beneficial organisms, including microbial agents.

Efforts should be made to document the indigenous technical knowledge of farmers and establish the scientific basis and efficacy of these practices. If found effective, these can be popularized. Indian farmers have traditionally used extracts of neem, turmeric, garlic, etc. over the centuries. Practices widely followed by farmers such as application of cow's urine, ash, chilli, powder, extracts of various plants, etc. should be studied and standardized if found effective. In China, thousands of rice farmers in Yunnan province to mix rice (*Oryza sativa*) varieties grown over two years. There was a 94% decrease in fungal rice blast disease as compared with monoculture plots growing only a single rice variety. The resistant crops also required low or negligible pesticide inputs. This method was soon adopted as an integral part of organic farming practices followed worldwide.

Cultural measures are the cultural practices associated with crop production to make the environmental less favorable for survival, growth or reproduction of pest species. Cultural practices if followed in the manner suggested provide relief from pests without incurring any major cost.

CONCLUSIONS

Pest control is an acceptable and necessary part of modern agriculture. However, some of the methods used during the past three decades have produced some undesirable side effects. The present system of pest control in today's exploitive agriculture which relies so heavily on use of pesticides, is not a viable one from ecological, sociological and economical viewpoints. If the non-chemical methods are not as effective as chemical methods, it is basically due to lack of understanding of the host-pest-environmental relationships. Hence, there is an acute need for basic and fundamental research, which can provide back-up technology for formulating ecological pest management programs for various pests. The different methods may interact with one another and should be brought together to form an integrated control programme, which exploits the strengths and compensates for the weaknesses of individual components. Knowledge of ecological causes of pest outbreaks is an essential component of pest management strategy which needs to be generated.

It is time to look at these so called *age-old practices* from a new angle and without prejudice, for they hold greater promise in the context of sustainable agriculture. Their necessity is strongly felt in the third world countries as many of these non-chemical methods are simple, easy to follow and economical, with least fear of environmental pollution.

CHAPTER 12

Neem: The Panacea for Organic Farming

The neem tree (*Azadirachta indica*) is native to tropical South East Asia. It is fast growing, can survive drought and poor soil keeps its leaves all year round. It is a tall tree, up to 30 metres high, with leafy spreading branches. Many white flowers, which smell of honey appear for the first time when the tree is 2 to 3 years old, and the tree bears fruit at 3 to 4 years. The ripe fruit are about 2 cm long and oval shaped. Inside the fruit there is a light-coloured seed about 1.5 cm long.

Although thousand-year-old Sanskrit medical writings mention neem's usefulness, the tree's exciting potential for controlling insects has only recently become clear.

First report about Neem's ability to repeal insects was is in the scientific literature in 1928 and 1929. Two Indian scientists, R.N. Chopra and M.A. Husain, used a 0.001-percent aqueous suspension of ground neem kernels to repel desert locusts. Not until 1962, however, was the real significance demonstrated. That year in field tests in New Delhi, S. Pradhan ground up neem kernels in water and sprayed the resulting suspension over different crops. He found that, although locusts landed on the plants, they refused to eat anything, sometimes for up to 3 weeks after the treatment. Furthermore, she noted that neem kernels were even more potent than the conventional insecticides then available and that neem's repellency was as important as its toxicity. In neighboring insecticide-treated fields, for instance, the insects also died, but not before consuming the crops.

Neem based pesticides are suitable for use in developing countries because the useful chemicals can be easily removed from the neem without the use of expensive and complicated equipment.

USES OF NEEM

Almost every part of the tree is used.

1. The wood of the neem tree is strong and resistant to termite damage. It is also good for firewood and making charcoal.
2. The neem tree is good for shade and is often planted for shade on roadsides.
3. Neem extracts are used to treat many health problems. People bathe in neem water to relieve heat rashes and boils. Neem oil is used against stomach ulcers and rheumatism Neem bark contains a strong antiseptic and neem is used to make soap and toothpaste. Neem twigs are used to

clean teeth.

4. Neem leaves can be used to make soil less acid.
5. Extracts from the neem tree are also used as mosquito repellents, fertilizers, diabetic food and animal feed.
6. Neem cake, which is left over when oil has been removed from seeds, and leaves, can improve the soil and add plant food and acts as nitrification inhibitor.

Although it has many used, the most important use for neem products is to fight against crop pests and diseases.

Worldwide approximately one third of crops in the field and store are lost to pests each year. The worst affected are maize and rice in Africa and Asia.

Neem for Pest Control

Different parts of the neem tree can affect more than 200 insect species and some nematodes, fungi, bacteria and viruses. Neem contains several active chemicals, which work in different ways; as a result of this, pests are unlikely to become resistant to neem. The most well known chemical in neem is **azadirachtin**. Neem is easy to prepare and use, and is environmentally safe and not harmful to man and animals.

Neem protects itself from the multitude of pests with a multitude of pesticidal ingredients. Its main chemical broadside is a mixture of 3 or 4 related compounds, and it backs these up with 20 or so others that are minor but nonetheless active in one way or another. In the main, these compounds belong to a general class of natural products called *triterpenes*; more specifically, *limonoids*.

Limonoids

Nine neem limonoids have been found effective to block insect growth, affecting a range of species that includes some of the most deadly pest of agriculture and human health. New limonoids are still being discovered in neem, but azadirachtin, salannin, meliantriol, and nimbin are the best known and, for now at least, seem to be the most significant.

Azadirachtin

One of the first active ingredients isolated from neem, azadirachtin has proved to be the tree's main agent for battling insects. It appears to cause some 90 percent of the effect on most pest. It does not kill insects-at least not immediately. Instead it both repels and disrupts their growth and reproduction. Research over the past 20 years has shown that it is one of the most potent growth regulators and feeding deterrents ever assayed. It will repel or reduce the feeding of many species of pest insects as well as some nematodes. In fact, it is so potent that a mere trace of its presence prevents some insects from even touching plants.

Azadirachtin is structurally similar to insect hormones called *ecdysones*, which control the process of metamorphosis as the insects pass from larva to pupa to adult. It affects the corpus cardiacum, an organ similar to the human pituitary, which controls the secretion of hormones. Metamorphosis requires the careful synchrony of many hormones and other physiological changes to be successful, and azadirachtin seems to be an *ecdysone blocker*. It blocks the insect's production and release of these vital hormones. Insects then will not molt. This of course breaks their life cycle. On average, neem kernels contain between 2 and 4 mg of azadirachtin per gram of kernel. The highest figure so far reported-9 mg per g-was measured in samples from Senegal.

Meliantriol

Another feeding inhibitor, meliantriol, is able, in extremely low concentrations, to cause insects to cease eating. The demonstration of its ability to prevent locusts chewing on crops was the first scientific proof for neem's traditional use for insect control on India's crops.

Salannin

Yet a third triterpenoid isolated from neem is salannin. Studies indicate that this composed also powerfully inhibits feeding, but also not influence insect molts. The migratory locust, California red scale, striped cucumber beetle, houseflies, and the Japanese beetle have been strongly deterred in both laboratory and field tests.

Nimbin and Nimbidin

Two more neem components, nimbin and nimbidin, have been found to have antiviral activity. They affect potato virus X, vaccinia virus, and fowl pox virus. They could perhaps open a way to control these and other viral diseases of crops and livestock.

Nimbidin is the primary component of the bitter principles obtained when neem seeds are extracted with alcohol. It occurs in sizable quantities-about 2 percent of the kernel.

Others

Certain minor ingredients also work as anti-hormones. Research has shown that some of these minor neem chemicals even paralyze the *swallowing mechanism* and so prevent insects from eating. Examples of these newly found limonoids from neem include deacetylazadirachtinol. This ingredient, isolated from fresh fruits, appears to be as effective as azadirachtin in assays against the tobacco budworm, but it has not yet been widely tested in field practice.

Chemical structures of neem's main ingredients: The complexity of these compounds demonstrates that nature is still the greatest chemist. Of the numerous pesticidal agents isolated so far from neem kernels, azadirachtin is the most active against insects. In addition to inhibiting their growth, it interferes with their powers of taste. Many leaf-eating insects are repelled by plants to, which even small amounts azadirachtin have been applied.

Azadirachtin, salannin, and nimbin all have the same basic limonoid structure. This differs from, but is not unlike, that of the stocks to, which the insect hormones (*ecdysones*) belong. An insect ingesting traces of these compounds is deeply affected because these *hormones mimics* block the parts of the brain that produce the hormones necessary to growth and development. In many cases, for instance, the insect's body may be ready to change while the hormones to complete the molt are not available. These deep-seated hormonal effects are the reason for neem's subtle, powerful, and yet insect-specific influences. Two compounds related to salannin, 3-deacetylsalannin and salanol, recently isolated from neem, also act as antifeedants.

Mode of Action

The fact that the extracts can be taken up by plants (and thereby confer protection from within) is one of neem's most interesting and potentially useful features. As has been noted, however, the level of this systemic activity differs from plant to plant and formulation to formulation. Extracts without oil, with a little oil, and with much oil exhibit different levels of systemic action.

The systemic activity differs with the insect as well. It is not effective on some aphids, for instance. They fed in phloem tissues, where (for reasons yet unknown) the concentration of azadirachtin is very low. Phloem is the plat's outermost layer of conductive tissues and insects such as these, whose mouthparts cannot penetrate past it, are little affected by neem treatments. On the other hand, leafhoppers and plant-hoppers, that feed at least half the time on the deeper layer of conductive tissues (called the xylem), get knocked down.

The growing accumulation of experience demonstrates that neem products work by intervening at several stages of an insect's life. The ingredients from this tree approximate the shape and structure of hormones vital to the lives of insects (not to mention some other invertebrates and even some microbes). The bodies of these insects absorb the neem compounds as if they were the real hormones, but this only blocks their endocrine systems. The resulting deep-seated behavioral and physiological aberrations leave the insects so confused in brain and body that they cannot reproduce and their populations plummet.

Increasingly, approaches of this kind are seen as desirable methods of pest control: pests don't have to be killed instantly if their populations can be incapacitated in ways that are harmless to people and the planet as a whole. In the 1990s this is particularly important: many synthetic pesticides are being withdrawn, few replacements are being registered, and rising numbers of insects are developing resistance to the shrinking number of remaining chemical controls.

But, for all the uncertainty over details, various neem extracts are known to act on various insects in the following ways:

1. Disrupting or inhibiting the development of eggs, larvae, or pupae;
2. Blocking the molting of larvae or nymphs;
3. Disrupting mating and sexual communication;
4. Repelling larvae and adults;
5. Deterring females from laying eggs;
6. Sterilizing adults;
7. Poisoning larvae and adults;
8. Deterring feeding;
9. Blocking the ability to "swallow" (that is, reducing the motility of the gut).
10. Sending metamorphosis awry at various stages; and
11. Inhibiting the formation of chitin.

As noted earlier, neem extracts have proved as potent as many commercially available synthetic pesticides. They are effective against dozens of species of insects at concentration in the parts-per-million range. At present, it can be said that repellency is probably the weakest effects, except in some locust and grasshopper species. Antifeedant activity (although interesting and potentially extremely valuable) is probably of limited significance; its effects are short-lived, and highly variable. Blocking the larvae from molting is likely to be neem's most important quality. Eventually, this larvicidal activity will be used to kill off many pest species.

Effectiveness

Neem is most useful against pests that feed on plants. Neem extracts control some insects well and others not so well, and are better at controlling pests on some crops than other crops.

Cold pressed neem oil and seed cake can both the used for pest control. The leaves can also be used to control pests, but there is less of the useful chemicals than in the seed.

Good Control

Neem extract is usually most effective against beetle larvae, butterfly and moth caterpillars. Examples of these are Mexican bean beetle, Colorado potato beetle and diamondback moth.

Neem is very effective against grasshoppers, leaf miners and leaf hoppers, for example variegated grasshoppers, green rice leafhopper and cotton jassid.

When neem is taken up by the plant, it will usually affect leaf hoppers and plant hoppers as they feed from the part of the plant, which carries the azadirachtin around inside the plant. Grasshoppers will stop eating almost immediately after neem extract is applied but caterpillars may not stop eating for 2 or 3 days.

Moderate Control

Neem is fairly good at controlling adult, beetles, aphids and white flies. White fly are less likely to settle and lay their eggs for some time after spraying. Beetles that feed on plant material as adults, eg brown leaf beetle, may sometimes avoid plants treated with neem extracts. Aphids avoid plants, which are sprayed regularly. When neem extracts are taken up by the plant it will not control aphids as they feed on the outer layer of the plant, which contains very little of the neem extract.

Poor Control

Neem gives only poor control of mealybugs and scale insects, adult, bugs, fruits maggots and spider mites.

Nontarget Species

As has been mentioned, neem extracts proved to be *soft* on unintended targets. Further examples are as follows:

Earthworms

In greenhouse studies, when neem leaves and seed kernels were incorporated into potting soil containing earthworms (*Eisenia foetida*), the number of young worms produced increased 25 per cent. In field trials there were no differences in the number of worms, but the average weight of each worm was highest in neem-treated plots. Thus, it seems possible that neem products can favor earthworms, at least under certain conditions.

Beneficial Insects

Neem seems remarkable benign to spiders, butterflies, and insects such as bees that pollinate crops and trees, ladybugs that consume aphids, and wasps that act parasites on various crop pests. In the main, this is because neem products must be ingested to be effective. Thus, insects that feed on plant tissues succumb, while those that feed on nectar or other insects rarely contact significant concentrations of neem products.

All this is coming clearer from recent research. For example, only after repeated spraying of highly concentrated neem products onto plants in flower were worker bees at all affected. Under these extreme conditions, the workers carried contaminated pollen or nectar to the hives and fed it to the brood. Small hives then showed insect-growth-regulating effects; however, medium-sized and large bee populations were unaffected.

Under laboratory conditions, the larvae of ladybugs and lacewings have shown some insect-growth-regulating effects from neem picked up from the bodies of to the insects. However, in greenhouse trials in Florida, neem products proved essentially nontoxic to predators and parasitoids of the cotton aphid and the sweet potato whitefly. Neither the amount of predation nor of parasitism was notably reduced.

A census of natural aphid enemies collected from seven different field trials indicated that neem has no detrimental effects on either predators (coccinellids, chrysopids, syrphids) or parasitoids (ichneumonids, braconids). The aphids in the neem-treated plots were actually carrying more parasites then were those in either the control plots or the plots treated with the insecticide pyrethrum.

PREPARATIONS FOR PEST CONTROL

Although bioactive compounds are found throughout the tree, those in the seed kernels are the most concentrated and accessible. They are obtained by making various extracts of the kernels and, to a lesser extent, of the press cake. Although the active ingredients are only slightly soluble in water, they are freely soluble in organic solvents such as hydrocarbons, alcohols, ketones, or ethers.

Methods of Application

Neem extracts can be applied in many ways, including some of the most sophisticated. For example, they may be employed as sprays, powders, drenches, or diluents in irrigation water-even through trickle- or subsurface-irrigation systems. In addition, they can be applied to plants through injection or topical applications, either as dusts or sprays. Moreover, they can be added to baits that attract insects (a process used, for instance, with cockroaches). They are even burned. For example, neem leaves and seeds and dry neem cake are ingredients in some mosquito coils

No new or unusual technology is required for any of the processing. It can be done using either simple village-scale technology or high-technology methods and industrialized facilities. The most common procedures are summarized below.

Water Extraction

The simplest technique (and the most widely employed today) is to crush or grind the kernels and extract them with water. They may, for example be steeped overnight in a cloth bag suspended in a barrel of water. For reasons not yet understood, this process is less effective than pouring the water into the bag and collecting the extract as it emerges. The resulting crude suspension can be used in the field without further modification. It can also be filtered and employed as a spray-able emulsion.

Water extracts of ground neem leaves are also very useful. Because neem is an evergreen, they are obtainable throughout the year.

Hexane Extraction

If the kernels are grated and steeped in the solvent hexane, only the oil is removed. The oil is not considered an active pesticide. However, new results show that it is an especially interesting material, which in certain cases can be used to kill the eggs of many types of insect, the larvae of mosquitoes, and various stages of certain pests (such as leafhoppers) that are often hard to control by other means.

The residue left after hexane extraction still contains the main active limonoid ingredients, and subsequent extractions with water or alcohol produce them in large amounts, clean and uncontaminated by oil.

Pentane Extraction

Pentane extracts of seed kernels are effective against spider mites. They reduce the fecundity (number of eggs) of *Tetranychus urticae*, for example. The active principles in the extracts differ from azadirachtin.

Alcohol Extraction

Alcohol extraction is the most direct process for producing neem based pesticidal materials in concentrated form. Limonoids are highly soluble in alcohol solvents. The grated kernels are usually soaked in ethanol, but sometimes in methoanol. The yield of active ingredients varies from 0.2 to 6.2 per cent.

Although water extracts are effective as pesticides, neem compounds are not highly soluble in water; the alcohol extracts are about 50 times more concentrated. They may contain 3,000 parts per million (ppm) or even 10,000 ppm azadirachtin.

Formulations

As noted, the simplest neem pesticide is a crude extract. However, for more sophisticated use, modifications can be made. These advanced formulations may convert neem extracts into the form of granules, dust, wettable powders, or emulsifiable concentrates. Aqueous extracts can also be formulated with soap for ease of application against skin disease.

Other formulations may involve the addition of chemicals or even the chemical modification of the neem ingredients themselves. These changes may be made to increase shelf stability and reproducibility, and for ease of handling or of scaling up the process. They may also reduce phytotoxicity, the damage to sensitive plants.

One particularly valuable class of additives are those that inhibit ultraviolet degradation. These include sesame oil, lecithin, and *para*-aminobenzoic acid (PABA).

When neem products are exposed to light they begin to lose their ability to control pests. For this reason the commercial neem based insecticide, Margosan-O, that is sold in the USA, contains a sunscreen.

Additives

Mixing neem extracts with other materials can boost their power 10-20-fold. Among these so-called "promoters" are sesame oil, pyrethrins (a type of insecticide mostly extracted from chrysanthemum flowers, see sidebar page 91), and piperonyl butoxide. They are used to produce a quicker kill.

Combinations with synthetic pesticides also can work well-they add rapid "knockdown" to neem's ability to suppress the subsequent rebound in the pest population. The effectiveness of neem extracts can even be boosted with the insect-killing bacterium *Bacillus thuringensis* (Bt) to provide a multifaceted pesticide.

PRACTICAL METHODS FOR PREPARATIONS

Control of Stored Grain Pest

Dried Neem leaves are mixed with storage grains. When neem is used in this way pests can be kept from the grain for a whole year; but if the grain is already infected with pests the protection will not work.

Method 1

- A ½ cm (1/2 inch) layer of fresh neem leaves, is placed in the bottom of a storage container.
- A layer of sun dried grain (up to 30 cm) is placed on top of this followed by another layer of neem leaves
- These layers can be repeated until the container is full, finishing with a good layer of leaves.
- Neem leaves for grain storage

Method 2

- Dry neem leaves in the sun so that the leaves stay green. Grind them into a powder. Mix the powder with clay and water. Plaster the inside walls of the storage container with the mixture and allow to dry
- Place a layer of neem leaves, which have been dried in the shade, on the bottom of the container.
- Fill the container with grain
- Place a layer of dried neem leaves on top and close the storage container.

Method 3

If grain is being stored in sacks, neem leaf powder can be mixed directly with the grain, (1 or 2 kilos of powder to 100 kilos of grain).

Uses of Neem Extract

There are many different ways to apply neem extracts. Some of the most common methods are described in this section.

Preparing Crushed Neem Seed

1. The ripe fruit pulp should be removed from the seed as soon possible after harvest, otherwise the seeds may become covered in mould. In some areas birds or fruits bats eat the pulp leaving the seed on the ground.
2. The seeds should then be laid out in a thin layer in the sun to dry out for a few days. The dried seeds should be stored in containers with plenty of air to stop mould growing.

3. The shells have to be removed using stones or a big mortar. The shells can then be removed by winnowing in the same way as with cereals.
4. The kernels are then ground in a mill or in a mortar.

Neem to Control Stem Borers on Young Plants

1. A small amount of crushed neem seed powder should be mixed with the same amount of dry amount of dry clay or sawdust.
2. The mixture is placed in the funnel of young maize of sorghum plants.
3. Rain will gradually dissolve the active chemicals in the neem seed.
4. This treatment may need to be repeated every 8 to 10 days until the plants flower.

Extracting Neem oil

To press neem oil by hand the kernels of the neem seed should be crushed as described in the previous section

- Add a small amount of water until the mixture forms a firm paste that can be kneaded.
- Knead the paste until oil drops form on the surface.
- Press firmly to extract the oil.
- The kneading and pressing should be continued in turn until the maximum amount of oil is removed. (The oil content of the seed kernel is about 45%).

In some areas there are traditional ways of removing oil from other seeds such as sesame or groundnut. It is a good idea to try these methods with neem.

If the oil is heated in the process, its ability to control insects will not be affected.

Controlling Bruchid Beetles in Stored Beans

Neem oil is used to control Bruchid beetles, which are small beetles whose larvae eat into stored beans and other legumes.

- Mix a small quantity of neem seed oil with the beans or seeds before storing.

The oil has a bitter taste but is not reported to change the taste of stored beans for humans to eat

Control of Soil-Borne Pests

The neem cake, which is left after the oil is extracted from the seed, is also useful for controlling several pests, which live in the soil, particularly nematodes.

Neem Water Extract for Plant Protection

- Prepare 500g of crushed neem seed kernels as described previously.

Mix crushed neem seed with 10 litres of water. It is necessary to use a lot of water because the active ingredients do not dissolve easily. Stir the mixture well and leave to stand for at least 5 hours.

- Spray the neem water directly on vegetables using a sprayer or straw brush.
- The effect lasts for 3 to 6 days. If kept in the dark neem water will be effective for 3 to 6 days.
- It has been estimated that 20-30 kg neem seed (an average yield from 2 trees), can normally treat one hectare.
- If crops have to be watered, water should go directly on the soil because water running over the leaves of sprayed plants may wash off the extract.
- If alcohol is available, 50 times more azadirachtin can be dissolved and extracted.

Water Based Neem Spray to Control Cutworms

Water based neem spray is most effective against pests such as Cutworms. During the day the caterpillars stay on the ground and fed on plant roots. At night they eat young stems.

Plants most affected include many vegetables and other affected plants include maize. Tobacco and coffee.

SUCCESS STORIES

Neem's effects vary with different insects. Some effects on a small selection of major pests are summarized below.

Desert Locust

Recent laboratory research has shown that neem oil causes *solitarization* of gregarious locust nymphs. After exposure to doses equal to a mere 2.5 liters per hectare, the juveniles fail to form the massive, moving, marauding plagues that are so destructive of crops and trees. Although alive, they became solitary, lethargic, almost motionless, and thus extremely susceptible to predators such as birds. Neem affects grasshopper nymphs similarly.

This discovery differs from earlier ones on locusts. Those first approaches used alcoholic extracts and were aimed at disrupting metamorphosis or at stopping adult locusts from feeding on crops. The new approach uses neem oil enriched with azadirachtin to prevent locusts from developing into their migratory swarms. It apparently blocks the formation of the hormones and the pheromones needed to maintain the yellow-and-black gregarious form, which plagues arid Africa and the Middle East. In an interesting aside, it has been shown that neem oil destroys their antennae, even when applied to the abdomen.

Neem trees grow well throughout the locust zones of Africa and the Middle East, and thus, in principle at least, the means to control the plagues could be locally produced.

Cockroach

Neem kills young cockroaches and inhibits the adults from laying eggs. Baits impregnated with a commercial preparation of neem-seed extract proved to retard the growth of oriental, brown-banded, and German cockroaches. First-instar nymphs of all three species failed to develop, and all died within 10 weeks. Last-instar nymphs exhibited retarded growth, and half of them died within 9 weeks. After 24 weeks, only 2 out of the 10 surviving German-cockroach nymphs had reached adulthood.

In a "taste test", American cockroach adults preferred neem-treated pellets over untreated ones, but neem-treated milk cartons repelled them.

Brown Planthopper

Neem cake (the residue left after oil has been removed from the kernel) has proved so successful that Philippine farmers are already using it on a trial basis against the brown planthopper (and other rice pests). Neem oil is being employed as well. Five applications of a 25-percent neem-oil emulsion sprayed with an ultra-low-volume applicator is said to protect rice crops against this increasingly severe scourge. It has been estimated that one neem tree provides enough ingredients to protect a hectare of rice. This use alone exemplifies the economic importance of further developing the neem tree for pest control.

Stored-Product Insects

Neem shows considerable potential for controlling pests of stored products. This is one of the oldest uses in Asia, and the literature contain many references to its benefits. In the traditional practice, neem leaves are mixed with grain kept in storage for 3-6 months. The ingredients responsible for keeping out the stored-grain pests are not yet identified-but they work well.

In this connection, repellency seems of primary importance. For instance, treating jute sacks with neem oil or neem extracts prevents pests-in particular, weevils (*Sitophilus* species) and flour beetles (*Tribolium* species)-from penetrating for several months. For this use, the degradation problem caused by sunlight is less of a concern because the products are mostly away from the sunlight, inside jars or other containers.

Neem oil is an extremely effective and cheap protection for stored beans, cowpeas, and other legumes. It keeps them free of bruchid-beetle infestations for at least 6 months, regardless of whether beans were infested before treatment or not. This process may be unsuited for use in large-scale food stores, but it is potentially valuable for household use and for protecting seeds being held for planting. The treatment in no way inhibits the capacity of seeds to germinate.

Neem has also been used in India to protect stored roots as well as tubers against the potato moth. Small amounts of neem powder are said to extend the storage life of potatoes 3 months.

Armyworm

Azadirachtin has proved an effective prophylactic against armyworms at extremely low concentration-a mere 10 mg per hectare.

The U. S. Environmental Protection Agency has approved a neem-seed-extract formulation for use on leaf-miners. This commercial product, now available almost nationwide, is expected to be especially useful against those leaf-miners that attack horticultural crops: Added to the soil, neem compounds enter the roots and move up into the crop's leaves so that leaf-miners munching on the leaves get their molting-hormone jammed, and they end up fatally trapped inside their own juvenile skins.

Mosquitoes

The larvae of a number of mosquito species (including *Aedes* and *Anopheles*) are sensitive to neem. They stop feeding and die within 24 hours after treatment. If neem derivative are used alone, relatively high concentrations are required to obtain high mortality. Nonetheless, the use of simple and cheap neem products seems promising for treating pools and ponds in the towns and villages of developing countries. In one test, crushed neem seeds thrown into pools proved nearly as effective at preventing mosquito breeding as methoprene, a rather expensive pesticide that is usually imported in developing countries.

Aphids

In the Dominican Republic, water extracts of neem seed provide effective against *Aphis gossypii* on cucumber and okra and against *Lipaphis erysimi* on cabbage. This was in direct-contact sprays.

As noted earlier, neem extracts applied in a systemic manner (that is, within plants) usually have little effect on aphids. Apparently, this is because aphids feed only on the phloem tissues, where, for some unknown reason, neem materials accumulate least.

Fruit Flies

Fruit flies (including the notorious medfly) are among the most serious horticultural pests. They cause millions of dollars in damage in to fruits, and their very presence in the tropics is keeping dozens of delicious fruits from becoming major items of international trade. But, at least in experiments, the medfly is proving susceptible to neem. This insect pupates underground, and in trials in Hawaii, spraying dilute neem solution under fruits trees resulted in 100 percent control.

More important, the neem material were compatible with the biological-control organisms (braconid wasps) used to control fruit flies. When neem was applied to soil at levels that completely inhibited the pest from emerging from pupation, the parasites developing in these pupae emerged and exhibited normal life spans and reproductive rates. Thus, neem is compatible with biological control of fruit flies. Diazinon, the current soil treatment for fruit flies, kills not only fruit flies but their internal parasites as well.

Nematodes

Neem products affect various types of nematodes. This may be significant because certain of these thread worms are among the most devastating agricultural pests and are also among the most difficult to control. In addition, an increasing number of synthetic nematicides have had to be withdrawn from the market for toxicological reasons.

Today, there is a small but increasing body of evidence that neem might provide useful replacements. Certain limonoid fractions extracted from neem kernels are proving active against root-knot nematodes, the type most devastating to plants. They inhibit the larvae from emerging and the eggs from hatching, and in at least one test they have done so at concentrations in the parts-per-million range. Water extracts of neem cake (the residue remaining after the oil has been pressed out of the seeds) are also nematocidal.

In a careful trial in Aligarh, India, amending soil with sawdust and neem cake dropped the root-knot index to zero and, of all the treatments tested, gave the greatest growth of tomatoes, a crop that is very sensitive to these nematodes.

Cardamom growers in South India are already using neem cake to control nematodes. Of 19 growers interviewed recently, 17 said that nothing works as well. These were sophisticated farmers who monitor world cardamom prices regularly and use synthetic chemicals for controlling other pests in their fields. In other words, they weren't using neem out of ignorance or poverty. They incorporate 100-259 kg per hectare of neem cake in their cardamom fields every year. About 3,000 tons of neem cake are now used annually in India's Cardamom Hills. It is sold by pesticide dealers, who transport it from 250-300 km away.

Snails

Various neem extracts kill snails. This appears to be beneficial in some cases.

In laboratory test, for example, ethanol extracts proved toxic to the aquatic snail (*Biomphalaria glabrata*), a species that is necessary to the life cycle of the parasite causing schistosomiasis (bilharzias). The extracts killed both the adult snail and its eggs. This raises the possibility that neem products may find a role in controlling schistosomiasis, a horrible scourge that infects some 200 million people in the tropics.

In another test, an aqueous solution of neem fruits resulted in a 100-percent kill of *Malania scabra.* This snail, common throughout the Orient, is a vector of lung flukes, a parasitic flatworm that encysts in the lungs of livestock, wildlife, and people, causing debilitation and sometimes death.

Crustaceans

Little is known about neem's effects-beneficial or detrimental-on crustaceans. However, in one intriguing set of experiments in the Philippines, it proved beneficial.

In rice paddies, the ostracod *Heterocypris luzonensis* feeds on the blue-green algae that fix nitrogen from the air. This minute aquatic crustacean thereby reduce a source of fertilizer for the crop. Killing this tiny creature thus would indirectly boost the nitrogen available and probably increase rice yields. Aqueous neem-kernel extracts have killed it very effectively under laboratory conditions.

Fungi

Neem has demonstrated anti-fungal activity. Should this prove widely applicable, the availability of a natural fungicide that can be grown, extracted and applied by farmers themselves could be of great consequence to worldwide agriculture and food supply. Fungi attack crops in countless numbers and forms. They are constantly evolving enemies of farms and forest. Many can reach epidemic proportions, a few have no cures, and some can make certain crop impossible to grow. And, despite the best of modern science, they still threaten wheat, corn, rice, and other plants that feed the world.

Not a lot is known about neem's practical use against rots, smuts, wilts, mildews, die-backs, and other fungal plant disease. However, several tests have indicated considerable promise.

In one test, neem oil protected the seeds of chickpeas against the serious fungal disease *Rhizoctonia solani, Sclerotium rolfsii,* and *Sclerotinia sclerotiorum.* It also slowed the growth of *Fusarium oxysporum* but did not kill it. In addition, neem cake incorporated into the soil completely blocked the development of the resting forms of *R. solani*-thereby interfering with the long-term survival of this devastating fungus.

In another, neem-seed extracts showed beneficial effects against leaf fungi. Spraying crude neem oil on lilac bushes, when done before any sign of outbreak, prevented powdery mildew from breaking out for the rest of the season. This protestant also gave essentially 100 percent control on hydrangeas in greenhouses-better than Benlate® (benomyl), the standard mildew treatment in much of the world.

In the case of bean rust, neem extracts have given 90 percent control when applied before the plants were exposed to the fungus. However, they worked poorly once rust was established.

In addition to affecting root-knot nematodes, treating soil with neem can reduce the populations of pest fungi in the rhizosphere that attack and feed off plant roots.

Aflatoxin

A truly unusual and potentially notable connection between neem and fungi has recently been reported from Louisiana. In trials there, neem-leaf extract failed to kill the fungus *Aspergillus flavus,* but, against all expectations, it completely stopped it from producing aflatoxin (see sidebar).

Plant Viruses

Plant viruses pose some of the most severe threats to world agriculture. Because they invade the crop's cells and cloak themselves with the plant's normal life processes, they are far more difficult to control than free-living organisms such as bacteria, protozoa, or fungi. At present, we an only try to halt their spread-something nearly impossible to achieve under even the best of circumstances-because viruses "hitch rides" in insects such as aphids, as well as on dirty tools, blowing dust, or spreading floodwaters.

A few virus-inhibiting chemicals are know for treating human and animal disease (AZT for AIDS, for instance), but at present, none are available for treating plants. Neem might be the first. Crude extracts seem to bind certain plant viruses effectively, and so limit infection.

However, for the moment at least, neem seems most effective at interfering with the transmission of plant viruses carried by insects. This conclusion is drawn from several successful tests of neem's effects against insect vectors of plant viruses.

These tests include the following

- A trial in the Philippines where rice fields sprayed with neem oil had significantly lower incidence of the ragged-stunt virus, which affects rice and is transmitted by the brown plant-hopper;
- A second trial in the Philippines where mixtures of neem oil and custard-apple oil interfered with the transmission of tungro virus, another rice pest;
- Experiments in India where neem-leaf extracts reduced the transmission of tobacco mosaic, a virus that seriously affects several vegetable crops;
- Field trials in the Philippines where fields treated with urea and neem cake were found to be lower in viral disease than those treated with urea alone;
- Enzyme-linked immunosorbent assays showing that rice seedlings grown in soil treated with neem cake were significantly freer of rice tungro viruses (transmitted by green leaf-hopper) than those in untreated control plots.

On the other hand, not all trial have been this successful. In the United States, daily applications of neem leaf extracts over a month's time to turnip plants infected with cauliflower mosaic virus did not reduce viral infection.

PROPAGATION AND PLANTING OF NEEM

Climatic Requirements

Rainfall

Neem trees can be grown in areas, which have between 405 mm and 1500mm of rain each year.

Temperature

Neem trees will survive very hot temperatures, up to 44°C and as low as 4°C. Some people report neem trees surviving light frost.

Raising Seedlings

The seeds of need do not live long and are usually planted as soon as possible after the fruit ripens and usually within three months. To help the seeds live longer the fruit pulp should be removed by hand and the seeds dried in the shade. Properly air dried seeds survive for at least 12 months in a refrigerator at 4°C. The seed are sown in the poly-tubes in nursery and 4-5 months old seedlings are ready for transplanting.

Transplanting

Transplanting is usually done in rainy season on the pits 60X60 size filled with manure and soil mixture. For initial one year irrigation and protection from wild animals is essential.

CONCLUSIONS

The mode of action of the various neem-tree extracts on a given insect species are often difficult to pinpoint and rather it has integrated suppression effect on various activities of pest, which ultimately helpful for controlling the population of pest. Neem's complexity of ingredients and its mixed modes of action vastly complicate clarification. Moreover, the studies to date are hard to compare because they have used differing test insects, and formulations. Further, the materials used in various tests have often been handled and stored differently, taken from differing parts of the tree, or produced under different environmental conditions. Still the use and mode of action of Neem products is very near to the philosophy of organic farming where the aim is to manage the insect population in such a way so that these are not economically harmful as against in the intensive /conventional agriculture in, which killing of insect is the aim.

CHAPTER 13

Peripherals for Self Sustainability

In many respects, organic farming is a way of life as much as it is a method of farming. The profitability of organic farms depends on the higher prices that their products command in the market place. Organic agriculture is not based exclusively on short-term economics, but also considers ecological concepts. It utilizes appropriate technology and appropriate traditional farming methods. This form of farming can also be called sustainable agriculture. The principals of this method are:

1. Organize the production of crops and livestock and the management of farm resources so that they harmonize rather than conflict with natural systems.
2. Use and development of appropriate technologies based upon an understanding of biological systems.
3. Achieve and maintain soil fertility for optimum production by relying primarily on renewable resources.
4. Use diversification to pursue optimum production.
5. Aim for optimum nutritional value of staple food.
6. Use decentralized structures for processing, distributing and marketing of products.
7. Strive for equitable relationships between those who work and live on the land.
8. Create a system, which is aesthetically pleasing for those working in this system and for those viewing it from the outside, e. g., it should enhance rather than scare the landscape of which it forms a part.
9. Maintain and preserve wildlife and their habitats.

SELF SUSTAINABLE SYSTEM

If the production system is solely dependant on the external inputs as in the case of conventional agriculture, the balance of the system disturbs as the buffer capacity exhaust. The results of this dependence of conventional agriculture are clearly visible in terms of decreasing production and environmental degradation. Although the organic farming is self dependant in terms of nutrient supply and plant protection, yet in addition to this some technologies are incorporated in the system which decrease the dependence on external supply sources for other inputs like water, energy etc. the system sustainability increase several times and it become more eco-friendly.

In the development of a more self-sufficient and sustaining agriculture, a deep understanding of the nature of agro-ecosystems needs the following principles:

1. They should function and to be design in such a way that they can manage agro-ecosystems.
2. They should be productive.
3. They should be obtained after conservation of natural resources.
4. They should be culturally sensitive and socially acceptable.
5. They should be economically viable.

Plants and animals communities together make agro-ecosystems which interact with their physical and chemical environments. They are modified by people to produce food, fibre, fuel and other products for human consumption and processing. An area used of agricultural production, looks as a complex system in which ecological processes found under natural conditions, such as nutrient cycling, predatory/prey interactions, competition, symbiosis and successional changes. By understanding these ecological relationships and process agro-ecosystems can be manipulated to improve production and to produce more sustainably, with fewer negative environmental or social impacts, and fewer external inputs.

The design of such systems is based on the application of the following ecological principles.

1. Enhance recycling of biomass and optimizing nutrient availability and balancing nutrient flow.
2. Securing favorable soil conditions for plant growth, particularly by managing organic matter and enhancing soil biotic activity.
3. Minimizing losses due to flows of solar radiation, air and water by way of microclimate management, water harvesting and soil management through increased soil cover.
4. Species and genetic diversification of the agro-ecosystem in time and space.
5. Enhance beneficial biological interactions and synergisms among agro-biodiversity components thus resulting in the promotion of key ecological processes and services.

These principles can be applied by way of various techniques and strategies. Each of these have different effects on productivity, stability and resiliency within the farm system, depending on the local opportunities, resource constructions and, in most cases, on the market. The ultimate goal of agroecological design is to integrate components so that overall biological efficiency is improved, biodiversity is preserved, and the agro-ecosystem productivity and its self-sustaining capacity is maintained. The goal is to design a quit of agro-ecosystems within a landscape unit, each mimicking the structure and function of natural ecosystems.

DESIGN OF SELF-SUSTAINABLE AGRO-ECOSYSTEMS

Most people involved in the promotion of sustainable agriculture, aim at creating a form of agriculture that maintains productivity in the long term by.

1. Optimizing the use of locally available resources by combining the different components of the farm system, i.e., plants, animals, soil, water, climate and people, so that they complement each other and have the greatest possible synergetic effects.
2. Reducing the use of off-farm, external and non-renewable inputs with the greatest potential to damage the environment or harm the health of farmers and consumers, and a more targeted use of the remaining inputs used with a view to minimizing variable costs.
3. Relying mainly on resources within the agro-ecosystem by replacing external inputs with nutrient cycling, better conservation, and an expanded use of local resources;
4. Improving the match between cropping patterns, and the productive potential and environmental constraints of climate and landscape to ensure long-term sustainability.
5. Working to value and conserve biological diversity, both in the wild and in domesticated landscapes, and making optimal use of the biological and genetic potential of plant and animals species; and
6. Taking full advantage of local knowledge and practices, innovative approaches not yet fully understood by scientists although widely adopted by farmers.

The ultimate goal of agroecological design is to integrate components so that overall biological efficiency is improved, biodiversity is preserved, and the agro-ecosystem productivity and its self-regulating capacity is maintained. The goal is to design an agro-ecosystem that mimics the structure and function of local natural ecosystems; that is, a system with high species diversity and a biologically active soil, one that promotes natural pest control, nutrient recycling and high soil cover to prevent resources losses.

Ecological Processes to Optimize in Agro-Ecosystems

1. Strengthen the immune system (proper functioning of natural pest control).
2. Decrease toxicity through elimination of agrochemicals.
3. Balance regulatory systems (nutrient cycles, water balance, energy flow, population regulation, etc.).
4. Enhance conservation and regeneration of soil-water resources and biodiversity.

The concept of sustainable agriculture is a relatively recent response to the decline in the quality of the natural resource base associated with modern agriculture. Today, the question of agricultural production has evolved from a purely technical one to a more complex one characterized by social, cultural, political and economic dimensions. The concept of sustainability although controversial and diffuse due to existing conflicting definitions and interpretations of its meaning, is useful because it captures a set of concerns about agriculture which is conceived as the result of the co-evolution of socioeconomic and natural systems. A wider understanding of the agricultural context requires the study between agriculture, the global environment and social systems given that agricultural development results from the complex interaction of a multitude of factors. It is through this deeper understanding of the ecology of agricultural systems that doors will open to new management options more in tune with the objectives of a truly sustainable agriculture.

Mechanisms to Improve Agro-Ecosystem Immunity

1. Increase of plant species and genetic diversity in time and space.
2. Enhancement of functional biodiversity (natural enemies, antagonists etc.).
3. Enhancement of soil organic matter and biological activity.
4. Increase of soil cover and crop competitive ability.
5. Elimination of toxic inputs and residues.

The sustainability concept has prompted much discussion and has promoted the need to propose major adjustments in conventional agriculture to make it more environmentally, socially and economically viable and compatible. Several possible solutions to the environmental problems created by capital and technology intensive farming systems have been proposed and research is currently in progress to evaluate alternative systems. The main focus lies on the reduction or elimination of agrochemical inputs through changes in management to assure adequate plant nutrition and plant protection through organic nutrient sources and integrated pest management, respectively.

PERIPHERALS FOR SELF-SUSTAINABILITY

Bio-Diversified Agro-Ecosystems

From a management perspective, the agroecological objective is to provide a balanced environments, sustained yields, biologically mediated soil fertility and natural pest regulation through the design of diversified agro-ecosystems and the use of low-input technologies intercropping, agroforestry and other diversification methods mimic natural ecological processes, and that the sustainability of complex agro ecosystems lies in the ecological models they follow. By designing farming systems that mimic nature, optimal use can be made of sunlight, soil nutrients and rainfall.

Agroecological management must lead management to optimal recycling of nutrients and organic matter turnover, closed energy flows, water and soil conservation and balance pest-natural enemy populations. The strategy exploits the complementarities and synergisms that result from the various combinations of crops, tree and animals in spatial and temporal arrangements.

In essence, the optimal behavior of agro-ecosystems depends on the level of interactions between the various biotic and abiotic components. By assembling a functional biodiversity it is possible to initiate synergisms which subsidize agro ecosystems processes by providing ecological services such as the activation of soil biology, the recycling of nutrients the enhancement of beneficial arthropods and antagonists, and so on. Today, there is a diverse selection of practices and technologies available, and which very in effectiveness as well as in strategic value. Key practices are those of a preventative nature and which series of mechanisms.

Various strategies to restore agricultural diversity in time and space include crop rotations, cover crops, intercropping, crop/livestock mixtures, and so on, which exhibit the following ecological features.

Crop Rotations

Temporal diversity incorporated into cropping systems, providing crop nutrients and breaking the life cycles of several insect pests, diseases, and weed life cycles.

Polycultures

Complex cropping systems in which tow or more crop species are planted within sufficient spatial proximity to result in competition or complementation, thus enhancing yields.

Agroforestry Systems

An agricultural system where trees are grown together with annual crop and/or animals, resulting in enhanced complementary relations between components increasing multiple use of the agro-ecosystem.

Cover Crops

The use of pure or mixed stands of legumes or other annual plant species under fruit trees for the purpose of improving soil fertility, enhancing biological control of pests, and modifying the orchard microclimate.

Animal Integration

Animal integration in agro-ecosystems aids in achieving high biomass output and optimal recycling.

All of the above diversified forms of agro-ecosystems share in common the following features:

1. Maintain vegetative cover as an effective soil and water conserving measure, met through the use of no-till practices, mulch farming, and use of cover crops and other appropriate methods.
2. Provide a regular supply of organic matter through the addition of organic matter (manure, compost, and promotion of soil biotic activity).
3. Enhance nutrient recycling mechanisms through the use of livestock systems based on legumes, etc.
4. Promote pest regulation through enhanced activity of biological control agents achieved by introducing and/or conserving natural enemies and antagonists.

Research on diversified cropping systems underscores the great importance of diversity in an agricultural setting. Diversity is of value in agro-ecosystems for a variety of reasons.

1. As diversity increases, so do opportunities for coexistence and beneficial interactions between species that can enhance agro-ecosystem sustainability.
2. Greater diversity often allows better resource-use-efficiency in an agro-ecosystem. There is better system-level adaptation to habitat heterogeneity, leading to complementarity in crop species needs, diversification of niches, overlap of species niches, and partitioning of resources.
3. Ecosystems in which plant species are intermingled possess an associated resistance to herbivores as in diverse systems there is a greater abundance and diversity of natural enemies of pest insects keeping in check the populations of individual herbivore species.
4. A diverse crop assemblage can create a diversity of microclimates within the cropping system that can be occupied by a range of non-crop organisms – including beneficial predators, parasites, pollinators, soil fauna and antagonists – that are of importance for the entire system.
5. Diversity in the agricultural landscape can contribute to the conservation of biodiversity in surrounding natural ecosystems.
6. Diversity in the agricultural landscape can contribute to the conservation of biodiversity in surrounding natural ecosystems.
7. Diversity in the soil performs a variety of ecological services such as nutrient recycling and detoxification of noxious chemicals and regulation of plant growth.

Diversity reduces risk for farmers, especially in marginal areas with more unpredictable environmental conditions. If one crop does not do well, income from others can compensate.

Integration of Livestock

Livestock play a vital role in most smallholder systems in developing countries. Cattle, sheep and goats graze areas, which can be cultivated only temporarily or not at all. In this respect, resources use for cropping and livestock-keeping is complementary. Apart from providing milk and meat, animals can provide manure for fertilizing cropland, draught power for cultivation, means of transport, e.g. water transport, and a savings account out of which larger expenditures for household or farm can be financed. Livestock can play a particularly important economic role for smallholders in semiarid areas where cropping risk is high, as animals provide security in cases of crop failure and can be sold to buy cereals and other staples when the family's harvest is insufficient.

Direct application of manure as fertilizer, also in combination with other forms of fertilizer, or the use of manure in composting is a key element of integrated animal husbandry in eco-farming systems. Although manuring does not create new nutrients, it transfers nutrients from rangeland to cropland, concentrations nutrients on selected areas of land, and recycles nutrients more quickly than through natural decay of vegetative material. The use of livestock for draught can be a further important link in eco-farming, in the sense of producing with few external inputs. However, rather than intensifying production on small plots of land, this often leads to preparation of larger areas than could be cultivated by hoe. Animal traction is likely to make a greater contribution to increasing agricultural production for both subsistence and sale where human population density and, thus, labour availability is low. In more densely populated areas, its primary value lies in reducing human labour inputs and permitting more timely cropping operations.

Crops and livestock can interact not only within a farm but also between different production units, as illustrated by the examples of traditional manure/forage linkages given in the section 'Crop-livestock integration'. Particularly crop residues are an important source of fodder. In addition, cropping can create a more favourable environment for livestock production than non-manipulated (natural) areas. Where savanna woodland has been partially cleared for farming, the taste fly population has been reduced and the higher grass yield on fallow fields as compared with natural savanna provides more forage for grazing ruminants.

Such inter-farm integration of crops and livestock is found mainly in areas where land availability still permits fallow systems. In areas so densely populated or degraded that the existing system of resource use can no longer support the human population, a shift in emphasis from large ruminants to smaller species such as sheep, goats or rabbits may be necessary. Larger animals (cattle, buffalo, camels) may nevertheless have to be retained for draught purposes.

A change then becomes necessary from extensive grazing to systems of zero-grazing involving fodder collection and/or intensive forage production. In supporting such changes, it is important to recognize that smallholders tend to use potential fodder resources for several other purposes as well, e.g., woody plants provide fodder, food (leaves, fruits), fuel, fibre etc. Also soil conservation and amelioration can be achieved by promoting forage production via multipurpose measures. The possibility of using foliage as fodder to increase animal production can be decisive incentive for smallholders to invest in planting trees and shrubs which control erosion, or in planting fallows which speed up the process of restoring soil fertility as in the case of ley farming.

Integration of Aquaculture

In recent years, aquaculture-the keeping or breeding of organisms, which live in fresh or salt water – has become a subject of growing interest. Within the framework of smallholder agriculture at issue here, attention must be focused on fish farming in ponds and fish breeding in rice fields. Fish provides protein, which is of exceptionally high quality from the biological point of view; they are also highly efficient in converting feed. Moreover, fish production is an enterprise that – in contrast to many other forms of animal husbandry – competes little or not at all with the production of food for humans. All these factors give pond-based fish farming a position of considerable importance as an enterprise in smallholder agriculture.

Fish breeding in rice fields, as has long been practiced in Southeast Asia, is a close to ideal form of land use, with cereal and animal protein being produced from the same area of land. Depending on the way the rice is grown, the following production methods can be used:

1. One generation of fish following a single rice harvest per year (rice-fish rotation).
2. Fish farming between two cultivation periods, in the case of 2-3 rice harvests per year;
3. Fish farming simultaneously with rice growing (rice-fish culture).
4. In all cases, the fish culture remains secondary to the culture.

However, other forms of fish breeding have gained considerably greater importance. These range from extension management in natural waters (stocking with young fish, no additional feeding) to intensive fish farming in specially created ponds (high stocking rate, artificial feeding only). An interesting and important example is the carp polyculture of China. Various carp species with different feeding habits live in different zones of a pond fertilized with domestic refuse and sewage and animal excrements. The water plants floating on the surface as well as any green feed added are eaten by the grass carp. Silver carp filter the plankton out of the water, and common carp eat the insect larvae, worms and snails on the bottom of the pond. This system can also be combined with the keeping of ducks.

Although various techniques of low-external-input agriculture have been described here separately, it is essential to recognize that eco-farming involves the complex integration of several of these techniques in a manner, which makes the most efficient use of locally available resources. The combination of techniques, which is most suitable for a particular farm at any point in time depends on the specific situation of that farm, in terms of not only the natural conditions but also the sociological and economic situation of the farm family. As these conditions change, so too do the appropriate eco-farming techniques.

Indigenous Organic Farming Practices

Within an indigenous farming system, it is difficult to distinguish the individual organic-farming measures visualized by agricultural scientists, because man, animals, forests, grasslands and fields are inseparable components of a single agro-ecosystem. For example, agroforestry is a relatively new concept in agricultural science, but it is not at all new to smallholders in the tropics, most of whom regard trees as a integral part of their farming systems. Particularly in densely populated Southeast Asia and in the dryland of India, indigenous farmers have developed highly productive and sustainable land-use systems in which agroforestry, multiple cropping, use of biological symbionts, mulching, composting, integrated plant protection, integration of fish and other livestock, and other organic farming measures are inextricably combined. In less densely populated parts of the tropics and subtropics, the indigenous farming systems represent a less intensive form of resource use, but the techniques and strategies applied by the farmers to gain a more or less secure livelihood are equally dependent on their detailed ecological knowledge.

Various elements of organic farming in traditional systems, e.g. the burning of vegetative growth before planting to enhance soil fertility, are not immediately evident to an observer trained in formal agricultural science and may even be regarded as harmful until the underlying principles are understood. There has also been a tendency among western scientists to regard the farming practices of different ethnic groups as representative of stages in an evolutionary development of agriculture. Only now are scientists beginning to realize that techniques once regarded as "primitive" (e.g. minimum tillage) reflected local adaptations to rainfall intensity, soil texture and erosion risks. They are organic farming techniques; ecologically appropriate for a specific site.

Soil and Water Conservation

Soil and water conservation practices are based on the following principles:

1. **Soil Conservation measures** includes reduction of splash erosion, detachment and transport of soil practices; and Control of run-off – to check sheet, rill and gully erosion
2. **Rain water management** includes collection storage and conservation of rainwater

The choice and design of soil and water conservation measure for agricultural lands depend mainly on the soil, land slope, rainfall and wind characteristics of the area. The measures adopted are broadly classified as:

1. Agronomical and biological measures; and
2. Engineering measures.

Often, both engineering and biological methods are used together but differ for arable and non arable lands.

Arable Land Management

Agronomical Measures
Wind Erosion Control

1. Vegetable cover
2. Strip cropping
3. Stubble mulching
4. Tillage practices
5. Sand dune stabilization
6. Windbreaks and shelterbelts

Water Erosion Control Measures

1. Contour farming
2. Strip cropping:
   a) Field strip cropping
   b) Contour strip cropping
   c) Buffer strip cropping
3. Conservation cropping system (ley farming)
4. Mulching and crop residue management
5. Critical area planning
6. Vegetative barriers or live bunds
7. Land smoothening
8. Ridges and furrows/dead furrows

Engineering Measures

1. Contour/graded bunds
2. Stone wall terrace
3. Water ways with vegetative cover
4. Gully plugging

Non-Arable and Denuded Land Management

1. Drainage lines
2. *In situ* moisture conservation measures
3. Vegetative fencing (bio-fencing or hedge)
4. Trench and mound fencing
5. Stone wall fencing
6. Mud well fencing
7. Barded wire fencing
8. "Social" fencing
9. Vegetative cover-pastures and afforestation

Rainwater Conservation

Many of these practices have evolved from indigenous practices. The indigenous practices are, by and large, simple, and easily constructed with locally available tools and materials. Improvements to such practices became necessary, to make them more effective and durable, and so their maintenance would be relatively easy.

The following are some of the water harvesting practices to be undertaken in the watershed to conserve soil moisture and rainwater and utilize the excess water for increasing agricultural production. The treatment may be different in soils and light and heavy rainfall areas. Based on soil type and rainfall, the watershed plan can be prepared and implemented with involvement of the farmers in the watershed.

1. *In-situ* moisture conservation practices
   a) Contour cultivation
   b) Formation of basins, furrows, random tie ridges, broad based ridges and furrows
2. Soil conservation techniques – viz, contour bunding in low and medium rainfall areas; graded bunding in high rainfall areas in red soils; and broad-based structures or ridges with grass or vegetative barriers in black soils.
3. Construction of check dams and gully control structures
4. Establishing farm ponds in individual farmer's fields
5. Construction of percolation ponds
6. Construction of irrigation tanks
7. Conjunctive use of ground water to supplement rainfall

Mulches

It is any kind of material that is applied on top of the soil, such as straw, compost, manure, chipped bark and plastic sheets but there are many other highly effective materials which are sometimes overlooked. Its function vary considerably. It is mainly used for soil enrichment and weed suppression. It also help in moisture retention, temperature modification (above and below ground level), soil texture, pest and disease protection and appearance.

Essentiality of Mulching

These are among the commonest reasons for mulching, especially with manures or composts, although to a lesser degree with many other materials. When manure or compost is incorporated into the soil the stimulated microlife breaks down the material and releases much of its nutrient value and provides perfect conditions for a newly planted tree or a heavily feeding vegetable during the growing season. This may cause leaching of fertility, especially in wet weather and whenever there are insufficient active root systems to absorb it. A mulch using the same material is likely to break down more slowly, and any leaching has to pass through the entire depth of the soil, thus considerably reducing the chance of its being lost.

Mulch and Microlife Activities

The presence of a mulch increase micro-life activity in the soil because it provides warmth and moisture. The increased microbial activity then enriches the soil as the microbial biomass and leaves more by-products. This activity also unlocks nutrients bound into soil particles; in particular the amount of exchangeable potassium in the soil is increased. Mulching practice for each year increases the depth of soil enriched.

Activity of Earthworm

The burgeoning life is easily seen when a mulch is removed: the numerous earthworm holes and droppings testify to their activity. These animals improve the drainage and aeration with their system of burrows, and so further aid other micro-life and fertility. Worms work up into the mulch and leave many droppings, often deposited on top, especially on sand mulches. These casts are extremely rich in nutrients and have a water-stable texture.

Weed Suppression

Mulches help in weed suppression. Use of a shallow mulch on top of established weeds may look neat immediately, but after a very short time they will have grown through. Worse, they will be even more vigorous, as the mulch will suit them and eliminate a few of their weaker competitors. If the weeds are small, low-growing and do not include vigorous spreaders like ground ivy, nettles, thistles, *Equisetum* (horsetail) or bindweed then thick mulches will kill them. Straw is easiest to apply in bulk and the cheapest too. Spread at least 1ft (30 cm) thick in spring it will eradicate many of the weeds, but any that do emerge must be pulled and destroyed straight away to prevent them rampaging. Extra layers of mulch applied every time the weeds reappear will be effective at suppressing them, but may require enormous amounts of material. Impenetrable mulches are much easier.

Birds and Mulch Disturbance

For a careless horticulturist, birds are the main cause of mulch disturbance. Birds kick the mulch backwards and work towards obstacles. If the edge of the path or lawn is made 4-6 in (10-15 cm) deep and the bed is chamfered away from it the birds will face the perpendicular surface, kicking the mulch up onto the bed.

Mulch and Retention of Moisture

Mulch help in retention of soil moisture and help in water economy of soil. During increasing frequency of summer droughts and day winters means that mulches are being used more and more to ensure that as little water as possible is lost. It helps in following ways:

1. In absence of mulch, water is continually pulled from the subsoil by capillary action and evaporates from the surface.
2. It interrupts the flow by hindering the evaporation, trapping the water in its air spaces.
3. Mulch ensures that the soil remains moist all the way to the surface, thus increasing the volume available for micro-life and plant activity.
4. The rough surface of a mulch further reduces evaporation by breaking up the airflow, in the same way as a hedge does on a larger scale.
5. The mulch breaks the impact of heavy rain and then allows it to percolate, it prevents soil impaction and keeps the surface permeable, preventing excessive run-off and soil erosion.

The water saving is great: where bare soil will take three to five days to evaporate ½ in (1 cm) of water, a mulched soil takes six weeks. If there is heavy rain there is little problem, as it will quickly drive through many mulches, but when brief showers predominate it is necessary to rake the mulch aside and allow the rain to reach the soil.

Increase in Crop Yield

The yields of wheat crop by mulching with straw was increased in acidic red-loam, sierozem and laterite soils by 14.3, 24 and 29 per cent over un-mulched treatment under different conditions. Mulching showed better effect on pea crop in acidic red-loam, medium black and laterite soils and increased the grain yield by 77.6, 43 and 40 per cent, respectively. Trials were repeated in different agro-climatic conditions (Bangalore red sandy loam, Ranchi acidic red, Poona medium black, Hissar sierozem, Delhi and Barrackpur alluvial soils) on wheat, pea, *moong* and maize crop (Gaur, 1975; Gaur and Mukherjee, 1980). The previous year's findings were confirmed showing the beneficial effect of wheat or paddy straw or *karanj* leaves mulch at 10 tonnes/ha on the crops. The grain yield of wheat was increased by 66 to 67 per cent in sierozem and Barrackpur alluvial soil, by 31.6 per cent in red sandy loam, 14.3 per cent in acidic red loam and 4.2 per cent in medium black soil. Pea grain yield was increased by 77.6, 29.4, 13.4 and 9.6 per cent in acidic red loam, red sandy loam, medium black and Barrackpur alluvial soil, respectively. The grain yield of *moong* and maize was increased significantly in Delhi alluvial soil due to wheat straw mulch by 31.7 and 110.5 per cent, respectively The incorporation of mulched material used on previous crop – maize, in soil increased the grain yield of wheat crop by 16.1 per cent. Nitrogen uptake by both these crops was increased and mulching had favourable effect on the population of soil bacteria, fungi and actinomycetes. The population of nitrogen fixers and phosphate solubilizers were increased several-folds.

The direct utilisation of organics low in nitrogen, i.e. high C/N ratio material may not always be feasible in intensive agriculture and particularly in culture of cereal crops. The utilisation of organic materials in soil in case of cereals requires proper understanding of the technology by the farmer. While growing legume crops which have got host-*Rhizobium* symbiosis system i.e. fixing of nitrogen in the nodules by rhizobia for the supply to plant, is not affected with nitrogen immobilization processes which generally occurs when organic matter low in nitrogen is ploughed in the soil. On the other hand, this benefits the process of nitrogen fixation. More over, the beneficial effects can be obtained by improvement of soil conditions, nitrogen immobilization and greater supply of carbon dioxide to plants for production of sufficient photosynthates. This method of recycling can be carried out by ploughing straw one week before sowing with small dosage of nitrogen (10-20 kg N/ha) and phosphate (90 kg P_2O_5/ha) as recommended for leguminous crops.

Nodulation in groundnut crop was improved due to application of wheat straw at 2,5 and 10 tonnes per hectare in Delhi alluvial soil. Ped yield was significantly increased by 66 and 95 per cent due to application of 2 and 5 tonnes straw per hectare by this method. Growth of plants was favorably influenced by 5 tonnes straw application. The residual effect of straw incorporation after groundnut increased the wheat grain yield by 3.9

to 36.8 per cent. Maximum residual effect was obtained with 10 tonnes straw applied per hectare. The practice can be followed in legume-cereal rotation. The straw gets decomposed during the growth period of legume crops and residual effect can be observed on the following crop. This has been confirmed with other legume crops (lentil, peas and *moong*).

Control of Temperature

Mulch monitors temperature of the soil (above and below mulches). Mulch has critical effect, plants, as it works as an insulating blanket. It protects roots in the soil from frost. Where bare soil will freeze to a depth of 2 in (5 cm), the same thickness of mulch will prevent freezing entirely - even a thin layer of litter gives considerable protection.

TABLE 13.1: Effect of mulches on crop productivity in drylands

Station	Crop	Mulch adopted	Productivity (qtls/ha)	
			With mulch	With mulch
Hoshiarpur	Maize	Basoti mulch	42.0	30.9
	Wheat	Basoti mulch	30.0	21.2
Rewa	Wheat	Paddy straw	17.2	11.5
Varanasi	Maize	Wheat straw	5.7	4.7
Akola	Cotton	Vertical mulch	10.3	8.0

Soil under a mulch has a more even temperature range and is warmer for most of the year. This insulation limits heat loss at night and keeps the average soil minimum temperature slightly higher than in un-mulched soil throughout most of the year. At the same time, the average soil maximum temperature remains lower, especially through the heat of summer, thus creating those more stable conditions of warmer soil at night and cooler during the day which most plants require.

Mulches, therefore, are good at protecting roots from frosts but may prove fatal to top growth. An understanding of this heat exchange process enables you to devise your own permutations.

Protection Soil against Erosion

Mulches protect the soil against erosion in several ways. They break the impact of heavy rain and prevent drying, stopping the granular soil that has formed being broken down into finer particles which would be more easily washed away. While the mulch remains the soil cannot be blown about or form a crust. It also helps prevent the formation of rivulets, which would cause further erosion.

The texture of soil is improved more deeply as mulching continues over several years and as more material breaks down and becomes incorporated. Even inorganic mulches such as sharp sand can improve texture.

Control of Pest and Disease

Mulches provide extra benefits in disease prevention and control soil microlife. The warm moist conditions under the mulch also break down many of the dormant overwintering organisms.

TABLE 13.2: The temperature effect of mulches

Effect	Gain or loss and reason	Winners and [losers]
More stable temperature under mulch	+ plants do not like big fluctuations	everything
Frost protection by mulch for soil	+ roots protected from freezing	bulbs, root vegetables, herbaceous and tender shrubs that regrow from below ground such as fuchsias
	- colder above mulch as less heat radiated from soil	[plants that cannot regrow from the roots, such as grafted plants]
[blossoms lost which would have been protected by bare soil warmth, especially strawberries]		
Slower warming of soil in spring	- cooler soil holds back early crops	[potatoes, peas and most crops sown before mid-spring outside]
	+ retarded crops may avoid late frosts	strawberries and other fruits especially
Slower cooling of soil in autumn	+ root activity continues longer	everything
Less cooling from evaporation of water	+ warmer soil plus water retention	everything
Higher night-time temperature in soil during growing seasons	+ increased temperature promotes more root and microlife activity	everything
Lower day-time temperature in soil during growing seasons	+ top layer of soil not baked, so promoting root and microlife activity	everything, especially crops that depend on both
	- less heat radiating back holds back ripening	[low-growing fruits, especially grapes]

After a mulch is removed the soil will be in a much better condition than if it had been left bare. The combination of worm and other activity, favourable physical conditions and a lack of rain impaction gives most soils that crumbly brown-sugar texture beloved by gardeners and plants. Mulches on empty ground as a winter cover against soil erosion and leaching will also produce a good tilth without the digging and raking otherwise required.

Many diseases such as **rose blackspot** overwinter on layer of mulch is applied on top of this infective material after leaf-fall it is sealed in and cannot be splashed back onto the plant to restart the cycle in spring. Thus disease prevention may be an unnoticed side-effect, as when horse manure is applied to roses, and when strawberries and gooseberries are strawed. Mulching can also protect indirectly against **mildews**, which are often aggravated, if not caused, by insufficient water; a moisture-retaining mulch will reduce attacks.

Apart from encouraging the soil micro life with nutrients, warmth and moisture a mulch also forms a good home for many larger creatures. *Woodlice*, for example, process the material and break it down; they then provide a basic food source for shrews, birds and hedgehogs, which help to control other more harmful bugs.

Mulches will further assist in the control of some pests such as **per midge**, **gooseberry sawfly** and **raspberry beetle**. Many pests such as these can be brought to the surface to be picked off or left for the birds, if the area is well watered and then covered overnight with a sheet of black plastic or old carpet. Sheets of impenetrable mulch can be even more effective. If they are put down before the pupae emerge, the pests are trapped underneath, die and are converted to soil fertility. Sward the next day, especially in hot conditions, unless you wish to kill the grass as well.

Appearance

Gardeners use a mulch for its appearance. An attractive covering is certainly pleasing to the eye and acts as an excellent foil to plants. The problem of cost against aesthetic value can be answered by using a thin layer of an attractive mulch on top of a cheaper or more effective kind. Mulches also come into their own where it is desirable to keep plants clean. Thin fabric or horticultural paper laid on the ground are excellent materials to protect salaaming, especially spinach, from soil splash. Gravel can also be used for this purpose.

Drawbacks of Mulching

Mulches do have a following few drawbacks.

1. Certain pernicious ones are found in the mulch habitat. Slugs and snails like the moist conditions beneath, though they are reluctant to move around on top of a dry mulch. Many bugs are a mixed blessing, such as woodlice and millipedes, which attack crops as well as processing dead material.
2. Warm moist conditions may encourage fungal or other diseases, so that the crowns of herbaceous plants rot or the emergent growths suffer from damping off or a neck rot. To prevent neck or crown rot in susceptible plants leaves a clear space free of mulch about the plant. To prevent the mulch encroaching on a valued plant use a ring of wire netting as a miniature snow fence.
3. Too thick or unperforated an impermeable mulch may restrict soil aeration; while this may not damage established plants during dry periods, wet anaerobic (oxygen-starved) conditions during the dormant season can rapidly cause root die-back and death. Anaerobic conditions can also change the composition of the soil micro life, leading to decreased beneficial activity and reduction of fertility.
4. The most direct damage to plants from mulches generally comes

from birds spreading them on top of low-growing thymes and other susceptible carpeting plants.

5. A hidden danger is where the sites of dormant plants are not marked and shoots might unwittingly be trodden on.

Types of Mulch

There are two basic types: loose organic and non-organic, such as peat or gravel; and impenetrable or impermeable sheet types such as carpet or plastic. The mulches in these two groups are listed below alphabetically, with notes about advantages or problems specific to each.

Loose Organic and Non-Organic Mulches

Bark Products: They are rapidly replacing peat for ornamental mulches, mainly because they are a sustainable resource, whereas peat is not. Bark of white pine, silver maple, walnut, sitka spruce and Douglas fir is processed into different grades, from a powder replacing peat in compost to large pieces the size of two fingers. The larger-textured bark chips are very long-lasting, provide plentiful holes for creatures to live in and allow aeration of the soil underneath more than do the finer grades. The problem, other than cost, is that bark products contain phytotoxic oils that poison many plants, and these have to be removed by careful composting before application. Added dangers from raw bark are animal pests and honey fungus (*Armillaria*), which is destroyed by shredding and composting. Bark products contain some nutrients, less nitrogen and magnesium but more calcium than sphagnum peat; the hardwood barks contain more nutrients than softwood. Softwoods are more acidic than hardwoods: both are in the range 3.5 to 6.5 pH, but the hardwoods require no lime when composted, as the pH rises, while the softwoods require the addition of chalk at 6 lb (3 kg) per cubic yard/metre.

The short-term effect of **composted bark** pH and nutrients on the soil is slight where the pieces are large, but with the finer grades the breakdown is rapid enough to add fertility, and also to incur some nitrogen robbery if the bark is mixed into the soil. It is fairly expensive but best for the permanent mulching of ornamental areas.

Cocoa bean shells are very expensive and have much the same use as coarse-textured bark, being most suited to undisturbed ornamental areas. In wet climates they can cake and become covered with mould, so may not be as attractive after a few weeks, though it helps to mix them half and half with bark of the same grade.

Coffee grounds and tea leaves are both unavailable in quantity except near factories. They dry out quickly, are similar to peat in appearance, acidic and with higher levels of nutrients. They can be used around most plants at an inch or so thick and incorporated afterwards, where they will be loved by the worms.

Tea leaves contain high levels of aluminium, traditionally utilized for blueing hydrangeas.

Dust mulches are the top layer of the soil, frequently and thoroughly hoed to produce a powdery surface. You should make sure that all cracks and crevices are filled. These mulches are primarily intended to keep the soil moist by preventing evaporation; weed control comes from the hoeing. Though they cost nothing, dust mulches can be labour-intensive and rain will cause crusting of the surface, requiring more hoeing. They are useful for early crops such as asparagus.

Feather waste is an unusual material, available only near poultry-processing farms, though small quantities can be obtained from recycled bedding. It contains 14% nitrogen in a fairly slow-release form, which can be beneficial but may overstimulate some plants. It blows about badly and needs to be sprayed with water to settle it. The worst problem is the horrible smell from wet feathers in hot weather. Feather waste may cause disease problems: if packed up in a wet state against bark or stems it can cause rotting.

Flax. An alternative to peat, fibrous straw is ground and composted with bark to produce a relatively sterile material, pleasant to handle, with a pH of 6.5. Nutrient levels are not high but the amount of potassium is significant and may make this material of use for vegetable-bed mulching and under soft fruit.

Grass clippings are one of the most easily available mulches, underrated principally because of people's experience with smell rotting masses when clippings are continually dumped in one spot. Grass clippings can be used satisfactorily in most areas of the garden. They provide their own nitrogen if incorporated fresh, but may cause nitrogen robbery after long drying.

Gravel is pleasant to look at, easy to apply, long-lasting, cheap, and less prone to bird disturbance than many other materials. It may contain lime or even salt, so check this before applying it to sensitive plants. It allows drainage and plentiful aeration. Gravel is most suitable as a permanent rather than a temporary cover. Gravel is surprisingly good at germinating seeds that fall on to it. Once they emerge as seedlings they should be flame-gunned or raked; left longer, they will be harder to eradicate.

Hop waste can be used as an attractive mulch resembling a finer grade of bark. As it dries quickly and is then light it may be kept in place with some chipped bark. The waste does not contain much intrinsic nutrient value, though its protein and carbohydrate content means it rapidly decays and adds greatly to the microbial biomass.

Leaves and leaf-mould should be secured as much as possible. Leaves make good mulches but blow around, so they need anchoring with grass clippings, sand or chipped bark, or must be covered with a net or some form of sheet mulch. They are probably the best mulch for protecting dormant plants through the winter, keeping them warm and dry. Leaf-mould resembles peat and is very pleasing; it contains many plant foods in the most nutritious form for mulching. Different leaves have different value. Evergreen leaves, which are resistant to decay, can be used as a mulch, though as they often have phytotoxic effects they are best used under their own kind in the shrub border. Holly leaves are supposed to repel moles and mice, but will also repel fingers for many years. Yew is so poisonous that it should not be left where children may pick it up. *Leaf-mould may introduce weed seeds* *even after it has been stacked for years.*

Manures are best in the spring, as when spread in the autumn they lose nutrients through leaching. Covering the mulch with plastic sheet will prevent this and aid its breakdown. Putting the garden to bed in the autumn like this gives good soil conditions for the spring but precludes green manuring. The high nutrient level means it is possible to overfeed the crop, so manures may be better combined with materials of lower value such as straw. Orchard trees, asparagus, blackcurrants and raspberries, leeks and potatoes will gain most from these rich mulches.

Mushroom compost is also used in all areas of the garden. It is probably the most practical all-round mulch because it is widely available, usually free of weed seeds, inexpensive, visually not unappealing and pleasant to handle. As lime and gypsum are added during mushroom-growing it may vary from an acid to a very alkaline pH. Care needs also to be taken against the risk of pesticide residues, as many chemicals are used against mushroom flies.

Peat has little nutrient value. It is an effective soil conditioner and has good moisture-holding properties. The finer-textured sedge peat is less acidic than *Sphagnum* or moss peat.

Pine needles are also used as mulch for strawberries, as they reputedly improve the flavour and may discourage slugs and snails.

Sand is one of the cheapest mulches, though rarely used. Sharp sand is preferable to the finer builder's sand. It allows more aeration. In many ways sand is similar to gravel, with no nutrient value and little effect on pH unless it contains some chalk.

Sawdust contains only half the nutrients of straw, is slow to break down and causes nitrogen robbery, so should not be incorporated into the soil unless and until it has broken down to a brown 'soil' and worms are found in it. Softwood sawdusts take longer than hardwoods to decompose.

Stones are also used like gravel. They are more effective at condensing dew and throwing up heat. If the stone is crushed and has sharp surfaces then it will discourage slugs and snails. The larger the stones the less effective they are at stopping weeds, unless there is a thick layer. They increase the availability of water and so tend to be used more in hotter climates. The stone get hot on top during the day, expanding and driving out the air from underneath; in the night, they cool rapidly, sucking the air back into the spaces where the dew condenses on the cool underside of the stone and runs down into the soil.

Straw (and hay) is the commonest bulk mulching material, especially for soft fruit, and for keeping strawberries clean. Straw is low in nutrient value but does not cause nitrogen robbery. Straw makes an excellent quick-drying pathway and is a mulch in most parts of the productive garden. In the vegetable garden it is more suitable for larger transplants, as the birds more it around and it can smother seedlings. When used as a temporary mulch it causes a weed problem afterwards

TABLE 13.3: Types and uses of mulches

Type	Uses	Pro and Con	Suited for	Apply
Bark	moisture retention soil enrichment weed control frost protection (roots) pest deterrence appearance	+ sterile - expensive + attractive	ornamental beds; paths; vegetable blanching	2-3 in (5-7.5 cm) deep with coarse layer on top
Bracken	moisture retention soil enrichment weed control frost protection (roots) pest deterrence	+ inexpensive -unattractive - weed problems	tender and herbaceous -winter protection; tomatoes, potatoes; strawberries; pathways	6 in (15 cm) thick of dry fronds 3 in (7.5 cm) thick of green fronds 3-4 in (7.5-10 cm) thick of dry fronds
Cocoa Shell	moisture retention soil enrichment weed control frost protection (roots)	+sterile +expensive	ornamental beds (best mixed 50:50 with bark)	2-3 in (5-7.5 cm) deep (goes mouldy in wet climate)
Coffee Grounds. Tea Leaves	moisture retention soil enrichment weed control frost protection (roots) appcarance	+sterile +attractive +adds fertility rapidly	ornamental beds; vegetable beds (tea leaves for hydrangeas)	2-3 in (5-7.5 cm) deep
Dust	moisture retention weed control pest deterrence appearance	+inexpensive +attractive	all areas; beat for protecting early blossoms and young tender growth	hoe at least fortnightly and after every rainfall
Feather	moisture retention soil enrichment weed control frost protection (roots)	+sterile -unattractive -disease problems +adds fertility rapidly	fallows and overwintering; under blackcurrants, globe artichokes, comfrey (best mixed with other mulches)	2-3 in (5-7.5 cm) deep; water sown the layer
Flax	soil enrichment	+attractive	vegetable and ornamental beds	2-3 in (5-7.5 cm) deep
Garden Compost	moisture retention soil enrichment frost protection (roots)	+inexpensive -weed problems +adds fertility rapidly	vegetable beds; around fruit trees and bushes; winter `bedding-down'	2-3 in (5-7.5 cm) deep as available
Grass Clippings	moisture retention soil enrichment weed control frost control frost protection (roots)	+inexpensive -weed problems +adds fertility rapidly	vegetable beds; ornamental beds; around fruit trees and bushes	1 in (2.5 cm) per week, in dry weather

Type	Uses	Pro and Con	Suited for	Apply
Gravel	moisture retention weed control pest deterrence appearance	+sterile +inexpensive +attractive	ornamental beds; paths and drives; under grapevines around Mediterranean herbs; Alpine and raised beds	2-3 in (5-7.5 cm) deep
Green Manures	moisture retention soil enrichment weed control frost protection (roots)	+inexpensive +adds fertility rapidly -weed problems	vegetable beds; around fruit trees and bushes	as available
Hop Waste	moisture retention soil enrichment weed control frost protection (roots) appearance	+sterile +attractive +adds fertility rapidly	vegetable beds; ornamental beds; around fruit trees and bushes, especially plums	2-3 IN (5-7.5 cm) deep
Leaves and Leaf-Mould	moisture retention soil enrichment weed control frost protection (roots)	+inexpensive +add fertility rapidly -weed problems -pest problems	vegetable beds; ornamental beds; around fruit trees and bushes	2-3 in (5-7 cm) deep
Manures	moisture retention soil enrichment weed control frost protection (roots)	+inexpensive -unattractive -weed problems -disease problems -pest problems +add fertility rapidly	vegetable beds; around fruit trees and bushes; winter `bedding-down'	2-3 (5-7.5 cm) after stacking or composting
Mushroom Compost	moisture retention soil enrichment weed control frost protection (roots) appearance	+sterile +inexpensive +attractive +adds fertility rapidly	best all-round mulch	2-3 in (5-7.5 cm) deep
Peat	moisture retention soil enrichment weed control frost protection (roots) pest deterrence appearance	+sterile +attractive -ecological cost of extraction	good all-around mulch	2-3 in (5-7.5 cm) deep
Pine Needles	moisture retention soil enrichment frost protection (roots) pest deterrence appearance	+sterile +attractive	supposedly add flavour to strawberries; good for paths	2-3 in (5-7.5 cm) deep
Sand	moisture retention weed control appearance pest deterrence	+sterile +inexpensive	sowing mulch; herbaceous plants, especially asparagus; for heavy clay soils	2-3 in (5-7.5 cm) deep

Type	Uses	Pro and Con	Suited for	Apply
Sawdust	moisture retention soil enrichment weed protection (roots) pest deterrence	+sterile -disease problems +inexpensive -unattractive	vegetable beds; around fruit trees and bushes-not raspberries	only if well mixed and composted, preferably with manure
Shredded Prunings	moisture retention soil enrichment weed control frost protection (roots) appearance pest deterrence	+sterile +inexpensive +attractive	soil-conditioning and weed-suppressing for ornamental beds; good for paths; provides valuable carbon balance in compost bins	2-3 in (5-7.5 cm) deep, or mixed with soft material (e.g. grass cuttings) in compost
Stones	moisture weed control appearance	+sterile +inexpensive +attractive	ornamental areas, especially alpine and rock gardens; under grapevines	2-3 in (5-7.5 cm) deep for fine grade; more for larger stones
Straw and Hay	moisture retention soil enrichment weed control frost protection (roots)	+inexpensive -unattractive -weed problems	fruit trees and bushes; rhubarb, potatoes and strawberries; winter `bedding-down'; paths	4-6 in (10-15 cm) deep, spreading high-nitrogen material first

Vertical Mulch

This involves opening of trenches of 30 cm depth and 15 cm width across the slope at vertical intervals of 30 cm. Sorghum stubbles/red gram/cotton stalks are stacked vertically in these trenches. The stalks should protrude up to 10 cm above the ground. These vertical mulches act as intake points and divert run-off water to sub soil layers.

This is ideally suited to soils with high clay content and low infiltration rate. Vertical mulch conserves moisture better than other methods during drought periods.

Live Vegetative Barriers

Under this mulching method, *subabul* is planted in the field and *Gliricidia* on the contour keylines. They act as vegetative barriers and effective mulch. When *subabul* and *Gliricidia* leaves are cut and spread in the field, it improves soil moisture and also supplies about 20 to 30 kg nitrogen per hectare.

Agroforestry / Alternate Land Use Systems

Alternative land use/agroforestry is defined as an *effective economic utilization of land without harming the natural resources structures based on land capability.* This involves the addition of a perennial component which has drought tolerance, can withstand the aberrations of monsoon and imparts stability to production.

The need of agroforestry based organic farming is not only to improve micro-climate and increasing availability of organic residue but also to bring millions of hectares of economically-marginal or environmentally-sensitive cropland, forest and waste lands into sustainable use.

Agroforestry or multi-story farming is a form of multiple cropping which essentially involves a vertical arrangement of crops (plants growing one above the other), whereas the dominant concept of intercropping is one of horizontal arrangement (plants growing beside each other). The integration of trees and shrubs with field crops eliminates the spatial separation between field and forest. A structural diversity of vegetation is sought which approaches the optimum for that site (structure of the climax vegetation). As in intercropping, the plants are combined in space and time in such a way that they compete as little as possible for nutrients, water and light, but rather complement each other in their differing requirements so as to achieve an optimal output per unit area. Scientific research in interrelationships and practical methods of agroforestry is still in its infancy.

Trees not only preserve soil fertility; they also make an important *contribution to the economy of the farm household.* As trees provide various products such as firewood, timber and fodder, the integration of trees into farming systems can reduce household expenditures for these products or can provide an additional source of cash income. If trees and shrubs are planted in home gardens and adjacent fields, their products can be collected quickly at the convenience of the household members. This considerably reduces the time and energy that particularly women and children must spend collecting firewood and increases the flexibility of timing this activity. The contribution of trees to improving working conditions should also not be underrated: the physical efficiency of workers under the climatic conditions created by trees is considerably higher than in the open field.

Organic farming requires the supply of crop nutrients and protection of crops with organic inputs. Therefore, the supply of nutrients is essentially through the application of organic matter in the form of farmyard manure, green manure and compost. Availability of nutrient-rich organic matter in adequate quantities to meet the crop requirements is a major constraint limiting large-scale adoption of organic farming. Whereas chemical fertilizers contain crop nutrients in a concentrated form, the nutrient content of most organic manures is low. Therefore, the quantity of any manure required to be applied to get a satisfactory crop yield is very high. It is usually in tons per ha as against the kg per ha of fertilizers required. Sources of organic matter are limited in dry areas of India. As the number of farmers practicing organic farming in any given locality is very few at present, they are able to collect farmyard manure and farm residues in their neighborhood. If more and more farmers take to organic farming, there is likely to be a serious shortage of organic matter. Therefore, on-farm production of organic matter and its careful use within the farm are crucial factors determining the viability of organic farming by a large number of farmers in a dry area. In this regard, trees, if included appropriately in the farming system, can be of immense benefit. They can (a) add nutrients and organic matter to the soil, (b) reduce nutrient and organic matter losses from the soil and (c) decrease the requirement of chemicals for crop protection by reducing the incidence of pests, diseases and weeds in crop fields.

Basic Principles

1. Selection of a suitable land-use model.
2. Identification of trees or shrubs that are not relished by livestock.
3. Level of competition between bushes and crop for soil water and light is minimal.
4. Consider the farmers' preference for trees.
5. Improved planting spot (dug out pit).
6. Undertake *in situ* water harvesting (individual basin) measures.

Types of Agroforestry Systems

Alley Farming

Alley farming is an alternate land use system where crops are grown in alleys formed by hedgerows of trees or shrubs. It involves three critical considerations, namely choice of crops, alley width and pruning heights. Short duration crops like pearl millet are advantageous to grow in semi-arid tropics

To avoid unhealthy competition between the trees and crops for moisture, it is more profitable to plant hedgerows on contours with 10 to 20 m spacing. To avoid competition for light, adopt a hedge cutting height to 15 cm.

Ley Farming

Ley farming aims to generate *in situ* fertility through rotation of legume forages with cereals. Inclusion of a legume fodder in rotation to improved soil quality and increased yield of sorghum in semi-arid tropics of India. It is equally important for increased fodder production, a critical component in rainfed farming.

Silvipasture

Marginal drylands, generally shallow and poor in nutrients, can also yield better returns through fuelwood and fodder. After six to eight year rotation, even arable crops can be grown on built-up soil fertility, without further fertilizer application.

Agri-Horticulture

In a better rainfall and other input facility agri-horticulture can be better option over sole fruit crop or sole food grain crop. Grain legumes are generally intercropped in this system to improve soil fertility Jujube- green gram is a successful system for the arid and semi-arid regions. Shade loving crops like Ginger, Turmeric and some medicinal plants are taken as inter crop in higher rainfall areas.

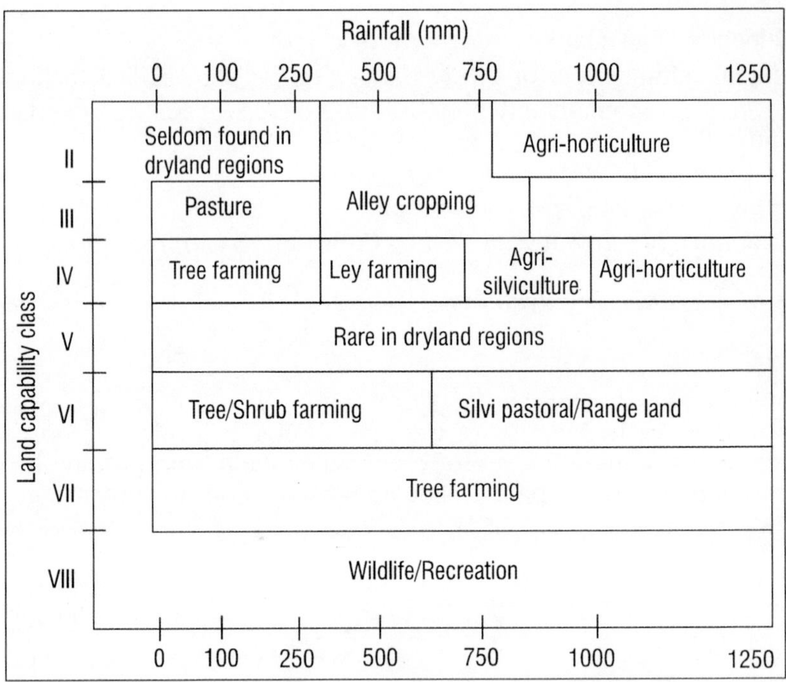

FIG. 13.1: Agroforestry/Alternate land use systems according to land capability classes and rainfall availability

Windbreaks and Shelterbelts

Windbreaks are narrow strips of trees, shrubs and/or grasses planted to protect fields, homes, canals, and other areas from the wind and blowing sand. Shelterbelts, a type of windbreak, are long, multiple rows of trees and shrubs, usually along sea coasts, to protect agricultural fields from inundation by tidal waves. Where wind is a major cause of soil erosion and moisture loss, windbreaks can make a significant contribution to sustainable production. There is a long tradition of using windbreaks in semiarid temperate regions of North America, Europe, and Asia for crop and soil-protection from wind and wind erosion as well as in the semiarid tropics (Vandenbeldt, 1990). Shelterbelts have also been traditionally used for a long time in several places, most notably on the Bay of Bengal coast of India and Bangladesh. The temperate-zone windbreaks are discussed in; the discussion here is limited to tropical windbreak systems.

When properly designed and maintained, a windbreak reduces the velocity of the wind, and thus its ability to carry and deposit soil and sand. It can improve the microclimate in a given protected area by decreasing water evaporation from the soil and plants. It can also protect crops from loss of flowers, as well as reduce crop loss due to sand-shear of seedlings. In many cases windbreaks have been shown to increase the productivity of the crops they protect. In addition to these soil- and water-conservation effects, windbreaks can also provide a wide range of useful products, from poles and fuelwood to fruit, fodder, fiber, and mulch.

Interactions between Trees and Crops, Useful for Organic Farming

1. Trees control soil loss and runoff water. They improve soil properties through efficient nutrient recycling and nutrient conservation.
2. A combination of annual crops and trees/bushes can increase biomass production because of differences in their rooting depths.
3. Trees/bushes are less prone than annual crops to extreme soil and climatic conditions – they ensure at least returns even if the crops fails.

Effects of Trees on Soils

Agroforestry has considerable potential as a major land-management alternative for conserving soil as well as maintaining soil fertility and productivity in the tropics. Some facts for supporting this statement are-

1. from time immemorial, farmers have known that they will get a good crop by planting in forest clearings:
2. soils that develop under natural woodland and forest are known to be well structured, with good moisture-holding capacity and high organic matter content;
3. unlike agricultural systems, a forest ecosystem is a relatively closed system in terms of nutrient transfer, storage, and cycling;
4. the ability of trees to restore soil fertility is illustrated by experiences in many developing countries, which indicate that the best way to reclaim degraded land is through afforestation or a similar type of tree-based land-use;
5. the conversion of natural ecosystems to arable farming systems leads to a decline in soil fertility and a degradation of other soil properties unless appropriate, and often expensive, corrective measures are taken; and
6. the microsite enrichment qualities of trees such as *Faidherbia (Acacia) albida* in West Africa and *Prosopis cineraria* in India have long been recognized in many traditional farming systems.

Beneficial Effects

Soil Conservation

1. **Rainfall erosivity**: Erosivity refers to the rainfall factor (R) of USLE. It is usually expressed as the EI_{30} index, which is the product of the energy of the storms multiplied by their maximum 30-minute intensity for all storms of more than 12.5 mm. There is a widely-held assumption that agroforestry systems can reduce the rainfall erosivity.
2. **Soil erodibility**: The major influence of agroforestry practices on the soil erodibility (K) factor is through the effect on soil physical properties, mediated by soil organic matter. It has been widely observed that soil structure is of a higher grade under forest than under cultivation; this

includes increased stability, lower detachability, and higher infiltration capacity

3. **Reduction of runoff**: This is based on the barrier approach to erosion control, in which runoff and soil loss are checked by means of barriers. Where trees are planted on soil-conservation works, including grass strips, bunds, and terraces, runoff and erosion are reduced; no specific additional effect, however, can be attributed to the presence of trees. Barrier hedges such as in alley cropping are effective in limiting runoff.
4. **The ground surface cover**: Soil cover formed of living and dead plant material including herbaceous plants and perennial cover crops, crop residues, and tree litter and prunings can effectively check raindrop impact and runoff, and the potential of this "cover approach" for reducing erosion is greater than that of the "barrier approach" discussed earlier. Agroforestry can contribute to maintenance of such effective ground covers for longer periods of time in a number of ways.
5. **Protection from erosion**: The most serious effect of erosion is loss of soil organic matter and nutrients, and the resulting reduction in crop yield. Forest cover reduces erosion to low levels, primarily through ground-surface litter cover and understory vegetation; the protection afforded by the tree canopy is relatively slight. Trees and shrubs can also be employed, through proper planting arrangement and management, as effective barriers to control soil erosion (e.g., hedgerows for soil-erosion control).
6. **Maintenance or improvement of physical properties**: The enhancement of such properties as soil structure, porosity, moisture retention, and erosion resistance under forest cover is well documented, as is the decline of these properties without forest cover. There is much evidence of the influence of physical properties of tropical soils on crop growth, independent of nutrient or other effects
7. **Modification of extremes of soil temperature**: Studies of minimum tillage show that high soil temperatures adversely affect crop growth and other biological properties including microbial population. Furthermore, ground-surface litter-cover greatly reduces the high ground-surface temperatures of bare soils in the tropics, which sometimes exceed 50°C. It is likely that leaf litter cover and shade produced by trees would have a similar effect.
8. **Improved water balance.** Trees and shrubs increase the water absorption and retention ability (infiltration rate and field capacity) of the soil. As a result of the greater availability of water to the plants, the growing period can be extended into the dry season, total plant production increased, and the soil protected by vegetative cover for a longer period of the year. Competition between individual crops for water cannot be avoided; however – in relation to total production – water utilization is generally more efficient: less water is consumed per kilogram of dry matter produced

Soil Fertility

1. **Maintenance or increase of organic matter**: The main avenues of organic matter addition to soils by trees is believed to be through continuous degeneration or sloughing-off of roots of standing (live) trees.
2. **Nitrogen fixation**: This has been prove, both indirectly through soil nitrogen balance studies and directly by observation of nodulation and tracer studies.
3. **Nutrient uptake**: Trees are more efficient than herbaceous plants in taking up nutrients released by the weathering of deeper soil horizons. Potassium, phosphorus, bases such as calcium and magnesium, and micronutrients are released by rock-weathering, particularly in the B/C and C soil horizons which tree roots often penetrate. Thus, nutrients in deeper soil horizons, that are unavailable to shallow-rooted crops, are taken up by deep-rooted trees.
4. **Atmospheric input**: Atmospheric deposition makes a significant contribution to nutrient cycling, more so in humid regions than in dry regions. It consists of nutrients dissolved in rainfall (wet deposition) and those contained in dust (dry deposition). Trees reduce wind speed considerably and thus provide favorable conditions for dry deposition.
5. **Reduction of acidity**: Trees tend to moderate the effects of leaching through the addition of bases to the soil surface.
6. **Reduction of salinity or sodicity**: Afforestation has been used successfully to reclaim saline and alkaline soils. For example, under *Acacia nilotica* and *Eucalyptus tereticornis* in the Karnal region in India, a reduction of topsoil pH from 10.5 to 9.5 over five years and of electrical conductivity from 4 to 2 dS m^{-1} has been reported with tree establishment assisted by additions of gypsum and manure. In this type of reclamation, the improvement in the soil's chemical properties undoubtedly is aided by improved drainage caused by construction of ditches, which leads to better leaching.
7. **Nutrient retrieval (Enhanced nutrient-use efficiency)**: This is related to the nutrient uptake mentioned earlier. It is commonly supposed that tree-rot systems intercept, absorb, and recycle nutrients in the soil that would otherwise be lost through leaching, thereby making a more closed nutrient cycle. Evidence for this mechanism comes from the relatively closed nutrient cycles found under forest ecosystems. The mycorrhizal systems associated with tree roots are an agent in the nutrient cycling process; they penetrate a large proportion of the soil, facilitating the uptake of nutrients, which can move only short distances by diffusion. The efficiency of mycorrhizae is demonstrated by the sometimes dramatic effects of mycorrhizal inoculation on plant growth.
8. **Effects of shading**: Shade lowers ground-surface temperatures, which may reduce the rate of loss of soil organic matter by oxidation.
9. **Improved microclimate.** The presence of dispersed trees in fields and pastures also changes the climatic within the plant communities.

Slight shading increases air humidity and reduces diurnal temperature fluctuations; reduced heating of the soil slows down humus decomposition. At sites with high solar radiation and/or high temperatures, multistory cropping can result in better utilization of solar energy and prevents excessive generation of heat within the stand.

Management of Adverse Effects of Trees

Trees, both as individual plants and when grown in association with herbaceous plants, can have adverse effects on soils. These adverse effects need to be managed either by adjusting tree crop geometry or supply of additional inputs e.g. organic manure/water, used for biological yield of trees.

1. **Loss of organic matter and nutrients in tree harvest**: A major concern in forestry is the depletion of soil resources by fast growing trees, and the effect of this depletion on subsequent forest rotations. Trees accumulate large quantities of nutrients in their biomass, part of which is removed in harvest. The problem is greatest where there is whole-tree harvesting (for example, the gathering of fine branches and litter by local people after a timber harvest).
2. **Nutrient competition between trees and crops**: This problem is most likely to be serious when trees or shrubs have an established root system that dominates that of newly planted annual crops. Ideally, the rooting systems of trees in agroforestry systems should have deep penetration but limited lateral spread. Whereas lateral spread of the canopy can be controlled by pruning, root pruning is generally to expensive to be practical.
3. **Moisture competition between trees and crops**: In the semiarid and dry savanna zones, this is possibly the most serious problem encountered in agroforestry.
4. **Production of substances which inhibit germination or growth**: Some *Eucalyptus* species produce toxins which can inhibit the germination or growth of some annual herbs. It has also been suggested that the production of allelopathic substances by tree roots could present a problem in agroforestry, but there is little evidence of this.

Beneficial as well as adverse effects of trees on soils and there management is necessary for successful of agroforestry systems. Where the growth of crops or other associated species located near or beneath trees is inhibited, it is important to establish the degree to which this is caused by one or more of the above factors, or if, in fact, whether it is caused by other factors. Trees and shrubs, when appropriately incorporated into land-use systems and properly managed, can make a significant contribution and improve the fertility and overall productivity of the soil beneath them.

Management of Agroforestry for Organic Farming

In organic farming the most important thing is to maintain nutrient cycling and agroforestry systems provide an opportunity for modifying nutrient cycling through management which results in more efficient use of soil nutrients, whether added externally (such as fertilizers) or made available through natural processes (e.g., weathering), when compared to agricultural systems. The underlying mechanisms that contribute to efficient nutrient cycling, as well as other nutrient cycling considerations in agroforestry systems, are -

1. There is potential for enhanced uptake of nutrients from deeper soil horizons (where they might be available as a result of rock weathering or percolation past herbaceous plant roots). The deep root systems of trees may reach these sites, which are not often attained by roots of common agricultural crops. The magnitude of this process, which is commonly known as - nutrient pumping, is not known; is believed to be a significant factor of soil fertility improvement in agroforestry systems.
2. The nitrogen fixation plant can directly fertilis the soil and indirectly neighbouring plants through above- and below-ground litter, through root exudates and leakage from leaves and roots, and through the excreta of grazing animals. When nitrogen fixing plants are grown in mixed plantations with non-nitrogen fixing plants, an increase in both growth and yield of the non-nitrogen fixing plants have often been found. Moreover, the legume with the non-legume intercrops, on average, yield more efficiently per unit land with higher area-□-time equivalence ratios than either intercrops with two legume species or two non-legume species. The nitrogen fixing plants (donor) and non-nitrogen fixing plants (receiver) could provide a good system for investigating nitrogen transfer, which was often termed the process of **nitrogen decomposition** and uptake in a neighbouring plant, where neighbouring plants might have differing nitrogen status. Nitrogen transfer could lead to increase productivity of intercrops and forests without increasing the use of nitrogenous fertilisers if properly managed and quantified. However, the roles of mycorrhizal hyphae in direct N transfer were somewhat inconclusive, and it is not yet clear if the transfer can be large enough to contribute significantly to the N status of the receiver in agricultural or natural ecosystems. Gains from symbiotic N_2-fixation by trees can be enhanced through tree-species selection and admixture. However, it is important to distinguish between nitrogen fixation, an input into the plant-soil system, and nitrogen addition through litter or pruning, which may result in an internal transfer within the system.
3. Nutrient release from tree biomass can be synchronized with crop requirements by regulating the quality, quantity, timing, and method of application of tree prunings as manure or mulch, especially in alley cropping. Different shrubs used in alley-cropping systems vary in the quantity, quality, and decomposition dynamics of leaf biomass. The timing of hedge pruning in alley cropping (and therefore, application of leaf biomass as a source of manure to the planted crop) can be regulated in such a way that the nutrient (especially N) release through decomposition of biomass is synchronized with the peak period of the crop's nutrient demand.

4. Management practices that lead to improved organic matter status of the soil will lead inevitably to improved nutrient cycling and better soil productivity. Although the recognized principal benefit from tree biomass in agroforestry systems is nutrient-related, there are other advantages stemming from organic matter addition to the soil: the importance of soil organic matter on crop and soil productivity is well-known.
5. Another major management consideration in agroforestry is the possibility of reducing nutrient loss through soil conservation. The role of agroforestry in soil conservation and related management strategies is need to be kept in mind.
6. Nitrogen fixing trees directly add nitrogen, an important plant nutrient, to the farming system. Depending upon the species, environment and management, the quantity of nitrogen fixed by trees can vary from less than 10 kg to more than 100 kg per ha. The nitrogen requirement of most food crops is around 100 kg per ha. While a small quantity of the nitrogen fixed in the root nodules of trees reaches the soil directly, the bulk of it is found in the leaf biomass.
7. The substantial quantities of leaf biomass produced by trees, especially nitrogen fixing trees, can be used to enrich the soil. Species such as *Leucaena leucocephala* and *Gliricidia sepium* yield 3.0-6.0 tons of dry leaf biomass per ha every year. This biomass can be directly incorporated into the soil as green manure. It can also be fed to animals as fodder and then the farmyard manure applied to the soil. The leaf biomass of most nitrogen fixing species, when used as green manure, decompose rapidly to release nutrients which may not be fully utilized by crops. Such biomass can be composted with fibrous by products like straw to slow down decomposition. This also enhances the break down of straw-like material.
8. Organic matter addition to the soil by the decay of tree roots. This is the reason for the luxuriant growth of plants often seen under old trees. If a farmer can plant short-duration trees such as *Cajanus cajan* or *Sesbania sesban* on a part of his land for about three years, that area is likely to have an improvement in soil organic matter as a result of root decay. Shifting this tree fallow area within the farm every three years can be a strategy for fertility improvement without the application of chemical fertilizers.
9. Leaf litter fallen from trees forms a ground cover that acts as a mulch. The moderating influence of this mulch on soil temperature reduces the rate of decomposition of organic matter and thereby the loss of nutrients. When the litter decomposes eventually, it releases nutrients and organic matter that enrich the soil. Another benefit of the litter mulch is the protection of soil through the reduction in erosion. Reduced erosion saves valuable soil nutrients and organic matter.
10. In addition to the beneficial effects on organic matter and nutrient status of soils, trees may also be useful in crop protection. Bird presence is usually increased if there are trees in a field because of the availability of branches for birds to perch on. Birds prey on pests, mostly caterpillars,

and reduce their population. However, these birds can themselves become pests if they feed on grain and other crop parts. Trees such as neem may also have a pest repellant property. Control of the parasitic weed Striga is achieved with *Sesbania sesban* in Africa, and *Tephrosia candida* is believed to control nematodes in tea plantations.

Thus, it appears there are several management options for exploiting the advantages of efficient nutrient cycling in organic farming based on agroforestry systems. Management practices are the key to organic matter maintenance in agroforestry systems: if a major part of the tree and crop residues are returned to the soil, soil organic matter levels can not only be maintained, but can even be improved. Instead of individual effort, a community approach, as in a watershed, will maximize benefits.

CONCLUSION

Sustainability in a system can be achieve in real sense by making the system self-dependant for input supply It also has several other advantages like-

1. Least affected by the high cost input supply which is fast changing due to globalization.
2. Prediction of water scarcity in future, high rate of deforestation and several other effects of modern lifestyle can be minimized.
3. It increases the diversity of products from the production system and this provide resilience to the system for the variability of climate and market.
4. Thus, the peripherals make infrastructure for organic farming and they are necessarily be incorporated at the initial stage of the organic farming system.

CHAPTER 14

Adoption and Success

In most of the developing countries, the subtropical and tropical climate makes it practically difficult to maintain organic matter status at higher levels under traditional agricultural practices, which necessitates the orientation towards organic farming. The important reasons which emphasis the need for organic farming in these countries are as follows:

1. Most of the arable soils contain organic carbon below the threshold level.
2. Majority of the farming community is resource poor and purchase of fertilizers and chemicals in adequate quantities is beyond their capacity.

In this system, a large fraction of farm by-products of plant and animal origin is utilized for non-farm use that is, fuel or other domestic purposes. Small and scattered land holdings of a large farming communities compel them to leave the crop residue in the farm itself rather than recycle it for composting. Lack of location specific technology to recycle organic wastes and lack of awareness to recycle organic waste in agriculture are the main reasons for its slow adoption even though it is native technique for the farmers which got lost during the period of green revolution.

Use of chemical in agriculture is considered by our planners and also by a sizeable section of elite farmers as the most efficient way of augmenting food production. The need to produce more and more food to satisfy the increasing population has led man to adopt new method of agriculture which have now proved to be (1) harmful to the soil in the long run and (2) harmful to the people who consume its produce. This distorted, system of agriculture is called *Modern Method of Farming*.

The harmful practices adopted in the modern farming are:

1. Excessive tilling
2. Excessive use of water
3. Use of inorganic fertilizer
4. Use of chemical pesticides
5. Neglect of natural allies of farming.
6. Use of varieties inconsistent with local factors.

Since, 1950s, farming methods derived from the West are being followed. These techniques are adopted regardless of whether they suit our climatic, economic and social conditions or nor. Use of Hybrid seeds is one such technique. These seeds may yield a bountiful crop but their nutritive values, taste, and satisfying capacity is doubtful

The adoption of conventional farming obviously did succeed in achieving higher production rather than with traditional farming but it is capital intensive. The cost of inputs are so high that sometime it includes as much as 80% of the realizable values of the produce. When farmers work on such a think margin of profit, even one or two adverse seasons can put them into a tremendous debt burden. It is out considered opinion that if farmers switch over to *Organic Farming Methods*, they will never suffer an economic loss over a period of time.

ORGANIC AGRICULTURE AN ATTRACTIVE ALTERNATIVE

Organic practices are more environmentally sustainable than most conventional practices. Emissions of toxic substances and dependence on non-renewable resources are limited. Organic practices are inherently better-adapted to prevailing agro-ecological conditions.

Export-oriented organic agriculture is attractive because of premium prices paid by North American and Western European consumers for certified organic products such as coffee, tea, vegetables, and cotton. The export income of these farms and estates is supplemented by a cash income from the sale of some products in the local market, primarily animal products introduced into the farming system to cover the need for manure. Furthermore, export-oriented organic farms produce food for home consumption, s well as by-products. Because of this increased product diversity, dependence on export income is lower than in comparable conventional export crop production.

Subsistence-oriented organic agriculture is attractive for marginalized farmers who primarily produce food crops for home consumption because food self-sufficiency and self-reliance for soil fertility and pest management are improved. Several cases are located in marginal areas which are in various stages of degradation. Organic practices combine well with efforts to regenerate soil, control erosion, or harvest water because of the many positive ecological effects of simple methods.

Compost for example, improves soil fertility, nutrient recycling, water retention, infiltration, and soil life. Contour farming with leguminous shrubs and trees replaces NPK fertilizers while at the same time improving soil structure and the microclimate and reducing weeds and erosion. In those situations organic agriculture allows sustainable productive use of marginal lands, combining improved food security and food self sufficiency with minimal capital investment.

The available material tends to support the conclusion that both in high potential areas and marginal lands, organic agriculture offers agronomically feasible includes a positive microeconomic effects for most of the producers involved, and possibility leads to positive effects on regional and national economics as well.

NOT A MIRACLE SOLUTION

However, Organic Agriculture Systems do not offer miracle solutions for all problems related to agricultural development. Social and economic problems like marginalization, prices, gender conflicts, and land tenure largely surpass the extent of this survey. Further marginalization may result from organic practices if the poorest sections of the population cannot afford the costs or labor needed for conversion. Better-off farmers might gain most when the approach towards organic farming is purely technocratic, just as they did earlier from Green Revolution technology. However, organic agriculture offers a better perspective for equitable development because it tends to increase farmers' self-reliance for food and inputs, and contributes to their autonomy and self-confidence. Furthermore, it increases biological and labor diversity, thus spreading risks in the agronomic as well as the financial sense. Unequal development of richer and poorer farmers in organic agriculture systems is not inherent to the technology, but remains related to the given *resources endowment* (quality and acreage of land).

ADOPTION IN THE FARMING SYSTEMS CONTEXT

The history of agricultural development over the last several decades contains many examples of failures in adoption or sustenance of technologies that were judged scientifically sound. The resource management technology may be failure due to water-management schemes in the inland, minimum tillage practices, poor take-up of improved clearing methods, soil conservation methods, and spread of alley cropping. The other reasons for failure to adopt or persist with technology are social, cultural or economic factors as to technical deficiencies. This necessitates an approach to technology development and transfer than integrates the criteria used by the farmer in deciding the actions, as well as those assumed in the scientists models.

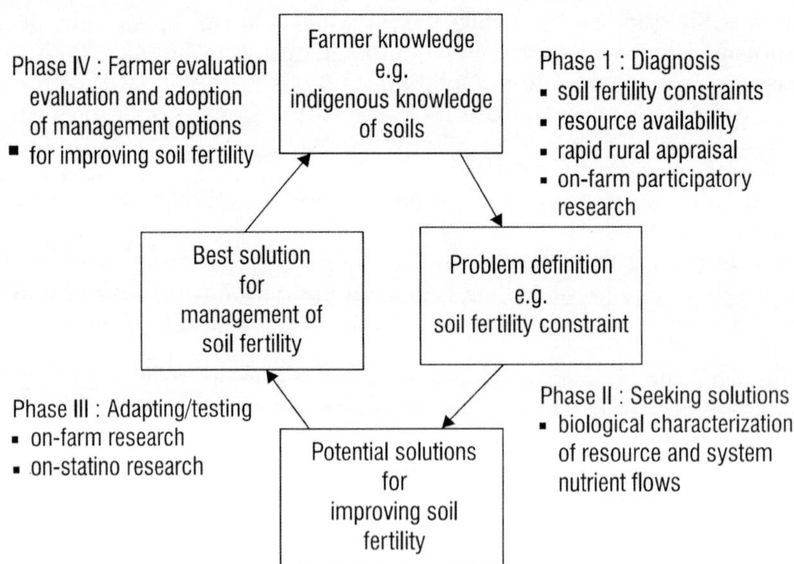

FIG 14.1: The *farmer-back-to farmer* approach to agricultural problem identification and solution (Rhoades and Booth, 1982; Moran, 1987) as applied to soil fertility research.

Soil management is only one small component of a farmer's spectrum of activity. The knowledge and perceptions of soil-based constraints should be combined with a holistic appraisal. Farmer's should have balanced approach between various alternatives based on whole range of available environmental, biological, and economic information.

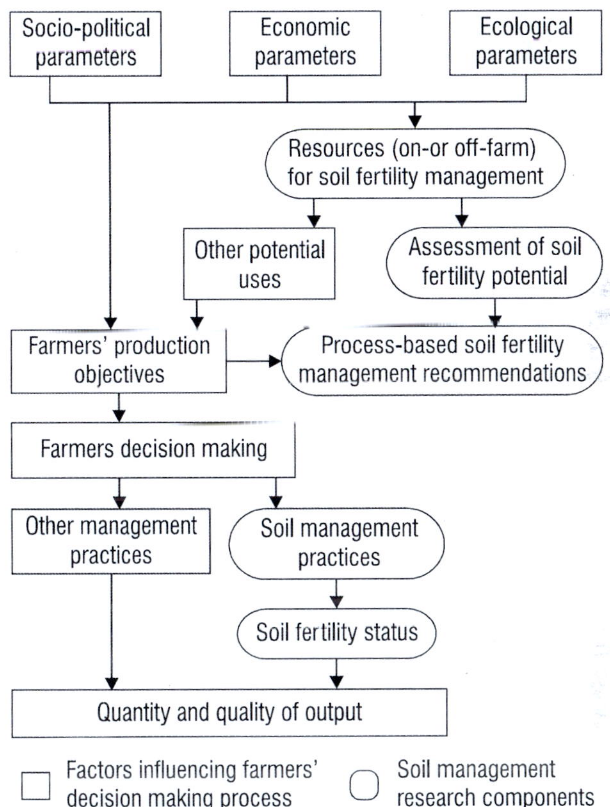

FIG. 14.2: Integration of the research results for biological management of soil fertility with system level information on farmers' decision making.

Technological change is thus more likely to succeed in terms of both adoption and of impact on production when it is adapted to farmers' circumstances rather than merely optimized in terms of scientists' criteria. This is most likely to be achieved when scientists and farmers work together throughout the research and development process.

Figure 14.2 illustrates stages in a research processes that is initiated by joint diagnosis of soil fertility problems. This information, based both on farmers' perceptions and on biophysical and socioeconomic characterization of the farming system. It should be used to guide scientific investigation of the opportunities for improvement in soil management practices.

The process-and-system approach places farmer decision making at the center of the technology development process, highlighting the need for detailed characterization of the environmental context that influences those decisions.

The capacity for increasing and sustaining productivity through improved soil fertility management is subject to a variety of limits of physicochemical, biological economic, social, and cultural dimensions. The determination of these limits, the establishment of interactions – including hierarchical relationships, and the opportunities for modifying them is one of the major purposes of system characterization

SOIL MANAGEMENT AND BIOLOGICAL PROCESSES

The range of management practices that a farmer can employ to regulate soil fertility is limited

TABLE: 14.1: Farmers Management Practices for influencing Soil Fertility Manipulation of Biological Process

Management practice	Constraints to use	Biological processes influenced	Soil fertility effects
Biological inputs	Cost		
Rhizobium inoculation	Availability of inoculum	N-fixation	Increased N-acquisition
Mycorrhiza inoculation	Environmental adaptation	Nutrient uptake by mycorrhiza	Increased efficiency of uptake of P and other nutrients
Soil fauna inoculation	Competition with fertilizer		Increased efficiency of H_2O uptake
	Competition with indigenous biota		Increased heavy metal tolerance
		Fauna burrowing	Soil structure/porosity
		Decomposition	Stimulation of nutrient release
Organic matter inputs			
Crop residues	Recycles nutrient only	Decomposition	Increased short-term nutrient availability
Root residues	Inaccessibility to management	SOM synthesis	
Weed residues	Crop impact from competition		Increased nutrient storage/exchange
Tree litters/ prunings	Land set aside		
Green manure	Land set aside		Soil physical structure improved
Farmyard manure	Livestock and fodder availability		Soil water regimes improved
Household waste	-		Acidity/toxicity diminished

Management practice	Constraints to use	Biological processes influenced	Soil fertility effects
Purchased organic input	Cost	Soil fauna/microflora growth	Macropore formation improved (macrofauna)
Pre-composting	-		Soil aggregation improved (microflora)
	All: Labor availability		
	Opportunity cost of other uses		
Inorganic fertilizer inputs	Cost Availability	Mycorrhiza inhibited	Direct transfer of nutrient to plant increased
		N fixation inhibited	Nutrient losses increased
		Mineralization/immobilization balance changed	Acidification
Tillage		(Intensive tillage)	(intensive tillage)
Conservation	Soilborne disease	Decomposition stimulated by OM incorporation	Short-term nutrient availability increased
Hand	Labor	SOM decay stimulated by aeration and particle size reduction	Root growth instilled layer promoted
Mechanized	Cost	Faunal and microbial populations diminished	Nutrient losses increased Long-term nutrient storage diminished
Pesticides	Cost Environmental and health impact	Nontarget organism populations diminished or eradicated	Destabilization of nutrient cycles Loss of soil structure

The above practices are not unique to biological management but common to farming practice the world over. The use of biological inputs such as N-fixing bacteria, mycorrhiza, or soil fauna as a means of enhancing the endemic biological activities are the means of biological soil management. Direct management is also achieved by the use of organic matter inputs – in effect a means of selectively feeding the heterotrophic biological populations of soil – a practice of very ancient origin but at times eschewed in modern agriculture. Equally direct but usually unintentional effects are also achieved by the use of pesticides, which may kill particular groups of soil organisms that are involved in processes of significance to soil fertility. Management techniques such as tillage and fertilization also influence the activity of the biota indirectly by altering the physical and chemical environment of the soil.

The success of biological management practices depends on following two preconditions which must be satisfied:

1. The availability of a management practice that is practically and economically acceptable to the farmer
2. The demonstration by the scientist that the practice leads to enhanced soil fertility.

BASIC RECOMMENDATION FOR INITIATING ADOPTION OF ORGANIC FARMING

Agronomic

Fertilization

1. Promote soil organic matter management, balanced fertilization, and soil protection as basic elements of soil management.
2. Facilitate an efficient, complementary use of mineral fertilizers, preferably of local origin such as phosphate, in the conversion period in areas where this is necessary to initiate the nutrient cycling.
3. Avoid the use of easily soluble chemical fertilizers, especially near sources of drinking water, in densely populated areas, and in areas of high ecological value.

Regeneration

1. Develop and implement organic matter management as the technique for regeneration in areas where soils are degraded.
2. Develop and implement organic matter management in combination with water harvesting and water storage as the primary techniques in areas where water is the limiting factor for agricultural production.
3. Enforce techniques such as alley cropping and contour farming in hilly and mountainous areas.

Pesticides

1. Phase out the use of synthetic pesticides, beginning with those most harmful to the environment and human health.
2. Replace synthetic pesticides by alternative pest management and techniques based on traditional and modern knowledge with a strong preventive character such as cultural practices, biological control, and selective botanical pesticides.
3. Immediately restrict the use of herbicides in areas subject to erosion.
4. Phase out the use of herbicides and replace them alternative strategies and techniques such as rotation, mulching, intercropping, and mechanical weeding.

Post-Harvest

1. Emphasize human and environmentally friendly local storage and conservation techniques for agricultural products.

Socio-Economic

Farm System

Promote mixed farms and diversification to improve resources efficiency, food self-sufficiency, and income security.

Human Resource Management

Develop education and training programs for extension officers and farmers for ecosystem management so that farmers can make choices based on a thorough understanding of their local agro-ecological production capacity and the market system.

Conversion

Assist farmers (especially women) intensively during the period of conversion towards sustainable agriculture systems (up to 5 years) and compensate them financially when necessary.

Processing

Promote on-farm or communal processing.

Marketing

Develop appropriate domestic and international marketing structures to ensure outlets for all products of organic farms.

FACTOR AFFECTING ADOPTION OF ORGANIC FARMING

The appropriateness and potential for adoption of organic farming practices depend upon a range of interrelated environmental and socio-economic factors. These are as follows:

The Degree of Intensification of Farming System

The potential for adoption of organic agriculture varies upon on the degree of intensification of the farming system. This, in turn, is related to population size growth and consequent land pressures.

Those farmers who practice unimproved traditional farming, where soil productivity tended to be sustained solely by fallowing. These farmers may not expending any serious effects from their current farming practices and are unlikely to be interested in any new techniques. However, they typically have limited access to agrochemicals and there may be scope for researching and promoting simple alternative organic techniques, especially for pest and disease control.

The second group of the farmers widely use agrochemicals. The farmers have reached limits to productivity through traditional means, and have had to adopt new technologies, the most common being use of chemical fertilizers. Traditional practices are still used, but they were supplemented with chemical inputs. Many of these farmers used a mixture of organic and inorganic techniques, but that they have generally not used chemicals to the extent of directly experiencing negative effects, and the medium to long term effects of using agrochemicals have not been explained to them. As such, the farmers tend not to appreciate any comparative benefits of using organic techniques, and are unwilling to adopt these unless artificial fertilizers and pesticides are unavailable or can not be afforded. Many of these farmers have previously used chemical fertilizers, but had stopped when subsidies were removed and the fertilizers had become too expensive. Thus there is an economic incentive to develop cheaper alternatives. Considerable potential exists for optimizing the use of available natural resources, and for a detailed exploration of the compatibility of organic and inorganic inputs. This group of farmers is also likely to experience many of the general constraints to the adoption of organic techniques.

The third group of farmers practice intensive agriculture. They rely heavily on agrochemicals, producing mainly for an urban market. However, they are often also the farmers who use the most sophisticated and intensive techniques of organic soil fertility management. They have adopted these strategies due to the need to intensify their production systems, and have been exposed to extension activities and introduced to new improved technologies. This group of farmers shows a relatively limited interest in using exclusively organic techniques, although observation of the negative effects of chemical inputs and the economic reality of increasingly costly inputs has served as incentives for the adoption of organic techniques. Providing that there is not a relative drop in fertilizer prices, it is likely that farmers would continue to use chemicals, and organic strategies for improving soil fertility management have considerable potential with this group. In particular, manure management can be substantially improved and there exists great possibility for urban waste utilisation.

Organic pest and disease control methods are considered ineffective by this group, and therefore not viable alternatives to pesticides. This area requires further research to prove its effectiveness before any serious uptake by farmers can be expected. However, in the short and medium-term, if environmental aspects, food safety and health prove to be considerations at least for consumers, there is a need for the promotion of safer pesticides and pesticide use, and for the use of more benign techniques such as biological control and the use of resistant varieties.

Significant potential to raise productivity through the optimized use of locally available natural resources exists in farming systems at all three levels of intensity.

Agroclimatic Factors

Specific organic farming techniques may be expected to be more suitable and common in some climatic zones than in others, and the constraints to sustainable husbandry are also likely to differ according to agro-climate. For example, the extent of manure use is dependent on livestock numbers and management, and these vary with agroecological zone.

The wet and humid tropics are characterised by nutrient poor soils and by high oxidation and mineralisation rates, and this results in difficulties in ensuring the presence of sufficient nutrient and organic matter inputs in cropping systems. Transport of organic materials is difficult due to heavy rainfall and inaccessible paths. Although vegetation growth is prolific, the rate of decay is fast and the management system such that the quantities of crop residues are too low for farmers to rely on these for sustainable organic matter inputs.

In the semi-arid areas where, livestock numbers are high and the management regimes of relatively low intensity. With biological processes being slower than the humid tropics, biomass production is lower and many of the sustainable technologies are less efficient. Farmers experience a lack of organic materials for soil fertility. This is aggravated by the fact that manure is used for fuel in some areas. The collection and distribution of organic material is especially time-consuming in the drier areas, where the farming systems tended to be extensive and cropped plots widely dispersed. Investment in land rehabilitation may not seem like a rational option when climate-related risks are high and returns to investment unpredictable. However, of the rehabilitation techniques available, organic manures are sometimes preferred by farmers in dry areas since chemical fertilizers require water to prevent crops scorching, and crops perform better without fertilizers under drought conditions. Organic manures also increases the water holding capacity of the soil.

Labour Inputs

A frequently cited constraint to organic soil fertility management is that of labour. This is true both in general and for individual organic techniques. For example, farmers identified lack of labour as a constraint to the adoption of composting. Organic soil fertility maintenance may also involve manure collection and spreading, composting, tree pruning and green manure incorporation. These techniques are likely to require more labour then either no maintenance traditional farming or modern high input mechanized agriculture. In areas with low natural resource potential, the labour requirement tended to be the highest and returns to the investment took longer to be realized. Cash poor farmers who could not afford to hire labour, and farmers who depended on off-farm work to supplement their farm income, may not have been able to adopt any agricultural practices which increased the labour burden; particularly if short-term increases in farm productivity is not guaranteed. Hired labour are primarily used for weeding, and as the cost of labour increased, so herbicides become an economic alternative. It is not just to availability of labour *per se*, but also unwillingness to spend the necessary time. Views such as 'the practice of organic arming is tedious to undertake' and that 'collecting and preparing ingredients needed for organic farming is tedious and time-consuming'

Environmental Concerns/Negative Experiences with Agrochemicals

In the more intensive farming systems, where artificial fertilizers have been promoted and used since the 1960's, there has been some experience of the negative impact of agrochemicals. There, farmers have an appreciation of the comparative benefits of organic soil inputs and many wish to reduce their fertilizer used. There is a widespread perception that chemical fertilizers are damaging both soil structure and soil moisture holding capacity and also creating an economic dependency on costly inputs. There is a clear understanding of the relationship between organic fertilizers and soil physical properties. A commonly known fact is that chemical fertilizer provide some concentrated nutrients for the plant but that manure or compost is required to feed and give structure to the soil, and that chemical fertilizers are effective for only one season whereas manure can last for two to four seasons. Manure is therefore seen as an investment.

Other farmer groups, who have not used fertilizers to a great extent, also have some reservations about artificial fertilizers for environmental reasons e.g. once they start using chemical fertilizer they have to continue to do so. With both limited cash and fertilizers supply in such areas, this creates feelings of fear and risk.

There is limited appreciation of alternatives to fertilizers, and lower availability and more constraints to manure use.

Land Size and Ownership

The regional land allocation/inheritance systems influence both tenure rights and the size and distribution of farms. This can directly or indirectly influence the current and potential development of organic techniques. Farmer willingness to adopt sustainable agricultural technologies which require investment in time and/or money and which will provide benefits in the long term, is directly related to land tenure.

Farmers have a choice of using manure or chemical fertilizer, rented land tends to receive less manure and more fertilizer than owned land. This is because the farmers do not wish to invest valuable manure for someone else's benefit, since the land could be repossessed by the landlord at any time. Furthermore, there is an incentive to use high chemical inputs on rented land to ensure a rapid return to cover the cost of land rental.

Cropping Systems and Sub-Systems within the Farm

Important differences also exist within the farm unit or among distinct holdings of one particular farmer. Many farmers operate a number of enterprises which are often spatially separated. For example, there is often more scope for sustainable soil management in home gardens, or in plots close to the homestead, than on the larger/more distant plots where the main cash and field crops are grown. This appeared to be the case with the current use of organic techniques. Farmers practices a bush fallow system without fertilizers on the main farm plots, whilst women use organic inputs to improve the soil fertility on vegetable plots. There is selectivity on the type of crop being fertilized by organic manures.

Proximity to Urban Markets

Farmers in remote areas with poor infrastructure have limited access to chemical inputs. In this situation, there is little option to achieve a more productive agriculture than to optimize the local resources available through organic farming techniques. Near urban areas, farmers tended to be commercially orientated and could generally afford to purchase inputs.

The demand from urban markets increased the use of pesticides to attain a high standard of crop appearance, and the use of fertilizers for continuous intensive vegetable production. Off-farm employment in the city may be seen as an attractive alternative to farming, particularly when the farming system requires high labour inputs, long hours, or direct handling of agrochemicals

However, farmers living relatively close to urban centres do have access to agro-industrial or urban municipal organic wastes, offering greater potential for organic soil fertility management. Intensification of animal production through zero grazing, due to high demand and lack of space, may increase the opportunities to produce and manage manure as a soil input.

Economic Factors

There is a link between the degree of commercialization and the use of agrochemicals, with farmers who supply well-developed markets having both the economic incentive and the ability to access inputs. Farmers using credit facilities is significantly higher near the urban areas than in rural areas, and there is a significant positive correlation between credit adoption and fertilizer used. Having limited funds to purchase fertilizers is seems to be an important reason for adopting alternative soil management techniques.

Organic fertilizers themselves are not always readily available at a low cost. The recycling of plant residues is rarely adequate on small intensive farms, and unless management practices include the concentrated use of green manures, N-fixing crops and zero-grazing by products, then external sources of fertilizer are necessary. This results in a market for organic fertilizers and composts, and they may become economic inputs with the same implications as chemical fertilizers for poor farmers. However, organic inputs on the market are more likely to be produced locally and often traded within the local exchange/bartering network.

There is a market for animal manure, it is sometimes transported over considerable distances. The same applied to urban wastes and poultry manure from intensive poultry farms in several countries. In the majority of cases, these markets are informal and with the little information available on the costs and value of the available material. These costs and values would be relative to that of chemical fertilizers but would need to take into account the potential long-term benefits from their use.

Education/Lack of Knowledge

A lack of knowledge was by far the most widely cited reason for not adopting organic farming techniques. This included a lack of awareness of the existence of organic methods, a lack of training and detailed instructions on application methods and rates, and lack of examples of success.

It is often assumed that farmers have a comprehensive indigenous knowledge about soil fertility management and pest control. This is frequently not the case. Farmers are not optimizing the natural resources at their disposal, nor are they clear as to why they were carrying out certain practices. Farmers stress that the use of organic inputs is not a positive decision based on value for soil fertility but a 'passive' or 'pretty much accidental' use. These inputs and practices included poultry manure and wood ash thrown onto gardens, crop residues left in the fields and unplanned and informal composting of piles of sweepings. However, the farmers are unintentionally following the habits and customs of those around them and before them, whether they understood it or not, and they are likely to underestimate their innate ability as farmers.

Farmers commonly lack the appropriate skills to convert their farming systems to become more productive and yet sustainable. Furthermore, the predominance of agrochemical use over the last 20 years or so has directly encouraged the loss of traditional knowledge. Examples are specific techniques, such as pit composting, that have been used in the past but have not been adopted by the younger generation. The same is true for traditional pest and disease control methods. Also, non-traditional crops are widely grown which are associated with pests for which there is no traditional knowledge. Many farmers feel that there are no effective alternatives to pesticide use; that the crop varieties they were cultivating are 'foreign' and therefore not strong enough to grow in the local environment without the use of pesticides.

The long-term benefits of organic fertilizers are often not clearly appreciated by farmers. The majority of farmers are aware that manure is of benefit to crops but not of the longer term effects of manure on soil fertility. As a result there is little understanding of the effect of cessation of manure application and the implications of this for the maintenance of soil fertility. Generally, manure management was poorly developed in many parts of the developing countries. The understanding and promotion of organic technologies for soil fertility maintenance was generally very low amongst extension staff. There was also little evidence of NGO work on organic agriculture among the random sample of farmers interviewed.

Availability and Cost of Organic Materials

Unavailability of organic materials as being the prime constraint to utilizing organic soil inputs. The particular materials mentioned are animal manure and crop resides. This highlights a lack of knowledge on the wider range of techniques possible.

Farmers in many areas are faced with a choice of uses for available materials. In areas, where there is a shortage of fuel wood, cow dung is used for fuel, in some areas, crop stalks are collected as fuelwood, for construction and mat making. Crop residues are often too valuable as an animal feed to be used as fertilizer or for composting.

The cost of off-farm organic materials may prevent the poorest farmers from using them. The farmers perception of the benefits of manure increase, so too do the price, particularly as the cost of inorganic fertilizer has also risen. The use and cost of the various types of manure is dependent on the distance of the farmer from the source.

Culture

In many areas, there is an unwillingness to handle waste, particularly human waste. Considerable social stigma is attached to manure handling as it is often mixed with human excreta and rubbish.

Gender

Consideration of the influence of gender the distribution of land, activities, income and decision-making need to be taken into account in relation to the potential for organic farming strategies. Women are often responsible for cultivating the small vegetable gardens as a contribution to household food production or for the market. These gardens currently receive organic inputs and have considerable potential for the further development of organic techniques.

Aversion to the Use of Organic Methods

When converting from chemical to organic inputs, there is usually a fall in productivity over a transitory period, until fertility, structure and microbial activity of the soil have been restored and enhanced through organic matter inputs. If properly managed, this fall is usually thought to be no more than 20% of crop yield, although this difference may be greater if very high yields are previously being obtained from over-use of agrochemical inputs. In agroforestry, techniques are being introduced, the effects from trees planted to provide products and improve soil fertility will usually not be realized in the first few years. It may not be feasible for resource poor farmers to adopt alternatives to fertilizers unless short-term benefits can be realized. Farmers express a fear of changing to organic farming methods for fear of failure or risks of reduced yields, and a lack of 'immediate' results as a reason for not using organic inputs.

The problems of increased weeds, pests and disease is also perceived by the farmers. Farmers believe that animal manure encourages weeds'. It was a common that manure from animals which had been fed infected residues could carry diseases such as tomato and potato blight, or 'compost heaps attract very difficult pests and animals' or 'some claim the methods, like application of mulch, harbors pests which destroy crops'. These misconceptions again highlight the lack of sound knowledge on the subject.

Perceptions of the value of organic inputs is very important. Some farmers are reluctant to use poultry manure in vegetable production systems as they perceived it to have slow nutrient-release properties, to encourage pests and diseases within the soil and to increase decay in vegetables during post harvest storage. In contrast, some places poultry manure is highly valued as fertilizer, and it is often used on vegetables.

Adverse comparisons of organic inputs with non-organic ones are also expresses in phrases which indicates that the former are considered to be *old-fashioned*.

Active Promotion and Preference for Agrochemicals

There is concern among farmers of the high costs of agrochemical pesticides, and also on their safety. The emphasis is clearly on agrochemical pesticide use. This, combines with the facts that pesticides are readily available, considered effective, and give the product a marketable appearance, encouraged the farmers to view natural pest control methods as having limited potential.

The uptake and increased use of chemicals has been directly influenced by active promotion. The influence of government agriculture departments and, more particularly, the extension services, on choice of farming techniques is a reason for use of fertilizers. Comments such as 'Government extension service advocates modern farming methods' are common, and several replies referred to the fact that the extension services are trained to use, and therefore promote, chemically-based methods. Government encourages the use of chemicals in their radio agricultural programmes. The promotion, through subsidy, of fertilizers and pesticides which has contributes to farmers becoming used to non-organic methods over a long period of time.

ADOPTION IN INDIA

For centuries, Indian farmers practiced a balanced system of agriculture, which was based on a thorough understanding of environments, seasons, crop mixing, rotation and gene preservation. Yet, over time practices became ritualized, without concern for the concept of balance. Hence, the systems stagnated and failed to respond to the changing needs of the times. After independence in 1947, traditional agricultural systems in India failed to meet the needs of a burgeoning population and suffered setbacks through a series of natural disasters. The government took on the responsibility of improving agriculture, by making huge investments in irrigation development and fertilizer production. The focus of agricultural research was on high-yielding varieties. As a result, chemical-intensive and water-intensive agricultural practices were actively encouraged. Heavy subsides for chemical inputs induced farmers to practice chemical farming.

The fall-out of such practices was soon amply evident. It is in this context the organic farming came to the fore. Many farmers took it up because of the ill-effects of modern farming. Though still in its infancy, organic farming has a strong ideological foundation and offers many fresh ideas. It is a response to the growing eco-awareness in agriculture.

By the late' 80s, three types of people were into organic farming. The first type consisted of urban-educated technocrats. They embraced this new ideology, for the sheer pleasure of feeling the soil. But, in several cases their enthusiasm was short-lived. The second type consisted of farmer with a modern education. Their farming practices reflect their earnestness to understand the strengths of traditional farming as well as their ability to assimilate scientific knowledge. A third group consists of ordinary farmers, who by sheer observation, trial and error have tried to understand and practice organic farming. Though this third group is insignificant in number right now, the future of organic farming in India lies with them.

Two striking common factors underlie most of the success stories of organic farming in the country. Almost all the successful farmers have access to sufficient water, and enough organic inputs in and around their farms. Many of them therefore grow cash crops like sugarcane, areca, cocoa, coconut, pepper and various spices. These farmers have effectively allayed apprehensions that a switch-over to organic agriculture would affect yields and income. But small and marginal dry-land farmers are hesitant to change over precisely because they lack the resource advantages of the successful farmers (water and organic inputs close to the farms).

Causes of Low adoption

2. Chemicals are easy to use and less costly
3. The benefit of organic practices are not seen immediately
4. Large quantities of organic inputs are required
5. Difficult to get organic fertilizers
6. Unorganized market for organically grown produce
7. Preferential behavior of consumers towards organic food not yet established
8. Economic loss due to transition (from traditional agriculture to organic agriculture or conventional to organic)
9. No experimental evidence on the cost benefit ratio of organic farming
10. Government effort to propagate organic farming is very little
11. Scientific research programmes on organic farming are also scarce

The main element that is brought up time and time again is the fact that there is lack of knowledge about the above. The above points show the areas in which education is needed. The government also needs to be informed, as this is one of the first steps in telling people about the benefits of organic agriculture, because many think whatever the government says is right and best.

SUCCESS WITH ORGANIC FARMING

The Organic farming has all round favourable effects and these effects are documented which ultimately results in long term sustainability of the production system. To strengthen this fact some examples of the areas or farms where organic farming is being followed since long time, are as follows:

Improvement in the Environment

Organic farms require both to avoid all forms of pollution that may result from agriculture and to maintain the genetic diversity of the agricultural system and its surroundings, including the protection of plant and wildlife habitats. These principles are given equal status with those of production and economic viability, so that organic farming systems consider environmental protection as one of the key outcomes of farming practice. The complex interactions between farm management and the components of the farming system interact with natural eco-systems at the field, farm and regional scale.

Soil Health

Organic Matter Build-Up

Organic matter build up in organically managed soils is widely reported such increases reflect the inclusion of leys and green manures in crop rotations and the incorporation of crop residues and manures into the soil. Higher levels of Carbon have been measured in the *light fraction* of soils under organic management, which is thought to represent a more biologically active pool. Changes in organic matter drive/underpin many of the other changes in soil biological and physical properties.

Biological Activities

Higher levels of biological activity and increased microbial populations have been measured in cultivated soils managed organically. However, the differences between phases of the rotation in any system can be equally great; microbial activity was always higher under a ley than a wheat crop. Bacterial biomass was similar in long-term grasslands, whether managed organically or conventionally measured an increased phospholipid fatty acid (PLFA) content in organically managed soils which was correlated to biomass C contents, but could not show any difference in the taxonomic composition or physiological status of the microbial biomass. However, increased abundance and diversity of the microbial community was observed in soils managed organically.

In long-term grassland the size of the active fungal population was increased under organic compared to conventional management. In cultivated soils increased fungal diversity is also seen, including more zygomycetous fungi and more species which are antagonistic to plant pathogens. The species richness of *Fusarium* was increased in soils under organic management, with the proportion of antagonistic *Fusarium* species higher than under conventional management. Increased numbers of nematodes have also been measured in long-term grassland soils. In cultivated soils in organic farming systems the use of composts and manures increases the numbers of non-parasitic nematodes.

An increase in earthworm abundance and activity was measured in cultivated organically managed soils. Increased in the biomass, density, and population diversity of earthworms we observed in organically managed soils. However, in long-term organic grassland. Rotational phases can have a large effect, especially in crops where intensive tillage is required such as potatoes.

Physical Conditions

Aggregate stability and porosity was increased under organic management.

Such changes in soil structure largely result from changes in the management of crop residues and organic matter inputs and may lead to increased aeration and water holding capacity of soils under organic management Lower rates of run-off and soil erosion have also been measured in organic systems. Increases in the depth of the A horizon of the soil have been recorded for organic systems, which may result from decreased bulk densities and/or decreased soil losses through erosion. However, several research studies have found no differences in the soil physical properties between organic and conventional farming systems and no differences in soil particle detachment between organic and conventional systems. Measurable changes in soil physical properties can take decades to establish.

Chemical Properties

Available phosphorus (P) and potassium (K) have been shown to increase and decline in soils under organic management. Increased soil pH has been measured in some organic systems. In New Zealand lower levels of total S and total P was measured in soils under organic management, as conventionally managed soils often received large inputs of these nutrients. Most changes in soil chemical properties could be predicted from a consideration of the inputs and outputs of nutrients to the system.

Nutrient budgets are a useful tool to summarize the nutrient status of complete systems. Depending on the farm management and the balance of imports and exports of nutrients, nitrogen, phosphorus, and potassium budgets range from deficit to surplus in organic farming systems. Balanced budgets are often more easily achieved in mixed farming systems; where imports of manure were large in a horticultural system, N, P, and K budgets were very positive.

Water Quality

Where nitrate leaching losses are measured directly, they are generally significantly lower than comparable conventional systems measured leaching losses between 7 and 52 kg N ha^{-1} at all points in the rotation on organic farms and suggest that overall farm losses would average c. 20 kg N ha^{-1} $year^{-1}$. However, where nitrate leaching is expressed relative to crop production levels rather than on an area basis, differences between conventional and organic systems are reduced. Lower nitrate concentrations was measured in porous cups under organic crops.

The management of residual nitrogen from legumes is a key to the minimization of nitrate leaching. Large leaching losses of N have been measured following the cultivation of grass-clover leys. About 33% of the N accumulated above ground in cut-mulched green manures was lost by leaching, when they were incorporated in the autumn. The timing of incorporation and the nitrogen demand of subsequent crops must be carefully considered. Storage and composting of farmyard manure on non-paved surfaces can also result in significant nitrate leaching. Covering manure or pre-composting on paved areas can reduce these losses. Leaching losses from manures and green manures can also represent significant losses of potassium, which while they may not be environmentally significant represent agronomically important removals of potassium from the farming system.

Measurements of soil mineral N concentrations tend to indicate no increase and often a decreased leaching risk in organic farming systems. Similar nitrate concentrations was measured in the soil in autumn in organic and conventional farms and suggested that there was no inherent difference in the leaching risk from conventional and organic farming systems

Nitrogen budgets are generally positive for organic systems. There is surplus nitrogen, which may be lost by nitrate leaching. Lower nitrogen surpluses was measured on organic than conventional mixed dairy farms. However the nitrogen surplus of organic pig production was greater than that of conventional systems. In livestock systems, the nitrogen surplus increases with increasing stocking density.

The water quality risks associated with the pesticides used inorganic farming systems remain largely unstudied. However, the permitted active ingredients, pyrethroids and rotenone, are not persistent and are also highly non-mobile in soil. It is only, therefore, by spillage or mishandling that these might enter water resources. The absence of conventional synthetic pesticides in organic farming systems suggests that the risks of water contamination are significantly lower than in conventional systems. Prohibition of the routine use of antibiotics organophosphates, in-feed supplements such as copper and zinc in livestock production reduce the impact of organic farming systems in comparison with conventional livestock systems. However, certain products such as synthetic pyrethroids permitted for ecto-parasite control in organic livestock, do represent a threat to water quality if spilled. The risks of contamination of water with pathogenic organisms are reduced in organic farming systems due to the reduced intensity of livestock grazing and housing.

Air Quality

There has been little work carried out on gaseous losses from organic farming systems. Calculated CO_2 emissions for crop and dairy systems suggest that emissions are generally lower in organic farming systems. CO_2 emissions and energy use are usually closely linked. Nitrogen surpluses in organic farming systems, discussed in tend to be lower than those for conventional systems, which indicate a lower potential for gaseous nitrogen emissions. Estimated NO_2 emissions, which also contribute to the greenhouse effect, show no clear differences between organic and conventional farming systems. Reduced emissions of ammonia from livestock housing in organic systems is likely due to the increased use of straw for bedding. However, handling and spreading manure can result in significant losses of ammonia. Lower stocking rates are likely to result in lower gaseous nitrogen emissions per hectare, even though emissions per animal are likely to be similar in organic and conventional systems.

Research on methane emissions is very scarce in organic farming systems. Methane emission per hectare is likely to be lower in organic farming systems, due to lower stocking rates.

Energy Use

For arable crops, lower energy consumption in organic farming systems in commonly calculated. Lower energy input to direct crop protection operations in organic farming systems was partially offset by greater investment of energy in cultivations and showed that although three times as much machine energy was required to produce an organic crop, when the external energy inputs of fertilizer and pesticide production were taken into account organic farming systems required only half the energy input of a conventional system. For crops requiring repeated mechanical weeding, e.g., potatoes, the energy consumption per unit output may be the same or greater than that of conventional farming systems. For livestock farms, 70% lower energy consumption, mainly due to reduced imports of feed. Variation between organic farming systems in terms of energy consumption is high, which is controlled by the soil type, proportion of fodder production on the farm, and proportion of weed control achieved by secondary cultivations.

Biodiversity

Flora

The use of cultural controls and the targeting of weed control to specific periods in organic systems results in a greater abundance and diversity of weed species in organic systems than conventional arable systems. Weeds were able to complete their development in organically managed fields of winter barley, while in conventional systems, weeds were confined to margins and adjacent areas.

Un-cropped areas [sown grass strips (beetle banks), grass margins, un-cropped wildlife and flower strips, hedges, ditch and bank habitats] are intrinsic in organic regimes where their management is central to the regime. These areas tend to be the reservoirs for both faunal and floral diversity, and therefore, the nature and extent of these habitats are the key in determining the overall biodiversity of agricultural areas showed that organic farms had a greater diversity of landscapes and in general had “more forms, colors, smells, sounds and spatial experiences.” Because of the diversity of land-use types and the adaptation of land-use to the diversity of physical features, organic farms allowed a greater diversity of landscape elements.

Hedges tend to be higher and wider, field boundaries contain more trees and field sizes are smaller on organic farms. In N. America the scarcity of hedgerows and herbaceous field margins has reduced much of the potential for wildlife integration within agricultural landscapes, which remains in much of Europe. However, field margins in organic farming systems did not contain significantly more diverse plant species than conventional margins. Floral species richness, and abundance of weed species in field margins is heavily influenced by management. In the short term at least, management techniques which increase biodiversity in the field margin do not result in increased weed ingress to the adjacent field edge

Management of grassland, particularly the intensity and balance between cutting and grazing and the level of external inputs, rather than the system to which it belongs controls potential to provide more stable, less disturbed environments for the establishment of biodiversity. Established organic pastures showed more early flowering species. However, there were few major differences in flora between organic and conventional mixed lowland grassland. Experimental grass leys managed for biodiversity conservation by sowing with a species-rich seed mix and without nutrient inputs were less productive and tended to harbor fewer pernicious weeds than conventional leys, which gradually became dominated by rye-grass.

Landscape

Issues relating to scale and landscape structure (e.g., patch fragmentation and heterogeneity) are increasingly important in ecology. They are particularly pertinent to arable farmland, which is uniquely characterized by a mosaic of distinct habitats and by severe and abrupt perturbations e.g., harvesting, cultivation. In comparing organic and conventional farming systems we are looking at large scale landscape processes. These are often also the hardest to study and indeed most of the studies reviewed here which inform us about system differences focus at a much smaller level on differences between individual fields. Overall the combination of management and structural practices adopted under an organic farming regime clearly have the potential to enhance the biodiversity of arable farmland. However, in terms of the regional biodiversity, effect of organic farms is small since they currently occupy less than 1% of agricultural land.

Sustainability of Food Production System

The global population is expects to reach 8.5 billion by the year 2025. A frequently expressed concern is whether, if organic farming were adopted on a large scale, global food requirements could still be met. This issue is highly complex. It is, of course, a different question than whether malnutrition will be prevented, as it is clear than even in countries with ample food supply quantities and of appropriate quality, malnutrition may still exist. Malnutrition has many roots: inadequate food supply, limited purchasing power, poor health conditions, and incomplete knowledge about nutrition. Food supply is also somewhat divorced from the technical ability of farming systems to provide crop and animal products but is largely controlled by the systems and infrastructure for distribution, markets, finance, and political structures.

The limited studies that have been carried out to date to study the implications of increasing use of organic farming systems on global food production and food security suffer from serious problems with underlying assumptions, availability of data, and the limited range of factors analyzed. For example, in Europe, we know that crops yields are commonly 20-40% lower in organic systems compared with conventional systems. Rotational constraints prevent crops being grown so frequently on arable farms, so that the overall yield reduction for cereal crops may be 40-50%. In northern Europe domestic food demand could be met with organic methods, but food would be more expensive. Similar conclusions were reached in the studies in the U.S. The proportion of grain legumes grown would increase substantially, while some studies suggest the vegetables crops would stay stable or increase slightly, and other crops such as oilseeds and sugar beet would stay stable or increase slightly, and other crops such as oilseeds and sugar beet would decline by as much as 75%. But most of these assumptions reflect current demand patterns for organic products, which place more emphasis on horticultural crops and less emphasis on crops for processing such as oilseeds and sugar beet. The extent to which these crops are produced organically will depend on the market for them, and this is still very difficult to predict.

Future livestock production levels are equally difficult to determine. It is reasonable to suppose that the production of ruminant livestock might decline by 10-30% and that pigs and poultry production levels might be substantially lower as a consequence of reverting to land-based, extensive systems. In addition, an implicit assumption is often made that we actually need to maintain current patterns of crop use. Organic farming, with its emphasis on farm-produced feed for livestock, is likely to result in significant reductions in the total quantities of cereals and other crops used to feed livestock, emphasizing instead forages produced as part of maintaining soil fertility or in areas not suited to crop production. By relying more on home-produced proteins, organic farming is also likely to reduce the need for protein crops, like soya, produced in developing countries, thus releasing resources to meet domestic food needs. Consumer demand patterns may also change to favor low meat diets as awareness of the environmental, welfare and health costs of intensive livestock production systems increases.

In addition, the assumption is often made that the yield reduction associated with organic farming in Western Europe will also apply in other parts of the world. The evidence is to the contrary. Where resource-poor farmers are unable to afford purchased inputs, the information-intensive, agro-ecosystem management approaches which characterize organic farming may be critical to achieve production increases. However, such approaches require changes from teaching and dissemination approaches to agricultural extension to the development of methods which develop the farmers capacity to think for themselves and develop their own solutions.

Many authors have argued that it would be possible to feed double the current global population only if we continue to intensify production through increased use of fertilizers, energy, bio-technology and water. Yet it is increasingly clear that the pressures on some of these resources, particularly water, and the negative environmental impacts of intensive input use on soil, water and air quality, are such that further intensification is not a sustainable option. Instead a strategy that increases total food supply over time, while doing this with lower levels of resource use and environmental pollution per unit of food produced, is a fundamental requirement–increasingly organic farming is demonstrating its capacity to do this.

BALANCE SHEET TO DECIDE ADOPTION

1. Most studies find that organic agriculture requires significantly greater labour input than conventional farms. This is especially true in areas of low ecological potential. However, when labour is not a constraint organic agriculture can benefit underemployed labour in rural communities. Furthermore, the diversification of crops typically found on organic farms, with their various planting and harvesting schedules, may distribute labour demand more evenly which could help stabilize employment. Land tenure is also critical to the adoption of organic agriculture. It is highly unlikely that tenant farmers would invest the necessary labour and sustain the difficult conversion period without some guarantee of access to the land in later years when the benefits of organic production are attainable.
2. Soil-building rotations need to be designed both from the economic and the technical points of view – uses must be identified for all the crop and livestock products produced. As in all agricultural systems, diversity in production increases income-generating opportunities and can, as in the case of fruits, supply essential health protecting minerals and vitamins to the family diet. It also spreads the risks of failure over a wide range of crops. It is possible that, even on those farms where organic crop yields are lower than those produced under systems which use high levels of inputs, the overall economic yields of the farm will be competitive since organic systems benefit from market premiums and sometimes lowered input costs.
3. In the situations where farmers cannot produce competitively for the market or are resource poor, the case income they obtain from farming is too small to allow them to buy external inputs. High transport and credit costs are amongst the factors that make these inputs expensive and risky to use. Where ecological conditions are not particularly favourable for farming or the soil has been degraded, using synthetic fertilizers is relatively inefficient. In these conditions, synthetic fertilizers do not generate enough profit to make them a viable option in compensating nutrient losses. Increasing numbers of farmers are being confronted with these problems as the relative cost of synthetic fertilizer increases and soil degradation accelerates. If these farmers continue to produce for the market and export/loose nutrients in quantities higher than they can afford seen their options for compensation of nutrient losses, further nutrient depletion, soil degradation and yield decline will be unavoidable. At the moment this process can be seen in many regions in Africa and elsewhere in the tropics. Organic farming based on ecologically sound practices, therefore, seems to be the only sustainable basis for livelihood within the reach of these farmers. This does not entirely exclude production for the market but must be kept to levels that farmers can afford.
4. The insights generated by organic farmers in their search for site-specific production strategies can be of great benefit to non-organic farmers interested in expanding their management options. However,

organic farmers still face huge uncertainties. A lack of information is an obstacle to organic conversion (e.g. surveys find that 63 percent of sub-Saharan African farmers and 73 percent of North American (US and Canada) organic farmers cite a lack of knowledge as the greatest barrier to adoption). Extension personnel rarely receive adequate training in organic methods and studies have shown that they sometimes discourage farmers from converting. Furthermore, institutional support in developing countries is scarce. Professional institutions with capacity to assist farmers throughout the production, post-production and marketing processes are non-existent in many developing countries. While there are helpful research results that immediately could be extended to farmers, much more are needed. In 1990, FAO sponsored a conference at which organic research needs were identified (e.g. economics of stockless farms, animal husbandry, nitrogen cycling) however these challenges have largely gone unmet.

5. The explicit goal of organic agriculture is to contribute to the enhancement of sustainability. Nevertheless, negative impacts may occur and organic agriculture is not an exclusive method for sustainable farming. The soil and water protection and conservation techniques of sustainable agriculture used to combat erosion, compaction, salinization and other forms of degradation are evident in organic farming. The use of crop rotations, organic manures and mulches improves soil structure and encourages the development of a vigorous population of soil microorganisms. Mixed and relay cropping provide a more continuous soil cover and thus a shorter period when the soil is fully exposed to the erosive power of the rain, wind and sun. Terracing to conserve moisture, and soil are used in appropriate situations and particular attention is paid in irrigated areas to on-farm water management. Properly managed organic farming reduces or eliminates water pollution and helps conserve water and soil on the farm (although improper use of manure can seriously pollute water). A few developed countries compel or subsidize farmers to use organic techniques as a solution to water pollution problems (e.g. Germany, France).
6. Organic farmers rely on natural pest controls (e.g. biological control, plants with pest control properties) rather than synthetic pesticides which, when misused, are known to kill beneficial organisms (e.g. natural parasites of pests, bees, earthworms), cause pest resistance, and often pollute water and land. Reduction in the use of toxic synthetic pesticides, which the World Health Organization (WHO) estimates to poison three million people each year, should lead to improved health of farm families.
7. Organic farmers aim to make the maximum use of the recyclable fertility in on-farm crop residues (straws, Stover's and other non-edible parts) either directly as compost and mulch or through livestock as farmyard manure. Eliminating the use of synthetic nitrogenous fertilizer greatly lowers the risks of nitrogen contamination of water. Crop rotation is a widely used method of fertility maintenance and pest and

disease control, which is used in large- and small-scale farming in both developed and developing countries, especially under intensification. Fodder legumes are well-known fertility for the dual purpose of feeding livestock and adding nitrogen to the farm fertility cycle. Grain legumes may also produce a reasonable amounts f nitrogen to the overall farm system through biological fixation; other nitrogen-fixing plants such as *Azolla* may also be used.

8. Biological nitrogen fixation is a powerful technique but it often requires some addition of minerals to the soil, especially phosphorus. Most certification programmes restrict the use of mineral fertilizers which may be necessary to supplement the organic manure produced on the farm. Natural and organic fertilizers from outside the farm are used (e.g. rock phosphate, potash, guano, seaweed, slaughterhouse by-products, ground limestone, seaweed, wood-ash). While most certification programmes prohibit the use of sewage sludge and night soil they are still used in some places. However, sludge may contain many contaminants including heavy metals which can have a deleterious and cumulative effect on the soil, while night-soil contains human pathogens and must be carefully composed before use.
9. Crop rotations encourage a diversity of food crops, fodder and under-utilized plants; this, in addition to improving overall farm production and fertility may assist the on-farm conservation of plant genetic resources. Integrating livestock into the system adds income through organic meat, eggs and dairy products, as well as draught animal power. The crops and on-farm forestry integrated into the system provide shade and windbreaks while providing food, income, fuel and wood. Integrated agri-aquaculture may also be found within diverse organic agricultural systems. Economic objectives are not the only motivation of organic farmers; their intent is often to optimize land, animal, and plant interactions, preserve natural nutrient and energy flows, and enhance biodiversity, all of which contribute to the overall objective of sustainable agriculture to preserve natural resources and ecosystems for future generations.

Prospects

The increasing development of organic production systems and markets illustrates that this represents a valid alternative approach, often in direct opposition to intensive biotechnology agriculture. Certainly organic farming systems fit well with a view of the countryside where agricultural production is tempered with conservation and where less-intensive more environmentally friendly agriculture is practiced through the whole countryside rather than the North American model, where agriculture is separated into highly intensive bread-basket regions, leaving large nature reserves/wildernesses on the less productive areas.

Organic farming systems can deliver agronomic and environmental benefits both through the structural changes and the tactical daily management of the farming system. Organic farming aspires to a complex mix of agronomic, environmental, social and ethic objectives, which set the target for the development of a farming system in its particular location. In well managed organic farming systems are relevant both to developed nations (environmental protection, biodiversity enhancement, reduced energy use and CO_2 emissions) and to developing countries (sustainable resource use, increased crop yields without over-reliance on costly external inputs, environmental and biodiversity protection). With greater political will and investment in research, more of this potential could be realized.

To be the successful organic farmer an applied ecologist with a broad range of skills and knowledge is required. To provide and enhance those skills and knowledge, the development of research in organic agriculture has to focus on a much closer association with many aspects of ecological research to develop the tools necessary for understanding complex interactions at different systems levels. Such developments should help not only to further the development of organic agriculture but also inform and improve the sustainability of other forms of agriculture.

SUCCESS STORIES OF ORGANIC FARMING

Organic Food (Gujarat, India)

Organically grown food is now a popular concept everywhere particularly in the markets of the developed countries of the world. Bhaskar Hiraji Save, a farmer in a tiny village called *Dehri* in south Gujarat, has been well awarded of the benefits of organic farming four decades ago. While India was widely propagating the use of modern methods of agriculture, B.H. Save was making good progress on perfecting organic farming techniques.

B.H. Save, has not used fertilizers or pesticides in his farm 'Kalpavruksa', and still manages to earn a record profit of nearly 400%. The coconut and chiku plants in his farm have yielded more fruit than in farms using pesticides and fertilizers. In fact the rice grown in his farm fetched higher prices in the market than rice grown using chemical fertilizers owing to its superior quality.

B.H. Save started farming back in 1950 when he was a primary school teacher. He bought half an acre of land in Dehri. He took up full-fledged farming only in 1956. He began by using chemical fertilizers and pesticides like the others. *I not only used fertilizers and pesticides, but also had an agency through which I supplied the items to other farmers*, recalls B.H. Save. It didn't however, take him much time to realize that the input requirements as far as pesticides and fertilizers were concerned, were increasing steadily. In fact his cost of production was increasing at the rate of nearly 10% per year. He says, *I was quick to realize that if I continued to put in this extra amount, in 10 years my cost would go up by 100 per cent, leaving me with no profits*.

The setback prompted B.H. Save to look at other means of improving the yield. He then completely revamped his approach and conducted an experiment on his farm. He stopped using any kind of chemical fertilizers and pesticides. The immediate consequence of this sudden move was the production fell and the yield witnessed negative growth for a period of nearly five years. Due to this some people dubbed his new approach as a *mad experiment*.

However, after five years the yield stabilized and the production slowly started to rise. People had by now realized that B.H. Save had proved his point. They said that he had come up with a new method of farming. According to him, he had not done anything new and was only doing what his forefathers had done for centuries before him.

B.H. Save says that in agriculture, the emphasis should not be on increasing production. Instead one should focus on reducing the cost of production. The four things, which are considered by him as unnecessary for his method of farming are fertilizers, pesticides, weeding and unnecessary tilling. An important aspect of agriculture in B.H. Save's opinion is that farmers should learn to work with nature rather than attempt to control it through conventional methods involving mechanization and inorganic substances.

In addition to B.H. Save's style of agriculture being low on production costs, it also reduces the water requirement of the area under cultivation by as much as 70%. The nutrient requirements of the crops are met by using dead leaves and organic waste. The process also utilizes biological pest control. While earthworms and red ants carry out the tilling in his farm, the grasshoppers and frogs carry out the job of insect control.

Economic Viability

In his "Kalpavruksha" farm he switched over to "Organic Farming" which did not show spectacular results in the very first year. The soil was recovering from the negative effects of chemical farming to which it was subjected earlier. Once again the soil regained its life, production increased and the cost of input nose-dived. The coconut trees which were planted 35 years ago and which were yielding 200 to 250 coconuts per year are now yielding 350 to 400 coconuts per year. "Kalpavruksha" farm was started with "Organic Farming" after full development he has moved over to "Natural Farming".

Recently, he has started "Sanghavi Farm" where he is willing to experiment every possible method of organic, irrespective of the risk involved, and is based solely and fully on the principles of organic Farming. That was a unique experiment for the first time in India 4 years ago. Today, one just has to visit this farm and get convinced about the efficiency of organic Farming.

In any type of farming there is a principle of keeping certain distance between two plants. Normally, coconut trees are planted on 25' x 25' basis. So that 4 trees occupy 625 sq. feet of land, but they require sunlight falling on 625 sq. feet space.

He therefore, planted 2 trees together leaving 50 feed land on one side and 25 feet on the other. This way the trees were getting sunlight on 625 sq. feet of space. Benefits of such plantation however, were many.

1. Per acre we could accommodate 84 trees as against 70 in the normal method.

2. Because of the two trees being together, we could save expenses for maintenance, planting watering, etc.
3. On the vacant land of 50 feet between two rows of trees, we could plant vegetables and pulses (which improves nutrient content of the soil) and also planted Banana trees on the vacant land of 25 feet between two columns of coconut tree.
4. This cosmopolitan plantation of short-living, medium-living and long-living trees & plants helps all the trees & plants. Both the quality & the volume of production improves. Normally coconut trees yield first fruits after 5 to 7 years. But at "Sanghavi Farm", we got fruits in a period of 4 years. Isn't that a big hit!

The quality & volume of bananas grown at "Sanghavi Farm" is very good. Such products also have a much longer shelf life.

His experiences in this regard are:

	Weight When grown Natural Farming Way	**Weight when grown Modern way**
First Bunch of Banana on a tree	18 kg.	25 kg
Second Bunch	30 kg.	30 kg.
Third Bunch	25 kg.	20 kg.
Fourth Bunch	15 kg.	Tree dies
Total	**88 kg**	**75 kg.**

He sold this @ Rs. 2.50 per kg. Rs. 1.75 per kg.

Total Income	Rs. 220.00	Rs. 131.25
Less: Expenses	Rs. 66.00 Water, labour, transportation)	Rs. 105.00 (Fertilizer, pesticides, Labour Water, Weeding, Transportation
Net Profit per tree	Rs. 154.00	Rs. 26.25

Secondly, the satisfaction gained by consuming such products. Enjoying such fruits satisfies the hunger sooner, thus leading to overall savings of individuals and also leading to their better health.

Having touched upon the beneficial features of Organic farming, he now move over to the principles of the system.

1. The soil needs no tilling. However, he would like to quality this statement to the extent that in any type of farming, initial tilling is necessary to clear the site. But subsequent tilling is harmful. Initial tilling is to facilitate the spread to roots. Any tilling done after sowing is, however, harmful as it cuts off the roots. Further, even as regards the initial, mandatory tilling, it must be done using light implements as the use of very heavy tractors tends to compact the soil, making it less arable.
2. The most effective tilling is carried out for the farmers by earthworms through their burrowing action. Earthworms multiply very rapidly in millions per hectare in 2 or 3 months time. Earthworms eat organic debris along with soil continuously and excrete digested material at

frequent intervals on the soil surface. For excretion they have to come to soil surface about 8 to 10 time a day. In this process they mix various soil components and improve thereby the soil structure. The earthworm castings are very rich plant nutrients. The population of earthworms will be high in the soil receiving plenty of organic manures.

3. Chemical fertilizers are not at all used in organic farming. They have no beneficial influence on the activities of soil organisms. Only organic manures, bulky and concentrated, are used.
4. By adopting the 'platform and trench' method, both organic manures and water are kept at regulated distance from the plants/tree. They are never "fed" directly to the plants as manure is not the food to plans/ trees but it is the food of the living organisms present in the soil like earthworms, bacteria, etc. These organisms make the soil rich and arable. The humus made by them also enables the soil to absorb and retain moisture.
5. When the plants/trees are being nourished organically, they develop strong roots which protect them against strong winds. Secondly, because no chemical fertilizers are used, their water requirement is also less to the extent to only 30% of the normal requirement.
6. In organic farming, weed growth in plant basins is suppressed through organic mulching. In other areas weeds are not controlled as they provide surface protection against high temperature and the water run off.
7. No chemical pesticides are required in organic farming. It is not that pests do not attack plants if been grown by the way of organic farming, but the use of organic pesticides such as tobacco decoction, diluted cow urine etc. is effective against pest infestation. Wherever possible, biological control using parasites and predators could be practiced.
8. As the principle of organic farming states "care of plants and not cure" it is possible to manage various activities with less workforce. Displacement of under employed work force from traditional agriculture to more productive sectors of economy becomes possible when organic farming is more widely practiced.

Having enunciated the principles of organic farming, he hasten to add that the above are illustrative and not exhaustive. In fact, he strongly believe that regular workshops be conducted by and for the actual farmers, right at the farms, to enable the idea of organic farming to be propagated.

Personally speaking he is getting minimum 400% profit from his farms (without even treating farming as a business) even if he take in to account cost of all the inputs like seeds, water, organic manure (which is purchased only in the beginning of the farms) labour, electricity, transportation etc.

It is very important to understand that this increasing population is reducing the area of soil available for agriculture. Hence, unless we move to methods which renew the soil fertility even while getting maximum production from it, we can not hope to solve the problem of food for all the people in the years to come. As we have always been saying, Oil may last till 2050, soil will not. Leading to people's better health, economic gain for our nation, for the society, for the individuals & for the farmers themselves and to help understand and propagate the ideas of Natural farming.

The Organic Way

1. Natural farming is a path, not a technique.
2. Weeds are a blessing, chemical fertilizers an interference in the life of the soil.
3. Co-operation is the basic principle of nature.
4. The practical philosophy for maintaining environmental balance is live and let live.
5. The life of all plants and creatures is dependent on the condition of the soil. Farming should be done with minimum interference in the ways of nature.
6. Nature's tillers are earthworms living earth.
7. Inorganic fertilizers are a threat to humus.
8. Humus play a vital role in growing tasty and whole some food.
9. Only healthy, living fertilizers can nurture healthy plants.
10. Only healthy plants can provide health to humankind.
11. Living soil is the foundation of plant growth.
12. Co-operate with the organisms of the soil.
13. Trees do not require much. Just keep the soil alive.
14. Trees need moisture, not a great deal of watering.
15. Natural farming is only possible through compassion.

Narayan Reddy's Farm

This 5-hectare farm is a half-hour drive from the city of Bangalore, in southern India. Reddy grows the same kinds of crops that his neighbours do: coconuts, fruits, vegetables, mulberry, finger millet, and rice. Also, like many other farmers in the area, he keeps dairy cows and rears silkworms.

Speciality: His goal is to make his farm self-sufficient. He and his family grow or make nearly everything, they need. They try out new ways of doing thing, to see if they can do the tasks faster or better or at a lower cost. Sometimes they find that the old method still work best. And they are always willing to share what they have learned.

Reddy's father began cultivating the land more than 20 years ago. He used chemical fertilizers and pesticides because that was the modern way. At first this worked well. But after a few years, he found he was losing money. The chemical were not always available. They were expensive. Crop yield were decreasing. What's more, the soil was deteriorating.

So Reddy decided to use different farming method. He knew about the old, traditional ways of cultivation. He felt that if he combined the best of the traditional methods with modern technology, he could improve his farm. And that is just what he did.

Reddy tries may ideas: Reddy learns all he can about modern scientific ways of farming. He chooses and tries out those ideas that he thinks will work for him. But he also makes sure that the methods will not pollute the soil and water or make him too dependent on outside sources for his supplies.

For example, one of the first things Reddy did was to stop buying chemical fertilizers and pesticides. At first this meant that he had to bring more organic materials onto the farm. For instance, he gathered animal manure from other farms. He collected leaves on roadsides and in the forest. He started to make lots of compost. He grew green manure crops, such as Sesbania (*Sesbania sesban*) and sunn hemp (*Crotalaria juncea*). And he used the leftover leaves and stalks of harvested crops for compost or mulch instead of burning them.

He also reduced pesticide use. Reddy now makes his own insecticide from ground neem seeds, cow urine, and other materials available on the farm. His crops look perfectly healthy, and the fruit trees are laden with fruit.

During the time that Reddy was changing from chemical farming to organic faming, he needed more labour on the farm. And his yields went down a little, but because his costs also went down, his profit was about the same. With time, as the soil improved, and he started growing a number of different crops each season, his overall profit increased.

Reddy has many crops and enterprises: The main crops Reddy grows are coconuts, fruits – especially sapota (*Achras sapota*), mango, and watermelon – finger millet (*Eleusine coracana*), and mulberry (*Morus spp.*). He also harvests two crops of rice a year. In between the coconut trees he grows vegetables to sell in the city.

All the food for the family is grown on the farm. They have their own cereals – rice and millet – and coconuts. They get plenty of milk from their cows. Even the coffee they drink at breakfast is their own, grown on a row of shiny coffee bushes just outside their house.

Over time Reddy has added other enterprises. One of these is rearing silkworms. In nature, silkworms live and feed on mulberry trees. They can also be reared indoors by farmer if they are fed the right kind of mulberry leaves regularly. Silkworms are valued for the cocoons they spin. These cocoons are reeled into silk yarn, which is used to make costly silk cloth.

On one side of Reddy's farmhouse is a large shed with shelves from floor to ceiling. Here the silkworms are kept in flat round bamboo trays and fed mulberry leaves. In about 4 weeks they spin their cocoons. These cocoons sell for a high price.

Rearing the silkworms does not cost much. Because Reddy has planted a large area of this farm with mulberry, he always has mulberry leaves to feed the silkworms. He has also planted several bamboo clumps. The bamboo is split and woven to make the trays where the silkworms live.

Reddy has planned his farm enterprises carefully. Each one links with and supports the others. The many different farm activities ensure that Reddy and his family have a secure income all year round.

From loss to profit: An organic farming success with imagination and hard work, he combines modern technology and age-old farming methods. He and his family grow or make nearly everything they need on their 5-hectare farm. They are always trying ways to lower costs and work more efficiently. And every enterprise on the farm is carefully planned to serve several purposes.

One of Reddy's most important enterprises is livestock. He owns 14 excellent dairy cows. They are specially bred to yield a lot of milk. They cost more to buy at first. But he makes a good profit from selling the high-quantity milk.

To keep up that quality, Reddy looks after the cows well. They are not allowed to graze free in the fields where they might damage the crops. Instead, they are kept in clean stalls in a cowshed and fed properly. They are also vaccinated against diseases.

Reddy built a biogas plant: The cows supply valuable manure. As every farmer knows, cow dung makes excellent crop fertilizer. But in this part of India, because firewood is scarce and costly, it is common to dry cow dung and use it as cooking fuel. This is a waste of good free crop fertilizer. Reddy learned that there was a way of using livestock manure to provide both fuel and fertilizer. Before it is spread in the field, it can be used to make biogas, a low-cost cooking fuel.

So, Reddy built a biogas plant near his hours. This was costly to do. But it was only a one-time expense. The biogas plant has paid for itself many times over. It now costs very little to maintain, and it turns a waste material, cow, dung, into two essential products – fuel and fertilizer.

First, all the cow dung from the dairy is put into the underground biogas pit. As it decays there, it gives off gas. This gas is piped into the kitchen and supplies all the cooking fuel for the household. So there is no need to burn dried cow dung or cut down trees for fuel.

The best part is that the cow dung can still be used as fertilizer. The leftover material from the biogas plant, called slurry, is still full of good plant nutrients. Reddy adds this to his compost bins. From there, it is taken to fertilize the crops.

So Reddy's cows supply both milk for sale and manure for the crops. And, because Reddy has built a biogas plant, the dung from his cows also provides biogas fuel for cooking.

Reddy links all his farm activities. He uses the biogas slurry to fertilize Napier grass. This in turn is used as fodder for the dairy cows. And the whole cycle begins again. So you see how everything is interlinked? This is a good example of how Reddy has used new ideas to save money and farm better.

Reddy also makes compost to use as fertilizer. He uses earthworms to speed up and improve the composting process.

To make the compost, Reddy has built stone bins from granite slabs, which are easily available in his area. The bins are about 5 meters long, 1 meter wide, and 1 meter wide, and 1 meter high. In these, he layers the farm and kitchen wastes as they collect. He then covers them with a layer of soil and adds them earthworms. The worms feed on the waste and break it down quickly.

This compost is one of the main fertilizers now used on the farm. It nourishes the crops. It builds up the soil. It is safe to use year after year. Because it is made from wastes, it costs nothing! And it returns to the earth whatever was taken out of it.

These are some ways the Reddy family uses to keep their farming costs low. They also make sure they do not have to depend on anyone else for their materials.

Of course some of the equipment Reddy uses is costly to install. But only at first. He only chooses equipment that will save him money and work in the long run, such as the biogas plant.

Reddy did not have much formal schooling. He faces the same problems as any other farmer in the areas. But he has taught himself new ways of looking for solutions. He finds out where he can get information. He learns how to do things himself. And, perhaps best of all, this pioneer spirit has encouraged other farmers to feel more confident and try new ways of farming.

Breaking out of the Pesticides Cotton Wedlock

Cotton is cultivated in over 60 countries on more than 80 million hectares which is equivalent to 5 per cent of all cultivated land, and its production uses 25 per cent of all pesticides consumed annually by the world's agriculture, with an estimated value of US$ 7.4 billion.

Of all the 300 million kilograms of pesticides used in Third World countries every year, half is used in cotton. In many African countries, over 70 percent of pesticide imports are used on cotton; in India 45 percent of pesticides are used in cotton production. Considering the importance of the fibre in their economy, for some countries cotton accounts for 70 percent of export earning, it is no wonder that poor countries spend so much of their foreign currency on the import of pesticides for use in cotton. Being a non-food crop cotton has attracted the most deadly pesticide formulations, classified under toxicity Classes Ia, and II of WHO.

Pesticides such as monocrotophos, dichlorvos, quinalphos, dimethrin, endosulfan among many others are still in use, including herbicides such as parquet and diuron. In the last ten years the use of synthetic pyrethroids such as cypermethrin though to be safer, created problems with resistance building up in the whitefly (*Bemisia tabaci*). In Senegal, the government used to supply farmers with a formulation containing a mixture of endosulphan, DDT and methyl parathion, the empty containers are still to be seen dumped around small farmers' households even in the remotest locations. This season Senegalese farmers have complained that the pesticide Cypercal has led to the burning and shedding of cotton leaves. In Zambia, the multinational, Lonrho, supplies farmer son credit with a range of potent organophosphate pesticides which end up used in maize crop.

A range of pesticide is used in the control of a whole variety of insect pests attacking the plant during its vegetative stage (thrips, aphids, jassids and leaf-eating caterpillars such as *Sylepta derogate* in Africa, *Spodoptera litura* in Asia). But it is in the reproductive state that the problem is greater since damage to capsules directly affects the yields. Species involved are larvae and adult caterpillars for instance, of the Noctuidae family several species of *Earias* sp. (Spotted and spiny bollworm), *Pectinophora gossypiella* (pink bollworm), and most important of all *Helicoverpa armigera* the famous American bollworm. Due to the high toxicity of the formulations and the regularity of application, cotton areas the world over have seriously reduced the natural predator population. In many cases imbalances are so acute that there is a need to re-introduce entomb-parasites and entomopathogens in order to hasten the biological recovery. The range of pests are clear symptoms of biological and plant nutrient/physiological imbalances originating from improper rotation and soil fertilization practices, such as the one-sided use of soluble nitrogen, especially urea, which led to increased attacks of sucking pests such as aphids and whitefly.

Other indirect problems arise when, as is the custom in most smallholder systems cattle put to graze on the leaves of cotton after the harvest or when cattle fed on de-oiled seed cake, a high protein concentration made from cotton seeds. This practice leads to contamination problems. In 1994, an entire consignment of Australian beef was rejected and returned by the American authorities after residues of pesticides used in cotton were found in levels higher than permitted by the Codex. In Senegal, farmers have noted the intoxication and consequent death of livestock after feeding on the leaves of cotton. If high levels of residues were found in samples of meat from Australia (where pesticides regulations are strict) one can only imagine what is happening with residue levels in meat and milk from countries like India, Pakistan, and many African countries. Cotton cake is also exported for animal feed to Europe's dairy 'factors'. No data is available on residue levels in such products.

The list of environmental, health, socio-economic and structural problems associated with cotton production is exhaustive. The drying of the Aral Sea for cotton irrigation, hunger in Sudan while cotton occupies the best soils, total depletion of soils and contamination of the environment and workers in Colombia due to constant aerial sprays, are just a few examples. Similar problems arise out of cotton processing (blenching, mercerizing, dyeing, etc.).

These problems are also seriously affecting the economic viability of the crop. Costs of production have risen steadily, a rise not being matched by an increase in yield or income. Pest outbreaks have led China to buy cotton on the international market for the first time in the past four years, which sent prices soaring upwards, only to fall again, destabilizing many textile industries which had already sold their yarn ahead of time with fixed prices and had suddenly to by expensive cotton to complete their orders.

All of this is making cotton farming a more risky business with decreasing margins for the farmer and a long-term problem for the industry which tries to meet the new demand for natural fibres at affordable prices.

The Organic Solution - Maikaal Project, M.P., India

Conversion to organic methods can reverse this process, has been proved in the Maikaal Project. Rather going in detail the technological strategies are being given here.

Technical elements of a cotton-based organic farming system developed for the Maikaall Project, M.P. state in India

1. **Cropping system:** rotations and intercropping systems have been

designed to improve soil fertility, guarantee the economic viability of the process, but also to increase bio-diversity and homeostasis by reducing pest pressure.

2. **Environmental reconstruction:** multi-purpose shrub and tree planting around field boundaries for fodder, microclimate and biodiversity improvement, reduction of contamination risk.
3. **Soil fertility management:** mycorrhiza favourable rotation. Azotobacter or *Azospirillum* and PSB (Phosphorus Solubilising Bacteria) as seed inoculants. Locally made bio-dynamic humus preparation. FYM compost enriched with rock phosphate applied on the basis of soil analysis, top-dressings with a special compost made of oil-cakes and bio-fertilizers treated with CCP (Compound of bio-dynamic compost preparations). Nome-made bio-fertilizer foliar sprays. Bio-dynamic silica spray for fibre quality improvement.
4. **Pest monitoring and mass trapping:** training for pest identification and pest counts to establish economic threshold levels. Pheromone traps for monitoring and mass trapping of bollworm complex.
5. **Pest controls and management:** environment management for the increase of natural predators, release of entomo-parasites and entomopathogens when necessary, commercial and locally identified botanical pesticides and formulations. Natural predators develop in bio-farms such as *Chrysoperla carnea,* ladybird-beetle, spiders, birds and become indicators for the restoration of the ecological balance. Soil fertility, pest management and natural controls result in minimal presence of aphids, and whitefly, reduced attack of jassids, presence of bollworms, at an economically tolerable level.
6. **Weeding:** cross cultivation using oxen and hand weeding.
7. **Harvest:** hand-picked, segregation of damaged capsules.
8. **Yields:** organic seed cotton average: 500 kg/acre, equal to 1200 kg/ha. Best organic farmer's yield registered at the Maikaal project in India using the above recommendations 3000 kg/ha (irrigated)

Eco-Farming of Medicinal Plants in the arid Tropics of India

The world's largest arid tropical region is spread from Northern Africa to the Western part of Asia. Its eastern end which covers about 12 percent (32 million ha) of India is popularly known as the *Thar* desert. Low (100 to 300 mm/yr) and erratic rainfall, high wind and solar regime and sandy soils make the Moisture Deficit Index – 85 to 90 percent in this region, which generates high probability of drought and desertification. Under such conditions the production of conventional agricultural/food grain crops is highly uncertain resulting in poor economic conditions for the majority of marginal farmers.

Within this ecosystem there exists a large degree of heterogeneity in terms of soil, climate and other economically relevant natural systems, which provides a rich biological diversity and, within species, a range of genetic diversity suitable for different conditions. About 657 species of plants belonging to 351 genera and 87 families, are found in this region, with 8 per cent of these endemic to the region. About 194 species are used for economic or medical purpose.

Medicinal Plant Production

Increase in human and animal populations in the region have put pressure on the natural vegetation of the ecosystem, causing deforestation, overgrazing of rangelands etc. Furthermore, some developmental activities, e.g. mining of sandstone, granite, marble etc., cultivation on marginal lands, increasing use of tractor in agriculture and the irrigation network of the Indira Gandhi Canal, are rapidly destroying natural habitats. Some introduced plants like *Prosopis juliflora* are becoming biological invaders.

Thus destruction of natural habitats and over-exploitation without replanting are threatening the population and diversity of medicinal plants (MP), in particular, for example, *Commiphora wightii* and *Asparagus racemosus*. It is thus imperative to protect and increase the production of medicinal plants.

Ecological Production of Medicinal Plants in the Thar Desert of India

The plants which are used for medicinal purposes should not have any residues of fertilizer or pesticides – otherwise they become more of poison than a medicine. Moreover, the MPs have several active ingredients (biochemical) and growing them with fertilizer, pesticides and irrigation may alter or deteriorated their quality. Thus, as far as possible, they have to be grown in near-to-near condition so that both the quality of MP and the eco-balance can be maintained. For that there are two possible ways: (i) *in-situ:* to make conditions suitable in their natural habitat to increase population; (ii) *ex situ*: to cultivate as an organic crop.

In-situ production

Nature has immense potential for restoration of an ecosystem provided the conditions are favorable. The arid ecosystem has been sustained since time immemorial and it is only the over exploitation of natural resources which makes the system unsustainable. In general, man-made reforestation/ replanting is difficult, as one would need to study the habitat and micro-environmental requirements of each and every species for their cultivation. Given these problems *in situ* conservation projects have been started, for example, a project near Jodhpur. The area has been protected by stone wall fencing, watershed soil and water conservation measures are in place and the participation of local people has been gained. The naturally regenerated plant population of *C. wightii* has increased by 813.3 percent in 7 years. *C. wightii* is an endangered medicinal plant in the region, however, despite much research into macro/micro propagation, experiments in cultivation have not been comparable with the *in situ* conservation project.

Agriculture Production

Although it is better to increase the production of MP by conserving them *in situ* there are plants which are required or demanded in large quantity such as Senna and Psyllium. For meeting this demand, cultivation is necessary. Under the arid tropical conditions cultivation of medicinal plants has economic advantages for farmers, helps the conservation of natural resources and can ultimately contribute to the sustainability of this fragile ecosystem.

Harnessed the Potential by N.D. Prajapati

Arid regions have huge possibilities and potential for the cultivation of medicinal plants and Mr. Narayan Das Prajapati an innovative farmer of the *Thar* desert has very well understood and concluded that

1. The MPs of this region are drought hardy by nature, thus their production is less affected by drought compared to food grain crops. As the root, stem and leaf are the useful parts in most MPs, rather than the fruits, the yields of these plants are less affected by drought.
2. The soils of the region are comparatively untouched by fertilizers, pesticides and irrigation (FPI). The use of FPI is highly risky due to climatic conditions, moreover the marginal farmers are unable to use costly inputs. On the other hand the farming systems of these regions are dominated by animals, and the *excreta* of animals can improve the properties of sandy soils. These, which are seen as limitations for food grain production, are opportunities for MPs cultivation. Furthermore, in organic faming quality (i.e. no residue of pesticides/fertilizers) is more important than the quantity.
3. A major part of the region is currently wasteland (12 million ha) due to various land forms e.g. deep sand/sand dunes, gravely, rocky areas, saline soils/water etc. where cultivation of food crops is not possible. MPs can, however, be grown in these areas, e.g. *Cassia angustifolia* (senna) in deep sand/sand dunes; *Commiphora wightii* on rocky/ gravely land; *Plantago ovata* and *Vinca rosea* in saline soils. Using wastelands for cultivation will not only increase the production of these plants but will also help halt desertification due to increased vegetative cover.
4. At present the farming community remains idle for 7-8 months, after the rainy season (Nov-June). Year round cultivation of MPs would provide more work and additional income, thus providing support to the economies of drought prone regions.
5. Because of the climatic uncertainty agroforestry systems have developed in the arid tropics, MPs can easily be accommodated in these systems.

There are many MPs which have potential for cultivation in this region. After several years of experimentation Mr. Prajapati has found that Senna is the most suitable plant for sandy arid lands because it grown well in bright sunshine and dry climates, its deep tap root system helps it to thrive in deep sand and no animals graze it.

Now only because of his efforts senna is being cultivated with completely organic metho in about 3000 ha area of the *Thar* desert. The *Plantago ovata (Psyllium)* is already commercially growing in saline soil/water areas, contributing about 60 per cent of the world's production.

Sekem in Egypt: Bio-Dynamic Production and a Way of Life

The SEKEM farm in Egypt is situated in the desert near the south tip of the Nile delta, not far from Cairo. It was established in 1977. The total farm area is 55 ha, of which about 30 ha are under cultivator. Additional farms, which supply products on a contract basis to the main farm cover about 500 ha. And this area is steadily expanding agriculture. The products are certified in accordance with the Demeter standards for biodynamic agriculture. The main crops grown can be divided into three categories: fruit trees, herbs and spices, and vegetable consumption in Western Europe. Spices and herbs are designed for both domestic consumption and exports. The farm also includes animal production.

From the very beginning SEKEM aimed at supplying local people with high quality organic food. SEKEM has contributed to getting organic production onto its feet in Egypt. It has established several organic shops in Cairo and delivers its products to about 7,000 pharmacies and 2,000 shops. In a tea-drinking country like Egypt it is significant that certified organic herb teas from SEKEM are amongst the most popular in the country. Because of the success of the SEKEM products on the West European market, exports have now risen to 40 percent of the total turnover instead of the 10-20 percent planed initially.

There is little doubt that the SEKEM initiative is a show is a show case of holism within the organic movement. Well-rooted in anthroposophic and bio-dynamic philosophy, all SEKEM's projects and enterprises have developed very rapidly. Today, SEKEM has more than 120 people in full-time employment and co-operation with over a hundred bio-dynamic farms marketing their products through SEKEM provides another 250 employment places for Egyptian men and women. SEKEM has also played a pioneer role in developing organic cotton production.

What began as a farm on sandy desert soil on the outskirts of Cairo, has now become a center of cultural and social development. The SEKEM kindergarten and school provides education for 350 children some of who are handicapped. A health clinic provides medical care and a mobile ambulance station reaches those who are unable to get to the clinic. Art and culture within the SEKEM community are expressions of SEKEM's approach to life and the design of its building and the conditions and environment reflect these principle.

AREAS, NEED ATTENTION TO POPULARIZE ORGANIC FARMING

1. Adopt the interests of the farmers and their families as the most important criteria when developing sustainable agriculture.
2. Put more emphasis on securing food production in the long-term rather than the short-term. Emphasis should be on a source-directed policy,

which prevents problems rather then trying to solve them in a non-sustainable way.

3. Secure long-term land tenure rights. This would demonstrate the government's interest and facilitate the farmer's interesting soil protection.
4. Develop a differentiated policy about the introduction of Organic Agriculture Systems in ecosystem variability and ecosystem users.
5. Introduce Organic Agriculture Systems in ecologically sensitive, biologically and genetically diverse, and nature-protected areas.
6. Promote the development of organic agriculture projects by facilitating organic food imports from developing countries.
7. The technique of mineral and organic matter balance being in equilibrium or positive for certified organic products should be enforced by the organic agriculture movement if it wands to substantiate its claim that organic agriculture is sustainable.
8. Promote the consumption of certified organic food and non-food products.
9. Research projects for formulating organic farming practices may have to be framed and implemented as national projects.
10. Incentives for production of good quality organic manure, bio-pesticides, biofertilizers and green manuring crops may be strengthened.
11. Development of pesticides of plant origin (such as neem) and use of agents especially under Integrated Pest Management Systems need to be promoted.
12. The indiscriminate use of inorganic fertilizer and pesticides needs to be discouraged.
13. Steps are needed to reduce hazardous chemical residues in seeds, fodder, food products and milk.
14. District-wise inventory of recyclable wastes (both on-farm and off-farm) needs to be prepared in order to know the potential of organic agriculture as also to develop location specific technologies.
15. Depending on the nature of wastes available, the following composting techniques need to be standardized and publicized among the farmers.
    a) Rapid composting of crop residues with suitable microbial strains such as *Trichoderma* spp. Or other microbes having cellulolytic and lignolytic actives.
    b) Vermi-composting using earthworms.

The nutrient content and microbial population of organic manures should be enhanced using rock phosphate and pyrites: biological agents such as N-fixes, P-solubilizers for preparation of good quality compost material.

Cropping system: Efforts should be made to maintain yield stability and suppress the weeds and pests through adoption of appropriate cropping systems and minimizing weeds and pests incidence through adoption of appropriate the term crop rotation systems such as:

- Introduction of legumes as break crops in intensive cereal-cereal system resulting in yield stability and restoration of soil fertility.
- Periodical substitution of wheat with a fodder crop in rice-wheat system is a remedy for *Phalaris minor*.
- Inter cropping of garlic with sugarcane minimizes incidence of shoot borer.

Alternative land use: Location specific alternative land uses should be identified and popularized for adoption in marginal lands as well as the lands under fragile eco-systems. Depending on the situations, these land uses may be agri-horticulture, agro-forestry, silvi-pastures etc.

Governmental Efforts

1. The changing scenario of increasing cost of fertilizers gradual removal of subsidy on fertilizers and increasing concern of likely pollution of ground waters by fertilizers through leaching from soils calls for proper use of fertilizers.
2. The encouraging results obtained by conjunctive use of fertilizers, organic manures, and bio-fertilizers are providing leads to future strategies of rational use of fertilizers and manures for continually enhancing productivity without being detrimental to the environment.
3. To promote the use of organic manures, the existing potential of about 650 million tonnes of rural compost and 16 million tonnes of urban compost may be exploited intensity. As per the existing estimates, only about 135 to 240 million tonnes of rural compost and 4.8 to 6.5 million tonnes of urban wastes are being used as manures.
4. To demonstrate and popularize the scientific method of compost making and for balanced fertilization though soil testing, a scheme of Balanced and Integrated Use of Fertilizers is being implemented by the Government of India. Under this scheme, Rs. 18 crores was spent during the Eight Plan. The outlay has been enhanced to Rs. 46 crores during the Ninth Plan for this scheme which is expected to be further increased.
5. Another new area, which requires special thrust, is the promotion of biofertilizers. Starting from the Seventh Plan, a total of 58 bio-fertilizers production units have been established with Government support, having a production capacity of 8300 tonnes per annum. Besides this, 29 biofertilizer units have been set up by private entrepreneurs and State Departments of Agriculture and others having an estimated production of 8,000 tonnes. Thus the total biofertilizer production capacity is about 16,000 tonnes/year.
6. Continuous Governmental emphasize on the usefulness of promoting integrated use of organic and chemical forms of nutrients for better results.
7. Need of continuous efforts to popularize organic farming with the main aim of ensuring sustained productivity, environmental protection and making available food and food products raised without chemicals of any king. The limitation of the system is the difficulty in supplying the

required amounts of nutrients to the crops in large areas only through organic sources and hence the integrated nutrient management system is a better option. However, organically grown food is expected to fetch higher price and this can offset any loss due to lower yields.

International Efforts

Policy

1. The environmental and economic benefits of organic agriculture have captured the attention of several countries; however, only a small number have enacted policies to assist the organic sector. Most assistance has developed in the private sector, especially by NGOs. This private sector infrastructure is not only recognized by countries, but also encouraged.
2. Farmers and consumers in almost all countries rely on a system of private self-organized producer organizations and independent certifiers which have, over the years, provided an economically-efficient mechanism of certification. The degree to which private organization are significant is indicated by the fact that IFOAM has some 650 individual and institutional members in over 100 countries, 75 percent of which are in developing countries. However, the network of private certifiers needs to expand; as mentioned earlier, many developing countries still lack certification organization.
3. Organic programmes, maintain evolving "input lists" of acceptable inputs for organic production, processing aids, and ingredients. Countries, in consultation with civil society organizations, can propose their own input lists to Codex and negotiations will probably take place over what ultimately is acceptable in the international marketplace. Many developing countries have not drawn up material lists and generally lack the resources and training to effectively participate in international standard-setting processes.
4. Engaging in international trade requires quantities of goods well beyond the production capacity of individual small organic farmers. NGOs have supported farmers in their efforts to establish appropriate organizations for collecting, processing and marketing organic products. Countries could support NGOs in building organizations which play a key role in disseminating best practices and ensuring that farmers use adequate production and post-production techniques.

Technical Support

1. Northern developed countries have invested by far the most in organic agriculture research but even so, the contribution is minimal compared to overall research agriculture (e.g. less than 0.01 percent of the US Department of Agriculture research budget is directed to organic agriculture). The lack of extensive formal organic research with the highly site-specific nature of organic agriculture, suggests that it would be most advantageous for farmers themselves to participate in locally-

based, applied field research. Experience with FAO-initiated Integrated Pest Management (IPM) Farmer Field Schools and community forestry projects has shown that farmers, whether owners or tenants, large or small, can practice good scientific methods if they are given orientation and technical support. Countries could also undertake market research. The USA, for example, analyzed the market for organic goods in more than 20 countries, with a focus on opportunities for US exports.

2. Most national and private organic standards, disallow genetically engineered organisms (GEOs), including transgenic crops, in organic production. By providing consumers with a non-GEO produced alternative organic agriculture has secured an important market, at the same time relieving much of the pressure to impose out-right bans on GEO-produced food. While current biotechnology applications have been rejected for organic production, it is possible that future scientific development and furthers discussions may produce technologies acceptable to consumers (e.g. the value of tissue culture in producing disease-free planting materials).

Direct Assistance

1. Interest in environmental protection and the preservation of small family farms has led developed European countries to subsidize organic production to varying degrees. The subsidies can provide significant encouragement (e.g. UK provides up to $450 per ha during the conversion period for designated land areas; European Community subsides have helped 15 percent of Irish organic farmers develop dairy operations). Interested communities would have to be advised about organizational aspects and on producing and commercializing the necessary inputs as well as on marketing any surplus produce (e.g. seed and planting material of adapted varieties, suitable livestock genitors).
2. Consumer confidence in the integrity of organic claims is essential if goods are to be sol at a premium. Because organic products cannot be distinguished from conventional products, consumers depend entirely on certifiers to truthfully distinguish organic from non-organic gods. Countries could better enforce organic standards by punishing those who engage in fraudulent activities as well as undertaking systematic tracking and measuring of fraud and its impact on the market.

INTERNATIONAL ORGANIZATIONS

Scientific Congress on Organic Agricultural Research (SCOAR)

Scientific Congress on Organic Agricultural Research (SCOAR) is a collaboration of producers, scientists, and others. Its mission is to plan and promote research and information exchange for understanding and improving organic agricultural systems.

The Goals of SCOAR

1. Cultivate a peer-level process of mutual learning and democratic collaboration among a diverse community of organic farmers & ranchers, scientists, extensionists, consultants, consumers and others for pursuit of the project Mission.
2. Create a shared understanding of organic agriculture as an ecological approach to managing agricultural systems.
3. Identify and characterize research and information-exchange priorities pertaining to the Mission, through various activities and documentation of those activities.
4. Advance the state-of-the-art of organic systems research, including on-farm, participatory, whole systems, and multi-farm studies.
5. Increase access to the knowledge, skills, and practices of successful organic producers by developing and implementing models for networking and information sharing among organic producers and scientists.
6. Encourage the use and implementation of SCOAR's results.
7. Potential SCOAR Themes: Core Research Needs and Some Big Issues
8. Scientific interpretation and elaboration of traditional *Best Management Principles:* (i) Feed the soil to feed the plant, (ii) Building health in the soil and animals builds healthy farms
9. Importance of conducting research in a bona fide organic context.
10. Emergence vs. extraction of beneficial organisms /behaviors: Systemic biological control paradigms.
11. Applied microbial ecological management: regional soil food web labs, micro-livestock husbandry, and *recombinant ecologies*.
12. Biological capital formation: lowering the costs of soil organic matter.
13. Getting past weeds vs. soil: reducing/mitigating impacts of tillage in systems without herbicides.
14. Explicit strategies for both conversion and post-conversion.
15. Applications of agricultural *rocket science* in an ecological context.
16. Surviving success: sustaining profitability as price premiums to farmers fall requires optimization of small-medium size production systems for long-term cost-efficacy.

International Federation of Organic Agricultural Movements (IFOAM)

Organic agriculture alone can ensure that we are assured a future that is both ecologically sound and economically sustainable. Basic Standards of Organic agriculture and Food Processing have been set by International Federation of Organic Agricultural Movements (IFOAM). Its main function is to coordinate the network of organic movement around the world.

The Organic Farming Research Foundation (OFRF)

The Organic Farming Research Foundation (OFRF) is a producer-directed organization dedicated to the improvement and adoption of organic farming methods. As analysts of U. S. policy for food and agricultural issues, we appreciate the opportunity to comment on FDA's policies and performance regarding *bioengineered foods*. OFRF not categorically opposed to every use of *bioengineering* in agriculture, but OFRF believes that the federal government has fundamentally failed to provide an adequate regulatory regime for transgenic technologies and products. In particular, the division of responsibilities among FDA, EPA and USDA ensures that there is no comprehensive, public interest oversight governing the release of transgenic organisms into the environment and the food supply. The environmental, agricultural, food safety and food security issues cannot be considered in isolation from each other, and sound public policy cannot be formulated or implemented in such a fragmented system. Furthermore, under the existing regulatory framework there is no definite point of accountability for the consequences of this regulatory failure. There is nobody ultimately in charge except the biotechnology industry itself. The Administration should immediately begin the process of forging a new, comprehensive regulatory process, with extensive participation from consumers, farmers and the public interest community.

Organic Farming in the Indian Sub-Continent

Background

Organic farming involves an approach to food production. It seeks to crate a sustainable agricultural system that relies first and foremost on ecological interactions and biological processes rather then chemical inputs. The aim of organic farming is to produce optimum yields with minimum harm to the surrounding environment.

Organic farming is a production system which avoids or largely excludes the use of synthetically compounded fertilizers, pesticides, growth regulators and livestock feed additives. To the maximum extent feasible, organic farming systems rely on crop rotations, crop residues, animal manures, legumes, green manures, off farm organic wastes and aspects of biological pest control to maintain soil productivity and tilth to supply nutrients and to control insects, weeds and other pests (Lampkin, 1990).

Farming before the western world colonized the developing world had, in many ways, been organic as there were no artificial inputs.

History

The ISC was involved in sustainable farming for several thousand years. Farming was not just practised for food production or income. It was a way of life. Agricultural activities and techniques were tailored to suit the soil conditions and climate. People were content with what they had, the land was natured with care, pesticides were unheard of. It was self-sustaining. Farming before the arrival of the western world into the east was at one with nature.

Half a century ago, Sir Alfred Howard, the father of modern sustainable farming wrote in this classic, An Agricultural Testament that; *in the agriculture of Asia we find ourselves confronted with a system of peasant farming which, in essentials, soon became stabilized. What is happening today is in the small fields of India and China took place many centuries ago. The agricultural practices of the Orient have passed the supreme test, they are almost as permanent as those of the primeval forest, the prairie or the Ocean.*

Status of Organic Farming in India

Green Revolution in India was mainly realized with the introduction of high yielding varieties of various crops and by following intensive cultivation practices with the use of fertilizers, pesticides and other inputs. The fertilizer consumption in the country accordingly has increased from 66,000 tonnes in 1951-52 to about 12. 16 million tonnes in 1992-93. In other terms, fertilizer consumption in India was only 0. 5 kg/ha in 1951-52, and increased to 67 kg during 1992-93. Same situation was with pesticides. All these agro chemicals are produced from non-renewable fossil fuel. In future, diminishing availability of these inputs may not only cost heavily on our foreign exchange but may also limit agricultural production.

With the increase in the country's own population, compulsion would be not only to stabilize agricultural production but to increase it further in sustainable manner. The scientists have realized that the green revolution with high input use has reached a plateau and is now sustained with diminishing return and falling dividend. The intensive use of inputs has not only polluted the soil, water and the environment causing their slow degradation but also affected the human beings. It is because of these the health conscious countries like, Germany, Japan, USA, Netherlands etc. and the affluent domestic population are ready to pay a premium for food raised without the use of chemicals. Thus, a natural balance need to be maintained for survival and well being of the human beings and plant and animal kingdom. The obvious choice for that would be adaptation of organic farming practices without compromising agricultural production.

Reports of deleterious effects arising with the use of agro-chemicals was being observed by the Government of India for quite sometime. This situation, coupled with the degradation of cultivable land and increasing agricultural pollution has created an unhealthy situation in the country. In order to balance this situation, it was felt to have a viable programme which may not only check soil degradation, underground water pollution, high residues of hazardous chemicals in food, but may sustain the food production and improve the soil health without effecting crop production. Organic farming, which aims at cultivating the land and raising crops in such a way as to keep the soil alive and in good health may be an alternative to the present system of farming solely depending on chemicals.

To study the viability of organic farming in India, a Technical Team constituted by the Ministry of Agriculture visited several research centres and farmers practicing organic farming during 1993 and made the following observations:

1. The country at present is not in a position to completely eliminate the use of chemicals especially fertilizers. However, it would not be difficult and unrealistic to phase out the use of these chemicals systematically.
2. The doses of fertilizer need to be gradually reduced and be balanced by increasing the use of optimum quantity of organic manures and biofertilizers.
3. Commercial sale of organic manure is uncontrolled and needs proper quality control standards so that the poor cultivator may not be cheated on one hand and on the other the eco-system may not be disturbed.
4. The use of farm yard manure even in their package of practices.
5. Development of knowledge in farmers about the hazards of pesticide use. They should minimize its use.
6. Neem and other materials of plant origin for pesticide should be used and should be readily available.
7. Integrated Pest Management (IPM) practices should be followed. Though, the plant extracts have been used by various farmers for controlling insects and diseases with sufficiently high efficiency and various combinations may be tried.

The Extension Staff of many States should be aware of the beneficial use of organic manure and deleterious effect of agro-chemicals.

Major meetings on organic farming have been held towards the end of the 80s. In 1994, a sum of 20 lakhs each is to be given to 10 medium sized municipalities to be identified by state governments to set up composting units.

Nodal Agency of IFOAM-India

Institute for Integrated Rural Development (IIRD)

IIRD is involved in various national and international networking programs. The Institute is an active member of The World Conservation Union (IUCN), International Federation of Organic Agricultural Movements (IFOAM), Asian South Pacific Bureau of Adult Education (ASPBAE), Associated Country Women of the World (ACWW), Environmental Liaison Centre International (ELCI) and World Rural Forum (WRF), Approtech Asia and ARISE. The Institute has about 65 full time staff and 103 part time staff. Besides, IIRD has about 300 volunteers in the project area.

IIRD and Organic Agriculture

The Institute has promoted organic agriculture through awareness programmes, training of farmers, community action programmes on organic agriculture, networking with like minded organizations for the cause of organic movement and lobbying for organic agriculture.

Organic Agriculture Training

Since IIRD has involved women as primary actors in development processes, IIRD's organic agriculture training programmes are primarily addressed to women farmers and leaders of rural women's groups.

The *Vikas Sevikas* (development animators) are women who organize *Mahila Mandals* (women's groups) in the villages covered by IIRD. At present 72 *Vikas Sevikas* attend every week, two days of issue based participatory leaning sessions at IIRD. In these training sessions, various issues relating to agriculture like soil fertility management, cultural practices, crop management and animal husbandry are solved by using organic agricultural practices.

IIRD with the involvement of women's groups in Paithan *Taluka*, has organized 48 *Pariyavaran Semites* (environment clubs) consisting of women in different villages. Most women in these *Pariyavaran Samitees* are small and marginal farmers. As an effort to restore their local environment in order to sustain them, these *Pariyavaran Semites* resolved to adopt organic agriculture in their farming practices. In May 1996, 15 *Pariyavaran Semites* decided to send 15 women organic farmers for a one year full time course on organic agriculture to IIRD'S newly established School for Organic Agriculture.

IIRD's School for Organic Agriculture is in Babhulgaon village in Paithan *Taluka*. The school offers a number of short need-based training on organic agriculture like composting, soil fertility management, livestock management etc. to farmers and a one year full time course for women organic farmers. The training sessions in this full time course include.

Community Action Programmes on Organic Agriculture

1. Home Gardens
2. Establishment of *Pariyavaran Semites*
3. Eco Development Programmes
4. Marketing of Organic Produce

IIRD has played a significant role in the development of National Standards for Organic Agriculture along with other IFOAM-India members. Presently, IIRD is involved in developing a domestic marketing structure suitable for the area, based upon *Teikei* principles.

Networking

IIRD is an active member of International Federation of Organic Agriculture Movement (IFOAM) and is the seat of the Asia regional office of IFOAM. IIRD has conducted a number of networking workshops for the members of IFOAM-India and is constantly involved in lobbying the government agencies for formulating favorable policies toward organic agriculture.

Current Organic Farming Practice in the Indian Sub continent

Bangladesh

There are 75 NGOs working towards an organic farming basis, they are however, facing difficulty in getting policy changes at government level for the promotion of organic agriculture.

Nepal

Organic agriculture began with post-political changes in 1990 and a workshop was held in 1991. The concept of sustainable agriculture is appreciated but it is not high on the agenda of most NGOs. There are only 4 NGOs actively promoting organic agriculture NESCOS and CWDS e Organic Farming Source Book pp 327.

Pakistan

The Pakistan Organic Farmers Association follows the approach of Sir Albert Howard, stressing self-sufficiency from on-site resources. There are a total of 30 farms all over the country producing cotton, sugarcane and vegetables. Their claims of success are however being ignored by the scientific community. (The Organic Farming Source Book pp 327).

Sri Lanka

Since 1960 funds from the World Bank, IMF and FAO have been importing agrochemicals to assist farmers. Attitudes towards agricultural practices are now changing. As subsidies on fertilizers have been withdrawn farmers can no longer afford to rely on these and there is now an increasing interest in compost making.

Organic teas by the Stassen Group has promoted other tea plantations to convert to organic methods. There are 3 direct marketing co-operatives established for organic produce.

CONCLUSIONS

There is considerable demand and scope for development of 'organic technologies', either individually or as a package, without necessarily aiming at full adoption of organic system. The potential to increase productivity by either traditional soil fertility management techniques, is limited. Modern organic techniques have, however, shown more potential than traditional techniques and long term benefits. With the organic system working on the principle that the whole is greater than the sum of its parts, the optimum output achievable through organic agriculture may not be reached in situations where organic techniques are combined with non-organic ones.

With the present technologies as well as availability of organic manures, plantation crops can be solely brought under organic farming, while in agriculture crops, organic agriculture can be practiced on a limited scale and in pockets. Practicing exclusively organic farming will have the limitation of being a low input-low output system

A Chinese experience of success in achieving high growth in agriculture with high amounts of organic sources of plant nutrients like pig manures and crop residues is often cited as a useful model. Farmers know about the benefits of organic manures but in the present context, a lot of scientific support is needed to make it a success. A well established composting technique to ensure mineralization of nutrients from all types of high and low cellulose containing bio-degradable materials including composting of urban waste has to be popularized. The production and supply of bio-fertilizers having efficient soil and crop specific strains of bacteria need special attention and popularization.

Each organic techniques holds a range of potentials and constraints that differ between agroecological zones. Individual organic techniques require to be supported and promoted by both basic and adaptive research in order to obtain information for more specific usage recommendations. Organic pest and disease control strategies are a major priority for basic research, while organic soil fertility techniques require both strategic and adaptive work.

CHAPTER 15

Economics and Marketing

Prices for organic foods reflect many of the same costs as conventional foods in terms of growing, harvesting, transportation and storage. Organically produced foods must meet strict regulations governing all these steps so the process is often more labor and management intensive, and farming tends to be on a smaller scale. There is also mounting evidence that if all the indirect costs of conventional food production (cleanup of polluted water, replacement of eroded soils, costs of health care for farmers and their workers) were factored into the price of food, organic foods would cost the same, or, more likely be cheaper.

ECONOMIC VIABILITY

Studies have shown that organic agriculture is economically viable, that farmers can achieve more income as a result of premiums and hat they need fewer inputs to maintain returns. Organic systems are based on the optimum use of local resources and technologies and can give farmers greater independence and more control over their means of production. However, more comprehensive monitoring is needed to analyze their sustainability and impact.

Many farmers enter organic production because they want to farm in a more holistic way. Other major stimuli for developing organic agriculture include environmental and social concerns, economic necessity, a lack of chemical inputs, and market demands. Small farmers are also encouraged to take up and stay in organic farming by the prospect of being able to produce more food at the subsistence level, having a larger surplus for local sale, or being abele to cultivate a product of significant export value. Farmers are most responsive to organic agriculture when they have not been exposed to the 'chemical message' and their farming systems involve traditional or nil inputs. When production is relatively labour intensive and if farmers have the chance of developing the organic concept themselves they are also more inclined to convert to organic agriculture.

Farmers are less likely to take up organic farming in situations of high labour cost and labour scarcity and where there has been an over exposure to the chemical message. Farmers with relatively mechanized farms and a commitment to high input, high output strategies are also less likely to convert. Insecure land tenure means small farmers will be reluctant to plant permanent crops.

Local and foreign NGO's local and foreign buyers, government agencies and large private farmers are generally the main initiators of developments in organic agriculture. Development often occurs at one focal point and spreads outwards to link a widening range of organizations. In Egypt, the Dominican Republic and India, pioneering work by individuals has lead to the development of local organization capable of representing the organic sector and to government recognition programme provides access to premium prices if certain practices are followed. It functions as an effective extension method with the capacity to support sustainability.

The Challenge of Going Organic

Organic agriculture relies on natural predators and an understanding of local soil and environment. It is knowledge intensive and, from the beginning, requires more design and management. When farmers only known chemical solutions, implementing control of pests and diseases through measure such as rotation, composting and time of planting represents a major change. Organic agriculture requires time and well-trained extension workers. Active organic management is required for a least twelve months before organic status is conferred. Benefits will not be immediate and small farmers will require considerable support in the first years. High input farmers will require financial support such as capital grants or annual area payments to offset the financial problems associated with conversion. Studies show that, in Western Europe, farmer's enthusiasm for organic agriculture correlates closely with the size of the conversion grants available.

It is probably no accident that coffee, cacao and tea are so prominent among the organic products supplied by developing countries. Conversion often does not involve radical change. These crops are often planted with shade trees even in conventional systems and, as such, they support a diversity similar to that encouraged in organic systems. Organic cocoa, for example, in contrast to conventionally cultivated plantation crops such as citrus, supports biodiversity and because the ground is not disturbed or cleared during planting and the root structures of other trees remain intact, soil and water resources are also conserved. However, an annual cropping system on bare, semi-arid land where there is little diversity and where few farmers can afford livestock will require much more time before it becomes a viable organic system. These factors represent a significant challenge to the small farmers' motivation to converting to organic production.

Organic produce is increasingly popular in Europe and a wide variety of products are available. All products are subject to inspection according to the guidelines of the current European regulation. Under this system, organic produce is deemed to comply with these standards if it can be proven that it has been subject to comparable guidelines and monitoring procedures. This makes it possible to monitor products originating in a wide variety of climatic conditions and agrarian structure. Monitoring small farmers (normally co-operatives) in Latin America, Africa and Asia, who are the main producers of organic coffee, for example, has proven to be an extremely arduous task. We are now trying to develop a practical and sophisticated monitoring procedure that is applicable to different farming structures and that meets stringent EU requirements for a 100% inspection rate and spot checks.

Farm Production and Profit

Absolute yield levels under organic management are increased over time, but at a slower rate than for comparable conventional systems, and are significantly higher than yields achieved in pre-1950s agriculture, to which organic farming is sometimes likened. Absolute yield are, however, subject to considerable variability due to a number of factors, including variety selection and plant breeding, soil type, climate, rotation design and nutrient management, length of time under organic management, management ability and developments in scientific knowledge and technology.

Yields relative to comparable conventional systems are directly related to the intensity of the prevailing conventional system. This is not only the case for comparisons between regions, but also between crops within a region, and for individual crop over time. Crop yields under organic management are somewhat lower than for comparable conventional systems. Yield reductions of 10-20% in some cases and similar or higher yields in others. In Europe, arable crops yields in organic systems are 60-80% those of conventional systems. Yields of forage crops are much more similar to those of conventional systems with reductions of 0-30% reported, feed quality for stock was not different between systems. For vegetable and fruit crops the proportion of marketable yield may be affected more severely than the total yields, and yields may be 50-80% of conventional systems.

In developing countries, organic farming methods seems to provide similar outputs, with less external resources, supplying a similar income per labor day as high input conventional approaches. Large increase is observed where local farmers adopt organic farming systems, up to 400%, reaching levels similar to those of high–input systems. Direct comparisons of yields are difficult because of the differences in the farming systems adopted under high-input or organic management.

Microeconomic Aspects

Output Mix

The output mix tends to widen, mainly as a result of conscious efforts to maximize the synergetic effects of organic farming. This is also the case in operating aimed at a single crop, such as Estate Tea-India and Estate Coffee-Mexico, which both have dairy production as a side activity. In these cases, the dairy herd produces not only a new output, but also a valuable internal input (manure).

The wider mix has the not yet quantifiable advantage of reducing dependency on one crop, and thus spreads risk. Even if the total farm area remains unchanged, the "new" products do not necessarily displace any of the cultivated area of the original crop (for example, Estate Tea-India). If they do, for example in Contour Farming Philippines and Alley Cropping-Tanzania, the drop in acreage is offset by higher yields. Market outlets for the new products will be one of the factors that determine overall profitability of the new output mix, as shown in Cotton-Turkey.

Output Value

The total output value of the organic farm is higher than in conventional farming. Often this is the result of a wider output mix, sometimes because of higher yields (for example Contour Farming-Philippines and Community Coffee-Mexico). In a few cases the increases is due to both factors (Alley Cropping-Tanzania). In Cotton-Turkey and Sugarcane-Brazil, despite a virtually unchanged output mix and lower yields, total output value was higher due to the premium for organic quality.

Output quality is in many cases perceived to be higher. Usually this is the case for export crops, where certified organic quality translates to pay high premium. The fact that Western consumers are willing sometimes to pay high premiums for certified organic foods has been the reason to begin a number of organic projects. In Vegetables-Indonesia and Vegetables-Senegal a premium is also paid on the local market.

In Estate Tea-India and Soya-Paraguay, the market for organic products is still too limited to take all the product, but in Community Coffee-Mexico, the coffee is not only marketed as organic, but also as "socially just," another quality aspect valued by a sector of Western consumers.

It is considered a very important advantage of Organic Farming Systems that diversification leads to a greater self-sufficiency in food, fertilizer, and fodder. The presence of animals provides the farm population with milk, meat, and a cash income in Estate Tea-India, Alley Cropping-Tanzania, and Dates-Morocco. These are important aspects, but hardly mentioned and certainly not quantified in the cases. Emoluments are not calculated by farmers, and thus evade economic analyses for this moment.

Input Mix

The input mix is invariably quite different from that prevailing under conventional methods. In many cases, it is also more diverse, in particular when the output mix is widened (for example, Peanuts-Paraguay). Two elements are outstanding:

- There is an overall shift away from off-farm inputs in favor of internal inputs and inputs available from nearby farms (for example manure versus chemical fertilizers).
- Labor requirements are generally higher.

Input Value

The overall input costs of organic farming is in many cases considerably higher than the customary alternatives in the area: Sugarcane-Brazil reports their costs, including labor, to be nearly four times higher, whereas the figures for Estate Coffee-Mexico show costs to be nearly twice as high as the HEIA (High External Input Agriculture) alternative, and nearly four times as high as the traditional alternative. Estates Tea-India estimates production costs to be twice as high as under the HEIA alternative.

Generally speaking, the larger part of these costs are labour. In one case, Soya-Paraguay, the costs of the non-labor inputs are higher.

Organic inputs are cheaper than the conventional ones in Vegetables-Indonesia, and Bananas-Dominican Republic).

Requirements of inputs other than labor tend to be lower in cases where synergistic effects occur, such as less soil erosion (for example Peanuts-Paraguay), or nitrogen fixing and nutrient recycling (for example Farm Systems Research-India). Where these synergistic effects do not occur (Bananas-Dominican Republic, Sugarcane-Brazil), input levels are high.

Investment levels are lower and of a different nature. Generally speaking they involve more labor and less capital. This is an advantage for motivated starters.

Labor Costs

With labor as a dominant input, the level of total input costs depends on the price tag attached to labor. In estate situations, labor is rightly treated as a cost. In family farm situations, the emphasis is on monetary outlays, and family labor is often not yet explicitly treated as a cost. The case studies would have gained in comparability and clarity if all labor had been valued in some way.

Two types of labor requirements can be distinguished. The first one is needed to install the system. For example, the establishment leguminous trees in Corn and Beans-Nicaragua, or trenching in Estate of Tea-India. The benefits of these inputs span a number of years. The second is required to maintain the system, for example compost making in Estate Coffee-Mexico and maintenance of alleys in Contour farming-Philippines, Alley Cropping Tanzania, and Corn and Beans-Nicaragua.

Benefits for Farmers

Compared to conventional methods, the output value of organic systems is in most cases higher, but the same is true for inputs if the labor input is properly values. It would appear that in six instances the net results are higher, sometimes considerably higher, then under conventional methods. These are Vegetables-Indonesia (2.5x), Vegetables-Senegal, Alley Cropping-Tanzania, Vegetables-Mexico (1.5x), Soya-Paraguay, and Peanuts-Paraguay (2.5x). In two cases, net results of organic farming are comparable to those of conventional farming (Farm Systems Research-India and Estate Coffee-Mexico), but in two of the 21 cases organic farming appears thus far to be a losing proposition (Estate Tea-India and Sugarcane-Brazil).

Employment Generation

Labor requirements of organic farming are generally higher. Many observes consider this one of it principal attractive aspects in developing countries, next to regenerative, environmental and output quality considerations. This is particularly true if the labor input is evenly spread over the year. Although not always specifically stated, both benefits, employment generation and evening out the labor requirements, occur in most of the cases. However, in Dates-Morocco and Corn and Beans-Nicaragua, the higher labor requirements are a constraint for the adoption of organic farming because of seasonal off-farm employment opportunities (for example tourism and logging). It is not clear from the avàilable reports how the higher labor requirement affect gender distribution.

Total Concept Approach

The introduction of organic farming systems included clear social objectives in Estate Tea-India, Horticulture-Egypt, Estate and Community Coffee-Mexico, and Sugarcane-Brazil. In most of these cases, the social improvements were initiated from the top down, but in the case of Community Coffee-Mexico they are initiated from the grassroots level. In that case, organic farming was seen as an instrument in the social struggle (because no credits were necessary to buy synthetic inputs).

Rural and Community Development

In Contour Farming-Philippines, Vegetables-Senegal, and in Community Coffee-Mexico, the implementation of organic techniques is done at the community level. This group forming is seen as important, not only because it increases the chances of successful conversion, but also because it improves the community's self confidence and structure.

In Corn and Beans-Nicaragua, it was expected that farmers would not have to migrate to new areas because the soil would not degenerate under agroforestry conditions. In the Alley Cropping-Tanzania case, erosion control was combined with improved production per hectare whereby the land could sustain more families then before. A similar result was observed in Contour Farming-Philippines. In Water Harvesting-Burkina Faso, migration stopped because of the success.

QUALITY OF ORGANIC PRODUCT

Food produced using organic methods tastes better and contains a better balance of vitamins and minerals than conventionally grown food. However, there is no clear scientific evidence with some studies showing increases in vitamin C, minerals and proteins and others not. The vitamin and mineral content of crops is controlled by a complex interactions of factors, including soil type and the ratios of minerals in added composts, manures and fertilizers. It is therefore difficult to separate the influences of the environment and farming system. Holistic quality methods, such as picture-developing methods, have been used successfully to distinguish between crop products from different farming systems. Organic wheat usually has lower protein levels than conventionally grown crops. However, milling quality in organic wheat can be achieved through careful choice of variety and management, such as cutting or grazing in early spring No differences is measured in mycotoxins or pesticide residues between wheat produced organically or conventionally.

PRODUCT PRICES

Premium prices are important for the financial success of organic farms, but their availability between countries varies. In the European Union Premiums are more widely available in Northern European countries and for crops, whereas for livestock produce and in the Southern European countries access to premiums is more limited. Typical farm gate prices for winter wheat range from 50-200% above conventional prices. An increasing number or market outlets and better co-ordination of supplies for livestock products has led to increased access to premium prices for livestock products, and premiums for milk and meat ranged between 8-36%. The situation of different in other parts of the world, where premiums for non-horticultural organic crops intended for export are less widely available and farmers consider that their organic systems should be able to function profitably without premium prices.

Level and availability of premiums is closely related to the choice of marketing outlets. In some cases, particular efforts on the side of the farmer are made to realize premiums, such as farm shops or other forms of direct marketing or on-farm processing and the actual price realized depends on the quantities sold through the various sales channels. The importance of different sales channels varies between countries, even though some common trends apply. For example, the share of produce sold through alternative marketing channels was commonly high for potatoes, whereas cereals were commonly sold through wholesalers. The costs of these activities, in particular, the increased labor and capital requirements, are not usually specifically identified, so that it is difficult to judge whether the emphasis on premium markets is really worthwhile. In the absence of higher prices to compensate for reduced yields, good financial performance of the organic system depends entirely on maintaining output and cost reductions.

Premium prices are largely a consequence of demand exceeding supply, but the share that the farmer received is also influenced by distribution and marketing costs. As the supply of products increases, prices to the consumer are likely to fall, but economies of scale in distribution and marketing may mean that prices to producers can be maintained. As the market grows, short-term oversupply may be encountered, leading to lower prices, until a critical mass of products is available to allow new traders and processors to enter the market and prices are restored. The sustainability of premium prices for producers seems secure in the medium to long-term.

THE ORGANIC MARKET

Growth

Recent years have seen very rapid growth in organic farming, particularly in Europe and the United States, but also in many other regions of the world including China, Latin America, and Africa. In the European Union, certified and policy-supported organic production accounted for just 100,000 ha on 6300 holdings in 1985, or less than 0.1% of the total utilizable agricultural area

(UAA). By the end of 1998, this had increased to more than 2.8 Mha (nearly 2.1% of total UAA) on 113,000 holdings, a 30-fold increase in 13 years. There are differences between farm types with a larger proportion of mixed farms managed organically than specialized dairy or arable units.

Consumer demand for "healthier" food has been considered one of the major factors influencing increased consumer demand for organic food. This perception is probably reflected in the lower consumption of livestock products among the consumers most likely to buy organic food. A survey of consumer choice in the United Kingdom showed that apart from health concerns, organic food was bought for environmental and animal welfare reasons. Approximately 70% of consumers buying organic food mentioned these two factors.

Alongside the increase in the supply base, the market for organic produce has also grown. Statistics on the overall size of the market for organic produce are still very limited. Recent estimates have suggested that the retail sales value of the European market for organic food was of the order of US$5-7 billion in 1998. In Canada, the growth of the market for organic produce was approximately 25% per annum. There has been a significant change in the development of the market for organic products, from supply-driven to demand led. In response, the role of multiples and multi-nationals in the processing, sale, and marketing of organic produce is increasing and the proportion or organic food purchased from specialist organic operations and local marketing schemes is decreasing as a result.

The trade in organic products is growing rapidly and becoming a reality throughout the world. Growth rates in the sector show organic products, that a few years ago supplied niche markets, have now entered mainstream marketing channels. Already 30,000 organic farmers are certified in Italy. In Scandinavian countries like Sweden and Finland and in Switzerland some 8% of farming is (certified) organic. Austria leads the world with 10% organic farming and in some Austrian provinces such as Salzburg and Tirol, the proportion is almost 50%. Although some in the organic movement might question whether we are straying from the path of 'healthy natural' growth, it cannot be denied that we are heading for a boom.

Respected organic market analysis, such as Professor Ulrich Hamm in Germany, have forecast in annual growth rate of between 20 to 30%. In some countries, this might be as high as 50 percent. The largest organic trader in the UK predicts that the organic market world wide will increase from US$ 11 billion to US$ 100 billion over the next ten years with the USA, Europe and Japan leading the way. In this context, Denmark's plans to make organic products account for 20% of the total domestic food market by the turn of the century seems far from being a green dream.

Approximately 1% of the U. S. food supply is grown using organic methods. In 1996, this represented over $3. 5 billion in retail sales. Over the past six years sales of organic products have shown an annual increase of at least 20%. Organic foods can be found at natural foods stores, health food sections and produce departments of supermarkets and at farmers' markets, as well as through grower direct-marketing such as C. S. A. s (Community Supported Agriculture). Many restaurant chefs across the country are using organic produce because they desire its superior quality and taste. Organic food is also gaining acceptance on a worldwide basis, with nations like Japan and Germany becoming important international organic food markets.

Though only a small percentage of farmers are expected to become organic producers, consumers demand for organically produced food and fibre products provides new market opportunities for farmers and businesses around the world. For many years, and with great success, the private sector alone has developed the concepts and markets for organic products. However, the surge in consumer interest has created new interest from the public sector, and developing countries are particularly in need of good information.

The export opportunities offered by the so-called developed world have clearly been a major stimulus in getting organic farming established in many countries. It is unlikely that organic agriculture would have developed so successfully in the northern hemisphere if there had been no demand pressure. Supplying a market in which demand outstrips supply is a producer's dream. This has been the case with organic farming for a long time and figures show that this trend is set to continue. Yet, the sector has shown it is also vulnerable and that prices can collapse if there are sudden and exceptional increases in the supply of particular products. Therefore, marketing with a clearly targeted strategy is a must for organic products. The organic market is a special one, partly because premium prices are involved. It requires a special effort. Attention should be given to meeting the requirements of a guarantee system that will ensure organic quality and allow consumers to develop their preferences for organic products with a feeling of trust.

The World Food summit Plan of Action recognized the importance of appropriate input technologies, farming techniques and other sustainable methods, such as organic farming, to assists farming operations to be profitable, with the goal of reducing environmental degradation, while creating financial resources within the farming operation.

Constraints and Opportunities

1. The demand for organic products has created new export opportunities for the developing world. While some consumers express a preference for locally-grown organic foods, the demand for a variety of foods year-round makes it impossible for any country to source organic food entirely within its own borders. As a result, many developing countries have begun to export organic products successfully (e.g. tropical fruit to the European baby food industry, Zimbabwean herbs to South Africa, six African nations exports cotton to the European Community). Typically, organic exports are sold at impressive premiums, often at prices 20 percent higher than identical products produced on non-organic farms. The ultimate profitability of organic farm varies, however, and few studies have assessed the long-term potential for such market premiums. Nevertheless, under the right circumstances the market returns from organic agriculture can potentially contribute to local food

security by increasing family incomes.

2. Farmers are denied access to developed country organic markets for two to three years after beginning organic management since such countries will not certify land and livestock as organic before that time, arguing that it is necessary for the purging of chemical residues. However, products produced on land under organic management for at least one year but less than the two-three year standard can be sold as *transitional organic*, although few markets have developed for such products.
3. In most cases farmers and post-harvest businesses seeking to sell their products in developed countries must hire an organic certification organization to annually inspect and confirm that these farms and businesses adhere to the organic standards established by various trading partners. The cost for this service can be expensive, although it varies in relation to farm size, volume of production, and the efficiency of the certification organization (e.g. IFOAM certification costs a maximum of 5 percent of sales value). Few developing countries have certification organizations within their borders, and even when sufficient resources are available to pay for certification farmers often lack the information to find credible inspectors.
4. While most developing country have focused on export markets in the developed world, domestic market opportunities for organic food or eco-food may also be exploited. In China, for example, there is a growing market for "green food" which, according to government grading standards, is produced without certain pesticides and fertilizers and with biological methods. Chinese farmers also produce organic food for export (e.g. tea to the Netherlands, soybeans to Japan).
5. Whether the intent is to sell organic products domestically or abroad, reliable market information is difficult to obtain. There is virtually no systematic production or market survey data being collected with which to assess the rate and pattern of organic market growth. In particular, no projections for the market in the developing world have been made, nor have markets systematically been identified for developing country exports. Estimates of the public's willingness to pay premiums, the impact of regional and tastes, and the incidence of market fraud have not been undertaken.
6. There is a growing consumer demand in the developed countries for healthy food. This change in the consumer preference reflects the increasing concern about the dangers associated with the consumption of food containing small quantities of food additives and pesticide residues. As the organically produced food is devoid of such harmful additives and residues, it is regarded as healthy food and, as such, enjoys special consumer preference. To the average consumer in the developed countries, organically produced food is more acceptable than the food produced under the conventional agricultural system because of the special qualities possessed by the former in regard to taste and nutrition.
7. In the developing countries organically produced food has not yet

gained consumer acceptance to any appreciable extent. This is despite the presence of harmful chemical residues at higher concentrations in vegetables, fruits and other farm produce. In India, through one meal comprising of chapattis, rice, dal, vegetables, salad (mixture of onions, tomatoes, carrot and cucumber), curd and banana/orange or grapes as dessert, one consumers 0.5 mg of DDT and BHC plus some malathion and endosulphan. This is 40 times more than the amount what an average American or European consumers. These pesticide residues lead to all kinds of chronic diseases of the heart, brain, kidney and liver including cancer.

8. There is lack of awareness among the people of the developing countries about poisons in the food they are consuming daily. Even for those who are aware of the dangers of poison in food, they do not have easy access to organic products. Wherever such products are available, they are costlier than the conventionally produced items mainly because organic agriculture does not enjoy the benefit of subsidy. However, there are still parts of India and other developing countries where people have access only to organic (chemical-free) foods because agro-chemicals have not yet penetrated into interiors and hinterlands. Unluckily, this fortunate situation may not prevail for long as agriculture is being developed as an industry without regard to the carrying capacity of the natural resource base.
9. On the other hand, in the developed countries there is a rapid awareness with regard to organically grown and processed foods and related issues. In Europe and U.S.A. in addition to Reform Houses on also sees special counters in departmental stores exclusively for organic products. In the west, supermarkets are generally considered as a threat by environmental lists, but they may still have a role to play. By introducing a wholly different segment of the consumer population to organic foods, supermarkets can help increase sales of organic products, improve the consistency of supply by creating greater demand and increase overall customer awareness of organic issues. These supermarkets can plough the vast masses of consumers, turning non-users into light user, and light users into regular customers of organic specialty stores. A problem for specially stores and supermarkets is consumer resistance to the higher price of organic products. In spite of this, consumers who are aware are willing to pay higher price for such products not only because they taste better, but also because of the realization that, in the process, they are supporting organic farming and thus protecting the environment
10. Organic agriculture is not only leading the way to viable alternatives to modern development models, but it is also part of the solution to global environment problems. Now, more than ever, opportunities exist for the organic market to expand from less than 1 per cent of overall food sales to between 10 and 20 percent by the turn of the century. There are, however, critical and sometimes very sensitive issues to be addressed in order to sustain a growing market for organically produced food.

Unfair Trends in the Market

Over the last ten years organic products have been marketed in increasingly larger quantities outside the area in which they are grown. Whilst seasonal production and regional markets remain an important objective in organic farming, there is nevertheless plenty of export opportunities for such products from the South as coffee, tea, cacao, bananas, spices, herbs and other subtropical and tropical products.

'Ecological dumping' is becoming more common and is a practice strongly resisted by the organic movement. There is a very real difference between production in keeping with the holistic principals of the organic movement and the purely commercial production of 'natural', 'biological', or 'organic' products such as organic cash-crops in monoculture. It is not enough to exclude what is known as 'ecological dumping' from the organic sector. Given the holistic nature of our movement, we also care about social aspects. In joining forces with the Fair Trade movement we also hope to be able to seriously challenge the 'social dumping' that involves the exploitation of cheap labour and child labour.

Bio-colonialism is one of the major challengers facing the organic movement. Given the much greater economic wealth and 'buying power' of the North, many products of organic quality find their way into these premium markets and so become unavailable to local and regional populations. Much more effort and creativity is needed if organic food is to become accessible to more people throughout the world. The SEKEM initiative is a good example of how this can be achieved.

Community supported agriculture with direct links between producer and consumer, in Japan are Asia known as TEKEI, offer a way of getting out of the bio-colonialism trap. Today, there are organic shops in India and Malaysia, farmer markets in Brazil, and organic produce in supermarkets in Argentina. Box schemes and home delivery of organic products are also being set up in several southern countries.

Very often organic farming systems begin with the production of 'cash crops' and gradually, with the knowledge gained in producing for export, the whole farming unit is converted to organic systems. Thus, planting shadow trees in coffee plantations and using compost is often followed by introducing organic methods into the cultivation of staple foods such as corn and beans.

There has always been agricultural trade. Trade remains a necessity whether it is barter between neighbors or long-distance trade facilitated by money. Trade makes it possible for house-holds to supplement their own production and satisfy their need for food and money whilst exploiting comparative advantages. In this century, there has been an enormous expansion in trade supported by cheap fossil energy, agro-chemicals, refrigeration and increasingly sophisticated forms of transport. At the same time, the buying power of urban populations in both the North and South have combined with trade liberalization and globalization to give a powerful stimulus to markets every-where.

This trade expansion has brought benefits to many agricultural producers and consumers. However, it has also been responsible for an uncontrolled concentration of power and wealth in the hands of a few countries, companies and persons, most of whom a be found in the industrial West. Control over capital and knowledge-intensive technologies, unfair trade practices including the dumping of cheap subsidized food and the manipulation of small farmers whose market choices are already limited have seriously effected small farmers everywhere. At the same time the capacity of industrialized countries to protect their own agricultural markets and the worsening terms of trade between agricultural and industrial products have also contributed to the decline and poverty of many agricultural communities. As small, market-oriented farmer lose out to larger farmers and the non-farming professions, and trans-national companies gain control over international commercial flows and agricultural production, many leave their land to search for jobs elsewhere.

This not only leads to an increasing gap between rich and poor but also to the loss of food security, ecological degradation, pollution and the decline of indigenous knowledge and cultural diversity. Market-oriented agriculture and agricultural trade an lead to a serious depletion of soil fertility, soil erosion and a loss of biodiversity. Agriculture under such circumstances is not ecologically sustainable.

However, agricultural trade and market-oriented agriculture do not have to be a trap for the small farmer. Fair and green trade offers important economic and ecological opportunities for sustainable development.

FAIR TRADE

Alternative approaches are being developed in reaction to the free trade model. Since the 1970s, the Fair Trade movement has been developing fast. Important organizations buy direct from Southern producer partners, pay a fair price and establish long-term working relationship. Initially, health food and Fair Trade shops were important alternative trade outlets. Today, however, Fair Trade products can be found in most supermarkets. Trade organizations that respect the principals of Fair Trade and accept external monitoring by labeling organizations, are awarded a seal of approval. The fundamental principal of Fair Trade is equal partnership between Southern producers, Northern importers, labeling organizations,

Fair trade is always associated with production in the developing world. However, we need fair and real cost covering prices for farmers all over the world. And here the organic market has something to offer all farmers. Organic production often requires more effort and may result in lower ('optimum rather than maximum') yields. Also, as millions of consumers all over the worlds have come to realize – 'nature has its price'. Therefore, growing numbers of consumers are prepared to pay premium prices because they know they are getting premium quality products in return. In calculating the price of organic products more factors are involved than in the pricing of conventional products. Demand, for example, plays a much bigger role. The complex of aspects involved makes it difficult for us to discuss price finding and setting mechanisms in the organic sector here. However, we can say that premium prices ranging from 20% to 30% seem to be normal and when combined with often higher fair trade premiums, they make it possible for small farmers to continue to work their land, feed their families and enjoy real development.

Fair Trade and Trade Development

'Fair trade' covers a wide range of ideas. Fair trade is seen as a conscious decision to help small producers in the Third World become viable exporters of their own produce and to support their learning process. Fair Trade does not mean, however, that other forms of trade are unfair.

Fair trade, of course, is not agricultural commodities and agricultural producers. All disadvantaged producers of any marketable product fall within the definition of the fair trade movement. In practice the fair trade labeling movement and organizations such as Equal Exchange in the USA and ATJ in Japan are mainly concerned with coffee, cocoa, tea, bananas, honey and sugar. The older alternative trading organizations work on a smaller scale and deal with thousands of food and non-food products. In this article we focus on small, co-operatively organized, agricultural commodity producers.

Small Farmers Disadvantaged

Major international agriculture commodities are frequently produced by small farmers who have no secure title to land. Their land holdings are small and they are unable to assert their civil and economic rights even when these are protected by law. Their access to market often depends on long and exploitative chains of intermediaries. Not only do they lack bargaining power, they have no access to information on world market prices. It is difficult for small farmers to acquire the technology and technical skills and the cost of financial services are frequently beyond their reach. As many farm in fragile, near subsistence economies, their choice of survival strategies are not necessarily rational from the economic and ecological point of view. Due to imperfections in the market, lack of transparency and politically inspired distortion, small farmers tend to receive lower prices for their produce.

Dilemma

There are many reasons why the problems of small farmers should be addressed. First, maintaining and boosting the economic viability of smallholdings puts a brake on migration to overcrowded cities incapable of offering employment alternatives. Second, creating an effective economic and institutional environmental makes it possible for people to exploit the market potential of their area. Finally, supporting farmers initiatives by encouraging co-operatives and associations enables them, collectively, to gain fair access to the (world) market: trade is better than aid. In recent years, fair trade labeling has increased the effectiveness of such interventions.

Fair Trade Labeling

Fair trade labeling aims to develop market niches where newcomers to the world market can learn the ins and outs of this type of trade in a relatively protected environment. In doing so, it is tries to provide small farmers' organizations with fair access to the (world) market, under conditions appropriate to their needs. The movement also supports farmers' organizations in their attempt to secure working capital and long-term loans for investments, market information, communications, management support and technical assistance.

Promoters of Fair Trade

Over the years, alternative trading organizations have promoted fair trade through specialize chains of (Third) World shops. The movement began about 30 years ago in the Netherlands and the *Fair Trade Organization* is still one of the most dynamic organizations in the field. Today, trading organization are found in almost all industrialized countries and most of them share supplies' databases.

In the eighties, the movement stagnated World Shops could not cope with the demand for markets coming from Third World suppliers. Mainstream distribution networks had to be accessed and pressure built up to professionalize the network of World Shops. This lead to the development of fair trade certification and tested in the model was first developed and tested in the Netherlands under the name Max Havelaar. Later is spread within and outside the European Union until today certification and labeling is carried out in some 15 countries. Some use the name 'Max Havelaar', others have chosen names and labels that reflect their own national and cultural identity. However, all are members of Fair Trade Labelling International. Coming up to standard

Action for Fair Trade

Fair trade organizations have been working with Southern producer organizations for decades to try and ensure that they receive a decent income for their work. Only recently, however, have fair trade products begun to reach mainstream Northern markets.

Coffee is probably the most successful fair trade product. It is also the most important traded commodity after oil and the main export for many Southern countries. Most comes from family-owned farms in developing countries. Fair trade organizations work with the producer organization, promote the sale of fair trade coffee, monitor buyers and roasters and guarantee that labeled products meet fair trade criteria. The ethical trading organizations bear the costs of inspecting producers.

The fair trade market was boosted by the development of recognizable fair trade marks awarded for products that met clear criteria. However, it was the development of a clearly recognizable brand. Café-direct, that brought fair trade coffee into mainstream retail outlets and within easy reach of consumers. Cafe-direct was developed by a partnership of four fair trade organizations that bought coffee from 14 producer organization in Africa and Latin America. Sales have increased dramatically over the last five years and Café-direct controls 3% of the roast and ground market and 2% of the instant market in the United Kingdom despite being 10% more expensive than most conventional brands.

As well as higher income and greater security, fair trade can also lead to environmental benefits. Fair trade can also lead to environmental benefits. Fair trade organizations often work with producers to improve the ecological sustainability of production.

Progress in Fair Trade Marketing

Fair trade certification has helped bring fair trade products into mainstream European distribution networks. Sales of coffee and bananas have increased tenfold. Today, small farmers are able to sell 14,000 tons of coffee with a trade value of some US$ 40 million in this way. In the Netherlands the market share of fair trade coffee has risen from 0.2% to 2% and in Switzerland to 5%, Mechanisms and alliances have been created to make substantial amounts of short and medium-term ethical bank loans available. While the impact to fair trade labeling appears to be unquestioned, its global impact should not be overestimated. In a global perspective fair trade is no more that a tiny niche market.

Although its impact is growing, results have been reached under very special market conditions and at a time when world market prices were persistently below subsistence level. Under these circumstances, fair trade offered a valuable alternative to those who had their name on the Fair Trade Labelling Procedures 'Register and who had secured an outlet for their product. However, not all organizations used (were able to use) this opportunity to effectively develop the capacity to survive under 'normal' market conditions. In 1994, coffee prices on the world market started to return to 'normal' levels, i.e. above the floor price established by the fair trade labeling organizations. For many farmers' organizations, the price premium of US$ 0.05 per pound above the market price was insufficient, while the inflexibilities inherent in the fair trade model eroded a substantial part of the price premium. A growing number of farmers' organizations who had been operating within the fair trade market ran into serious trouble despite higher-than-market prices, while others managed to survive and went onto develop into strong and went on to develop into strong and respected co-operative organizations, addressing the economic needs and rights to many thousands of small farmers and their families.

Protectionism

Unless alternative approaches to trade become mainstream, the inequity and environmental damage inherent in conventional trading practices will remain. There is need for a more appropriate economic framework to redirect the world economy towards sustainability. This is not a feasible option in the short-term. A more immediate approach would be a new type of protectionism in which governments formulate trade regulations based on socially fair and ecologically sound standards capable of creating enabling conditions for green and Fair Trade.

Priority to Local Economics

Even if such legislation were to be adopted, it would not be possible to eliminate completely the negative social and environmental effects of international trade. There will always be an important group of farmers who are unable to compete on the (inter) national market and whose farms fall in the margins of the globalizing economy. International trade is wholly dependent on fossil energy, itself an important source of pollution and, in this sense, International Trade is inherently unsustainable. To avoid the environmentally unsound transport of agricultural products over long-distances and a risky dependence on distant consumers, sustainable development and protection of the local economy. Particular attention should be paid to marginalized peoples and guaranteeing ecological and cultural appropriateness.

Strengthening the local economy fosters a community's sense of social and environmental accountability, self-reliance, food security, cultural and biological diversity and the commitment to conserving its natural resources base. Indigenous people in the Andes protect the cultural, ecological and economic integrity of their communities by sustaining barter - which as a system is probably more widely spread than officially recognized – as an important element of their economic and cultural system.

Strengthening Local Economics

In many communities economic activities function at an unnecessarily low level as money and local natural resources leak away into the wider market economy. Local exchanges of goods and services can be greatly enhanced by the development of local money systems. Local money system has successfully stimulated the local economy in Argentina. Many local groups and communities in the North and in the South are currently experimenting with these Local Exchange and Trade Systems (LETS).

Critical Factors

It would seem that in some cases fair trade procedures have not been adapted to the operational needs of farmers' organizations, especially under present unstable market conditions. In addition, the fair trade labeling organizations have not paid sufficient attention to the development and organization of the necessary management support services. Management is also a critical factor at farmer level. The minimum scale of operations needed to run an export business is far beyond the scope of small and often isolated farmers. Lacking the management skills needed to operate the business, they have to hire in managers over whom they may have little control. If there is no effective control and accountability is not enforced, the temptation to address private interests above co-operative interests can be irresistible and leads to serious problems. In this way a vicious circle develops and the co-operative is excluded from credit and loans because it has become un-creditworthy and its commercial liability has been further compromised by inadequate quality control.

Many co-operative organizations, however, have successfully exploited the opportunities of the fair trade market and built up and diversified their businesses to the benefit of their members. Critical success factors seem to be the availability of highly component and reliable management, sustained management support, and market diversification into conventional, fair trade and organic markets.

Challenges

There are several major challenges facing the fair trade and organic market. A strategy must be developed to facilitate the merger of ecological and fair criteria incorporating social fairness and ecological responsibility as preconditions for sustainable production. Criteria must be adapted to the operational needs and limitations to Third Worlds realities and international markets. At the same time the management and information gap must be closed and a solution found to the recurrent problem of credit. It is clear that the certifying organizations cannot solve these huge problems on their own. Strong global alliances with trade and industry, development organizations, ethical (and conventional) financial institutions, universities and consultants are needed in order to build a system that works and hence challenges 'business as usual'.

Trade Opportunities

In today's globalize world, export success is one of the major routes to economic progress for developing countries. But the conditions for success are changing as producers face rising environmental expectations in key export markets, resulting from tightening regulations, new corporate practices and changes in consumer values and lifestyles. There new expectations reflect the growing recognition that current patterns of consumption, particular in the richer, industrialized world, are not environmentally sustainable. Profound changes in the ways in which goods and services are produced, traded and consumed will be required, both to reduce the burden on the global environment and to ensure that a growing population has resources to meet its needs.

For these who can adapt to these requirements and start moving to anticipate trends, there are new opportunities to be found in sustainable trade that can generate financial, environmental and social benefits. Already growing numbers of farmers in the Sought are receiving higher prices and more long-term security by selling their products into environmentally friendly or fair trade markets in the North. Consumer demand for organic products is gradually increasing in responding to concerns about the environmental and health implications of industrial agriculture.

New Opportunities in a Growing Market

Latin American producers have been quick to tap into these markets. The organic sector in Mexico, for example, is now estimated to be worth US$ 500 million and Argentina is making serious efforts to development its organic sector with sales rising from US$ 1.5 million to US$ 20 million in the last six years. This rapid development has been supported by efforts to overcome the bureaucracy surrounding EU recognition imported organic goods. Argentina became the first developing country to obtain a place in the EU provisional list, thereby gaining market advantages.

New products have no established market; emerging markets suffer from either insufficient supply or over-supply of products. This contradiction also rules emerging markets for eco-friendly produce, necessitating a balancing act between supply and demand. Investments have to be made; initial losses, though inevitable, need to be pegged down, through judicious promotional campaigns and production planning, to avoid a demand-supply gap. No compromise is possible regarding timely production targets, both qualitative and quantitative, to satisfy and maintain the trust of an emerging clientele. Production has to be so organized that producers are able to conventional inputs. Here the issues are the short shelf-life (of eco-friendly inputs such as bio-fertilizers and bio-control agents which are living organisms) and the crop-specific or pest-specific action of many of these. The fact that these inputs are in great demand and in large quantities, but only during specific times –grow in season – complicates production planning and marketing.

Third, being eco-friendly also implies minimizing environmental damage in the course of transport and storage of the produce. Since energy-intensive operations have to be curtailed, transport and cold storage of eco-friendly foods for the right market become controversial. To overcome such dependence on non-renewable energy, production systems tuned to local markets and seasonal products may be well worth exploring.

However, wherever local production potential is limited, the required product range, be it for a healthy diet or a healthy, has to be met through export from high-potential areas.

Alternative Markets

Experiments to develop parallel markets have shown that it is not easy to do so. The problems encountered range from regularity of supplies and storage problems (space, shelf-life and eco-friendly storage methods) to retaining the goodwill of consumers. The tiniest upset can result in clients dropping out. There are several realities about alternative marketing that are gradually coming to light.

Role of the Trader

It is almost impossible for a small organic farmer to sell all his/her produce in the village. This calls for a whole chain of arrangements, including the presence of a trader. Fair trade considerations mean that a balance of power has to be ensured between producer and trader. To achieve this, producers' collective must be organized.

More often than not, in our villages, the trader is a key socio-economic actor – as money lender, input supplier, transport provider and linking agent with the larger market. By providing flexible and timely loans, assured transport and access to markets, he becomes an ally of small-scale producers, despite their indebtedness to him. He influences farmers' crop choices by being their financier and by sharing his intimate knowledge of markets; in turn he also creates a system for loan recovery.

Quality Guarantee

Any opportunity to earn quick money can increase the chances of producers and marketers cutting corners. It is not uncommon to come across organic produce with pesticide residues – eco-friendly inputs that are either spurious or have crossed the prescribed expiry date – being marketed. Such problems arise because of some unethical elements operating among producers and traders. Hence, quality guarantee through inspection and certification by independent bodies becomes inevitable. Besides quality, fair trade also implies that profit margins in the production-trade chain need scrutiny and certification.

Who should pay for inspection and certification? Reasonably speaking, the producer, trader and consumer should share the expenses. For effectiveness, a local-level certifying body could be constituted and governed by representatives from organizations of producers, traders and consumers. It could assume responsibility for defining standards, checking quickly, and even promoting the use of eco-friendly produce. To bolster this, the retail outlet should use an identifiable tag to indicate the marketing of certified produce.

The Consumer

In general, the average consumer does not take pains to buy eco-friendly produce. Considerations such as appearance, price, product diversity, retail points in the neighbourhood and easy mode of payment, prevail over assurance of product safety and quality. The consumer prefers products to be conveniently available, even if WHO standards of acceptable pesticide residue levels are violated. Any irregularity in supply, even those caused by unfavourable weather, makes the average customer walk away, confirming the saying-*Trust comes by foot and leaves on horseback.*

These are the realities that producers and traders have to take into account, to create suitable production and marketing systems for eco-friendly produce.

Retailing Arrangements

Various modes of retailing are being tried out to develop niche markets for eco-friendly produce. Be it direct marketing by individual farmers or NGOs, or retailing through intermediaries such as corporate houses and individual traders, each mode is faced with special problems, and the search for appropriate arrangements still goes on. Here are some possible reasons that may have affected the economics of these enterprises.

The retailing efforts of individual farmers were diverted by insufficient product range and higher prices. Exceptions were those farmers who had special non-perishable products in bulk, and had successfully linked up with international market. On the other hand, NGOs facilitating organic market development sourced product from multiple locations, offering farmers attractive prices. Erratic supply, inconveniently located outlets and limited clientele led to insufficient turnover, storage losses and higher prices, rendering such ventures economically unviable.

Corporate houses encountered problems in marketing their wares because they had opted for large-scale production of a limited number of products, but only in the normal seasons. This may have resulted in wildly fluctuating supplies, and consequent loss of consumer confidence.

Dilemma of the Farmer

The experience of *organic vegetable farmers* reflect the fact that growing and selling perishables is a complex process. The consumer expects variety round the year, even if some vegetables are off-season; but a repetitive mix of vegetables is not tolerated for more than a fortnight. The eco-market is small; thus quantities retailed have to be small. So, the producer has to engage in labour-intensive yet eco-friendly cultivation of vegetable on small plots, to maintain rotation and demand schedules. Organizing production in this manner increases the possibility of a steady market, which is essential for perishables, and can provide the producer with a small but steady income.

The eco-friendly *fruit farmer* produces large quantities at a time, making it difficult to find a large enough alternate market for his/her produce. With a shelf-life of about a week, the urgency to sell is very high. So, the producer has little choice but to sell in the open market.

Eco-friendly *cereal and pulse growers* opt for bulk-selling their produce to traders for want of storage facilities. Though traders have the necessary infrastructure, they get caught in problems of preservation, since consumers tend to buy moderate quantities at a time. The issue pertain to using eco-friendly storage method, preserving the palatability of the produce, and managing expenses and storage losses, to maintain profitability.

Processing

Various attempts at agro processing, either shelf-life or to add value and increase profits have floundered, because processing is not tailored to market realities. Entrepreneurs are pure prone to embarking on such ventures without thoroughly assessing and understanding al aspects of eco-friendly processing, as well as demand and scale of operation. Inappropriate products pile up, cost estimations go awry, qualities suffers and the enterprise closes down. Besides, there is the added complexity of providing the "environment-friendly" guarantee.

MARKETING OF PERISHABLES ORGANIC PRODUCE – STUDY IN BANGALORE, INDIA

Many farmers around Bangalore cultivate and market organic produce-either on their own initiative, or an NGO motivation. The product range includes milk, cereals, fruits and vegetables.

Fruits and Coconuts

Yields are seasonal and daily marketing is not necessary. Fruit harvesting is labour intensive and this pushes up costs, but merchants are willing to pay more for assured quality and quantity of organic produce.

The main marketing problem of these products is the large volumes to be harvested and sold. To cite an example, an average farmer has to sell nearly 82,000 coconuts from 565 trees or 650 kg of guavas from one harvest. Organic shops cannot handle such large volumes in one lot. Instead of organized harvesting, it the coconuts are allowed to ripen and fall, a farmer can dispose of 10 to 15 at a time. But then, overheads go up and profits come down. An advantage of coconut is that it can be stored in a husked form until prices get more profitable. Middlemen serve as useful intermediaries for small farmers, helping with credit and marketing.

Milk

Being perishable, milk has to be marketed every day. Government milk co-operatives offer a ready outlet. But "organic" milk enjoys no price advantage over "inorganic" milk. Milk is also marketed through local markets. Here too, the price is the same as that of regular milk.

Potatoes

Two NGOs had motivated small farmers to take up potato cultivation on small plots. After facing problems with procurement of quality organic manure, inputs and good seed material, they had to tackle the question of volumes and prices. First, the quantum produced by each farmer was not big enough to be marketed independently. Second, the exclusive outlets for organic produce (discussed below) could not handle the total output of 88,000 kg of potatoes, and had to scout around for other secondary organic outlets. Eventually, one NGO marketed its entire yield (72,000 kg) in the open market, while another opted to market the produce from its farmers (16,000 kg) in the organic market. Staggering the sale by offering only a part of the yield for sale every week meant higher overheads, as well as loss through prolonged storage. Better crop planning by the NGOs, based on study of market demand, would have resulted in a more organized marketing system, with better returns.

Exclusive Outlets for Organic Products

Econet, an NGO-run organic shop, offered its consumer members a price comparable to general retail market prices. Farmers' produce was collected and transported to the outlet, as an incentive to them. However, poor scheduling and irregular volumes affected the confidence of producers and consumers alike.

Consumers Collective (CoCo), another direct marketing initiative had a limited clientele, and faced supply-demand irregularities. This affected economic viability, and operations had to be shelved in the short run.

In both cases, the low scale of operations critically affected the quantum the kind of products that these outlets could deal with. So, the ability of small farmers to market through an exclusive outlet was constrained.

Lessons Learnt

Production planning without understanding the nature of organic markets, led to inconvenient production schedules and unmanageable quantities of produce, thereby affecting the confidence of producers and consumers alike. Poor market planning meant that the understanding of consumer needs was weak, and was not reflected in the production cycles of the participating farmers. Not enough attention has been given to developing consumer awareness about organic products, thus affecting the growth of the market.

Both self-motivated farmers and well-meaning NGOs have embarked on organic production without studying the market. The answer to stable organic markets probably lies somewhere between individual consumers and bulk purchasers. In Bangalore, therefore, some obvious starting points are student hostels, hotels and nursing homes. These bulk consumer could be motivated into going organic, their consumption patterns over a year worked out, and production planned accordingly. The marketing of organic produce demands a thorough knowledge not only of sales and marketing, but also of the technical, managerial and certification aspects of production.

CERTIFICATION OF ORGANIC PRODUCE

Organic farming has a major difference from other approaches to sustainable agriculture, is the existence of production standards. Certification procedures and, in many countries, a legislative basis giving a clear dividing line between organic and other farming systems, primarily for marketing purposes.

Organic farmers have imposed limits on the range of acceptable production practices and technologies. As organic farming has developed, acceptable production practices have been recorded in technical guides and handbooks. However, organic farming systems operate at a financial disadvantage relative to other producers, because they seek to deliver environmental and sustainability benefits and internalize externalities within the farming system. It was therefore necessary to look to consumers to help compensate through a price premium for organic produce. The market for organic products therefore developed as a means to an end, rather than an end in itself.

Where marketing of produce was locally based and small scale, production standards might be guaranteed by word of mouth. However, as the market for organic produce expanded, it was necessary to codify commonly accepted production practices. Since the early 1970s, considerable efforts have been put into the development of production standards for organic production (e.g., IFOAM, 1998), in order to create a differentiated market for organic products. Production standards are based on the over-arching principles of organic farming but may be expressed in great detail, related to local conditions and may indicate recommended, restricted and prohibited practices and inputs. Production standards therefore represent a blend of ethics, tradition, experience, scientific knowledge, and pragmatism. They are constantly evolving, reflecting the need to respond to the appearance of new technologies (e.g., genetic manipulation) and new evidence of environmental impact (e.g., the removal of Bordeaux mixture as restricted pesticide for organic farming). Of key importance is that the definition of organic agriculture is in terms of standards for production and processing, not the certification of the products themselves. It is the way food and fiber is produced, not the end product, which is the focus of organic standards.

Clear production standards has also required the introduction of processes of farm certification and inspection, to ensure that the standards are adhered to. A period of several years, known as the conversion or transition period, is needed to change a farm from conventional to organic management. The length of the conversion period depends on the certification scheme, but under many schemes a period of two years conversion is required before a farm, or part thereof, is certified as organic. However, there is wide agreement that the development of a fully-functioning organic farm takes much longer than two years and is dependent on the previous management of the farm and the soil and environmental conditions.

The Standards

Setting up the organic standards, rules and (governmental) regulations that clearly define the practice of organic farming is the first step for marketing. From the beginning the organic farming movement has been more than a set of ideas. It is firmly rooted in a normative description of farming in harmony with nature.

Standards, whether international, regional linked to a specific philosophy, are not intended to be a handbook for farmers working in the field. Indeed they make rather 'dry' reading. What they do provide is a guide to organic production and the means to develop a truly ecologically sound farming system. They define our systems and technologies and give us a huge advantage as we face the urgent need to distinguish organic farming from 'half-way approaches' such as integrated pest-crop management. Standards are not simply a collection of prohibitions describing what is not allowed in organic farming. Standards reflect clearly the positive approach and definitions of organic farming by emphasizing what ought to be done in order to farm organically and stressing the avoidance of synthetic chemical substitutes that have lead conventional agriculture to its present 'dead end'.

Implementing organic standards requires inspection. A body that is independent of standard setting and control should evaluate facts and find lings and where farmers and products are shown to have complied with organic standards, it should be able to issue a certificate a product 'certified organic'. A logo or seal that functions as a trademark is used to translate this into a statement of quality that the consumer can understand.

It is still a fact of organic life that a lot of product inspection and certification carried out in the southern hemisphere are done by institutions based in the North. However, efforts are being made in many parts of the world to develop regional and national organic product guarantee systems. Such systems not only reduce costs but also give organic producers in the 'developing' world more independence and self-control. IFOAM and many of its member organizations actively support this decentralization strategy.

However, even if regional and national guarantee system are established, there will still be a need for producers to work in close co-operation with certification programmes in the North if products are to be exported to these markets. There are about 300 organic trade marks on the market. Few consumers have an overview of this 'seal jungle'. There is obvious need to establish international 'control' and supervision of organic certification. Therefore, IFOAM has established an accreditation system for certifying institutions where, on the bases of mutual recognition, it is possible, for example, to allow licensing agreements.

Trading

Besides the usual information on transport logistics, custom regulations, tariffs and price mechanisms necessary in exporting any product, trading within the organic sector requires a particular type of knowledge. Most of this cannot readily be found in the literature on economics and marketing. Perhaps one of the most effective ways of developing an understanding of how the organic market works is to attend one of the several international organic fairs. The most international, prominent and by far the strictest of these 'organic' fairs is the Biotech which takes place in Germany every year

A number of donor organizations including SIDA and GTZ and development agencies in the North have established promotion and training programmes to foster export opportunities for organic products. These projects have already helped to encourage producers in developing countries to enter the rapidly growing market for organic products.

Serious Barriers

Small- and medium-sized exporters often face serious capacity constraints in responding to the challenges presented by sustainable trade. Conventional trade barriers in the industrialized world, such as restrictive trade policies in the agricultural and textile sectors, perverse subsidies and bureaucratic regulations are also serious barriers to Southern producers. Partnership along the product chain is essential in overcoming these problems. NGO's government agencies and by years in the North canal play an important role by providing technical or financial assistance, long-term security or in helping to develop producer organizations. Positive policy choices also help expand sustainable trade opportunities.

Governments in industrialized countries should ensure that policy making processes are transparent to exporting countries and that new regulations are phased in such a way that producers have time to make the necessary changes. Whilst Northern governments could play a much more proactive role in developing the market and improvement access for sustainably produced goods, governments in developing countries can do much to promote sustainable trade by integrating environmental factors into their export promotion strategies.

Meaning of Certified Organic

Certified organic refers to agricultural products that have been grown and processed according to strict uniform standards, verified annually by independent state or private organization. Certification includes inspection of farm fields and processing facilities. Farm practices inspected include long term soil management, buffering between organic farms and any neighboring conventional farms, product labeling, and record keeping. Processing inspections include review of the facility's cleaning and pest control methods, ingredient transportation and storage, and record keeping and audit control.

An organic label indicates to the consumer that a product was produced using certain production methods. In other words, organic is a *process claim* rather than a product claim. An apple produced by practices approved for organic production may very well be identical to an apple produced under other agricultural management regimes.

Partnerships are Needed

Organic certification can be a slow, laborious and relatively costly process and a particular challenge to smaller producers. A common solution is to form co-operatives to share the load. In Uganda, for example, the Lango Co-operatives Union has made the transition to organic cotton production with the support of the Swedish International Development Agency (SIDA). SIDA provided technical assistance, organized crop finance, and ensured organic certification. It also supported the training of local certifiers thus reducing the cost of third party inspection. The way prices were set was another innovative feature of this project. Before conducting business, all partners in the chain list their costs and claim a fixed margin based on open books. A fair trade organization then looks for the best possible price on the market. Any excess funds are paid out to the producers, either as extra premiums or as development find contributions. For the 5,500 farmers involved in organic cotton production in Lango, this has meant a 20% average increase on farm gate prices.

Links between members of the supply chain has meant that Lango solved one of the major problems facing Ugandan exporters: banks that refuse to pay crop finance. With the help of the Dutch government, the project received a loan, at commercial rates, and SIDA arranged that risks were covered by the Dutch HIVOS/Triodos Fund that provides funding for environmental and social projects in developing countries.

Another organic cotton partnership that has paid dividends is that between the Swiss cotton trading company Remei, and Maikaal Fibers, a spinning mill in Madhya Pradesh, India. Eighty-five villages produce cotton for Maikaal, which is certified by a Swiss company. The project guarantees to buy cotton and provides extension services, interest-free credit and a price premium of around 20% An estimated three-quarters of the cotton is sold through the Swiss Co-op, which launched a new range of clothes- **Natura**

Line -in 1995 and aims to be entirely organic by the year 2000. It should be noted that the Co-op has kept the price of Natura-Line products at the same level as conventional brands, despite the premium price paid to cotton growers. Partnerships are needed

Organic certification can be a slow, laborious and relatively costly process and a particular challenge to smaller producers. A common solution is to form co-operatives to share the load. In Uganda, for example, the Lango Co-operatives Union has made the transition to organic cotton production with the support of the Swedish International Development Agency (SIDA). SIDA provided technical assistance, organized crop finance, and ensured organic certification. It also supported the training of local certifiers thus reducing the cost of third party inspection. The way prices were set was another innovative feature of this project. Before conducting business, all partners in the chain list their costs and claim a fixed margin based on open books. A fair trade organization then looks for the best possible price on the market. Any excess funds are paid out to the producers, either as extra premiums or as development find contributions. For the 5,500 farmers involved in organic cotton production in Lango, this has meant a 20% average increase on farm gate prices.

Links between members of the supply chain has meant that Lango solved one of the major problems facing Ugandan exporters: banks that refuse to pay crop finance. With the help of the Dutch government, the project received a loan, at commercial rates, and SIDA arranged that risks were covered by the Dutch HIVOS/Triodos Fund that provides funding for environmental and social projects in developing countries.

Organic Farmers and Export Markets: The Role of Co-Operative - Case Study form India

Magosan Exports, a company launched in 1993, which obtains international certification for organic farmers (provided they are 100% organic). He also makes specific suggestions about action to put organic farming on a sound footing.

As of now in India, even successful organic farmers lack awareness about marketing. Many local micro initiatives in cities lack quality control and do not link with other networks. Both these factors are necessary to operate on a large scale.

With the increasing preference of consumers in developed countries for certified organic foods, the demand has been rising by about 12% to 17% every year, but the availability certified organic foods worldwide is only in the range of 3% to 8%. This indicates the excellent potential for exporting organic foods, which can perhaps be tapped best through group initiatives in organic farming.

Magosan Exports (ME) was launched in 1993, in Karnataka, following contacts between the promoters and organic traders in Europe. It is perhaps the only officially licensed export company in India that has been able to negotiate international certification on behalf of its member farmer. We have a tie-up with the certifying agency SKAL in the Netherlands, which has been inspecting and certifying all our farms and products since 1994. The SKAL inspector visits each farm twice a year to scrutinize and gather samples for analysis. Small farmers can overcome the prohibitive cost of certification, if they work collectively and avail of the certification facility. It gives them access to consumers in advanced countries who are willing to pay higher prices for products that bear the pro-environment EKO mark.

A statutory requirement for certification is that a farm has to be 100% organic – i.e. no dependence on chemical fertilizers and pesticides for at least the previous three years and only restricted farm inputs allowed thereafter. The compost pit and water source should be located at least six metres away from the border of any other conventional farm. Meticulous book-keeping of inflow and outflow of inputs, and maintaining a clean and congenial environmental – thereby creating natural habitats – are essential for organic certification.

Farmers' groups are kept informed of such requirements, and a minimum floor price is assured for their produce. They not insist on specific products. The farmers are allowed the flexibility to select and grow the crops that are best suited to their circumstances, and this arrangement works out very well.

Magosan is linked to 2,500 organic farmers and farmers' organizations in several districts of Karnataka and Kerala, and in the Sundarban area of West Bengal. The combined area under organic farming is more than 1,000 ha.

Magosan collects all details and product samples from its Registered Member Growers (RMG). It assumes all the costs for certification, facilitates inspection and quality tests and does the base work for an RMG to be able to meet international quality standards. RMGs are not restricted in their choice of crops, but have to ensure that quality is maintained.

ME collects and disseminates knowledge about various successful farming techniques and post-harvest technology to the farmers when necessary, at nominal cost. International consumers have given Indian organic foods a high rating, in terms of taste, appearance and aroma. At present, organic spices, pulses and cotton are being exported by us; we soon hope to be able to broaden the product range.

Yet, the rampant corruption that permeates the entire system of agricultural marketing and export, undermines the sincere efforts of organic farmers. The authorities are not vigilant about quality control; we as exporters face problems in transit or shipment. Many a time, banned chemicals used for disease treatment contaminate our consignments of organic food, enroute.

Organically produced food shows distinct quality differences from non-organic food. These differences could be assessed through various techniques such as Kirlian photography and microbiological and bio-chemical tests. While such techniques are useful, a perfect production systems having specific standards have to be promoted for adoption by the farmers processors. This is necessary to verify the validity of the claim of farmers 'organic' and of middlemen and food processors that their produce for sale is organically grown and processed.

To satisfy this requirement, a sound system of certification and labeling of the produce by a complement agency is essential. The farming unit where food production is organized as per accepted standards has to be supervised and inspected at frequent intervals and at different stages of production before certification in order to ensure quality and authenticity. The certification agency has to adopt very reliable methods such as soil tests water tests and food quality tests besides taking note of the natural qualitative indicators so as to ensure credibility of the system and to prevent fraudulent labeling of the product.

While certification gives reliability to product as organic, the certification itself would need a guarantee. This is provided through a system of accreditation. Until now there has been no international system to guarantee the organic certification in different countries. But now International Federation of Organic Agriculture Movements (IFOAM) Germany has established the IFOAM Accreditation Programme.

IFOAM and Certification

Briefly IFOAM is a global umbrella network of farmers, scientists, processors, traders, agriculture extension workers and consumers, who share the common interest in developing farming and food production system that provide healthy produce without depleting natural resource base. The IFOAM has a Standards Committee (SC) which is responsible for preparing standard documents. During the last few years the Standards Committee of the IFOAM concentrated on the following 3 major tasks:

1. Updating the Basic Standards on plant production and animal husbandry
2. Introducing processing standards
3. Reflection and cooperation with various state and super state institutions on guidelines for organic production, i.e. FAO/WHO Codex – Alimentarius and the EEC regulation.

The standards documents prepared by the IFOAM as found in the booklet 'Basic Standards of Organic Agriculture and Food Processing' reflect the present status of organic agriculture and processing methods. Their standards are not to be seen as a final product, but rather as a contribution to the dynamic development of organic farming. The IFOAM Basic Standards include a chapter on inspection and certification but only in the form of organic agriculture.

4. The principal aims of organic agriculture
5. Conditions in which these standards apply
6. Crop production
7. Animal Husbandry
8. Storage, transport and processing
9. Labelling and consumer information
10. Inspection and certification
11. Revision of the Standards

In the IFOAM Third World Task Force Policy Paper, the scope of farming has been enlarged to cover the cultural and socially diverse way of life in addition to producing food and fibre. Hence, the definition of 'organic agriculture' should also include sustainability of the social and cultural aspects of human beings and their communities. It further goes on and states that although some farming systems may be technically and environmentally sound, they should not be acceptable as organic if they are socially unjust. Ethical trade takes local socio-economic and agro-ecological realities into consideration. Hence socio-economic criteria need to be also incorporated into the standards of production/processing and trade of organic products. The SC of the IFOAM, therefore, is drawing up following standards:

1. Fair Trade and Social Rights
2. Environment and Nature Conservation.

Further, production and / or processing guidelines for specific crops of products, like the already existing Coffee and Tea Guidelines, are also being drawn up.

Since 1980 the IFOAM has set Basic Standards for the production of organic foods. These serve as frame for countries to draw up their own national standards. The IFOAM Standards Committee with experts from various countries meets regularly to adapt the Basic Standards to the changing conditions, the recent emphasis being Fair Trade, Social Rights and Environmental Conservation. The inspection, evaluation and accreditation of certification organizations take place within the IFOAM Accreditation Programme Board.

IFOAM and Accreditation

Although the IFOAM is not engaged in trading, it has realized the importance of the certification of organic products for the positive development of organic farming – all over the world. A programme under the sponsorship of the IFOAM guarantees commitment, competence, transparency and independence from any particular interest. The IFOAM Accreditation Programme has the following to offer:

1. The producer can be assured that international competition is fair
2. The certification programmes are assured of the equivalency of other programmes without having to make their own checks
3. The exporter gains access to the world wide organic market
4. The importer can buy organic products from all over the world, with a powerful guarantee
5. The consumer can trust the organic quality whatever be the origin of the food.
6. National and supranational authorities get a cost-efficient and trustworthy solution for the acceptance of imported organic goods.

Farmers organizations, consumers organizations, marketing organizations and food processing organizations should link up with international forums for the success of the movement.

Organic Foods Certification in India

Proper standards of production and certification of organic foods have not been drawn up for application in India where agro-climatic conditions and farming systems exhibit wide variations. To commence with certification, the following steps have to be undertaken:

1. In order to reduce confusion, the use of the wood 'sustainable' can be made synonymous with 'organic' and 'natural'. These are the major terms used world wide meaning food is produced without the use of synthetic chemicals and with an awareness of sustainable soil management practices as they affect the ecosystem of the farm and surrounding environment.
2. Develop a working group of farmers to establish 'Organic Farmers' Certification Organizations. Discussion between farmers is necessary to share information on problems and solutions to growing crops without synthetic pesticides and fertilizers.
3. Define what is meant by 'Natural', 'Organic' and 'Sustainable' Farming. Organic farmers themselves must get together to decide on what they mean by sustainable farming practices.
4. Standardize farming practices and materials (farm inputs) which are acceptable to the organization. With the increasing participation of farmers in natural, organic and sustainable farming, the need to establish standardization of practices and materials (farm inputs) becomes necessary and has to be addressed.

Farm inputs such as fertilizers and pesticides labels 'Organic' need to be verified. The farmer should have some way of knowing that the materials purchased are acceptable brands to be used in sustainable farming practice. Guidelines need to be developed concerning acceptable and non-acceptable inputs.

Introduction of Certification in India for Organic Agri Exports

Today organic agri cultivation is fast picking up. The market for organically grown agri products is very big in Western Europe and USA.

To curtail exports of spurious organic products from the country, the Commerce Ministry at New Delhi has recently made it compulsory for exports to obtain valid 'organic certificates' issued by accredited authorities. According to a public notice dated June 11,2001 issued by the Director-General of Foreign Trade, an agricultural product will be allowed to packed under a valid Organic Certificate issued by a certifying agency duly accredited by the Agricultural and processed Food Products Export Development Agency (APEDA), Coffee Board, Tea Board or Spices Board. The new condition would be effective from July 1, 2001.

Suggestion

1. It is essential to set up and enlarge organic food raw material production bases to satisfy people's growing demand for organic raw material. For

that purpose, bases should be selected on the basis of the investigation and analysis of the environment quality background of the locations. Measures should be taken to ensure the production bases free from environment pollution and in sound agricultural environment.

2. It is important to develop organic food processing industry by setting up various types of organic food processors producing various organic foods, thus increasing the variety of organic foods in the market as well as the economic benefit by processing.
3. Actively opening up organic food market both at home and abroad will stimulate development of the trade of organic food. Market determines production. Only market can stimulate development of the production and processing of organic food and production of quality organic food in turn does development of the market. In Asian countries, most consumers do not know much about organic food, which affects developments of the organic market. It is advisable to take various measures to give publicity to organic food and to educate the consumers.
4. In developing the market for organic food, the world market demand will usually stimulate development of a country's organic food. Therefore, at the initial stage of the organic food market, to develop products demanded by the world market will help develop domestic market. But to build a sustainable organic food marketing system, it is essential develop domestic and world market simultaneously.
5. To ensure the quality of organic, it is essential to formulate organic food production and processing standards in line with the basic standards of IFOAM, to set up certifying organisations and to strictly enforce the procedures for inspection, and certification of organic food. Inspection and certification of organic food are the major links in building high-quality sustainable organic food supply system as well as the guarantee of the quality control of organic food. Most of the countries have not yet worked out their own country's standards nor set up any certifying organizations, which has significantly affected the development of organic food more attention should be paid to this work.

INIDIA NEEDS

To place organic farming on a sound footing in India, urgent action is necessary on the following fronts:

1. Knowledge development and training of farmers in grading, curing and post-harvest practices according to international standards to ensure quality outputs. Enhancement of labour efficiency by developing intermediate technologies. Value addition processes at farm-level or community level to improve the income of farmers.
2. Arrangements for easy credit for working capital, supported by buyback arrangement, well encourage farmers to take up innovative cropping system.
3. A separate Organic Development Wing under the Ministry of Agriculture could bring greater synergy to the effort to promote organic farming and

the export of organic products.

CONCLUSION

As we have seen, the booming development taking place in organic farming and marketing offers many opportunities. But there are challenges as well as and these have to be faced. However, if we do not give up our holistic principles, we will be able to go on contributing to the establishment of organic production systems and this could lead to changes in life style and consumption patterns that will reach far beyond food and nutrition.

CHAPTER 16

Research

Although organic farmers have successfully dealt with many aspects of organic manures, composting, crop rotation design, nitrogen fixation and nitrogen supply in crop rotations and weed regulation. Therefore, research projects in organic farming need emphasize on the following fields:

1. Improved utilisation and efficiency of use of organic materials, e.g., by minimizing nutrient losses from manures through the development of better collection and storage techniques, reducing labour inputs required for land applications, and exploiting the organic fraction of the urban waste stream, particularly in urban and peri-urban farming systems.
2. Further research on agricultural systems, particularly to:
   a) elucidate potential of cumulative and cross benefits of organic techniques
   b) explore the compatibility of organic and non-organic fertilizers/inputs;
   c) investigate the changes in soil fertility and crop production which follow a cessation of fertilizer use and its substitution with organic inputs, to both determine the features of the 'transition' period under different conditions and also to determine best practice for maintaining yields and income over this period.
3. Further studies on the policies of national governments and international agencies to organic farming, to determine the extent to which these currently influence the promotion of non-organic versus organic farming and the potential for change through policy making at these levels.
4. Comparative studies on the price, value, quantity and supply of organic and non-organic fertilizers, particularly in areas of intensive agriculture, taking into account the 'true' longterm value of organic inputs. Such studies would then lead into a full cost-accounting analysis of organic and non-organic farming systems in SSA, to include current externalized costs such as pollution damage, loss of biodiversity and negative health impacts.
5. Further studies on consumer perceptions of the food, health and environmental issues raised by organic farming, to determine and current and potential influence of consumer demand and public pressure. In addition, a case study analysis to ascertain the usefulness of necessity of a domestic market as a base to the setting-up of export trade in organic products.

A HOLISTIC INTERDISCIPLINARY APPROACH

Importance of a Holistic Base

While holistic methods may be more difficult to apply to laboratory and field experiments, a holistic outlook is critical in deciding the context in which research should be undertaken. The holistic view must inform:

1. What is the real research question?
2. The choice and application of the research methods chosen;
3. Interpretation and validation of results.

Specific experiments need to be derived from a full consideration of the research question, each forming a component towards greater understanding or solution of the problem. The process should be highly interactive, *top to bottom*, *farmer back to farmer* and cross-referencing the relevance of the research at all levels.

Comparison of Organic and Conventional Experiments

As whole, organic and conventional experiments tend to have a different blend of characteristics, some of which are listed in Table 16.1.

An organic approach is based more on an acceptance of complexity, as an inherent part of the system rather than seeking to reduce or avoid complexity altogether. Control treatments will tend to be defined by the system, rather than artificially imposed. Environmental effects may be used positively, for example, to drive selection in organic plant breeding. Due to the greater range and interdependence of factors, there is likely to be a greater element of value judgment of intuition when setting up the experiment, compared to the conventional researcher who tends to be less involved.

TABLE 16.1: Generalized characteristics of organic and conventional experiments

Organic	Conventional
Inter (multi) disciplinary	Discipline based
Interdependence of activities	Singularity of activities
Role of value judgment/intuition	Objectivity
Longer-term	Short-term
Environmental interaction	Environmental control
Greater range of parameter studies	Fewer parameters evaluated
Based on farmer goals	Involve 'commercial' goals
Applied	Fundamental

Application of Techniques

Techniques are broadly classified according to:

1. Appropriateness to organic and conventional research e.g. many analytical techniques.

2. Suitability to organic research but with some adaptation e.g. molecular biology;
3. Novel e.g. new statistical techniques.

The requirements of holistically based research, will determine the approaches to be taken and the particular techniques (modified or otherwise) to be applied. Some possibilities are given in Table 16.2.

TABLE 16.2: Approaches and techniques for organic laboratory and field experiments

Requirements	Approaches	Techniques
Multi-disciplinary	Cooperation Brian-storming	Inputs from diverse sources e.g. Social sciences Molecular biology
Interdependence of activities	Define Primary effects Secondary effects	Computer modeling Causal and flow analysis techniques
Value judgment/intuition	Consultation Observation	Non-traditional plant breeding techniques
Range of parameters	Qualitative and quantitative methods Macro and micro scale	Better developed food quality assessments Soil biological activity 'Picture building' techniques Developing of 'indices'
Experimental control	Experimental design Observation study	Novel statistical techniques
Environmental interaction	Multi-site experiments Multi-factorial design	GPS technology

To accommodate more complex, interacting issues, greater emphasis will be given to consultation and observation studies, in formulating the research question and the experimental approach to be taken. The experience of a range of participants (farmers, researchers, retailers) from a variety of backgrounds (production, economics, social sciences) may be brought to bear, for example, to propose novel approaches to a particular problem. New and developing approaches in computer modeling, statistics or global positioning by satellite may be used to analyses data or improve experimental design. Even technologies such as molecular biology, normally thought to conflict with organic ideology may be used for example, to chart population dynamics within a non-traditional plant breeding programme.

Only in the simplest biological systems can positivist realist scientific approaches explain cause and effect mechanisms. The complex and interacting nature of all the biological and decision-making processes within organic farming systems requires a systems approach to scientific study, which considers processes within the contest of the system. This does not rule out reductionist approaches to specific questions, but demands a more holistic approach to the framing of hypotheses and the interpretation of results

DISCIPLINARY RESEARCH

Socio-Economic and Policy Research

Socio-economic and policy studies covers a very broad range of research. Consequently, one cannot judge the strengths and weaknesses of a particular research method independently from the particular topic being studies.

1. Descriptive (studies that report what the current situation is in a particular aspect of organic farming, without any analysis, comparisons of considerations of alternatives.)
2. Comparative (studies that compare organic and conventional farming with regard to performance measures such as yields, economic returns or environmental impacts).
3. Improvement (studies that try to improve the performance of an organic system according to any of these performances measures).
4. Projects (studies that attempt to predict the future status of organic farming under various assumptions or in response to various policy scenarios).

Economic Research

1. Develop proper cost/benefit analyses and accounting models for the natural resource base with an emphasis on the environmental impact of the different farming systems.
2. Develop models, which indicate the impact of conversion to Organic Agriculture Systems on a regional and national level.

Research on Production System Management

Agroforestry

1. Effect of multistory farming on the microclimate and yield physiology of food crops (CO_2-assimilation, photosynthesis etc.), as well as on key parameters of soil fertility (organic matter, structure, pore space, biological activity, water balance etc.).
2. Suitability of promising tree species for use in agroforestry systems (light transmission, competition of roots with food crops, allelopathic compatibility etc.);
3. Locational suitability and ecological requirements of tree and shrub species valuable in agroforestry terms;
4. Biomass production (for mulch and fodder) and nitrogen production of various shrub legumes grown as hedges (*Leucaena, Calliandra, Sesbania etc.*)

Multiple Cropping

1. Further development of traditional multiple cropping systems by introducing additional crops, using improved varieties (e.g. lower-story

crops with a high degree of shade tolerance) and applying improved cultivation techniques (better plant or row spacing, contour planting etc.);

2. Performance of traditional and improved multiple cropping and sole cropping systems (labour productivity, output per unit area, physical yield, farm income etc.);
3. Interactions between crops (e.g. allelopathy, N_2 fixation of legumes as influenced by a companion cereal);
4. Plant protection by means of multiple cropping.

Green Manuring

1. Locational suitability of specific green manuring plants and mixtures (biomass production, fertilizer effect, nutrient and water requirements);
2. Under sowing of green-manuring crops, e.g. in maize, sorghum, upland rice (shade tolerance, degree of competition for water and nutrients, N_2 enrichment capacity etc.);
3. Integration of green manuring period into existing crop rotations;
4. Labour efficiency, particularly with regard to working green manure crops into the soil;
5. Combination of green manuring with, e.g. farmyard manure, compost and/or mineral fertilizers;
6. Long-term changes in soil fertility (humus content etc.) as a result of green manuring;
7. Selection and breeding of fast-growing green-manuring legumes;
8. Crop/green-manure geometry in intercropping and sequential cropping patterns.

Composting

1. Long-term effects on yield and soil fertility of compost on its own and combined with other fertilizers (green manuring, rock phosphate, lime, mineral nitrogen) in all agroclimatic zones;
2. Heap versus pit composts where both methods are possible: rotting process and quality of final product (aerobic/anaerobic fermentation, rate of conversion, N_2 assimilation, formation of humic substances etc.);
3. Dung/plant composts versus pure plant composts;
4. Economic efficiency of compost making and use (labour efficiency, opportunity costs of materials composted etc.);
5. Anti phytopathogenic effect of compost application to various food crops.

Mulching

1. Mulching versus composting using the same base materials (organic wastes): yield function, labour efficiency, nutrient balance and dynamics, biological activity of soil;

2. Decomposition rate of various mulching materials, chopped differently and applied in different quantities, in selected agroclimatic locations.

Plant Breeding

Organic agriculture depends on either old crop varieties or on modern varieties bred for use in non-organic agriculture, neither of which are bred for modern organic agriculture. Focused organic plant breeding would lead to major improvements in organic production, probably more so than from any other research area.

Organic breeding would select varieties and populations best fitted to organic soils and rotations and much better able to compete with weeds. Selection would also be directed to high performance of populations, mixtures and inter-crops, which are potentially capable of providing effective buffering against environmental variation.

Organic plant breeding should draw farmers back into the process from which they have been gradually excluded during this Century. Indeed, this should happen in all aspects of organic research, because of the dependence on farmer experience at higher levels of organization.

Enhancing Knowledge Level

1. Extend the fundamental research on parameter of the agro-ecosystem, for example mineral and organic matter cycles, soil life management, and species interaction and energy balance in order to understand the processes contributing to resource efficiency and problem prevention.
2. Implement fundamental research on ethics and attitudes towards nature, rural development, and agriculture in order to change agriculture research and extension methods towards sustainable agriculture.
3. Implement fundamental research on gender issues.
4. Implement research on the relation between choosing hybrids and non-local varieties and the effect on biodiversity.
5. Begin research comparing conventional farming with existing organic agriculture systems.

Development of Organic Agriculture

1. Develop site-specific organic agriculture systems and techniques for four different levels (subsistence, small holder, commercial farmers, and estates) based on farmer participation.
2. In the various agro-ecological zones of each country, develop a metal farm based on participatory development. External advice on the establishment of such farms should preferably come through a south-south exchange process rather than the more common north-south exchange.
3. At a later stage, develop model regions in various agro-ecological zones.
4. Adapt a flexible long-term process approach in the development of sustainable agriculture. Methods of participatory evaluation and

monitoring should be developed and implemented to secure maximum appropriateness and acceptability of the developed technology.

5. Facilitate and strengthen farmers' capacity for natural resources management and develop technology towards a sustainable ecosystem management.
6. Promote on-farm demonstration and a farmer-to-farmer exchange of experiences.

ONGOING F.A.O. ACTIVITIES RELATED TO RESEARCH

FAO has undertaken several activities specific to organic agriculture, most of which are in the early stages of development. An FAO Library bibliographic search identified 49 to FAO reports on organic agriculture. The FAO Regional Office for Europe has been supporting meetings of researchers working on organic agriculture, sponsored a conference in 1990 on Biological Farming in Europe, and, most recently, an expert Round Table in 1997 which established a working group on Research Methodologies in Organic Farming under the European System of Co-operative Research Networks in Agriculture (ESCORNA). The Regional Office for Asia and the Pacific has coordinated an Asian Bio and Organic Fertilizer Network that for more than a decade has issued annual bulletins on organic recycling in the region. Software has been developed by FAO's Land and Water Department Division to facilitate collecting data on the use of organic nutrients. The Codex Committee on Food Labelling is currently considering Draft Guidelines for the Production, Processing, Marketing and Labelling of Organically produced Foods. Perspectives and guidelines for post-harvest handling of organic fruits, vegetables, aromatics, and spices in developing countries are under development by the Agro-industries and Post Harvest Management Service.

In 1997, an internal E-mail network on organic agriculture was established to facilitate the exchange and evaluation of information and develop common understanding among FAO staff. A focal point on organic agriculture was nominated within the Environmental and Natural Resources Service. After obtaining formal liaison status within FAO in 1997, IFOAM met with FAO experts from all technical departments in March, 1998 to identify areas of potential collaboration.

REDESIGN AGRICULTURAL SYSTEMS

The demand for a more careful stewardship of abiotic and biotic resources is central to the philosophy of sustainable organic farming. A widely recognized concept, sustainable development is an essential element in future development. Agriculture plays a vita role in the concept of sustainable development. Without sustainable agriculture, no sustainable development is possible. Agriculture related problems such as environmental pollution of soil, water and air; resource depletion and nature degradation as well as socio-economic problems, are high on the agenda of various communities and interest groups (including those of farmers, environmentalists, nature-conservators, scientists, consumers and policy makers). This has resulted in the serious demand that today's farming systems be redesigned and transformed into more sustainable ones. However, this is not an easy task, because agriculture is a multifunctional and multiple objective activities. Agriculture has to supply food in sufficient quantity and quality and the supply itself must be stable, sustainable and accessible. It must provide employment and generate basic income and profit at farm, regional and national levels. In addition, agriculture also has to try and avoid and minimize the pollution of abiotic resources, protect and steward nature and landscape, as well as ensure the overall health and well-being of farm animals and rural and urban people.

Desired Improvements of Farming Systems and the Measurement of their Sustainability

Taking into account agriculture-related problems, the following improvements seem to be desirable in order to make farming systems more sustainable:

1. **Ecologization**: prevention and/or minimization of environmental pollution through the overall management and the use of agricultural inputs.
2. Production of more *nature* and (bio) diversity on farms.
3. Ensures stable and fair income for farmers allowing them to maintain their dignity and function in society.
4. A higher degree of ethical and political acceptability.
5. A farming system should be designed and function in such a way that complies, as far as possible, with the requirements, criteria and parameters of sustainable agriculture. In practice, no farm entirely fulfils all the requirements for sustainable agriculture, even if these have been theoretically determined. The overall sustainability of a farming system can best be expressed by the index (degree) of sustainability reached. This index can be derived from farm balances such as nutrients, soil organic matter, energy, labour, economic return, resources use and biodiversity value and from other relevant farm data.

Methodology of Designing Sustainable Farming Systems

Design is not a typical research methodology. In fact it is a process very different from classical research. Classical research investigates a particular problem, phenomena or a set of problems, in order to understand the mechanisms involved. Sometimes it seeks answers to practical problems. In most cases, a research problem or object is segregated into smaller *researchable* units and analysis and experiments are performed to prove causal relations between manipulation and phenomena. Design however, involves another process entirely, in design one synthesizes the knowledge into the larger units through which the *whole* functions in reality. In other words, in research one *discovers something*, while in designing one *creates something* out of its vision and synthesized knowledge. However, although design takes another route than the one followed in research, it does not exclude the need for classical research, since designing is enriched by each new research discovery. Agro-designing methodology involves several steps. It starts with an inventory of the needs and objectives of the stakeholders concerned. Objectives are then ordered hierarchical and rated. The most important of these are then transformed into a suitable set of parameters. In other words, in order to quantify selected objectives, a set of measurable key parameters should be developed. The next step is to establish appropriate agricultural methods and techniques serving more than one objectives, a set of measurable key parameters should be developed. The next step is to establish appropriate agricultural methods and techniques serving more than one objective, for example, intensive fertilization in general serves the objective of high yield and is detrimental to other objectives such as the environment and capable of bridging gaps between conflicting objectives. Finally, the set of multi-objective parameters and methods is linked in a general theoretical prototype based on agronomic, agro-ecological and economic considerations. Theoretical prototypes are then tested in practice and their shortcomings are used as the learning points for the next phase, (re) designing of the same or another farming system.

1. Strengthen existing networks (IFOAM groups, Regional networks);
2. Carry out organic needs and their studies for all countries;
3. Improve the dialogue between scientists;
4. Organize a national and transnational/international dialogue for farmers, advisers and researchers;
5. Make on-going research and research results available.
6. Install a central database with research needs, on-going research projects and research results.

CONCLUSIONS

In the conventional agriculture research most of the components of organic farming have been studied in detail but individually e.g. organic mater, carbon cycle in soil, soil microbiology, etc. The important aspect which need research in relation to organic farming are, to study the interactions among the components, factors affecting long term sustainability of organic farming system and for that an holistic interdisciplinary research approach is need to be developed.

Glossary

Accelerated erosion. Erosion much more rapid than normal, geological erosion; primarily as a result of the activities of humans or, in some cases, of animals.

Acid soil A soil with a pH value <7.0. Usually applied to surface layer or root zone, but may be used to characterize any horizon.

Acidic cations. Hydrogen ions or cations that, on being added to water, undergo hydrolysis resulting in an acidic solution. Examples in soils are H^+, Al^{3+}, and Fe^{3+}.

Acidulation. The process of treating a fertilizer source with an acid. The most common process is treatment of phosphate rock with an acid (or mixture of acids) such as sulfuric, nitric, or phosphoric acid.

Acre. An area of 43,560 ft^2 or 0.405 ha.

Actinomycetes. A non-taxonomic term applied to a group of organisms with characteristics intermediate between the simple bacteria and the true fungi.

Activation energy. The energy required before the reaction of concern will proceed. Often, heating will supply the needed energy. Enzymes or other catalysts lower the activation energy of a reaction.

Additive. Substance intended to improve the properties of a fertilizer or soil conditioner.

Adhesion. Molecular attraction that holds the surfaces of two dissimilar substances in contact, such as water and rock particles.

Adsorption. The process by which atoms, molecules, or ions are taken up and retained on the surfaces of solids by chemical or physical binding (e.g., the adsorption of cations by negatively charged minerals).

Aerate. To allow or promote exchange of soil gases with atmospheric gases.

Aeration, soil. The process by which air in the soil is replaced by air from the atmosphere. In a well-aerated soil the soil air is similar in composition to the atmosphere above the soil. Poorly aerated soils usually contain a much higher percentage of carbon dioxide and a correspondingly lower percentage of oxygen than the atmosphere above the soil. The rate of aeration depends largely on the size, volume, and continuity of pores not filled with water from the surface and within the soil.

Aerobic. 1. Having molecular oxygen as a part of the environment. 2. Growing only in the presence of molecular oxygen, as aerobic organisms. 3. Occurring only in the presence of molecular oxygen (said of certain chemical of biochemical processes, such as aerobic decomposition).

Aeroponics. It is a system for growing plants with their roots supplied with moisture in the air. The rooted plants are placed in a special type of box with their shoots exposed to air and roots inside the box with computer controlled humid atmosphere. It is, however, a recent method of root hairs. Plants like Citrus and olive have been successfully grown through this method.

Aflatoxins. Potent carcinogens in nature that are antibodies produced by fungal growth on many grains and nuts.

Aggregate. A unit of soil structure, usually formed by natural processes in contrast with artificial processes, and generally < 10 mm in diameter.

Agricultural waste. The characteristics of the agricultural waste materials make them important in respect of their use and economic value. Agricultural wastes can be put to use because (1) Farm development (improvement of soil health, soil fertility, soil physical conditions and plant protection), it is the source of energy and power, (3) it acts as animal feed and fodder, (4) it is the human food.

Agronomy. A specialization of agriculture concerned with field crop production and soil management. The scientific management of land.

Air dry. (1) The state of dryness at equilibrium with the water content in the surrounding atmosphere. The actual water content will depend upon the relative humidity and temperature of the surrounding atmosphere. (2) To allow to reach equilibrium in water content with the surrounding atmosphere.

Alkaline (or basic) fertilizer. A fertilizer capable of increasing the residual alkalinity or reducing the residual acidity of the soil.

Alkaline forming. A term applied to commercial fertilizers that leaves an alkaline or basic residue in the soil. The basic residual effect is expressed in terms of `equivalent basicity.' More the basic residue, more will be the equivalent basicity.

Alkaline soil. A soil that is alkaline in reaction throughout the root zone or for a major part of the root zone. Precisely, any soil having a pH value greater than 7.0. Practically, a soil having a pH above 7.3 is called alkaline.

Allelopathy. The action of some substance produced by or in one plant species that reduces the growth of another plant species. Mostly these are toxic organic materials produced either by the plant or during decomposition of plant residues.

Alley cropping/farming. Growing annual crops between rows of trees or shrubs, often leguminous. Pruned material from these is used as mulch around annual crops and also as fodder and fuelwood.

Alluvial fan. Fan-shaped alluvium deposited at the mouth of ravines or canyons as the slope lessens and water-carried debris settles out of the slowed water.

Alluvial. Sediments deposited by running water of streams and rivers. It may occur on terraces well above present streams or in the normally flooded bottomland of existing streams.

Amendment, soil. Any material, such as lime, gypsum sawdust, sulphur or soil conditioner, that is added to the soil to influence plant growth by changing the soil reaction or pH or by improving the physical condition of the soil. Strictly, a fertilizer is also an amendment, but the term is used most commonly for added materials other than fertilisers.

Amensalism. Production by one organism of a substance that is inhibitory to other organisms. Similar in use to the term *allelopathy.*

Ammonia fixation. Chemisorption of ammonia (NH_3), and possibly ammonium, by the organic fraction of the soil.

Ammonia volatilization. Mass transfer of nitrogen as ammonia gas from soil, plant, or liquid systems to the atmosphere.

Ammonium fixation. The process of converting exchangeable or soluble ammonium ions to those occupying positions similar to K^+ in the micas. They are counterions entrapped in the ditrigonal voids in the plane of basal oxygen atoms of some phyllosilicates as a result of contraction of the interlayer space. The fixation may occur spontaneously with some minerals in aqueous suspensions, or as a result of hating to remove interlayer water in others. Ammonium ions so adsorbed are exchangeable only after expansion of the interlayer space.

Amorphous clays. Lacking adequate crystal orientation to diffract x-rays. Without regular crystalline form in molecular sizes.

Amorphous material. Non-crystalline soil materials.

Anaerobic respiration. The metabolic process whereby electrons are transferred from an organic compound to an inorganic acceptor molecule other than oxygen. The most common acceptors are carbonate, sulfate, and nitrate.

Anaerobic. (1) The absence of molecular oxygen. (2) Growing in the absence of molecular oxygen (such as anaerobic bacteria). (3) Occurring in the absence of molecular oxygen (as a biochemical process).

Analysis. As applied to fertilisers, it shall designate the percentage composition of the product expressed in terms which the existing trade practice and laws require and permit.

Animalia. One of the five kingdoms categorizing all living organisms. This kingdom includes the animals, multi-celled organisms capable of either searching for food (mobility) or pulling a media containing foods to them (sponges circulating water through themselves).

Anion exclusion. The exclusion or repulsion of anions from the vicinity of soil particle surfaces because of the negative potential associated with most soil particles.

Anion. (ann'-I-on) Negatively charged ion; ion that during electrolysis is attracted to the anode.

Anthropogenic, anthropic. (Soils) Changes in soils caused by people such as plowing, fertilizing, and using for construction.

Ants. In both urban and rural environments ants are active in tunneling and bringing up subsurface soil to construct mounds of various sizes. Because ants can carry particles no larger than allowed by the gap between their open mandibles (mouth parts), the mounds contain no stones or gravel. *Formica cinerea* were originally common in the grasslands of the American prairie but are now confined by cultivation to undisturbed lands such as those along railroad tracks, cemetery edges, and wetland borders. The mounds are built largely of subsoil and are rich in organic materials the ants bring to the colonies from nearby vegetation. Exudate from plant aphids is fed to the young in the mounds. As a result, the soil there is brought to a high level of fertility. Every ten or twenty years a mound is deserted after the queen dies. The soil collapses and settles at that spot. New mounds form between the old ones. Eventually, the soil over an entire prairie can be worked and fertilized by this type of activity.

Aqua ammonia. Aqua ammonia (ammonia dissolved in water) normally contains 20 per cent nitrogen, through grades containing upto 26 per cent nitrogen can be made. The major advantages of using aqua ammonia rather than anhydrous ammonia are the simple requirements in handling and elimination of most of the hazards. Ordinary storage tanks are satisfactory as opposed to the costly stainless steel tanks that are required for storage of anhydrous ammonia. Aqua ammonia also need to be applied deep into the soil to prevent loss of nitrogen.

Aquic. A soil water regime, mostly too wet (reducing conditions, waterlogged) for parts of the year.

Aquifer. A geologic formation that transmits water in sufficient quantity to supply the seeds for a water development. Aquifers are usually saturated sands and gravels and fractured, porous, cavernous, and vesicular rock

Arable land. Areas of land whose ecology and environment are suitable for the production of cultivated crops.

Arid climate. Climate in regions that lack sufficient moisture for crop production. In cool regions, annual precipitation is usually less than 250 mm. It may be as high 500 mm in tropical regions. Natural vegetation is desert shrubs.

Arid. A term applied to regions or climates that lack sufficient moisture for crop production without irrigation. The upper annual limit for cool regions is 250 mm (10 in.) or less and for tropical regions as much as 380-510 mm (15-20 in.). Contrast with Semiarid.

Aridic. A soil water regime with long dry periods.

Ash. The mineral residues remaining after the destruction of organic material by burning.

Backslope. The slope component that is the steepest, straight then concave, or merely concave middle portion of an erosional slope.

Badland. An area generally devoid of vegetation and broken by an intricate maze of narrow ravines, sharp crests, and pinnacles resulting from serious erosion of soft geologic materials. Most common in arid or semiarid regions.

Beaches. Sandy, gravelly, cobbly, or stone shores washed and reworked by waves. The areas may be partly covered by water during high tides or storms.

Bedder. A tillage tool to form ridge or narrow beds.

Bedding. The mounding of soil into elevated strips performed with tillage tools such as sweep, shovel, disk, or moldboard.

Bedrock. The solid rock underlying soils and the regolith in depths ranging from zero (where exposed by erosion) to several hundred centimeters.

Bench terrace. An embankment constructed across sloping fields with a steep drop on the downslope.

Bentonite. A highly plastic clay consisting of the minerals montmorillonite and beidellite (Smectites), which swell extensively when wet. Also called volcanic clay, soap clay, and amargo site.

Bioassay. A method for the quantitative determination of a substance by its effect on the growth of a suitable microorganism, plant, or animal under controlled conditions.

Bioconcentration. It is the ability of microorganisms to store materials in the environment. The storage may be within cells, as in the case of sulfur, or outside the cells, as in the case of nitrate or sulfur. The metabolic activities of sulfur oxidizers have been used to concentrate low-grade mineral ores. Sulfur oxidizers, such as *Thiobacillus*, can grow in slag from copper mines; the material remains after ore is roasted to remove most of the copper. The sulfuric acid produced by the oxidizing activity of *Thiobacillus* dissolves the copper, and copper sulfate accumulates at the base of the slag pile. A similar process is used to concentrate uranium from ores in the form of uranyl sulfate. Bioconcentration has caused the formation of commercially attractive deposits of manganese, but also hazardous accumulations of pesticides and radioisotopes in nature.

Biodegradable. A substance capable of being decomposed by biological processes.

Biodiversity. All species of plants and animals, their genetic material and the ecosystems of which they are part.

Biological magnification. The accumulation of increasing amounts of a substance in organisms of a food chain as one progresses up the chain (i.e., low in plankton, higher in small fish, large fish, and highest in fish-eating birds).

Biomass. (1) The total mass of living microorganisms in a given volume or mass of soil. (2) The total weight of all microorganisms in a particular environment.

Biomass. The weight of material produced by a living organism or collection of organisms. The term is usually applied to plants to include the entire plant, or it may be qualified to include only certain parts of the plant, e.g., above-ground or leafy biomass. Biomass is expressed as fresh weight or dry weight.

Biome. A large, easily recognized community formed by the interaction of regional climates with regional biota and substrates. In a given biome the life form of the climatic climax vegetation in uniform. Thus the climax vegetation of the grassland biome is grass, although the dominant species of grass may vary in different parts of the biome.

Biosequence. A sequence of related soils that differ, one form the other, primarily because of differences in kinds and numbers of plants and soil organisms as a soil-forming factor.

Biosphere. The space area supporting living organisms. It includes atmosphere, animals, vegetation, soil and oceans.

Bisequum. One sola above another in the same profile.

Biwall irrigation: It is an extruded dual chamber micro-irrigation tubing, manufactured from LLDPE. In this system, holes are drilled on main tube at even distance over the long lengths. This system is suitable for all *closely spaced crops.*

Blown-out. Areas from which all or almost all of the soil and material has been removed by wind erosion. Usually barren, shallow depressions with a flat or irregular floor consisting of a more resistant layer and/or accumulation of pebbles, or a wet zone immediately above a water table. Usually unfit for crop production.

BOD (Biochemical Oxygen Demand). Biochemical oxygen demand (BOD) is the most commonly used parameter to define the strength of a municipal or organic industrial wastewater. Widest application is in measuring waste loadings to treatment plants and in evaluating the efficiency of such treatment systems. In addition, the BOD test is used to determine the relative oxygen requirements of treated effluents and polluted waters. However, it is of limited value in measuring the actual oxygen demand of surface waters, and extrapolation of test results to actual stream oxygen demands is highly questionable, since the laboratory environment cannot reproduce the physical, chemical, and biological stream conditions. BOD is by definition the *quantity of oxygen utilized by a mixed population of microorganisms in the aerobic oxidation* (of the organic matter in a sample of wastewater) at a temperature of 20°C.

BOD Sensor. Biosensors are the combinations of biologically active material displaying characteristic specificity with chemical or electronic sensor to convert the response into electrical signal. They are, in fact, biocatalysts and are used as purified enzyme, antibody or as whole microbial cell or as an organelle. The response of biosensor is measured in terms of substrate used or product form. They are of different types like, carbon electrode, glucose electrode, ion sensitive electrode, photocell, oxygen electrode, thermistor, adenosine electrode, etc.

Buffer compounds, soil. The solid and solution phase components of soils that resist appreciable pH change in the soil solution (i.e., carbonates, phosphates, oxides, phyllosilicates, and some organic materials).

Buffer. A buffer solution is one which is resistant to changes in pH on the addition of small amount of acid or alkali. This resistance is called the buffer action. The magnitude of the buffer action is called the buffer capacity (ß). Such solutions consist of a mixture of a weak acid and its base (acetic acid and sodium acetate), a weak base and its salt (ammonium hydroxide and ammonium chloride). A salt of weak acid and a weak base (ammonium acetate) also has a buffer action. Some of the common buffers are given below:

Buffering. 1. The capacity of the soil solids and liquids to resist appreciable change in pH of the soil solution. 2. The ability to maintain the approximate concentration desired of any ion in the soil solution.

Bulk density (loose). The mass per unit volume of a material after it has been tipped freely into a container under clearly specified conditions.

Bulk density (tapped). The mass per unit volume of a material tipped into a container and then compacted under clearly specified conditions. Or liquid in a non-packed from (to which a label cannot be attached).

Bund. A ridge of earth placed in line to control water run-off and soil erosion, demarcate a plot boundary, or other uses.

Buried soil. Soil covered by an alluvial, loessal, or other depositional surface visibly when treated with cold 0.1 M HCl. These soils usually contain from as little as 10 as much as 200 g kg^{-1} $CaCO_0$ equivalent.

Caking. The formation of a coherent mass from individual particles.

Calcic horizon. A diagnostic mineral horizon of carbonate accumulation. Indicated by letter *k*.

Calcium. It is absorbed as divalent Ca^{2+} easily available for plant growth, but acidic soils where high rainfall occurs are often fertilized with lime (a mixture of CaO and $CaCO_3$) to raise the pH.

Caliche. (1) A zone near the surface, more or less cemented by secondary carbonates of Ca or Mg precipitated from the soil solution. It may occur as a soft thin soil horizon, as a hard thick bed, or as a surface layer exposed by erosion. (2) Alluvium cemented with $NaNO_3$, NaCl, and/or other soluble salts in the nitrate deposits of Chile and Peru.

Calmodulin. Calmodulin has been isolated from lower and higher plants, the green algae *Chlamydomonas* and *Mougeotia*; the fungi, *Neurospora cressa*, the yeasts *Candida elbucans* and *Saccharomyces cerevisiae*, *Blastocladia emersonii*, two *Basidiomycetes* and different sources of higher plants including peas, mungbeans, wild carrots and peanut seeds, barley shoots and roots, Zucchini hypocotyls, spinach leaves, corn, etc.

Capillary fringe. A zone in the soil just above the plane of zero gage pressure that remains saturated or almost saturated with water. The extent can be inferred from the retentivity curve and depends upon the size distribution of pores.

Carbon cycle. Carbon is the most important element in the biological world. It can be deemed as the foundation stone of the cell structures. Plants and microbial cell contain large amount of carbon approximately 40 to 50 per cent on the dry weight-basis. This huge amount of carbon which is present in biological world, comes from atmospheric carbon dioxide. The atmospheric carbon dioxide is converted to organic carbon, mainly by the action of photoautotrophic organisms which include higher green plants on land and algae in water. These photoautotrophs supply the organic nutrients in the form of complex carbon compounds, needed for heterotrophic animals and chlorophyll lacking microorganisms. Once the atmospheric carbon is fixed by photoautotrophs, in presence of light it become non-available for the generation of new plant-life. Therefore, it is essential for carbonaceous materials to be decomposed and returned to the atmosphere, otherwise the stock of carbon in nature would be exhausted. The return of carbon is brought about through the decomposition of organic carbon in nature. In fine, the world-wide circulation of carbon atom is brought about by living organisms. Essentially, green plants including water algae, convert atmospheric carbon dioxide into organic compounds during photosynthesis. Three compounds are then broken again into carbon dioxide during respiration and by decay. They are produced by soil bacteria and fungi after death; or the plants are eaten by heterotrophic herbivorous animals which may be finally eaten by carnivorous. Likewise, compounds are broken into carbon dioxide by respiration or death or decay of the animals. This is the general outline of carbon cycle in nature, which may be represented by the following scheme.

Carbon. The special role and function of carbon in plant life cause carbon to be discussed in the literature under the heading as being an essential element. The uptake of carbon by the plant is different from that of inorganic elements. Inorganic element uptake is mainly through the roots in the liquid phase, but carbon uptake is through the leaves and shoots in the gas phase. The toxicity symptoms of CO_2 in mature tomato and bean leaves, yellowing spots, bulkiness, etc., resemble the signs of deficiency of P and toxicity of Mg and could be related to the accumulation of starch.

Carbonation. A chemical weathering process combining the action of acidity and formation of soluble bicarbonates. The solvent is carbonic acid formed from carbon dioxide dissolved in water.

Carbon-nitrogen ratio. The ratio of the weight of total organic carbon to the weight of total nitrogen in a soil or in an organic material. C:N ratio of wheat straw is nearly 80:1 while that of soil is 10:1 When undecomposed straw with a high C:N ratio is applied to the soil, its C:N ratio is reduced through bacterial decomposition. To speed up bacterial decomposition, nitrogenous fertilizer is added to the soil at the time of turning under straw.

Cat clays. Strongly acidic clays resulting from aeration of anaerobic soils containing many sulfide minerals. The solfides are oxidized to sulfuric acid in aerated soil. In the Netherlands called *katteklei* (= cat's coat).

Cation. An ion carrying positive charge of electricity. The common soil cations are calcium, magnesium, sodium, potassium and hydrogen.

Cellulose ferments are of very great economic value. They are used for purifying water, disposing the waste and as a soil humus. Those microorganisms which utilize carbon may be referred as an intermediate between true autotrophic and heterotrophic forms. Bacteria which oxidize carbon dioxide, methane and other hydrocarbons may be aerobic or anaerobic in nature. Some of the aerobic and anaerobic forms of bacteria which oxidize these compounds as follows: *Carboxydomonas oligocarbophila:* These bacteria oxidize carbon monoxide to carbon dioxide. *Methanomonas methanica:* These bacteria oxidize hydrocarbons to form carbon dioxide and water. *Methanomonas aliphatica:* These bacteria oxidize n-pentane, n-hexane, n-octane and use paraffins, petroleum oils and methane as a source of carbon and energy. *Anaerobic, purple, sulphur bacteria:* These bacteria require simple fatty acids in order to absorb carbon dioxide and in presence of infra-red light they use hydrogen.

Cemented (indurated). Having a hard, brittle consistency, even when wet, because the particles are held together by cementing substances, such as humus, calcium carbonate, or the oxides of silicon, iron, and/or aluminum.

Chelates. An organic compound capable of holding the plant nutrient in a form which prevents it from getting tied with other elements in the soil, thus keeping it more or less in available form for the plant. The term refers to the claws of a crab illustrative of the way in which the atom is held.

Chelation. (key-lay'-shun) The formation of strong bonds between metals and organic compounds. Some are insoluble, such as in soil humus. Those chelates used in fertility work are soluble and help keep the nutrient metals (iron, zinc, manganese, and copper) mobile in the soil, thus aiding plant uptake.

Chemical oxygen demand (COD or **C.O.D.).** A measure of the oxygen-consuming capacity of inorganic and organic matter present in water or wastewater. It is expressed as the amount of oxygen consumed from a chemical oxidant in a specific test. Chemical Oxygen Demand (COD). The estimation of COD is of great importance for water having unfavourable conditions for the growth of microorganisms, such as presence of toxic chemicals. In such water BOD can not be determined accurately. However, COD too is not a perfect index of organic compounds present in water because, in this reaction, many inorganic compounds are also oxidized, and at the same time some organic compounds remain unaffected. There are both physical and chemical conditions that define a healthy environment for aquatic organisms. Among the physical conditions are temperature, pressure, osmotic pressure, light, and turbidity. Although a few organisms can live in hot water, most prefer a moderate temperature. Because there are so many varieties of aquatic life, it is impossible to generalize about an optimum temperature for a healthy aquatic environment. Among bacteria, some prefer warm water, some cold, and others grow best at intermediate temperatures.

Chemical weathering. The breakdown of rocks and minerals due to chemical activity, primarily due to the presence of water and components of the atmosphere.

Chemigation. The process where fertilizers and pesticides are applied into irrigation water to fertilize crops and control pests.

Chemodenitrification. Non biological processes leading to the production of gaseous forms o nitrogen (molecular nitrogen or an oxide of nitrogen).

Chilean nitrate of soda (Chile saltpetre). A product obtained by refining the crude nitrate deposits found in Chile and containing about 99 per cent sodium nitrate.

Chisel. A tillage implement with one or more shanks to which are attached chisel, spike, or narrow-shovel tools. When used to subsoil, and penetrate deeper than 0.4 m, the shanks are sturdier and spaced farther apart.

Chiseling. Breaking or loosening the soil, without inversion, with a chisel cultivator or chisel plow, usually below normal plow depth.

Chlorite. A layer structured group of silicate minerals of the 2:1 type that has the interlayer filled with a positively charged metal-hydroxide octahedral sheet. There are both trioctahedral (e.g., M = Fe^{2+}, Mg^{2+}) and dioctahedral (M = Al^{3+}) varieties.

Chlorosis. A loss of normal green color of the plant. Colors may be uniform pale green, brown spotted, yellowish over the entire leaf, or yellowish only between the leaf veins.

Chroma. The relative purity, strength, or saturation of a color; directly related to the dominance of the determining wavelength of the light and inversely related to grayness; one of the three variables of color.

Chronosequence. A sequence of related soils that differ, one from the other, in certain properties primarily as a result of time as a soil-forming factor.

Chute spillway. Open channels with steep slopes consisting of an inlet, vertical curve section, steep channel and an outlet. These can be constructed in all areas irrespective of rainfall and soil characteristics when run-off is to be carried along the steep slopes and sudden drops.

Citrate-insoluble phosphorus. That portion of P in fertilizer remaining after water and ammonium citrate extractions. Phosphorus content of fertilizers that is considered to be immediately unavailable to plants in the guaranteed analysis of the fertilizer.

Citrate-soluble phosphoric acid. It is that part of the total phosphoric acid (as P_2O_5) in a fertilizer that is insoluble in water but soluble in neutral N ammonium citrate solution.

Clay films. Coatings of clay on the surfaces of soil peds and mineral grains and in soil pores. (Also called clay skins, clay flows, illuviation cutans, argillans, or tonhautchen.)

Clay. (Soils) 1. A mineral soil separate consisting of particles less than 0.002 mm in eluivalent diameter. 2. A soil textural class. 3. A fine-grained soil that has a high plasticity index in relation to the liquid limits (engineering). 4. A specific mineral structure.

Claypan. A dense, compact layer in the subsoil having a much higher clay content than the overlying material, from which it is separated by a sharply defined boundary; formed by downward movement of clay or by synthesis of clay in place during soil formation. Claypans are usually hard when dry, and plastic and sticky when wet. Also, they usually impede the movement of water and air, and the growth of plant roots.

Cleavage. The smoothness of regular breaking surfaces of minerals, where "perfect" is very smooth and flat; "conchoidal" is convex and not smooth. Cleavage helps in mineral identification.

Climax. The most advanced successional community of plants capable of development under, and in dynamic equilibrium with, the prevailing environment.

Clod. A compact coherent mass of soil varying in size, usually produced by plowing, digging, and so on, especially when these operations are performed on soils that are either too wet or too dry and usually formed by compression, or breaking off from a larger unit, as opposed to a building-up action as in aggregation.

Closed formula mixed fertilisers. The fertilisers grade is disclosed on each bag of such fertilizer mixture, but the ingredients or straight fertilisers used in formulating the mixture are not disclosed. In India, fertilizer mixtures sold to the cultivators are usually of the closed formula type.

Coarse texture. The texture exhibited by sands, loamy sands, and sandy loams except very fine loam.

Coated fertilizer. Fertilizer, the granules of which are covered with a thin layer of a different material in order to improve the behaviour and/or layer of a different material in order to improve the behaviour and/or modify the characteristics of the fertilizer.

Cobblestone. Rounded rock and mineral particles of size between 8 and 25 cm diameter.

Cobbly. Containing appreciable quantities of cobblestones. (Said of soil and of land. The term angular cobbly is used when the fragments are less rounded.)

Cohesion. Holding together; force holding a solid or liquid together because of attraction between like molecules; cohesion decreases with rise in temperature.

Coliform. A general term for a group of bacteria that inhabit the intestinal tract of humans and other animals. Their presence in water constitutes presumptive evidence for fecal contamination. Includes all aerobic and facultatively anaerobic, gram-negative rods that are nonspore forming and that ferment lactose with gas formation. *Escherichia coli* are important members.

Colloid. A substance that, when suspended in water, diffuses not at all or very slowly through a semipermeable membrane; a substance in a state of fine subdivision with particles from 1 micrometer to 1 nanometer.

Colluvium. A general term applied to deposits on a slope or at the foot of a slope or cliff that were moved there chiefly by gravity. Talus and cliff debris are included in such deposits.

Cometabolism. Transformation of a substrate by a microorganism without deriving energy, carbon, or nutrients from the substrate. The organism is able to transform the substrate into intermediate degradation products but fails to multiply at its expense.

Commercial fertilizer. It has been indicated that the three elements most commonly deficient for crop production are nitrogen, phosphorus, and potassium. Therefore, a *complete* commercial fertilizer contains all three of these elements. The analysis (grade) of the fertilizer is required by state law to be shown on the bag and declares the percent by weight of nitrogen, phosphorus, and potassium in that order. The latter two may be expressed as oxides (P_2O_5 and K_2O). The analysis will appear as numbers such as 24-10-8 on the bag.

Community forestry. A form of social forestry, where tree planting is undertaken by a community on common or communal lands.

Community. All of the organisms that occupy a common habitat and that interact with one another.

Complete fertilizer. A product obtained by mixing different fertilizer stock materials, containing three major plant nutrients, namely, nitrogen, phosphoric acid and potash.

Complex fertilisers. The commercial fertilisers containing a least two or more of the primary essential nutrients. When such fertilisers contain only two of the primary nutrients, they are designed as incomplete complex fertilisers, while those containing all three primary nutrients are called as complex fertilisers.

Compost. Organic residues, or a mixture of organic residues and soil, which have been piled, moistened and allowed to undergo biological decomposition. Mineral fertilizers are sometimes added. Often called artificial or synthetic manure when produced primarily from plant residues.

Compost. Organic residues, or a mixture of organic residues and soil, which have been mixed, piled, and moistened, with or without addition of fertilizer and lime, and generally allowed to undergo thermophilic decomposition until the lime, and generally allowed to undergo thermophilic decomposition until the original organic materials have been substantially altered or decomposed. The final product can be easily worked into the soil, or can be used in or as a potting mix. Sometimes called artificial manure or synthetic manure. In Europe, the term may refer to a potting mix for container-grown plants.

Compound fertilizer. Fertilizer having a declarable content of at least two of the nutrients nitrogen, phosphorus, and potassium, obtained chemically or by blending or both.

Compressibility. The property of a soil pertaining to its susceptibility to decrease in bulk volume when subjected to a load.

Concentrated flow. The flowing of a rather large accumulated body of water over a relatively narrow course. It often causes serious erosion and gullying.

Concentration: The concentration of calmodulin in most plant tissues varied between 2 to 4 μm Mol Kg. Measurements of the sub cellular calmodulin based on enzymatic and radioimmunoassays of wheat leaf cells indicated that 89 to 93% of the cellular calmodulin occurs in the cytoplasm, while more than 1% to 9% is present in the organelles like chloroplasts, mitochondria and microsomes. The following table indicates the association of calmodulin with oat tissue, cells and cell fractions as measured by radioimmunoassay. The most important of all known Ca^{2+} modulated proteins is calmodulin, a highly conserved, heat stable, small molecular weight protein found in all eukaryotic cells. Containing 4 Ca^{2+} binding sites per molecule, calmodulin is composed of a single subunit and 148 amino acids. The binding of Ca^{2+} induces a conformational change resulting in increased alpha helicity. This Ca^{2+} induced conformational change allows the protein to interact with and alter the activity of calmodulin-dependent enzymes. The Ca^{2+} calmodulin complex has a wide range of biochemical activities, including effects on both coarse and fine metabolic regulation. Because of these characteristics, calmodulin has become the standard against which all other Ca^{2+} modulated proteins have been compared. Plant enzymes which have been reported to be directly regulated Ca^{2+} or Ca^{2+} calmodulin include NAD-kinase, β-glucan synthase, Ca^{2+} and H^+-ATPase, quinate; NAD^+-oxidoreductase and protein kinases.

Concretion. (Soil) A local concentration of a chemical compound, such as calcium carbonate or iron oxide, in the form of an aggregate or nodule, of varying size, shape, hardness, and color.

Conservation tillage. Any tillage sequence that reduces loss of soil or water relative to conventional tillage.

Conservation. The protection, improvement, and use of natural resources according to principles that will ensure their highest economic, social, and psychological benefits to people in perpetuity.

Consistence. (Soil) 1. The resistance of a material to deformation or rupture. 2. The degree of cohesion or adhesion of the soil mass. Terms used for describing consistency of soil material at various soil moisture contents.

Consumptive use. The total quantity of water transpired by vegetation plus that evaporated from soil, plus that in the plant material. Almost equal to evapotranspiration.

Contamination. Any substance added or accumulating in water or soil which makes the water or soil less desirable for people's use. (= pollution).

Contour border strips. These are strips on contour with no cross slope alone the width and are spaced at 0.3 m vertical interval. The levees in between the two strips measure 0.24 sq. m and are most suitable for the medium to deep soil with intake rates of more than 8 mm per hour and where the annual rainfall is around 750 mm.

Contour bunds. These are earthen embankments on contour. The bunds may be closed with hook up bunds at right angles to the contour bund or kept open and are used to break the long lengths of slopes to intercept run-off against the bunds for absorption.

Contour farming. A system of farming sloping lands in such a way that ploughing, preparing land, planting and sowing and cultivating operations are done on the contour. It creates numerous ridges and furrows, which help to conserve soil and moisture in low rainfall areas.

Contour stripcropping. Layout of crops in comparatively narrow strips in which the farming operations are performed approximately on the contour. Usually, strips of grass, close-growing crops, or fallow are alternated with those in cultivated crops.

Contour. 1. An *imaginary* line on the surface of the earth connecting points of the same elevation. 2. A *true* line drawn on a map connecting points of the same elevation.

Contour. An imaginary line connecting points of equal elevation on the surface of the soil. A contour terrace is laid out on a slope at right angles to the direction of the slope and nearly level throughout its course. A ditch is sometimes constructed along the contour to store and conserve water.

Coppice mound. A small mound of stabilized soil material around desert shrubs.

Coppicing. Cutting certain tree species close to ground level to produce new shoots from the stump. Also occurs, naturally in some species if the trees are damaged.

Coprogenic material. Remains of fish excreta and similar materials that occur in some organic soils.

Cotton seed cake. A by-product of the cotton seed crushing industry. When it is undecorticated, it is non-edible containing 3.9 per cent nitrogen, 18.1.8 per cent phosphoric acid and 1.6 per cent potash. When decorticated, cotton seed cake becomes edible oilcake containing higher percentages of plant nutrients. Decorticated cotton seed cake contains 6.4 per cent nitrogen, 2.9 per cent phosphoric acid and 2.2 per cent potash.

Coulter. A disk used in tillage to cut surface litter or open a slit a facilitate the passage of following chisels or knives in reduced tillage.

Cover crops. A close-growing crop grown primarily for the purpose of protecting and improving soil between periods of regular crop production or between trees and vines in orchards and plantations.

Cradle knoll. A small knoll formed by earth that is raised and left by an uprooted tree

Creep. (Soils) Slow mass movement of soil and soil material down relatively steep slopes, primarily under the influence of gravity but facilitated by saturation with water and by alternate freezing and thawing.

Crest. The slope component that is commonly at the top of an erosional ridge, hill, mountain, and so on.

Crop residue. The portion of a plant or crop that is left in the field after harvest, e.g., millet straw.

Cropping pattern. The yearly sequence and spatial arrangement of crops or of crops and fallow on a given area.

Cropping system. The cropping patterns used on farm and their interaction with resources, other farm enterprises and available technology which determine their makeup.

Crushing strength. The force required to crush a mass of dry soil or, conversely, the resistance of the dry soil mass to crushing. Expressed in units of force per unit area (pressure).

Crust. A soil-surface layer, ranging in thickness from a few millimeters to a few tens of millimeters that is much compact, hard, and brittle when dry than the material immediately beneath it.

Cryic. An annual soil temperature regime (0-8°C) in quite cold environments.

Cryogenic soil. Soil that has formed under the influence of cold soil temperatures.

Crystal structure. The orderly arrangement of atoms in a crystalline material.

Crystal. A regular arrangement of atoms in space.

Cultivation. A tillage operation used in preparing land for seeding or transplanting or later for weed control and for loosening the soil.

Cyanobacteria. New name for N_2-fixing microbes, which were previously called blue-green algae.

Cyclic salt. Salt derived from the sea or salt lakes deposited on the landscape from wind or rainfall.

Deflation. The removal of fine soil particles from soil by wind.

Deflocculate. (1) To separate the individual components of compound particles by chemical and/or physical means. (2) To cause the particles of the disperse phase of a colloidal system to become suspended in the dispersion medium.

Deforestation. The permanent removal of forest and undergrowth.

Degradation. The process whereby a compound is transformed into simpler compounds, although products more complex than the starting material may be formed.

Delta. An alluvial deposit formed where a stream or river drops its sediment load on entering a body of more quiet water.

Desert crust. A hard layer, containing calcium carbonate, gypsum, or other binding material, exposed at the surface on desert regions.

Desert varnish. A glossy sheen or coating on stones and gravel in arid erosion.

Desertification. Desertification is land degradation in arid, semi-arid and dry sub-humid areas resulting from various factors, including climatic variations and human activities.

Desertization. This word is used in connection with deserts. This process occurs in arid zone bordering the deserts under av. rainfall of 100-200 mm.

Desorption. The displacement of ions from the from the solid phase of the soil into solution by a displacing ion.

Detoxification. Conversion of an inhibitory molecule into a nontoxic compound.

Diagnostic horizons. (As used in the Soil Classification System of the National Cooperative Soil Survey in the United States) Combinations of specific soil characteristics that are indicative of certain classes of soils.

Diatomaceous earth. A geologic deposit or fine, grayish siliceous material composed chiefly or wholly of the remains of diatoms. It may occur as a powder or as a porous, rigid material.

Diatoms. Algae having siliceous cell walls that persist as a skeleton after death. Any of the microscopic unicellular or colonial algae constituting the class Bacillariaceae. They are abundant in fresh and salt waters and their remains are widely distributed in soils.

Dictahedral. An octahedral sheet or a mineral containing such a sheet that has two-thirds of the octahedral sites filled by trivalent ions such as aluminum or ferric iron.

Diffusion. The movement of an ion in water mostly by its own kinetic motion. Very slow, usually in nanometers per hour (10^{-4} cm/h).

Dinitrogen fixation. Conversion of molecular nitrogen (N_2) to ammonia and subsequently to organic combinations or to forms utilizable in biological processes.

Disperse. (1) To break up compound particles, such as aggregates, into the individual component particles. (2) To distribute or suspend fine particles, such as clay, in or throughout a dispersion medium, such as water.

Dispersion, soil. The breaking down of soil aggregates into individual particles. Generally speaking, the more easily dispersed the soil, the more erodible it is.

Diversion drain. Channels designed across the slope for the purpose of intercepting and diverting surface run-off to save arable lands from the damage due to overflow. These are adaptable in all areas irrespective of soil and climatic conditions. The channel is designed usually with trapezoidal shape and vegetated with suitable species.

Dolomite. A natural mineral composed of calcium and magnesium carbonate widely used as liming material and as an ingredient in fertilizer mixtures. Dolomite is used in fertilisers to render them non-acid forming. It also supplies available magnesium.

Dolomitic lime. A naturally occurring liming material composed chiefly of carbonates of Mg and Ca in approximately equimolar proportions.

Double sulphate of potash and magnesia. A commercial product containing not less than 21 per cent of potassium oxide, not less than 53 per cent of sulphate of magnesia and not more than 2.5 per cent of chlorine.

Drainage class. The description of the ease with which a soil drains off excess water by percolation. Terms such as *well drained, imperfectly drained, or very poorly drained* are used.

Drip irrigation. Irrigation whereby water is slowly applied to the soil surface through emitters having small orifices. It refers to the system of water application in small quantity literally drop by drop. Drip irrigation was developed originally as a sub-irrigation system but as a drip irrigation system, as it is today, this concept was developed in early 1940's, by Symcha Bless, an Israeli Engineer. Drip system essentially consists of main, sub main, laterals, drippers and few other accessories to deliver the required quantity of water at the root zone of the crop. Low pressure ranging from 0.35 to 7.0 cm. diameter pipes are used for mains and 2.5 to 3.75 cm. for submarines. For laterals, pipes of various ranging from 6 mm to 18 mm with a pressure rating of 4.o atm. are generally used. Two types of drip-irrigation systems are used in India. One is biwall irrigation system and another is micro-sprinkler and micro-sprayer irrigation.

Drop structures. Mechanical checks constructed across the nala or drainage channels to check the water velocity with a weir for conducting excess run-off. They are also constructed as drop inlet structures.

Drought tolerance. The capacity of plants to survive drought; specifically adaptation that enhance their power to withstand drought-induced stress.

Drought. The absence of precipitation for a period long enough to cause depletion of soil moisture and damage to plants.

Dry farming. The system in which field crops are raised with an annual rainfall of less than 25 inches, without irrigation facilities. Form plant nutrition point of view, lesser doses of fertilisers are recommended to dry farming areas for all field crops, compared to irrigate farming.

Dry weight. (Soils) The equilibrium weight of the solid soil particles after the water has been vaporized by heating to 105°C (221°F).

Dryland farming. It refers to the practice of crop production and related activities in arid and semi-arid areas where, precipitation is less than evaporation (rainfall

<750 mm), water table remains below the root zone, optimum growing season is short (60-120 days in general) and crops generally faces mild to severe moisture stress atleast once in the growing season.

Dry-mass content or ratio. The ratio of the mass of any component (of a soil) to the oven-dry mass (ODwt) of the soil.

Duff mull. A type of forest humus that is transitional between mull and mor and usually contains four horizons: Oi(L), Oe(F), Oa(H). and A.

Dune land. Consists of ridges and the intervening troughs made up of sand-size particles that shift with the wind. It is devoid of vegetation.

Dung. The semi-solid excrements of animals used as a fertilizer and soil conditioners.

Dust mulch. A loose, finely granular or powdery condition of the surface of the soil, usually produced by shallow cultivation when the soil is dry.

Dysic. Low level of bases in soil material, specified at family level of classification.

E horizon. A strongly leached layer, often referred to as an albic horizon.

EC (electrical conductivity). Measured in Siemens/meter.

EC_e. The electrolytic conductivity of an extract from saturated soil, normally expressed in units of siemens per meter at 25°C.

Ecology. The totality of relationships among organisms and their ambient (surrounding) environment.

Ecosystem. A community of organisms and the environment in which they live.

Edaphic. (1) Of or pertaining to the soil. (2) Resulting from or influenced by factors inherent in the soil or other substrate, rather than by climatic factors.

Edaphology. The science that deals with the influence of soils on living things, particularly plants, including human use of land for plant growth. (Term not in common usage).

Effective precipitation. The portion of the total precipitation which becomes available for plant growth.

Efflorescence. Accumulation of dried soluble salts on the soil surface, left as the water carrying the salts has evaporated.

Effluent. 1. The discharge or outflow of water from ground or subsurface storage. 2. The fluids discharged from domestic, industrial, and/or municipal waste collection systems or treatment facilities.

Eh. The potential that is generated between an oxidation or reduction half-reaction and the H electrode in the standard state.

Electodynomorphic soils: The soils with properties that have been influenced mainly by factors other than parent material.

Electrokinetic potential (zeta potential): The potential difference (fall of potential) between the fixed parts and diffused or freely mobile portions of double layer is known as electrokinetic or zeta potential.

Eluvial horizon. A horizon (layer) of soil that has been formed by the process of eluviation.

Endodynomorphic soils: Soils with properties that have been influenced mainly by parent material.

Environment. The term environment etymologically means surroundings and is generally used for an abstract concept. The surroundings may be natural or man-made, and may be physical, chemical or biotic. The environment is thus, a complex of so many things encompassing an organism that interact not only with the organisms but also among themselves. As a result of this it is difficult to isolate or alter any one of the them without affecting other components. The growth, behaviour and life history of an organism, are therefore, influenced by the environment in which they live. The preferred environment of an organism is usually referred to as habitat. The term, habitat denotes a more specific meaning of the surrounding than environment. On our planet earth, we find two major types of habitats, terrestrial (~ 174 million km²) and aquatic (~ 480 million km²) with different environmental complexes. Both of these habitats support innumerable forms of life, the nature and concentration of which vary considerably, although certain forms (amphibians) are dependent on both of them for completing their life cycle.

Enzyme. A protein mass with surface chemical groups arranged so that atoms around a certain kind of bond can attach to the enzyme, allowing the enzyme to act as a catalyst for a chemical reaction (such as splitting of the bond).

Eolian (aeolian) soil material. Soil material accumulated through wind action. The most extensive areas in the United States are silty deposits (loess), but large areas of sandy deposits, called sand dunes, also occur.

Equivalent acidity. The number of parts by weight of calcium carbonate (as $CaCO_3$) required to neutralize the acidity resulting from the use of 100 parts by weight of the fertilizer.

Equivalent basicity. The number of parts by weight of calcium carbonate (as $CaCO_3$) that corresponds in acid neutralizing capacity to 100 parts by weight of the fertilizer.

Erodibility. The state or condition of being erodible or the specific property of soil which refers to vulnerability of soil to erosion under specific circumstances.

Erosion classes. A grouping of erosion conditions based on the degree of erosion or on characteristic patterns. (Applied to accelerated erosion; not to normal, natural, or geological erosion.) Four erosion classes are recognized for water erosion and three for wind erosion. Specific definitions for each vary somewhat from one climatic zone, or major soil group, to another.

Erosion. (1) The wearing away of the land surface by running water, wind, ice or other geological agents, including such processes as gravitational creep. (2) Detachment and movement of soil or rock by water, wind ice or gravity. The following terms are sued to describe different types of water erosion:

Erosivity. The potential ability of water, wind, gravity, and so on, to cause erosion.

Essential chemical elements. Elements required by plants to complete their life cycles.

Essential element. (Plant nutrition) A chemical element required for the normal growth and reproduction of plants, people, or other animals.

Estimate agriculture. Agriculture as practiced on large plantations with modern management. Focusing on one product, and with profit and export-oriented strategy.

Eucaryote. (you-car'-ry-ot) Organism cell type that has the genetic material enclosed inside a nucleus. Includes all organisms except the bacteria and actinomycetes.

Euic. High level of bases in soil material, specified at family level of classification.

Eutrophic. Having concentrations of nutrients optimal, or nearly so, for plant or animal growth. (Said of nutrient or soil solutions and bodies of water.) The term literally means "self-feeding."

Eutrophication. Enrichment of waters with nutrients, primarily phosphorus, causing abundant aquatic plant growth. Eutrophication is the process of enrichment of surface waterbodies like lakes, reservoirs and streams with nutrients. Nutrient enrichment of waterbodies results in intense proliferation and accumulation of algae and higher aquatic plants in excessive quantities which can result in detrimental changes in water quality and can significantly interfere with man's use of the water resource. Surface waters also receive largest contribution of nutrients from non-point sources such as agricultural lands. Several studies have demonstrated that high levels of total N in surface waters are generally related to farming activities. A large part of the nutrient load generated from agricultural land is related to the natural nutrient content of the soil. Practices which do not allow for the proper incorporation of fertilizers applied at excessive rates can cause short term elevated nutrient concentrations in runoff.

E-value. This method is a direct application of the isotopic dilution principle and is an attempt to measure the amount of nutrient in the soil that is in equilibrium with the same nutrient in the soil solution (Russel *et al.*, 1954). Thus in the case of estimation of available phosphate, use is made of the principle that when ^{32}P is added to a system containing ^{31}P, the ^{32}P will distribute itself such that equilibrium ratio of ^{32}P to ^{31}P (specific activity) will be constant for all the phosphate participating in the equilibrium reaction.

Evaporites. Residue of salts (including gypsum and all more soluble species) precipitated by evaporation.

Exchange capacity. The total ionic charge of the adsorption complex active in the adsorption of ions.

Exchange reactions. Although colloids have a negative charge, no electrical charge in the soil goes unbalanced for very long. A soil colloidal system has a double layer of charges. The inner layer is the negative charge of the colloidal particle; the outer layer is formed by cations in the soil solution, which are attracted to the colloidal surface in proportion to the negative charges available. This means that a divalent cation such as calcium (Ca^{++}) or magnesium (Mg^{++}) will neutralize two negative charges of the colloidal particle, whereas monovalent ions such as potassium (K^+), sodium (Na^+), or hydrogen (H^+) will neutralize one negative charge each. In very acid soils aluminum ions ($A1^{+++}$), which may be combined with one or two hydroxyls (OH^-), can be attracted to colloidal surfaces. There may be many other cations attracted to the colloids in small amounts. Some of these are trace elements that are of great significance to growing plants.

Exchangeable aluminium. Trivalent positively charged ion that can be retained on the negative charge of the clay minerals or humus, and forming along with hydrogen the group of acid-forming exchangeable cations. The exchange acidity is due almost uniquely to aluminium ions though it is generally considered that exchangeable hydrogen is responsible for this acidity.

Exchangeable cation percentage. The extent to which the adsorption complex of a soil is occupied by a particular cation. It is expressed as follows: exchangeable cation ($cmol_c kg^{-1}$ soil)

Exchangeable sodium fraction. The fraction of the cation exchange capacity of a soil occupied by sodium atoms.

Exchangeable sodium percentage (ECP). The percentage of the cation exchange capacity of a soil occupied by sodium.

Exchangeable sodium ratio. The ratio of exchangeable sodium (cmol kg^{-1} soil) to all other exchangeable cations (cmol kg^{-1} soil).

Extensive farming. The fish feed on the natural food supply in the pond.

Extrusive rocks. Volcanic igneous rocks formed from molten magma that cooled somewhat rapidly (has glassy nature of small crystals) because it was exposed above the earth's crust.

Exudate. Low-molecular-weight metabolites that enter the soil from plant roots.

Exudation. The excretion or natural elimination of substances from the plant, usually from the roots and leaves, but can be from any plant part.

Facultative organism. An organism that is able to carry out both options of a mutually exclusive process (e.g., aerobic and anaerobic metabolism). May also be used in reference to other processes, such as photosynthesis (e.g., a facultative photosynthetic organism is one that can use either light or the oxidation of organic or inorganic compounds as a source of energy).

Fallow. Allowing cropland to lie idle, either tilled or untilled, during the whole or greater portion of the growing season.

Fallowing. Keeping the land free of a crop or weeds for a period of time. This is done to restore soil productivity mainly through accumulation of water, nutrients or both. Fallow during the kharif season is a common practice in wheat-growing regions which receive limited rainfall. During fallow, the field is cultivated to control weeds and to help the storage of moisture for the succeeding crop.

Family, soil. A category in soil classification between soil series and subgroup. It is the most useful category for determining plant growth and making land-use recommendations.

Fertility index. Fertility index is defined as the relative sufficiency expressed as a percentage of the amount of nutrient adequate for optimum yields. it is related with soil test values and crop response.

Fertility, soil. The ability of a soil to supply the nutrients essential to plant growth.

Fertilizer controlled-release (slow release). A fertilizer term used interchangeably with delayed release, slow release, controlled availability, slow acting, and metered release to designate a controlled dissolution of fertilizer at a lower rate than that of conventional water-soluble fertilizers. Controlled-release properties may result from coatings on water-soluble fertilizers or from low dissolution and/or mineralization rates of fertilizer materials in soil.

Fertilizer Pollution. Increased agricultural productivity through improved technologies has helped in maintaining the balance between food supply and demand. The supply of nutrients from the soil from recycled organic matter and biological fixation of N is not sufficient to ensure the present level of food production. Inorganic fertilizers are, therefore, used to provide additional plant nutrients to ensure crops produce their optimal yields while maintaining the fertility of the soil. According to an estimate, fertilizers accounted for 75 per cent Asian cereal growth during 1979-89. Chemical fertilizers, no doubt, constitute an invaluable input for producing required food crops.

Fertilizer requirement. The quantity of certain plant nutrients needed, in addition to the amount supplied by the soil, to increase plant growth to a designated level.

Fertilizer Use Patterns. Fertilizer consumption has increased tremendously all over the world in recent years. Worldwide, total application of N through fertilizers was around 60 million tonnes in 1981-82 and it has increased to more than 77 million tonnes in 1991-92. Thus, as developing countries like India strive to produce more food by applying large amounts of plant nutrients through fertilizers, it becomes increasingly important to consider seriously the fertilizer related environmental problems. Consumption of NPK in India during 1990-91 was million tonnes and that in the states of U.P. and Punjab, it was and million tonnes (FAI, 1994). Since 10 per cent or more of total nutrients applied through fertilizers in India is consumed in Punjab having only around 3 per cent of the net cultivated area of the country, the effect of fertilizer pollution shall be felt in Punjab quite early. There are, of course, specific crops and agricutural areas in a state or country receiving higher rates of fertilizers than the averages indicated above. Some farmers use much more fertilizer than their neighbours.

Fertilizer, inorganic. A fertilizer material in which carbon is not an essential component of its basic chemical structure. Urea is often considered an inorganic fertilizer because of its rapidly hydrolysis to form ammonium ions in soil.

Fertilizer, mixed. Two or more fertilizer materials blended or granulated together into individual mixes. The term Includes dry mix powders, granulated, clear liquid, suspension, and slurry mixtures.

Fertilizer, organic. A material containing carbon and one or more plant nutrients in addition to hydrogen and/or oxygen. Urea is often considered an inorganic fertilizer because of its rapid hydrolysis to form ammonium ions in soil.

Fertilizer, salt index. The ratio of the decrease in osmotic potential or increase in pressure of a solution containing a fertilizer compound or mixture to that produced by the same weight of $NaNO_3$ x 100.

Fertilizer, starter. A relatively small application of fertilizer applied with or near the seed for accelirating early growth of the crop.

Fertilizer, suspension. A fertilizer containing dissolved and undissolved plant nutrients. The undissolved plant nutrients are kept in suspension with a suspending agent, usually a swelling-type clay. The suspension must be flowable enough to be mixed, pumped, agitated, and applied to the soil in a homogeneous mixture.

Fibric material. Mostly undecomposed plant remains that contain large amounts of fibers that are well preserved and are recognizable. Bulk density is usually very low.

Field capacity, *in situ* (field water capacity). The content of water, on a mass or volume basis, remaining in a soil 2 or 3 days after having been wetted with water and after free drainage is negligible.

Field experiments. Experiments conducted in the field to determine the type and amount of fertilisers to suit particular soil type and crop.

Final sample. laboratory sample: A representative part of the reduced sample or, where no intermediate reduction is required, of the aggregate sample.

Fine texture. Consisting of or containing large quantities of the fine fractions, particularly of silt and clay. (Includes clay loam, silty clay loam, sandy clay, silty clay, and clay textural classes. Sometimes subdivided into clayey texture and moderately fine texture.)

Fine texture. Containing large portions of silt and clay fractions.

Firm. A term describing the consistency of moist soil that offers distinctly noticeable resistance to crushing but can be crushed with moderate pressure between the thumb and forefinger.

Fixed ammonium. The ammonium in soil that cannot by replaced by a neutral potassium salt solution (e.g., 1 M KCl).

Flagstone. A relatively thin fragment, 15.2 to 38 cm long, of sandstone, limestone, slate, shale, or rarely, of schist.

Flexible cropping. A nonsystematic sequence of growing adapted crops with cropping and fallow decisions at each prospective date of planting based on available water in the soil plus expected growing season precipitation.

Flocculation. To aggregate or clump together individual soil particles, especially fine clay and humus, into small clumps or granules that usually settle out of suspension quickly. Opposite of deflocculate of disperse.

Foliar diagnosis. Diagnostic technique based on analyzing plant tissues for the total content of nitrogen, phosphorus, potassium, boron, etc. Low values of nutrient percentages in plants under test compared to nutrient percentage in high yielding plants indicate deficiency of nutrients. The nutrients commonly tested by this method are nitrogen, phosphorus and potash.

Forest soil. The total forest area in India is estimated as 75 m ha which constitutes 18 per cent of the total geographical area. The major forest areas are covered by tropical deciduous, coniferous and tropical evergreen forests and occur in Himachal Pradesh, Jammu & Kashmir, Uttar Pradesh, Arunachal Pradesh, Nagaland, Assam, Meghalaya, Mizoram, Manipur, Orissa, Madhya Pradesh, Maharashtra, Kerala and Andaman & Nicobar Islands. While the climatic condition and altitude control the kind of forest species growing in different areas, the kind of vegetation and topography control the soil profile development. The major soils as observed in different forest areas are: Brown Forest and Podzolic (in the Himalayas) and Red and Lateritic soils (in the Deccan Plateau). The soils in the Himalayas are formed on sedimentary, igneous or metamorphic rocks under cool/cold (sub) humid conditions. The Deccan Plateau soils are formed from igneous and metamorphic rocks under (sub) tropical conditions.

Fragipan. A hard, dense, brittle-when-dry soil pan.

Fragmental. Stones, cobbles, gravel and very coarse sand with too little fine sand, silt, and clay to fill soil interstices larger than 1 mm.

Free iron oxides. A general term for those iron oxides that can be reduced and dissolved by a dithionite treatment. Generally includes goethite, hematite, ferrihydrite, lepidocrocite, and maghemite, but not magnetite.

Friable. A consistency term pertaining to the ease of crumbling of soils. **Friction cone penetrometer.** A cone penetrometer with the additional capacity of measuring the local side-friction component of penetration resistance. The resistance to penetration developed by the friction sleeve equals the vertical force applied to the sleeve divided by its surface area.

Frigid. A cool soil temperature regime (0-8°C) with warm summers.

Fritted trace elements. Sintered silicates having total guaranteed analyses of micronutrients with controlled (relatively slow) release characteristics.

Frost heaving. The raising of the surface soil due to the growth of ice crystals in the underlying soil. Such action often pushes plants out of the ground.

Fulvic acid. (1) The mixture of organic substances remaining in solution upon acidification of a dilute alkali extract from soil, in which case the expression fulvic acid fraction is often used, and (2) the colored material that remains in solution after removal of humic acid by acidification.

Gated pipe. Portable pipe with small gates (slide openings) installed along one side for uniformly distributing irrigation water to corrugations or furrows.

Genetic. Resulting from, or produced by, soil-forming processes: for example, a genetic soil profile or a genetic horizon.

Geomorphic surface. A portion of the landscape specifically defined in space and time that has determinable geographic boundaries and is formed by one or more agencies during a given time period.

Geophagy. The eating of soil by animals and people.

Gilgai. The microrelief of soils produced by expansion and contraction with changes in water content. Found in soils that contain large amounts of clay, which swells and shrinks considerably with wetting and drying. Usually a succession of micro basins and microknolls in nearly level areas or of micro valleys and micro ridges parallel to the direction of the slope.

Glacial drift. Rock debris that has been transported by glaciers and deposited, either directly from the ice or from the meltwater. The debris may or may not be heterogeneous.

Glacial till. Unstratified, nonsorted materials deposited from melting glaciers and consisting of clays, silts, sands, gravels, and boulders.

Glaciers. Large masses of ice that formed, in part, on land by the compaction and recrystallization of snow. They may be moving downslope or outward in all directions because of the stress of their own weight or they may be retreating or be stagnant.

Glaciofluvial deposits. Material moved by glaciers and subsequently sorted and deposited by streams flowing from the melting ice. The deposits are stratified and may occur in the form of outwash plains, deltas, kanes, eskers, and kame terraces.

Gley. Some layer of mineral soil developed under conditions of poor drainage (poor aeration), resulting in reduction of Iron and other elements and in gray colors and mottles (blobs of variously colored soils).

GNESIS. The foliated metamorphic rock formed from the igneous rock (Granite) by the process of metamorphism.

Grain size. The dimension which corresponds to the smallest sieve aperture size through which a particle will pass if presented in the most favourable attitude.

Granulation. Techniques using a process such as agglomeration, accretion or crushing, to make a granular fertilizer.

Granule. A natural soil aggregate or ped of relatively low porosity.

Gravel. Coarse soil particle 0.2 to 2 cm in diameter (Pomerol, Inf. et Bio.)

Gravelly. Containing appreciable or significant amounts of gravel. (Used to describe soils or sands.)

Gravitational potential. The amount of work that must be done per unit quantity of pure water in order to transport reversibly and isothermally and infinitesimal quantity of water, identical in composition to the soil water, from a pool at a specified elevation and at atmospheric pressure, to a similar pool at the elevation of the point under question.

Gravitational water. An obsolete term previously referring to water that moves into, through, or out of the soil under the influence of gravity.

Great soil group. A category of the U.S. Soil Taxonomy system. Soils are placed according to soil moisture, temperature, base saturation status, and expression of horizons.

Green Food. Green Food is a kind of pollution-free food with high safety, quality and nutrition which is certified by an authorized organization. The standard system of Green Food is composed of four parts: environmental standards; producing and operating standards; product standards; packaging and labeling standards. On the basis of adopting IFOAM Basic Standards and considering the actual conditions of China, Green Food is divided into two grades: grade AA and grade A. The former is full organic, and the later is transitional. This introduction will focus on grade AA Green Food. The environmental factors of the Green Food production site including air, soil and water must be tested and up to the first or second grade of national environment standards, and any pollution index can not be higher than the corresponding standards. In the producing process, all chemical fertilizers, chemical pesticides, synthetic food additives and other chemical inputs can never be used. All chemical residues must be undetectable O in grade AA Green Food products. The logo and attached words of grade AA Green Food are required to print in green color on white background, and it is just the reverse of grade A Green Food. In conclusion, the standards of grade AA Green Food is up to or more strict than the IFOAM Basic Standards.

Green manure crops. Crops that are plowed under to improve the fertility and physical condition of the soil are called green manure crops. The best for this purpose are legumes such as alfalfa and clover because they are high in nitrogen content. This is a result of the presence in root nodules of bacteria that are capable of taking nitrogen from the air. This is called nitrogen fixing.

Green manuring. A practice of ploughing or turning into the soil undecomposed green plant material for improving the physical condition of the soil or for adding nitrogen when the green manure crop is legume. Two types of green manuring are being practiced by the cultivators of India. These are (i) green manuring *in situ* and (ii) green leaf manuring.

Ground phosphate rock. Material obtained by grinding naturally occurring phosphate rock to a fineness so that 90 per cent of it passes through 150 micron IS sieve.

Groundnut cake. A by-product of the oil industry. This is an edible oil cake. It contains 7.3 per cent nitrogen, 1.5 per cent phosphoric acid and 1.3 per cent potash. In the Deccan tract of Maharashtra. Karnataka. Andhra Pradesh and Tamil Nadu, groundnut cake is applied to sugarcane and bananas on a very large scale.

Besides the nutrients, it supplies organic matter to the soil.

Groundwater recharge. The water that fills large soil pores moves downward under the force of gravity. We have already spoken of this gravitational water. It percolates until either it is absorbed by drier soil below and is very much slowed in its movement or it reaches the water table. The water table is the plane in the porous substratum below which all pores are filled with water. This may be within the surface soil or in the deep bedrock. Groundwater is the term for the water below the water table. The water table surface has high and low positions, as does the land surface, but with less vertical range. Wetlands, lakes, and streams are places where the water table comes to the surface and where discharge is taking place. Wetlands in humid regions are actually not recharge areas, for water is being lost from them in the form of stream flow and evapotranspiration. Porous rock layers such as sandstone may be saturated and can be important sources of well water. Such layers are called aquifers.

Groundwater. That portion of the water below the surface of the ground at a pressure equal to or greater that atmospheric.

Groundwater. Water that occupies pores and crevices in rock and soil below the surface and above a layer of impermeable material. The upper limit of groundwater is the water table.

Growth medium. Any material such as soil, peat, etc., used as a support for plant roots, that has a capacity for water retention and which may contain added or naturally occurring nutrients.

Guano. It includes many materials which vary in source composition and readiness for use.

Guarantee (of composition). Quantitative and/or qualitative characteristics with which a marketed product must comply for contractual or legal requirements.

Gully erosion. The removal of soil and water concentrated in deep, narrow channels.

Gully. A channel resulting from erosion and caused by the concentrated but intermittent flow of water usually during and immediately following heavy rains. Deep enough to interfere with, and not to be obliterated by, normal tillage operations.

Gypsic horizon. A diagnostic horizon having an accumulation of gypsum.

Gypsum requirement. The quantity of gypsum or its equivalent required to reduce the exchangeable sodium content of a given amount of soil to an acceptable level. However, for effective use, the gypsum applied in practice is 1.2 to 1.5 times the gypsum requirement, thus found out. Pure gypsum contains 18.6% S and commercial agricultural grade gypsum being of 70-80% purity contains 13-15% S. According to present standards mineral gypsum should contain not less than 70% $CaSO_4.2H_2O$, not more than 0.75% sodium (Na) and entire material should pass through 2 mm sieve but atleast 50% it should pass through a 0.25 mm (60 mesh) sieve. Its solubility is 2.5 g/L which appears to be low. That is the reason gypsum requires a lot of water.

Gypsum. A hydrated calcium sulphate ($CaSO_4$, $2H_2O$). The commercial material contains varying amount of impurities and is used as a soil amendment.

H bond. (Hydrogen bond) The bond of a hydrogen ion already strongly attached to one electronegative ion (such as oxygen, nitrogen, sulfur), less strongly to a second electronegative ion. H bonds are common between water molecules, between water and mineral oxygen, and between organic substance hydrogens, and mineral oxygen.

Habitat. The place where a given organism lives.

Half-life. The time for half of a substance to be destroyed or inactivated or, for radioactive substances, to lose half of its radiation.

Halomorphic Soil. It is a sub-order of the intrazonal soil order, consisting of salt affected soils (saline and sodic soils etc.) formed under imperfect drainage in arid regions.

Halophytic vegetation. Vegetation requiring or tolerating a saline environment.

Hardpan. A hardened soil layer in the lower A or a deeper horizon caused by cementation of soil particles with organic matter or with such materials as silica, sesquioxides, or calcium carbonate. The hardness does not change appreciably with changes in moisture content and pieces of the hard layer do not slake in water.

Harvest index. The quantity of biomass produced per unit of input of a plant nutrient.

Heat of wetting. When clay is wetted it tends to be sticky and is easily molded. On drying it absorbs considerable heat energy and on wetting again, it evolves the same amount of heat. This phenomenon is known as "heat of wetting".

Heavy Metals. Those metals of high atomic weight, having densities greater than 5.0 Mg/m^3. Many heavy metals are toxic when accumulated into animal bodies. The more common ones of concern in the environment are arsenic, cadmium, chromium, copper, beryllium, lead, manganese, mercury, nickel, and zinc, among many others. Lately, there has been growing concern on health hazards associated with heavy metals entering the food cycle through soil. Fertilizers contain heavy metals as impurities. The application of rock phosphate or its products to soil always implies the addition of a significant amount of lead and cadmium into the soil.

Hectare. (ha) A metric unit, 10,000 square meters or 2.471 acres.

Hemic material. Organic residues in organic soils (Histosols) having an intermediate stage of decomposition.

Herbicides. The usage of chemicals for controlling weeds started with the introduction of 2,4-D (2,4-Dichlorodiphenyl trichloro ethane) in the late forties. The usage of herbicides is higher than any other agrochemicals. The main use of herbicides in the world is in maize and soybean.

Heterotroph. An organism capable of deriving carbon and energy for growth and cell synthesis by the utilization of organic compounds.

Heterotrophic. An organism capable of deriving energy for life processes from the oxidation of organic compounds.

Hidden Hunger of Crops. It refers to a situation in which a crop needs more of a given element, yet has shown no deficiency symptoms.

High analysis fertilizer. A fertilizer containing not less than 25 per cent of the major plant nutrients, namely, nitrogen, phosphoric acid (as P_2O_5) and potash (as K_2O). Such fertilisers are urea containing 46 per cent nitrogen, triple superphosphate containing 45 to 47 per cent P_2O_5 and muriate of potash containing 60 per cent K_2O. High-analysis fertilisers have an advantage in the cost of bagging, handling and transportation per unit of plant nutrient.

Humic acid. The dark-colored organic material that can be extracted from soil by various reagents (e.g., dilute alkali) and that is precipitated by acidification of pH 1 to 2.

Humic substances. A series of relatively high-molecular-weight, brown to black substances formed by secondary synthesis reactions. The term is used in a generic sense to describe the colored material or its fractions obtained on the basis of solubility characteristics (e.g., humic acid, fulvic acid).

Humin. The fraction of the soil organic matter that is not dissolved upon extraction of the soil with dilute alkali.

Humus. It is the complex organic matter derived from decomposition of plant and animals tissue by the action of various soil microorganisms. Cellulose constitutes the major source of carbon and of energy for microbic-flora of soil. Humus is a brownish-black material of soil and is of great importance for plant growth. It improves the physical condition of soil like bettering texture and water holding capacity and forming reservoir of mineral nutrients. Chemically, humus is composed of dead organic matter alone, however, sometimes various kinds of salts may be associated with it, particularly iron and some of the elements which are required in small amount for life in general.

Hydraulic conductivity. (*K*) The proportionality factor in Darcy's law, indicating the soil's ability to transmit flowing water. Values of *K* commonly range from 1×10^{-1} cm/s in porous materials (sands) to 1×10^{-9} cm/s in tight clays.

Hydrogen bond. The chemical bond between a hydrogen atom of one molecule and two unshared electrons of another molecule.

Hydrogenic soil. Soil developed under the influence of water standing within the profile for considerable periods, formed mainly in cold, humid regions.

Hydrologic cycle. The circuit of water movement from the atmosphere to the earth and return to the atmosphere through various stages or processes, as precipitation, runoff, percolation, storage, evaporation, and transpiration.

Hydrolysis. That chemical reaction involving double displacement in which hydrogen of water combines with the anion of the mineral and hydroxyl of water combines with the cation of the mineral to form an acid and a base.

Hydrophobic soils. Soils that are water repellent, often due to dense fungal mycelial mats or hydrophobic substances vaporized and reprecipitated during fire.

Hydroponics (chemical gardening). Term hydroponic was used by Gerick (1940) to refer soilless growth of plants in water and sand cultures. The solution used in culture contains all essential elements in proper proportions for normal growth of plants. This practice has become quite popular during recent past even for large scale commercial growth of plants such as roses, tomatoes, lettuce, carrots and potatoes. Hydroponic cultures have certain advantages over *geoponics* (soil culture or agriculture).

Hydrous mica. (Illite) (Obsolete) A hydrous aluminosilicate clay mineral with structurally mixed mica and smectite or vermiculite, similar to montmorillonite but containing potassium between the crystal layers. It has a smaller cation exchange capacity than montmorillonite. Sometimes referred to as illite or mica No term given to replace these names. Still widely used.

Hydroxy-aluminum interlayers. Polymers of general composition $[Al(OH)_{3}-_{x}]_{m}^{+nx}$ which adsorbed on interlayer cation-exchange sites. Although not exchangeable by unbuffered salt solutions, they are responsible for a considerable portion of the titratable acidity (and pH-dependent charge) in soils.

Hydroxy-interlayered vermiculite. A vermiculite with partially filled interlayers of hydroxy aluminum groups. It is normally dioctahedral in both the interlayer and the vermiculite layer. It is common in the coarse clay fraction of acid surface soil horizons. It has intermediate cation-exchange properties between vermiculite and chlorite. Synonyms are *chlorite vermiculite intergrade* and *vermiculite-chlorite intergrade*.

Hygroscopic. A term used to designate materials that absorb moisture from the air.

Hymatomelanic acid. The alcohol-soluble fraction of humic acid.

Hyperthermic. Continuously hot temperature regime (15 22°C).

Hyriic soils. Soils that are periodically wet long enough to produce anaerobic conditions, thereby influencing the growth of plants.

Hysteresis. Phenomenon according to which some of the physical properties Y related to the independent variable X are irreversible even when the variations in X are carried out slowly, the curve obtained for increasing X does not correspond to that for decreasing X. This phenomenon is observed during drying (desorption) and moistening (sorption) of soils.

Identification of silicate clay minerals. Since different clay minerals have contrasting properties, it is important to identify them so that a soil's capabilities and limitations can be accurately predicted. Nearly pure deposits of a clay mineral are valuable as sources of industrial materials.

Illite. The mica component of a structurally mixed fine grained mica and smectite or vermiculite. Sometimes the entire mixture is referred to as illite. Illites are dioctahedral.

Illuvial horizon. A soil layer or horizon in which material carried from an overlying layer has been precipitated from solution or deposited from suspension. The layer of accumulation.

Illuviation. The process of deposition of soil material removed from one horizon to another in the soil; usually from an upper to a lower horizon in the soil profile.

Immobilization. The conversion of an element from the inorganic to the organic form in microbial tissues or in plant tissues.

Impact of Desertification: The impact of desertification is more severe and distinct in arid regions due to their inherent morphological vulnerability and climatic severity. In the western Rajasthan, different type of fluvial and aeolian fragile ecosystems occur and they are undergoing morphological and biological changes due to natural and accelerated desertification processes like wind erosion/deposition, water erosion/deposition, salinity/alkalinity and waterlogging. In the last few decades, the impact of human and livestock population on arid ecosystems resulting in the desertification/depletion of their biological productivity. Based on the above definitions of desertification, the studies on the wind erosion/deposition, water erosion/deposition, salinity/alkalinity and waterlogging hazards responsible for degradation of various type of arid ecosystems were conducted in different parts of the western Rajasthan using aerial, space and ground survey techniques. The causes, impacts and future trends of the desertification in western Rajasthan in general and in Shergarh tehsil of Jodhpur district and the Bandi catchment in particular as two specific case studies have been highlighted in this paper.

Impeded drainage. A condition that hinders the movement of water through soils under the influence of gravity.

Impervious soil. A soil through which water, air, or roots cannot penetrate.

Impervious. Resistant to penetration by fluids or by roots.

Indicator plant. A plant that indicates, in general, and often in a specific manner, a deficiency of plant nutrient in the soil.

Indigenous knowledge. Local knowledge unique to a given culture or society.

Indigenous. Native to a specific area; not introduced. Opposite of exotic.

Indurated. (Soil) Soil material cemented into a hard mass that will not soften on wetting.

Industrial agriculture. Agricultural management and practices which respond to market impulses (satisfaction of buying power) and aim at maximum production. Plant selection, crop rotation (if any), and corresponding amounts of energy usage, mechanization, and inputs of synthetic fertilizers and pesticides are directly related to market and maximum production. In this study, synonymous with High External Input Agriculture (HEIA), modern and intensive agriculture.

Infiltration. The downward entry of water into the soil through the soil surface.

Infiltration. The downward movement of water into the soil.

Inoculate. To treat (usually seeds) with microorganisms for the purpose of creating a favorable response. Most often refers to the treatment of legume seeds with *Rhizobium* to stimulate dinitrogen fixation.

Inoculation. The process of introducing cultures of microorganisms into soils or culture media, such as by adding *Rhizobia* bacteria to legume seed.

Instructive rocks. Rocks formed from molten magma that cooled slowly (has large crystals) because it was deep within the earth's crust. Also called plutonic igneous rocks.

Integrated agriculture. A type of agriculture with various objectives. This includes not only farm productivity, but also maintaining farm employment, income, environment, and democratization.

Integrated Nutrient Management. The use of carefully derived combinations of mineral and organic fertilizers in combination with complementary crop practices such as tillage, rotation, and moisture conservation.

Integrated Pest Management (IPM). A mixed strategy of selective use of agrochemicals, biological (predator/prey) methods, genetic resistance, and appropriate management practices. Pesticides are either an integral part or a last resort of pest control strategy.

Intensive fish farming. High stocking rate (10 kg fish/1000 1 water), high investments in technical facilities and exclusively artificial feeding permit intensive production; problems include waste water disposal and high susceptibility to disease.

Inter-cropping. Two or more crops are managed simultaneously in the same field.

Intercropping. Growing two or more crops in the same field at the same time in a mixture.

Internal soil drainage. The downward movement of water through the soil profile.

Interstratification. Mixing of silicate layers in a given stack. Interstratification may be regular or random.

Intrinsic permeability. The property of a porous material that expresses the ease with which gases or liquid flow through it.

Ion selectivity. The relative adsorption of an ion by the solid phase in relation to the adsorption of other ions. The relative adsorption of an ion by a root in relation to absorption of other ions.

Ions. Electrically charged particles. As used in soil, an ion refers to an electrically charged element, resulting from the breaking up of many substances like acids, salts, and alkaline. Positively charged ions are called `cations' and negatively charged as `anions' for further information

Iron - Microbial Transformation. Iron is one of the necessary nutrients of all the living organisms and is required in very minute amount for growth. It is found in all living cells. In a few species, however, only minute amount is present. It is always abundant in the terrestrial habitat and is one of the major constituents of the earth's crusts. The absence of iron in plants leads to the absence of greenness. Iron ore is often present in non-available form for plants. Therefore, it essential to convert this non-available form into the available one. This conversion is brought about by the activities of microorganisms and is brought about in various ways.

Iron Bacteria. In 1888, the eminent Russian microbiologist, S. Winogradsky, reported the isolation of iron bacteria- a strain of *Leptothrix* which has the capacity to oxidize the ferrous iron and precipitate ferric hydroxide in its sheath. These kinds of bacteria are present in iron containing spring water in non-iron containing waters, in raw and even in treated drinking water and derive their energy from the oxidation of ferrous compounds. These bacteria are responsible for producing foul smell in water pipes as a result of accumulation of large amount of ferric salts.

Iron oxides. Group name for the oxides and hydroxides of iron. Includes the minerals goethite, hematite, ferrihydrite, maghemite, and magnetite. Sometimes referred to as free iron oxides, sesquioxides, or hydrous oxides.

Iron-pan. An indurated soil horizon in which iron oxides are the principal cementing agent along with aluminum oxides. If from plinthite, it is called ironstone.

Ironstone. Hardened plinthite materials often occurring as nodules and concretions.

Irrigation application efficiency. Percentage of irrigation water applied to an area that is stored in the soil for crop use.

Irrigation efficiency. The ratio of the water actually consumed by crops on an irrigated area to the amount of water applied to the area.

Irrigation methods. The manner in which water is artificially applied to an area.

Irrigation. The intentional application of water to the soil.

Isomorphous substitution. The replacement of one atom by another of similar size in a crystal lattice without changing the crystal structure.

Jarosite. A pale yellow potassium iron sulfate mineral.

Jasper. Siliceous sedimentary rock, deriving from a silicified Radiolaria-bearing ooze (variety of Chalcedony), light red (cornaline) or brown (sardoine).

Kainite. An impure mixture of potassium and sodium chlorides, sometimes containing sulphate of magnesia, with not less than 12 per cent of potash (as K_2O).

Kame. An irregular ridge or hill of stratified galcial drift.

Kandic horizon. A diagnostic argillic horizon having mostly low activity (1:1 crystal), clays, such as Kaolinite.

Kaolin. A subgroup name of aluminum silicates with a 1:1 layer structure. Kaolinite is the most common clay mineral in the subgroup. Also, a soft, usually white rock composed largely of kaolinite. White and friable clayey rock the main constituent of which is kaolinite. It results mainly from surface weathering under hot, humid climate, or by hydrothermal alteration, of acid igneous rocks rich in potash feldspars. Being refractory it is mainly used in porcelain manufacture.

Kaolinite. A clay mineral of the kaolin subgroup. Its general formula is $Al_2Si_2O_5(OH)_4$. It has a 1:1 layer structure composed of shared sheets of Si-O tetrahedrons and Al-(O, OH) octahedrons. Its CEC is 10 (3-15) c mol (P^+) Kg^{-1}.

Labile pool. The total sum of a nutrient that readily solubilizes or exchanges, when the soil is equilibrated with a salt solution, to become available to plants during a season.

Labile. A substance that is readily transformed by microorganisms or is readily available to plants.

Lacustrine deposit. Sediments deposited in fresh (nonsaline) take water and later exposed either by lowering of the water level or by the elevation of the land.

Land capability class. One of eight classes of land in the land capability classification of the USDA-Soil Conservation Service, distinguished according to the risk of land damage or the difficulty of land use.

Land use planning. According to FAO, it is the systematic assessment of land and water potential, alternative patterns of land use and other physical, social and economic conditions, for the purpose of selecting and adopting land-use options which are most beneficial to land users without degrading the resources or the environment, together with the selection of measures most likely to encourage such land uses. Land use planning may be at international, national, district of local levels. It includes participation by land users, and covers educational, legal, fiscal and financial measures.

Land utilization. Geographical area of the country is 328.73 million hectares and whole of the area can not be put under agricultural use i.e. under cultivation or production of crops since portion of the area is under rocks, marshy lands, sloppy lands, rivers, roads and buildings, forests, etc. and as such only a portion of it can be utilized. According to an estimate, out of the total area of country, only 265 million hectares (80%) representing varying degrees of potential is available for production.

Landform. A three-dimensional part of the land surface, formed of soil, sediment, or rock that is distinctive because of its shape that is significant for land use or to landscape genesis, that repeats in various landscapes, and that also has a fairly consistent position relative to surrounding landforms.

Landforms of the Arid Ecosystem. Different landforms created by the exo genic and endogenic processes, particularly the sand dunes, are the spectacular features of the arid region of western India

Landscape patterns. The landscape is a mosaic of soil bodies that fit together as neatly as pieces in a jigsaw puzzle. Some landscapes have concentric, circular soil patterns; some are characterized by looped patterns; and others based on a soil map, each representing a square kilometer (0.4 mi^2) of land in south-eastern Wisconsin.

Landscape. All the natural features such as fields, hills, forests, water, and so on, which distinguish one part of the earth's surface from another part. Usually, that portion of land or territory which the eye can comprehend in a single view, including all its natural characteristics.

Landslide. (1) A mass of material that has slipped downhill under the influence of gravity, frequently assisted by water (i.e., when the material is saturated). (2) Rapid movement downslope of a mass of soil, rock, or debris.

Land-use system. The way in which land is used particular group of people with a specified area.

Land-utilisation index. It is defined as the number of days during which the crops occupy the land during a year divided by 365 and can be expressed as a fraction or as a percentage.

Layer charge. Magnitude of charge per formula unit balanced by ions of opposite charge external to the unit layer.

Layer silicate minerals. Minerals with the sheet silicate structures of the phyllosilicates.

Leaching fraction or requirement. The leaching fraction necessary to keep soil salinity, chloride, or sodium (the choice being that which is most demanding) from exceeding a tolerance level of the crop in question. It applies to steady-state or long-term average conditions.

Leaching of metals by microbes. It is only the last two decades that people came to know the important role of microbes in hydrometallurgical industries for the extraction of metals from their ore. They hydrometallurgical method of leaching of metals from low grade ores by passing through stacks of ores has been known from ancient times. The sulphide metals are oxidized to sulphate and sulphuric acid. The rate of this process can be increased by the activities of iron oxidizing bacteria. Biological leaching is brought about by the bacteria like *Thiobacillus thiooxidans, T. ferrooxidans, Ferrobacillus ferrooxidans, F. sulfooxidans,* etc. These bacteria have been reported to be helpful in the leaching of copper, nickel, molybdenum, zinc, arsenic, antimony and germanium from their sulphates.

Leaching. Removal of plant nutrients in solution by the passage of water through soil. This is one of the ways in which plant nutrients are lost from the soil. Among the major nutrients, nitrogen is lost in large quantities by leaching

Leaching. The removal of soluble materials from the upper soil layer by water moving vertically down

Legend. Soil map units are arranged in many ways. Any organized list of map units is called a legend. There may be several kinds of legends, based on nomenclature of map units.

Legume. A member of the legume or pulse family, *Leguminosae.* Includes many valuable food and forage species, such as peas, beans, peanuts, clovers, alfalfas, sweetclover, lespedezas, vetches, and kudzu, almost all legumes are nitrogen fixing plants.

Legume. Any plant species from the family Leguminosae (e.g., beans, peas) many of which have the ability to fix atmospheric nitrogen via bacteria in or on their roots.

LEIA (low external input agriculture). A strain of traditional, subsistence farming characterized by low incidence of any inputs from the farm/village (in case of total absence LIA) and little or no contact with other then local markets. Such systems may not now be sustainable, owing to increasing population pressure on the natural resource base.

LEISA (low external input and sustainable agriculture). A form of agriculture which makes optimal use of locally available natural and human resources (such as climate, landscape, soil, water vegetation, local crops and animals, labor, local skills, and indigenous knowledge) and which is economically feasible, ecologically sound, culturally adapted, and socially just. The use of external inputs such as mineral fertilizers, pesticides, hybrid seeds, and machinery is not excluded, but is seen as complementary to the use of local resources and has to meet the above-mentioned criteria of sustainability.

Lichen. A symbiotic relationship of a fungus and an alga whereby the fungus supplies water and dissolved nutrients and the alga photosynthesizes carbohydrates and fixes nitrogen. Lichens colonize bare minerals, rocks, and large trees, a first step in rock weathering and soil formation.

Ligand. An organic molecule that can bond to various metals through two or more bonds. The ligand-metal is called a *chelate.*

Light soil. A term used for sandy or coarse-textured soil. Leaching of nutrients is more in light soils, as such it is advisable to apply nitrogenous fertilisers in small doses. Wherever practicable, green manuring should be done to improve the physical condition of a light soil.

Lime (calcium) requirement. The amount of agricultural limestone required per acre to a soil depth of 15 cm (6 in.) or on about 910,000 kg (2 million pounds) of soil to raise the pH of the soil to a desired value under field conditions.

Lime chlorosis. A yellowing of foliage commonly occurring in soils high in lime or irrigated with water high in lime. It is caused by a deficiency of available iron, zinc or copper because insoluble calcium compounds are readily formed with these minerals.

Liquid limit. The minimum water mass content at which a small sample of soil will barely flow under a standard treatment. Synonymous with upper plastic limit.

Liquid manure. Liquid arising from animal urine and litter juices or from a dung heap.

Lithic contact. A boundary between soil and continuous, coherent, underlying material must be sufficiently coherent to make hand-digging with a spade impractical. If mineral, it must have a hardness of 3 or more (Mohs scale), and gravel size chunks that can be broken out do not disperse with 15 hours' shaking in water sodium hexametaphosphate solution.

Lithosequence. A group of related soils that differ, one from the other, in certain properties primarily as a result of differences in the parent rock as a soil-forming factor.

Loess. Material transported and deposited by wind and consisting of predominantly silt-sized particles.

Low-analysis fertilizer materials. The fertilisers containing a low percentage of plant nutrients. Such fertilisers are Chilean nitrate containing 16 per cent nitrogen and single superphosphate containing 16 to 18 per cent P_2O_5.

Lowland vs. upland soils. Terms commonly used in connection with rice (*Oryza sativa* L.) culture to denote flooded (paddy) vs. unflooded conditions.

Luxury uptake. The absorption by plants of nutrients in excess of their need for growth. Luxury concentrations during early growth may be utilized in later growth.

Lysimeter. (1) A device for measuring percolation and leaching losses from a column of soil under controlled conditions. (2) A device for measuring gains (irrigation, precipitation, and condensation) and losses (evapotranspiration) by a column of soil.

Magnesia (magnesium oxide) (MgO). A product consisting chiefly of the oxide of magnesium.

Magnesium sulphate. ($MgSO_4 7H_2O$) is commonly used to supply magnesium to the soil through fertilizer mixtures or foliar sprays. It contains 16% MgO and 13% S.

Magnetotaxis. Some microorganisms have evolved unusual mechanisms to remain in their optimum environments. One such organism is *Aquaspirillum* (ah-qua-spy-RIL-lum) sp. members of this genus can orient themselves to the earth's magnetic field and migrate into anaerobic sediments when disturbed by turbulence in the water and sediment. Cellular orientation is produced by a magnetic pull that causes individual organisms to move downward. This phenomenon, called *magnetotaxis*, involves structures called *magnetosomes*, which contain iron in the form of magnetite. A mechanism similar to the one observed in bacteria also appears to be responsible for the migration patterns of some animals able to detect and use geomagnetism as a cue for orientation. Magnetite has been detected in honeybees as well as in birds and fish.

Magnitude of Soil Loss. Soil has been described as a complex and valuable resource that must be protected from loss by erosion of cultivated fields. The top 4 to 8 inches (10 to 20 cm) of soil is the most valuable portion for crop production. It is here that most of the nutrients are found and the root systems are most extensive. After the topsoil is gone, crop growth may still be Possible. But this cannot be done as economically as before because the lost humus-enriched layer was the storehouse of most of the essential plant nutrients and the subsoil may have undesirable physical properties for crop production.

Even under good control measures, cultivated sloping land may be expected to erode at a rate of 3 to 4 tons per acre per year (7.4 to 8.6 metric tons/ha). The eroded soil is not necessarily washed or blown completely off the farm but may simply be moved from the high ground to the low ground. Eventually, the high ground becomes much less productive and crops on the low ground may be damaged by the sediment.

Maintenance application. Application fertilizer materials in amounts and at intervals to maintain available soil nutrients at levels necessary to produce a desired yield.

Major fertilizer elements. These elements are nitrogen, phosphorus and potassium.

Manuring. Application of animal dung, compost or other organic material used to fertilize the soil.

Marl. An earthy, unconsolidated deposit formed in freshwater likes, consisting chiefly of calcium carbonate mixed with clay or other Impurities. The clay content of marl may range from around 10% to over 70%.

Marsh. Periodically wet or continually flooded areas with the surface not deeply submerged. Covered dominantly with sedges, cattails, rushes, or other hydrophytic plants. Subclasses include freshwater and saltwater marshes. The differ from bogs in that circulation is good so that there is continuous supply of fresh oxygenated water.

Mass flow (nutrient). The movement of solutes associated with net movement of water.

Mass water content. The water content expressed as the weight of water in a soil divided by the oven-dry weight of soil. When expressed as a percentage, it is the mass water percentage or simply water percentage.

Massive. Lack of soil structure in coherent materials; structureless but holds together.

Matric potential. The amount of work an infinitesimal quantity of water in the soil can do as it moves from the soil to a pool of free water of the same composition and at the same location. This work is less than zero or *negative* work, thus reported in negative values. Matric potential nearly equals water potential in non-salty soils.

Mecron. A unit of linear measurement. One micron = 10^{-6} meter.

Medium-texture. Intermediate between fine-textured and coarse-textured (soils). (It includes the following textural classes: very fine sandy loam, loam, silt loam, and silt.).

Mesic. A soil temperature regime of intermediate ranges (8-15°C).

Metal Toxicity. High metal availability, arising from mining and industrial activities, disposal of sewage sludge or soil acidification, is an increasing problem in agriculture and forestry. Metal toxicity causes multiple, direct and indirect effects in plants which concern practically all physiological functions. The primary toxicity mechanisms of the different metal ions may be as different as their chemical properties, especially valence, ion radius and capacity to form complexes. Nevertheless, an excess of these metal ions or of soluble metal chelates may induce a series of biochemical and physiological alterations in plants which present some common characteristics. Membrane damage, alteration of enzyme activities and the inhibition of root growth are considered characteristic features of heavy metal stress. These may eventually lead to a large range of secondary effects such as disturbance of hormone balance, deficiency of essential nutrients, inhibition of photosynthesis, changes in photoassimilates translocation, alteration of water relations, etc., which further enhance the metal- induced growth reduction. In studies on metal toxicity in plants, the primary toxicity effects these secondary alterations which concern practically all physiological processes. Plant cell membranes are generally considered primary sites of metal injury. In response to metal ion stress, the plasmalemma gets leaky to K^+. Toxic heavy metal concentrations have an effect on K^+ and H^+ extrusion and decreased K^+ content in root cells. Aluminium is found to affect the K^+ concentration mostly in the tip region of pea roots. The K^+ leakage is not believed to be a consequence of a general loss of membrane integrity, or of a general decrease of energy (ATP) available for ion pumping, but due to a direct interaction of the metal ion with the cell surface causing metal - sulfhydryl bonds, which determine the effects on membrane leakage. Analysis of the influences of excess heavy metals on plant water relations have to distinguish between effects on water availability in soils, effects on root growth limiting water uptake and other phytotoxic effects which may influence water relations. Metal toxicity not only affects the length of the primary root, but changes the architecture of the entire root system. Toxic levels of Al, Pb, Cu, Zn, Mn, Cd or Cr have been found to enhance lateral root formation giving rise to a denser and compact root system. Root hair density is generally decreased. These morphogenetic changes indicate that metal toxicity alters the root hormone balance, but the changes of endogenous hormone levels underlying these morphogenetic alterations are virtually unexplored.

Microbial Activities. The pioneer American ecologist F. E. Clements has classified the activities in ecosystems as one of three kinds: action, reaction, or coaction.

Microbial Habitats. Microorganisms have been found in nearly all regions of the earth. In many cases the organisms are only transients in the environment, but some are definitely native microorganisms. These indigenous microorganisms are able to grow during those periods when environmental conditions are appropriate and to survive during periods when conditions for growth are unsuitable. Survival in some cases results from the formation of bacterial endospores or other dormant microbial structures.

Microclimate. (1) The climatic condition of a small area resulting from the modification of the general climatic conditions by local differences in elevation or exposure. (2) The sequence of atmospheric changes within a very small region. (3) Most plant, however, are more greatly affected by the environment closer to the ground which may deviate considerably from that a few meter above (2-3 m) for example natural surfaces around crops, forest, pond etc.

Microorganisms and Plant Growth. Microorganisms make available different kinds of nutrients to plants which are required for their growth. Following are the important roles of microorganisms in plant growth are: (1) certain microorganisms produce growth promoting substances, e.g., auxin and gibberellin which stimulate plant growth. (2) There may be a competition between microorganisms and plants for nutrients available in soil. (3) Certain pathogenic microorganisms excrete some toxic substances which may be harmful to the plants. (4) Viruses-specially bacteriophages also show indirect harmful effect on plant growth by destroying soil bacteria.

Microorganisms in Nutrient Cycling. Soil microorganisms (bacteria, fungi, nematodes, protozoa, actinomycetes, and others) decompose and dispose of plant and animal remains. In the process these organisms form humus, which is a more active component of soil than mineral clays. Microorganisms also perform important steps in various nutrient cycles and in solid, liquid, and gaseous phases of the soil-plant-atmosphere-hydrosphere system. Without these organic processes the cycles would stop, and living organisms, including people, would lose life support. There is a biological rule stating that the smaller the organism, the greater its number and influence. Thus the action of microorganisms in the soil is far more widespread and of greater importance than of insects and rodents.

Microrelief. (1) Small scale, local differences in topography, including mounds, swales, or pits that are usually <1 m in diameter and with elevation differences of up to 2 m. (2) Differences in topography altered by tillage operation, generally over an area of about 1 m^2 with elevation differences of a few centimeters or less.

Micro-sprinkler and micro-sprayer irrigation: This is a combination of sprinkler and drip irrigation. It is so, because water is sprinkled around the root zone of trees with a small sprinkler working under low pressure. Water is given only to root zone area as in case of drip irrigation but not to the entire ground surfaces as done in the case of sprinkler method.

Mineral Nutrient Depletion in Soil and Plant Growth. India has harvested 182 million tonnes of food grains during 1993-94. The increase in per hectare yields of crops is possible through scientific farming, including judicious use of fertilizers. Mineral nutrients play an important role in agricultural production, however, the problem of nutrient depletion with cropping is of paramount importance. Therefore, it is essential to develop those practices which can reduce nutrients depletion and increase crop production. There are three basic processes through which nutrient is depleted in soil, they are (i) Crop removal, (ii) leaching losses and (iii) Loss in gaseous form. *Crop Removal:* Roots absorb large amount of nutrient from soil. The NPK and micronutrient are required for crop growth. With the intensive cropping when more than one crop is grown during a year, the amount of nutrients removed correspondingly increases. Weed control, should therefore, be effective to save nutrient depletion. *Leaching Losses:* Out of major essential nutrients, only nitrogen is subject to leaching, essentially as nitrate due to its readily soluble nature. This kind of leaching depends on texture of soil and intensity of irrigation. Nitrogen losses are higher in light-textured soils and increase with heavy irrigation. Fertilizer management is essential to avoid such depletion. *Losses of phosphorus* is also a serious problem. It is immobile in soils and is subject to loss through erosion and runoff only. Practices which increase organic residues in soil reduce erosion and runoff and thereby, save phosphorus losses. *Losses in Gaseous Forms:* Nitrogen is subject to removal through gaseous forms from soil. Two processes are involved in this loss: (i) volatilization as ammonia and (ii) denitrification (under aerobic conditions, as in rice field).

Mineral Nutrition - Crop Productivity. For good cultivation of plants especially, high priced field crops soil nutrients should be available at appropriate stages of plant development. Unless all the nutrients are present in available forms, crop yields will be limited. Since mostly these nutrients are limited in arable soils under natural conditions, soil fertility can be controlled by supplying as fertilizers, those ions that limit plant growth. Crop yield can be increased by carrying out systematic investigations of fertilizer needs. Generally governmental agencies can provide data regarding schedule of fertilizing the soil, which nutrient has to be applied, when and how much. Thus deficiency of nutrients and as a consequence its negative impact on physiological processes can be avoided. The actual chemical fertilizer requirements of crop plants can be evaluated by exposing the plants to two tests- soil analysis and plant analysis. These two methods are complementary and each has a role in determining the nutrient situation for a given crop. While soil analysis reveals the low, or high status of nutrient supply, plant analysis on the other hand may be helpful in diagnosing a mineral nutrient deficiency. Thus in cases in which visual symptoms alone are not sufficient to pinpoint a deficiency plant analysis can be used to verify and specify deficiency.

Mineral Nutrition. This is a broad field which is concerned with the complex of biosynthetic events by which higher plants produce organic materials from inorganic materials absorbed from their environment. These events include the absorption process itself, the translocation of absorbed mineral nutrients from roots to stems, leaves, fruits and finally conversion of absorbed (inorganic) nutrients to organic form. The nutritional needs of plants were understood when the addition of farmyard manures, plant tissue debris, wood ashes, ground bones, composts, and other materials of plant and animal origin to soils were found to increase their productivity. In the early half of the nineteenth century it was realized that plant growth and development can proceed only when plants are supplied with certain chemical elements referred to as essential elements. These essential elements are mostly derived from mineral constituents of the soil and are absorbed principally by the roots as inorganic ions. Roots are the most important part of the plant as the nutrient uptake takes place mainly through them.

Mineral soil. A soil consisting predominantly of, and having its properties determined predominantly by, mineral matter. Usually contains < 200 g kg^{-1} organic carbon (< 120 to 180 g kg^{-1} if saturated with water), but may contain an organic surface layer up to 30 cm thick.

Mineralization. The conversion of an element from an organic combination to an inorganic form as a result of microbial decomposition.

Minimum tillage. The least amount of tillage required to create a suitable soil condition for seed germination, plant growth, prevention of weeds, and the control of erosion. The concept of minimum tillage was stored in USA during 70's. Research has shown that frequent tillage is rarely beneficial and often detrimental. This led to the development of concept of minimum tillage. Minimum tillage is aimed at reducing tillage to the minimum necessary (generally 10-15% of conventional tillage) for ensuring a good seedbed rapid germination, a satisfactory stand and favourable growing conditions.

Mixed cropping. Cropping system which involves raising of various crops.

Mixed intercropping. Growing two or more crops simultaneously with no distinct row arrangement.

Mixed inter-cropping. Two or more crops are wholly interspersed, maximizing their interaction. Can be considered at the within-species level (variety mixtures) or at the between-species level (grass-clover leys and other gramineous/leguminous species).

Moisture (soil). Soil particle holds water on its surface in the form of a thin film with a certain force. Soil pore space contains both air and water in varying proportions. Soil moisture expressed in percentage on oven-dry basis either on weight basis or on volume basis. The terms soil moisture and soil water are used generally as synonyms.

Moisture-retention curve. (Obsolete) A graph showing many soil moisture percentages versus applied tensions (or suctions, pressures, or soil water potentials).

Monera. (*moan-ee'-ra*) One of five kingdoms for organisms. Those in the monera kingdom have procaryotic cells (do not have genetic material enclosed in a membrane). Includes bacteria, actinomycetes, and cyanobacteria.

Monocropping. The repetitive growing of the same, single crop on the same land.

Monoculture – usually refers to the continuous use of a single crop species over a large area but can also be used with reference to single varieties or even single disease resistance genes. Avoided in organic agriculture.

Monoculture. Only one species is kept (e.g. trout); the fish are generally very demanding in terms of nutrition and environment.

Montmorillonite. An aluminum silicate (smectite) with a layer structure composed of two silica tetrahedral sheets and a shared aluminum and magnesium octahedral sheet. Montmorillonite has a permanent negative charge that attracts interlayer cations that exist in various degrees of hydration thus causing expansion and collapse of the structure (i.e., shrink-swell). Its general formula is $Si_4Al_{1.5}Mg_{0.5}O_{10}(OH)_2Ca_{0.25}$. The calcium is exchangeable. CEC = 80 (50-120) C mol (P^+) Kg^{-1}.

Mor. A type of forest humus in which the Oa horizon is present and in which there is practically no mixing of surface organic matter with mineral soil; that is, the transition from the Oa to A horizon is abrupt. (Sometimes differentiated into thick mor, thin mor, or felty mor.)

Moraine. An accumulation of glacial drift formed chiefly by the direct deposition from glacial ice. Examples are ground, lateral, recessional, and terminal moraines.

Morphology. The physical nature of the soil as exhibited by horizon differences and such physical properties as texture, porosity, and color.

Most profitable rate or MPR or optimum dose. Refers to the amount of fertilizer that produces the greatest profit per acre. If. More than the optimum dose is given, the profit will be sacrificed.

Mottles. Spots or blotches of different color or shades of color interspersed with the dominant color.

Muck soil. (1) A soil containing between 10-40% of organic matter. (2) An organic soil in which the plant residues have been altered beyond recognition. The soils are generally blue due to presence of ferrous iron under anaerobic condition.

Muck. Highly decomposed organic material in which the original plant parts are not recognizable.

Mulch farming. A system of tillage and planting operations resulting in minimum incorporation of plant residues or other mulch into the soil surface.

Mulch. A natural or artificial layer of plant residue or other materials, such as crop residues, leaves, sand, plastic, or paper, on the soil surface.

Mulch. Plant or non-living materials used to cover the soil surface with the object of protecting the soil from the impact of rainfall, controlling weeds or moisture loss and in some cases, fertilizing the soil.

Multiple cropping. Growing two or more crops in the same field in one year at the same time or one after the other.

Munsell color notation system. A system that specifies the relative degrees of the three simple variables of color: hue, value, and chroma. For example: 10YR 6/4 is a color with hue 10YR, value 6, and chroma 4.

Mutualism. An interaction between two species in which both species benefit. Mutualism, like the other forms of symbiosis, occurs widely among most of the principal groups of animals, plants, and microorganisms and includes an amazing range of physiological and behavioral adjustments. Mutualism is usually essential to the survival of the organisms involved because of the dependence they have developed for one another. In many situations, one partner cannot survive without the other.

Nala bunding. Construction of bunds of suitable dimensions across the nala or gullies to hold the maximum run-off water to create flooding of the upstream area temporarily for some days or weeks, with surplussing arrangements at suitable intervals to drain the water.

Nala. A rivulet or streams of shallow depth.

Natural erosion. Wearing away of the earth's surface by water, ice, or other natural agents under natural environmental conditions of climatic, vegetation and so on, undisturbed by humans.

Natural Inputs: Natural is commonly understood as anything that is non-synthetic.

Natural Nutrients. Plants may take forty or more different elements from the soil, only sixteen of which are considered essential for complete plant development. Most of the plant tissue is made up of carbon (C), hydrogen (H), and oxygen (O) that come from carbon dioxide in the air and water in the soil. The remaining thirteen elements are normally absorbed in ionic form from the soil, although small amounts can be taken in through the leaves if placed there in solution by precipitation or spray irrigation.

Natural vegetation. The vegetative cover that would exist in a given area without interference from humans.

Nematode. Small (0.5-1.5 mm long) nonsegmented worms. Some are parasitic on many crop plant roots; others parasitize destructive insects.

Neutral soil. A soil in which the surface layer, at least in the tillage zone, is in the pH range 6.6 to 7.3.

Neutralising index. It is also called effective calcium carbonate (ECC). The neutralising index of a liming material is defined as the product of its calcium carbonate equivalent (CCE) and the fineness factor. The fineness factor is the sum of the products of the percentage of material in each of the three size fractions (60, 20 and 8 mesh) multiplied by the appropriate effectiveness factor. N. I. = CCE x fineness factor.

Niche. (1) The particular role that a given species plays in the ecosystem; (2) the physical space occupied by an organism.

Night soil. Night soil is human excreta, solid and liquid In India, it is directly applied to the soil to a limited extent but converted mainly as town compost. In cities which have sewage facilities, sewage water and sludge are used directly to raise crops. On an average night-soil contains 5.5 per cent nitrogen, 4.0 per cent phosphorus (P_2O_5) and 2.0 per cent potash (K_2O) on oven dry basis.

Nitrate depression in soil. It is a relationship between organic residues of wider C:N ratio and the presence of nitrate nitrogen in soil resulting nitrate nitrogen depression due to higher rate of decomposition of organic residues. As long as the C/N ratio is wide, the general purpose decay organisms are dominant and at the same time nitrification process blocks due to inactiveness of the nitrifies which results depression in the amount of nitrate nitrogen in the soil.

Nitrate reduction (biological). The process whereby nitrate is reduced by plants and microorganisms to ammonium for cell synthesis (nitrate assimilation, assimilatory nitrate reduction) or to various lower oxidation states (N_2, N_2O, NO, NO_2^-) by bacteria using nitrate as the terminal electron acceptor in anaerobic respiration (respiratory nitrate reduction, dissimilatory nitrate reduction, denitrification).

Nitrate toxicity of forage. Forage containing more than 6000 ppm of nitrate (NO_3) may be toxic to cattle. Causes of high nitrate include drought, cool weather, cloudy weather, and heavier than recommended applications of manures, sludges, and nitrogen fertilizers.

Nitrogen activity. It indicates how readily the nitrogen in a fertilizer or manure would become available to plants after application to soil. It, therefore, offers a qualitative distinction between slow and quick acting nitrogenous fertilisers and manures.

Nitrogen Cycle. Nitrogen is one of the chief and important constituents of protein and nucleic acid molecules that play the basic role in cell metabolism, growth, reproduction and transmission of heritable characters. Therefore, without a constant supply of this unavoidable element, life can not go on. Gaseous nitrogen is the most abundant element of atmosphere (78.08% v/v), and seems to have a highly complex nutrient cycle in the terrestrial and aquatic ecosystems. There is an interrelationship between different elements required for plant growth in environment. These all are affected by microbes. Atmospheric nitrogen is chemically inert. Therefore, it can be used, as such by most of the living organisms. The common source of nitrogen in the nature is the nitrate ions (NO_3). The nitrate ion is often absorbed by the plants as a mineral raw material from the environment. The absorbed nitrate ions are then converted into amino groups ($-NH_2$). The conversion of $-NO_3$ to NH_2 group is only possible by plants and not by animals. Minerals accumulation in soil, dissolution of rock by water, addition of fertilizers in soil are also the sources of NO_3 in the soil. Apart from these sources, atmospheric nitrogen and oxygen combine together in presence of lightening which come down with the help of rain-water to the ground, and serve as NO_3 source on the earth. Absorbed nitrogen (in the form of NO_3) incorporates into organic structures of plants which remain intact until death. Animals also obtain nitrogen by eating plants. Finally, both plants and animals die or decay. Through gradual decay of all nitrogen of dead plants and animals, it gets concentrated into ammonia (NH_3). The so forming NH_3 is then utilized as a source of nutrition for nitrifying bacteria. Usually, these bacteria are of two kinds: 1. those bacteria which absorb ammonia and convert it into nitrite ions i.e., $NH_3 \rightarrow NO_2$. 2. Those bacteria which absorb nitrite ions and convert it into nitrate i.e., $NO_2 \rightarrow NO_3$. Thus, the environment again gets NO_3 by the activities of these nitrifying bacteria. The available NO_3 in the environment acts upon denitrifying bacteria which converts it into molecular, atmospheric nitrogen. The free molecular nitrogen is then, utilized by nitrogen fixing organisms in which few species of bacteria and blue green algae have been included. They are often present in soil and water. They absorb nitrogen and incorporate it into ammonia and protein. The process is called *Nitrogen fixation*.

Nitrogen fixation. Nitrogen fixation has always been a subject of great importance in biology, but since two decades the recent biotechnological development has lent its practical urgency. Nitrogenous fertilizers have become more and more expensive in terms of fossil energy, and hard cash; the population pressure have increased the need for high protein plant food. Biotechnologists are actively engaged in improving the efficiency of microorganisms to fix atmospheric nitrogen in fields. This will improve the functioning of the nitrogen cycle in biosphere which will ultimately increase plant productivity. The programme of improvement also includes the creation of nitrogen-fixing symbiotic association other than those existing between leguminous plants and bacteria of the genus *Rhizobium*. This will help to improve agricultural production.

Nodules. (1) Specialized tissue enlargements, or swellings, on the roots or leaves of plants, such as are caused by nitrogen-fixing microorganisms, (2) globules with an undifferentiated fabric; in the context undifferentiated fabric includes recognizable rock and soil fabrics.

NO-tillage. (No-till) A method of growing crops that involves no seedbed preparation other than opening a small slit or punching a hole into the soil in order to place the seed at the intended depth. There is no cultivation; no chemical for weed and pest control is used.

Nutrient antagonism. The depressing effect caused by one or more plant nutrients on the uptake and availability of another.

Nutrient balance. An as yet undefined ratio among concentrations of nutrients necessary for maximum growth rate and yield. An imbalance results when one or more nutrients are present in either or excess supply.

Nutrient. Single or compound element used as raw material for the development and growth of living organisms. For example, carbon, oxygen, phosphorus. The nutrients drawn from the soil by plants are often classified according to their role in the yield of crops into: indispensable or essential elements (microelements): N, P, K. Secondary elements (meso elements): Ca, Mg, Mn, S, Fe. Minor elements (microelements) or oligoelementos: Zn, Cu, B, Mo, Co.

Oilcakes. When oil is extracted from oilseeds, the remaining solid portion is the oilcakes are of two types, edible oilcakes which can be safely fed to livestock, and non-edible oilcakes which are not fit for feeding to livestock. Oilcakes are added to the soil as concentrated organic manures. They supply organic matter and all the three major plant nutrients (N, P_2O_5, and K_2O) but mostly they supply nitrogen.

Open-formula mixed fertilisers. Fertilizer mixtures of which the composition and the complete make-up are disclosed by manufacturers. Thus the fertilizer grade, the straight fertilisers and the filler used are disclosed on each bag. Cultivators and extension workers are in a better position to know the quantity and quality of nutrients contained in such a fertilizer mixture.

Organic agriculture. A production system which avoids or largely excludes the use of synthetically compounded fertilizers, pesticides, growth regulators, and livestock feed additives. To the maximum extent feasible, organic farming systems rely on crop rotations, crop residues, animal manures, legumes, green manures, off-farm organic wastes, and aspects of biological pest control to maintain soil productivity and tilth, to supply plant nutrients, and to control insects, weeds, and other pests. The concept of the soil as a living system that develops the activities of beneficial organisms is central to this definition.

Organic matter. The term *soil organic matter* is generally used to represent the organic constituents in the soil, excluding undecayed plant and animal tissue, their partial decomposition products, and the soil biomass. The end product of organic matter decomposition is humus.

Organic phosphorus. Phosphorus present as a part of organic compounds, such as Glycerophosphoric acid, inositol phosphoric acid, and cytidylic acid.

Organic soils. (Histosols) (As used in the Soil Classification System of the National

Cooperative Soil Survey in the United States). Saturated with water for prolonged periods unless artificially drained and having at least 12 or 18 percent organic carbon by weight, depending on the mineral fraction and the kind of organic materials. Never saturated with water for more than a few days and hiving 20 percent or more organic carbon by weight.

Osmotic potential. (Solute potential) The amount of work an infinitesimal quantity of water will do in moving from a pool of free water the same composition as the soil water to a pool of pure water at the same location. The effect of dissolved is not involved. Except in salty soils, the osmotic potential is usually very small.

Osmotic pressure. The pressure to which a pool of water, identical in composition to the soil water, must be subjected in order to be in equilibrium, through a semipermeable membrane, with a pool of pure water (semipermeable means permeable only to water). May be identified with the osmotic potential defined above.

Outwash. Stratified glacial drift, often high in sands and gravels, which is deposited by meltwater near glaciers.

Oven-dry soil. Soil that has been dried at 105°C until it reaches constant weight (oDwt).

Overburden. (1) Material recently deposited by a transportation mode that occurs immediately superjacent to the surface horizon of a contemporaneous soil. (2) A term used to designated disturbed or undisturbed material of any nature, consolidated or unconsolidated, that overlies a deposit of useful materials, ores, lignites, or coals, especially those deposits mined from the surface by open cuts.

Oxic horizon. A diagnostic horizon common to Oxisols. It is thick, low in weather able minerals, and with low CEC clays.

Oxidation. Combination with oxygen; addition of oxygen or other atom or group; removal of hydrogen or other atom or group.

Oxide clays. To this point we have considered only silicate clays, but oxide and hydrated oxide clay minerals are also present in soils. Normally, these are oxides of iron and aluminum and are found most abundantly in soils formed from parent materials rich in iron and aluminum in tropical and subtropical regions where weathering has removed much of the silica from the clay fraction. Oxide clays have little or no crystallinity and very low capacity to hold plant nutrients.

Oxisols. A soil order (highest category) in the US Comprehensive System of Soil Classification. The Oxisols are highly weathered soils, rich in oxides of iron and aluminium but depleted of silica.

P_2O_5. Phosphorus pentoxide; when phosphorus burns into oxygen it forms P_2O_5. Designation on the fertilizer label that denotes the percentage of available phosphate.

Paleosol, buried. A soil formed on a landscape during the geological past and subsequently buried by sedimentation.

Paleosol, exhumed. A formerly buried paleosol that has been exposed on the landscape by the erosive stripping of an overlying mantle of sediment.

Paleosols. Soils formed under (ancient) climates different from those that now exist.

Pan, genetic. A natural subsurface soil layer of low or very low permeability, with a high concentration of small particles, and differing in certain physical and chemical properties from the soil immediately above or below the pan.

Pan, pressure or induced. A subsurface horizon or soil layer having a higher bulk density and a lower total porosity than the soil directly above or below it, as a result of pressure that has been applied by normal tillage operations or by other artificial means. Frequently referred to as plowpan, plowsole, or traffic pan.

Pans. Horizons or layers, in soils, that are strongly compacted, indurated, or very high in clay content.

Paralithic contact. Similar to a lithic contact except that it is softer, can be dug with difficulty with a spade, if a single mineral has hardness < 3 (mohs scale), and gravel size chunks that can be broken out will partially disperse within 15 hours' shaking in water or sodium hexametaphosphate solution.

Parasitism. An association whereby one organism (parasite) lives in or on another organism (host) and benefits at the expense of the host.

Parent material. (Soils) The unconsolidated, chemically weathered mineral or organic matter from which the A and B horizons (solum) of soils may have developed by pedogenic processes. The C horizon may or may not consist of materials similar to those from which the A and B horizons developed.

Participatory rural appraisal (PRA). A set of facilitative and participatory technologies developed in the late 1980s by researchers and NGOs, to build upon local people's capabilities and to empower local people in the process.

Participatory Technology Development (PTD). Technological development building on existing farmers' knowledge and enhancing farmers' capacities to experiment and to communicate their findings.

Particle density. The density of the soil particles, the dry mass of the particles being divided by the solid (not bulk) volume of the particles, in contrast with bulk density. Units are g cm^{-3}.

Particle size analysis (granulometry) by sieving. The division of a sample by sieving into size fractions.

Particle size. The effective diameter of a particle measured by sedimentation, sieving, or micrometric methods.

Particle-size classes for family groupings. (as used in the Soil Classification System of the National Cooperative Soil Survey in the United States) Various particle-size classes are applied to soils.

Particle-size distribution. The fractions of the various soil separated in a soil sample, often expressed as mass percentages.

Pascal. A unit of pressure equal to 1 newton per square meter.

Pasture. A portion of land covered with grasses legume mixture suitable for grazing.

Peat soil. An organic soil containing % organic matter. Used in the United States to refer to the stage of decomposition of the organic matter, "peat" referring to the slightly decomposed materials.

Peat. Residual matter from plants grown and decayed in almost permanently waterlogged conditions and which may contain a limited quantity of naturally occurring mineral material.

Ped. A unit of soil structure such as an aggregate, crumb, prism, block, or granule, formed by natural processes (in contrast with a clod, which is formed artificially).

Pediment. The footslope component of an erosional slope; geomorphologically an erosional surface that lies at the foot of a receded slope, with underlying rocks or sediments that also underline the upland, which is barren of, or mantled with sediment, and which normally has a concave upward profile.

Pediplain. A geomorphic term for an outwash plain landform.

Pedogenesis. Caused by the natural processes of soil development, Synonyms: soil genesis, soil development, and soil formation.

Pedology. Pedology is that branch of soil science which deals with the survey, genesis and classification of soils for land use-planning.

Pedon. A three-dimensional body of soil with lateral dimensions large enough to permit the study of horizon shapes and relations. Its area ranges from 1 to 10 m^2. Where horizons are includes one-half of the cycle. Where the cycle is < 2 m, or all horizons are continuous and of uniform thickness, the pedon has an area of approximately 1 m^2. If the horizons are cyclic but recur at intervals > 7 m, the pedon reverts to the 1-m^2 size, and more than one soil will usually be represented in each cycle.

Peneplain. A "near" plain, usually the alluvium-covered low areas that have almost formed a level plain near the level of water drainage.

Penetrability. The ease with which a probe can be pushed into the soil. (May be expressed in units of distance, speed, force, or work, depending on the type of penetrometer used.)

Percolation, soil water. A qualitative term applying to the downward flow of water in saturated or nearly saturated soil in response to the pull of gravity.

Percolation, soil water. The downward movement of water through soil. Especially, the downward flow of water in saturated or nearly saturated soil at hydraulic gradients of the order of 1.0 or less.

Pergelic. A soil temperature regime (<0°C) averaging subzero temperature (permafrost).

Permafrost. (1) Permanently frozen material underlying the solum. (2) A perennially frozen soil horizon.

Permeability, soil. (1) The ease with which gases, liquids, or plant roots penetrate or pass through a bulk mass of soil or a layer of soil. Since different soil horizons vary in permeability, the particular horizon under question should be designated. (2) The property of a porous medium itself that expresses the ease with which gases, liquids, or other substances can flow through it, and is the same as intrinsic permeability k.

Pesticidal Pollution of Soil. ***Insecticides.*** Pesticides are now extremely used to control various pests which are harmful to the crops raised by man for food, feed and fibre production. They have played an important role in boosting agricultural production during the past from decades. However, most of the chlorinated pesticides are non-bio degradable and leave residues which are detrimental to human beings, animals and the environment. Some of the harmful pesticides have either been banned or restricted specially in developed nations. Residue studies in India have also revealed wide spread contamination of the food products and the environment as a whole. The presence of residues of insecticides in food commodities and other components of the environment is a matter of serious concern. Even small quantities of the residues ingested daily alongwith food can build up to high levels in body fat. There is, therefore, a constant fear that society is being slowly poisoned by the intake of food contaminated with pesticide residues. In case of wheat grains in India, main contaminants have been DDT, HCH and to some extent other organochlorine insecticides such as aldrin and heptachlor. A few samples from Ludhiana and Mysore were found to contain malathion ranging from 0.02 to 11.92 ppm which appears to be a case of direct spray of malathion on the gunny bags/grains in the storage godowns. The samples analysed during 1960s and 1970s contained relatively more amount of DDT (Tr-9 ppm), HCH (Tr-13.9 ppm) whereas samples analysed after 1980 showed contamination level to be much lower than earlier studies (DDT, Tr-1.75 ppm; HCH, Tr-3.92 ppm). The seasonal vegetables consisting of green leafy vegetables, tomato, potato, cauliflower, okra, cabbage, etc., have been analysed mainly for residue content of organochlorine insecticides such as DDT, HCH, aldrin, dieldrin, heptachlor and indene. Most of the samples have been found to be contaminated with DDT, HCH or both. However, a few of them showed the presence of aldrin, dieldrin, endrin, heptachlor and endosulfan. The contamination of bovine milk with residues of HCH has been reported to be quite alarming. About 90 per cent of 980 samples of milk samples analysed in the states of Andhra Pradesh, Maharashtra, Delhi and Punjab showed HCH residues at levels ranging from 0.12 to 40 ppm on fat basis. Samples analysed under the ICMR project also reveal that there is a widespread contamination of HCH residues. Human breast milk is also known to contain residues of insecticides, most commonly those of DDT and HCH. The mean level of total DDT, HCH and aldrin was 0.52, 0.2, 0.04 mg/kg in 50 samples collected from rural areas in two districts of the Punjab (India) also showed high levels of residues of DDT and HCH. *Fungicides.* Agriculture is now being transformed into dynamic productive and profitable sector because of the ever increasing demand for food and feed products. In order to cope up with the food production targets, more emphasis is likely to be given to agricultural inputs. By the end of the century, the average agricultural production in the developing countries shall have to be increased by 30% over the present level. The use of chemicals as plant protection practices acts as an insurance against the risks of loosing not only the crop yield but also the money spent on advanced technologies used as inputs in attaining higher production. Breakthroughs in the development of different agrochemicals and their widespread use at the farm level indeed have helped to boost the agricultural production. In all countries, fungicides constitute 10 per cent of the total consumption of pesticides in India and 17 per cent in the world.

Petrocalcic horizon. A diagnostic horizon of carbonate accumulation and hard cementation.

Petroferric contact. A boundary between soil and a continuous layer of indurated soil in which iron is an important cement. Contains little or no organic matter.

pH meter. The pH meter consists of two electrodes namely, (1) glass electrode and (2) calomel electrode. The glass electrode. The glass electrode consists of a very thin bulb (about 0.1 mm thick). The bulb is made of high conductivity glass which is sensitive to pH change. The glass electrode consists of silver chloride, silver and 0.1 N HCl. Its tip is covered by a special glass surface. It allows only hydrogen ions. Each of these B components exist in ionized state. Different type of glass electrodes are available. Reasonably rugged glass electrodes which operate from 0-80°C and upto pH of about 10. In more recently, combination electrodes (glass and reference electrodes) can be used. Combination electrodes may be used to determine the pH of small volume of solutions. The glass electrode can be used for almost any solution except that with is very acidic or basic. The other electrode is a calomel electrode. Here also, a series of ionizations take place resulting in the generation of electron but not hydrogen ion. If these two electrodes (glass and calomel electrode) are connected, the electrons will move from the more positive electrode to other. The electrodes are exposed to solution containing high concentration will not respond directly. But the equilibrium of reaction (3) in glass electrode will be affected, this in turn will trigger changes in reactions of (1) and (2). The ultimate effect of the increase in H concentration will change the electron concentration in reaction (1) of the glass electrode. The resulting current flow will be a measure of hydrogen ion concentration to the medium. It is seen that the measurement of the current between the glass and calomel electrodes will be a measure of hydrogen ion concentration in the solution. In pH meter, the current is fed into a calibrated dial. In such a way that the dial reading gives the pH of the solution.

pH, soil. The negative logarithm of the hydrogen ion activity of a soil. The degree of acidity (or alkalinity) of a soil as determined by means of a glass, quinhydrone, or other suitable electrode or indicator at a specified moisture content or soil-water ratio, and expressed in terms of the pH scale.

pH. *(p = Power, H = hydrogen ion concentration).* The term pH was introduced by a Danish biochemist, Soren Sorensen in 1909. The term pH may be defined as the negative logarithm of hydrogen ion concentration (moles/litre). In other words, pH is the negative logarithm of the hydrogen ion activity or function. pH may be expressed as pH = log $1/aH^+$ = -log aH^+; aH^+ is the activity of H^+.A term used to indicate the degree of acidity or alkalinity. Technically, pH is the common logarithm of the reciprocal of the hydrogen ion concentration of a solution. A pH of 7.0 indicates precise neutrality, higher values indicate increasing alkalinity, and lower values indicate increasing acidity. ***Measurement of pH.*** The measurement and study of changes in the hydrogen ion concentration by using indicators, comparators and pH meters. ***pH indicators.*** These are organic compounds of natural or synthetic origin. The colour is dependent upon the pH of the solution. Some indicators are commonly used to determine the acidity such as Bromothymol blue, phenolphthalein and methyl orange. These indicators when added to a solution respond to the H^+ ions present in the medium by changing colour. This is because the indicators are chemicals which are capable of ionizing in solution. *Comparators.* Various comparators like Lovibond comparator and Hellige comparator are available. Discs with different shades of colours for different pH values are supplied. The test solution is treated with the prescribed quantity of indicator specified or supplied. The colour developed is matched with the colours in the disc. The pH for the colour in the disc matching with the colour of the experimental solution is the pH of the test solution.

pH. The negative logarithm of hydrogen ion activity (concentration) of a soil. A pH of 7 is neutral; less than 7 is acidic; higher than 7 is alkaline.

Phase, soil. A subdivision of soil taxon, usually a soil series or other unit or classification, based on characteristics that affect the use and management of the soil; phases include degree of slope, degree of erosion, content of stones, and texture of the surface.

pH-dependent charge. The portion of the cation or anion exchange capacity that varies with pH.

Phytoliths. Inorganic bodies derived from replacement of plant cells; they are usually opaline.

Pica. An unnatural appetite of animals and people that results in eating soil, bark, bones, and/or hair.

Plain. A flat, undulating, or even rolling area, larger or smaller, that includes few prominent hills or valleys that usually is at low elevation in reference to surrounding areas, and that may have considerable overall slope and local relief.

Plane of atoms. A flat (planar) array of one atomic thickness. Example: in soil mineralogy, a plane of basal oxygen atoms within a tetrahedral sheet.

Plant analysis. Analytical procedures to determine the nutrient content of plants or plant parts.

Plant food ratio. The ratio of the numbers of fertilizer units in a given mass of fertilizer expressed in the order N-P-K. This ratio may be based on nitrogen as unity.

Plant roots and the rhizosphere. The rhizosphere is the volume of soil, water, and air (with associated organisms) immediately around the root of a plant. The roots may exude carbon (CO_2) and even oxygen, The CO_2 makes the soil solution slightly acid so that plant nutrients are dissolved from fine rock particles. The oxygen may favor precipitation of iron to form a film in the soil near the root. Outer layers of the root may slough off, enriching the soil with organic matter.

Plant tissue test. This refers to the rapid chemical testing of plant tissue, which is analogous to chemical testing and diagnosis in medicine. The basis of rapid chemical testing of plant tissue is the colorimetric determination of the levels of nitrate. Phosphorus and potassium determination of the levels of nitrate, phosphorus and potassium in the sap of fresh plant tissue. Sometimes, tests are also carried out for magnesium, calcium, manganese and zinc. Plant tissue tests, though only qualitative, are highly useful guides to interpretation of the relative supply of nutrients actually being taken by the plant. The underlying assumption is that an adequate supply of the elements is indicated by abundance in the plant sap.

Plastic limit. The minimum water mass content at which a small sample of soil material can be deformed without rupture. Synonymous with lower plastic limit.

Plastic soil. A soil capable of being molded or deformed continuously and permanently, by relatively moderate pressure, into various shapes.

Playa. An ephemerally flooded, vegetatively barren area on a basin floor that is veneered with fine-textured sediment and acts as a temporary or as the final sink for drainage water.

Plinthite. A nonindurated mixture of iron and aluminum oxides, clay, quartz, and other diluents that commonly occurs as red soil mottles usually arranged in platy, polygonal, or reticulate patterns. Plinthite changes irreversibly to ironstone hardpans or irregular ironstone aggregates on exposure to repeated cycles of wetting and drying.

Pocosin. A swamp, usually containing organic soil, and partly or completely enclosed by a sandy rim (e.g., the Carolina Bays of the southeastern United States).

Polder. A low-lying land area normally under water but reclaimed from the water and protected from resubmission by dikes, as in the Netherlands.

Pollution. Environmental pollution is a global problem. The decline in environmental quality as a consequence of pollution is evidenced by loss of vegetation cover and biological diversity, excessive concentration of harmful chemicals in the ambient atmosphere and in food grains, growing risks of environmental accidents and threats to *life support system*. Those substances which cause pollution are known as pollutants. Pollutant may be defined as any chemical (radionuclide, organo-phosphorus compound or trace gases) or geo-chemical substance (dust, sediment, grit, etc), biotic component or product (pollens or products of microbial activity), or physical agent (heat, sound, etc.) that released intentionally or inadvertently by man into the environment in such concentration that may have adverse, harmful or unpleasant and inconvenient effects (Agarwal, 1993). It is also defined as any solid, liquid or gaseous substance present in such concentration as may be or tend to be injurious to environment [Indian Environment (protection) Act, 1986]. The term pollutant is usually applied to non-living, man-made substance or other nuisances or excess in a specific location. The important pollutants are: Gaseous pollutants [Oxides of nitrogen, sulphur compounds (SO_2, H_2S), carbon compounds (CO_2, CO), ozone, halogens (chlorine, bromine, iodine, etc.], and chlorofluorocarbons, fluoride compounds, metals (Mercury, lead, iron, zinc, nickel, tin, cadmium, etc.), agricultural pollutants (pesticides, herbicides, fungicides and fertilizers), complex organic pollutants (benzene, benzopyrenes, acetic acid, ether, etc.), biotic component (pollens), deposited matter (soot, smoke, tar, dust, grit, etc.), solid waste, radioactive waste, noise, heat, etc. Soot is a black substance, essentially carbon from the incomplete combustion of wood, coal, natural gas, etc., as deposited on the inside of chimneys and other surfaces. Detergents, pesticides and biocides, chlorofluorocarbons, plastics and plasticizers, solvents, fuel, paints, dyes, medicines, fuel and food additives, etc., are some examples of the multiplicity of chemical products made and disseminated for the benefit of man. All these have the inherent capacity to disturb the life support system.

Polyculture. Usually refers to the use of a range of different species over time on a single management unit, which may be a farm or a field. **a. Sole (single) cropping:** different crops are used but are managed separately.

Polyculture. Various species with differing nutritional and environmental requirements are kept together; competition should be avoided.

Pore-size distribution. The volume fractions of the various size ranges of pores in a soil, expressed as percentages of the soil bulk volume (soil particles plus pores).

Potash. Term used to refer to potassium or potassium fertilizers and usually designated as K_2O.

Potential acidity. It is also called reserve or exchange acidity. It is defined as the acidity develops due to adsorbed hydrogen (H^+) and aluminium (Al^{3+}) on the soil colloids. The magnitude of this potential acidity is very high.

Powder. Solid substance in the form of very fine particle.

Precipitated bone phosphate. A by-product from the manufacture of glue from bones and is obtained by neutralizing the hydrochloric acid solution of processed bone with lime. The phosphoric acid is chiefly present as dicalcium phosphate.

Precipitation interception. The stopping, interrupting, and temporary holding of precipitation in any form by a vegetative canopy or vegetation residue.

Predation. A relationship between two organisms whereby one organism (predator) engulfs and digests the second organism (prey).

Primary nutrients. Refers to N, P, and K in fertilizers.

Primary plant foods. The primary plant, foods are nitrogen (N). phosphoric acid (P_2O_5) and potash (K_2O). These nutrients are also called major plant nutrients.

Prime agricultural lands. Soils capable of the highest production levels with the least hazard of erosion or damage. Mostly deep soils with adequate natural or developed water that are nearly level, without rocks, easily managed, and in productive climates.

Prokaryote. (pro-car'-ry-ot) Organisms whose cell types do not have nuclear materials enclosed in a nuclear membrane (bacteria, cyanobacteria, actinomycetes).

Productivity, soil. The capacity of a soil to produce a specified plant or sequence of plants under a specified system management. Productivity emphasizes the capacity of soil to produce plant products and should be expressed in terms of yield.

Profile, soil. A vertical section of the soil through all its horizons and extending into the C horizon.

Protista. One of five kingdoms into which all organisms are fitted. Protista includes mostly one-called organisms with eukaryotic cells: amoeba, sporozoan, slime molds, ciliates, and flagellates.

Pruning. Cutting back plant growth, including side branches or roots.

Puddling. Making a soil impermeable by manipulating and compacting it in standing water, which reduces its apparent specific volume thus facilitates transplanting.

Pyroclastics. A general term applied to detrital volcanic material that have been explosively or aerially ejected from a volcanic vent.

Quantitative measurement. No phase of the study of soil science is more far reaching than the exchange complex. The amount of exchangeable cation is expressed in terms of the equivalent weight (gram-atomic weight divided by the valence) of cations per 100 g of soil. The equivalent weight of ions varies according to atomic weights; but all are compared to hydrogen, which weighs 1 g/equivalent. If all exchange sites in 100 g of good agricultural soil in the temperate region were occupied by Ca^{++} ions, about 0.4 g of exchangeable Ca would be present.* This would represent about 8000 lb (3600 kg) of Ca^{++} in 1 acre (0.4 ha) of soil to a depth of 6 in. (15 cm).+

Quick soil tests. These are simple and rapid chemical tests of soils designed to give an approximation of the nutrients available to plants. Interpretation of results depends upon previous standardization of soil tests with responses to fertilisers. It varies for different kinds of soil.

Rainfall erosion index. A measure of the erosive potential of a specific rainfall event. In the universal soil-loss equation it is defined as the product of two rainstorm characteristics: total kinetic energy of the storm x its maximum 30-min intensity.

Reduced tillage. Less tillage than in "conventional" tillage.

Reduction. Atoms or ions that gain electrons. Often associated with very wet, waterlogged soil conditions.

Regenerative agriculture. Aims to restore an agricultural system, especially after the decline to a low productivity level, through the use of resource efficient and ecologically sound farming practices.

Remote sensing. In the broadest sense, the measurement or acquisition of information of some property of an object or phenomenon, by a recording device that is not in physical or intimate contact with the object or phenomenon under study [e.g., the utilization at a distance (as from aircraft, spacecraft, or ship) of any device and its attendant display for gathering information pertinent to the environment, such as measurements of force fields, electromagnetic radiation, or acoustic energy]. The technique employs such devices as the camera, lasers, radio-frequency receivers, radar systems, sonar, seismographs, gravimeters, magnetometers, and scintillation counters.

Residual effect of manure. This refers to the residual beneficial effect of application of farmyard manure on the succeeding crops. This beneficial effect is due to improvement in the physical condition of the soil, and also due to the unutilized plant nutrients. It is estimated that only one third of the nitrogen present in farmyard manure is utilized by the first crop. Similarly, about two-third of the phosphate is effective but most of the potash is available for the first crop.

Residual fertility. The available nutrient content of a soil carried over to subsequent crops.

Residual material. Unconsolidated and partly weathered mineral materials accumulated by disintegration of consolidated rock in place.

Retardation factor. The capability of a soil for slowing or retarding the movement of a solute, and is defined for solutes subject to equilibrium reactions with the soil matrix.

Reticulate mottling. A network of streaks of different color; most commonly found in the deeper profiles of Lateritic soils containing plinthite.

Rhizocylinder. The plant root plus the adjacent soil that is influenced by the root.

Rhizosphere. The zone of soil immediately adjacent to plant roots in which microbial numbers and kinds may be much different than in the bulk soil in general.

Rill erosion. An erosion process in which numerous small channels of only several centimeters in depth are formed; occurs mainly on recently cultivated soils.

River wash. Barren alluvial areas, usually coarse-textured, exposed along streams at low water and subject to shifting during normal high water. A miscellaneous area.

Rock outcrop. Consists of exposures of base bedrock, other than lava flows and rock-lined pits. Most rock outcrops are of hard rock, but some are soft rock.

Role of Microorganisms in Carbon Cycle. Decay, combustion and animal respiration are the chief steps of carbon cycle. The decay process involves the liberation of carbon and hydrogen as carbon dioxide, methane, water, etc. A huge amount of these substances are also evolved from mineral spring and volcanoes. Complex organic matter like simple carbohydrate and proteins are decomposed by most of the soil microorganisms but cellulose and nucleoproteins are difficult to decompose and relatively a very few soil organisms are able to do it. Usually, cellulose is digested by the members of general e.g., *Cellulomonas, Cellvibrio, Clostridium, Pseudomonas*. Actinomyces and some of the molds. The decomposition of complex organic compounds leads to the formation of following compounds: 1. Acid which reacts with insoluble material, changing them and making themselves available as plant food. 2. Gases like carbon dioxide, hydrogen, methane, etc. Some of the gases are released in the atmosphere.

Root nodule. A swelling formed on the roots of leguminous plants, caused by the symbiotic nitrogen-fixing bacteria, *Rhizobium.*

Rotational cropping. Production of different crops among fields and over years. This is the basis of organic farming systems, designed to provide efficient nutrient cycling and restriction of pests, diseases, and weeds.

Row inter-cropping. One or more of two or more crops are planted in alternating rows, thus reducing interactions. Commonly practiced in small-scale intensive organic horticulture. *Strip inter-cropping* – two or more crops are grown in strips; the strips are wide enough for independent management but narrow enough for some interaction among the crops. E.G., maize/beans; brassicas with clover; leeks with clover. *Relay inter-cropping* – intermediate between rotation and inter-cropping. It is often applied as a form of strip inter-cropping; two or more crops are grown simultaneously during part of the life-cycle of each. E.g., wheat/ maize; mustard/wheat

Runoff. In a productive soil the soil pores need to be large enough to permit water to enter and move freely. If the soil pores are very small because of soil compaction or a natural "tight" soil condition, much water will run off. This takes water away from plant roots. Runoff water may also erode, carrying soil with it.

Run-off. The portion of the precipitation on area that is discharged from the area through stream channels. That which is lost without entering the soil is surface run-off and that which enters the soil before reaching the stream is called "groundwater run-off" or "seepage flow" from groundwater.

Salination. The process whereby soluble salts accumulate in soil. It replaces the obsolete term *salinization.*

Saline seep. Intermittent or continuous saline water discharge at or near the soil surface under dryland conditions which reduces or eliminates crop growth. It is differentiated from other saline soil conditions by recent and local origin, shallow water table, saturated root zone, and sensitivity to cropping systems and precipitation.

Saline soil. A saline soil contains enough soluble salts so distributed in the soil that they interfere with growth of most crop plants. Ordinarily the saline soil is only slightly alkaline in reaction (pH 7.4 to 8.5) and contains very little absorbed sodium. Saline soils are often recognised by the presence of white salt crusts. As such, a saline soil is sometimes referred to as white alkali soil.

Saline-sodic soil (Salt affected soil). 1. A soil containing sufficient exchangeable sodium to interfere with the growth of most crop plants and containing appreciable quantities of soluble salts. The exchangeable-sodium percentage is greater than 15 (or SAR greater than 13), the conductivity of the saturation extract greater than 4 dS m^{-1} (25°C), and the pH is usually 8.5 or less in the saturated soil. 2. A saline-sodic soil has combination of harmful quantities of salts and either a high alkalinity or a high content of exchangeable sodium, or both, so distributed in the profile that the growth of most crop plants is reduced.

Salinization. (Obsolete) The process whereby soluble salts accumulate in soil.

Salt balance. The relation between the quantities of dissolved salts carried to an area in irrigation water and the quantities of dissolved salts removed by drainage water from the same area.

Salt flats. Undrained areas in closed basins in arid regions. In these areas, 10 to 75 cm of crystalline salt overlie stratified, very strongly saline sediments. The water table may be within 20 cm of the surface at some period during the year.

Salt tolerance. The ability of plants to resist the adverse, nonspecific effects of excessive soluble salts in the root medium. Salt tolerance is distinguished from tolerances to specific ions (e.g., Na^+ and Cl^-) or solutes (e.g., H_3PO_3) and from nutritional imbalances.

Saltation. Particle movement in water or wind where particles skip or bounce along the streambed or soil surface. Synonym: valuation.

Sand. (1) A soil particle between 0.05 and 2.0 mm in diameter. (2) Any one of five soil separates: very coarse sand, coarse sand, medium sand, fine sand, and very fine sand.

Sapric material. One of the components of organic soils with highly decomposed plant remains. Material is not recognizable and bulk density is low.

SARD (sustainable agriculture and rural development). Management and conservation of the natural resource base and the orientation of technological and institutional change in such a manner as to ensure the attainment and continued satisfaction of human needs for present and future generations. Such sustainable development (in agriculture, fisheries, and forestry) conserves land, water, plant and animal genetic resources, is environmentally non-degrading technically appropriate, economically viable, and socially acceptable.

Saturate. (1) To fill all the voids between soil particles with a liquid. (2) To form the most concentrated solution possible under a given set of physical conditions in the presence of an excess of the solute. (3) To fill to capacity, as the adsorption complex with a cation species (e.g., H^+-saturated, etc.).

Saturated flow. Movement of water through soil that is saturated with water as in irrigation or during a rainstorm; mostly gravitational water.

Saturated soil paste. A particular mixture of soil and water. At saturation, the soil paste glistens as it reflects light, flows slightly when the container is tipped, and the paste slides freely and cleanly from a spatula.

Saturation content. The water content of a saturated soil paste, expressed as a fraction of the dry soil mass.

Saturation paste extract. The extract from a *saturated paste,* a condition of wetted soil that will glisten and flow slightly when tipped.

Secondary fertilizer. These elements are calcium, magnesium and sulphur.

Secondary mineral. A mineral resulting from the decomposition of a primary mineral or from the reprecipitation of the products of decomposition of a primary mineral.

Sediment. Transported and deposited particles derived from rocks, soil, or biological material.

Sedimentation. The process of sediment deposition.

Self-mulching soil. A soil in which the surface layer becomes so well aggregated that it does not crust and seal under the impact of rain but instead serves as a surface mulch upon drying.

Semiarid. A term applied to regions or climates where moisture is normally greater than under arid conditions but still definitely limits the growth of most crops. Dryland faming methods or irrigation generally are required for crop production. The upper limit of average annual precipitation in the cool semiarid regions is as low as 38 cm (15 in.) whereas in tropical regions it is as high as 114-127 cm (45-50 in.). Contrast with arid.

Semi-intensive fish farming. (a) Stocking rate of 1-2 kg fish/1000 1 water; some supplementary feeding (fodder plants); omnivorous fish (especially plankton, algae and higher plants) are most suitable (this system is appropriate for low-input smallholder farms). (b) Young fish are bred on an intensive basis, e.g. in cages or tanks, then released and reared on an extensive basis without supplementary feeding.

Semi-organic fertilizer. Product in which declared nutrients are of both organic and inorganic origin obtained by mixing and/or chemical combination of organic and inorganic fertilisers.

Semi-organic soil conditioner. Product in which the useful substances and elements are of both organic and inorganic origin obtained by mixing and/or chemical combination of organic soil conditioners and the soil conditioners are calcium, magnesium and/or sulphur.

Sequential cropping. Two or more crops in sequence in the same field in 1 year.

Sesquioxides. (Metal oxides) A term for minerals containing 1.5 atoms of oxygen per atom of the metal, particularly Al_2O_3 and Fe_2O_3. Often TiO_2 is included, although it does not strictly fit the meaning of *sesqui* (= 1.5 times).

Shaly. (1) Containing a large amount of shale fragments, as a soil. (2) A soil phase, as for example, the shale phase.

Shear strength. The maximum resistance of a soil to shearing stresses.

Shear. Force, as with a tillage tool, acting at a right angle to the direction of movement of the tillage implement.

Sheet erosion. The removal of fairly uniform layer of soil from the land surface by run-off water.

Shrink-swell potential. Susceptibility to volume change due to loss or gain in moisture content.

Siderophore. A nonporphyrin metabolite, secreted by certain microorganisms, that forms highly stable bonds with iron, thereby mobilizing iron in soil. The two major types are catecholate and hydroxamate.

Siemen. A unit of electrical inductance; in SI metric units, electrical conductance is measured in siemens per meter ($S\ m^{-1}$). Decisiemens per meter ($dS\ m^{-1}$) is equivalent to millimho per centimeter (mMho/cm).

Silt. 1. A soil separate consisting of particles between 0.05 and 0.002 mm in equivalent diameter. 2. A soil textural class.

Silting. The deposition of waterborne sediments in stream channels, lakes, reservoirs, or on floodplains, usually resulting from a decrease in the velocity of the water.

Single grain. (Obsolete when called "structure") A term for a *lack of structure* in which individual particles do not cohere together. Example: coarse sands.

Sink. 1. Depression in the land surface. 2. A term for a material or reaction that acts as an infinite reservoir or removal mechanism. Example: Soil is a reservoir for carbon monoxide (CO).

Sinkhole (karst). Term refers to geomorphic component in a karst topography, and so on.

Site index. (1) A quantitative evaluation of the productivity of a soil for forest growth under the existing or specified environment. (2) The height in meters of the dominant forest vegetation taken at or calculated to an index age, commonly 25, 50, or 100 years.

Site. (1) In ecology, an area described or defined by its biotic, climatic, and soil conditions as related to its capacity to produce vegetation. (2) An area sufficiently uniform in biotic, climatic, and soil conditions to produce a particular climax vegetation.

Skeletal. A textural term used in soil "families" to indicate that the soil has greater than 35 percent coarse fragments (larger than sand).

Slaty. Containing a considerable quantity of slate fragments. (Used to modify soil texture class names, such as "slaty clay loam," etc.)

Slick spots. Small areas in a field that are slick when wet due to a high content of clay with high exchangeable sodium.

Slickensides. Polished and grooved surfaces produced by one soil mass sliding past another. Common in Vertisols. Contrast with Clay skins.

Slickspots. Areas of sodic soils.

Sodium adsorption ratio (SAR). A relation between soluble sodium and soluble divalent cations which can be used to predict the exchangeable sodium percentage of soil equilibrated with a given solution.

Soil Acidification: Mechanisms. Protons, which are the chemical agents of acidification of soils, arise from respiration and exudation of roots, metabolism of microbes, decay of plant and animal residues, oxidation of pyrite, application of acid-generating fertilizers, and deposition of acid rains. Therefore, the development of soil acidity is related to soil parent materials, the advancement of the stage of soil formation, intensification of agriculture, and industrialization. The stage of soil development advances with time, Industrialization is being promoted around the world since the Industrial Revolution in the latter part of the 19th century. We have witnessed the world-wide population explosion in the 20th century, the resultant necessity to assure plentiful supplies of foods, fiber and shelter for billions of people, and the concomitant escalation of human desire for material goods. Consequently, if there are no proper management practices, the extent of acidification of soils is deemed to increase as time goes by. The area of acid soils throughout the world can be estimated around 20%. Acid soils are localized in tropical areas, the western and central part of Europe, and the eastern and central part of America. In tropical regions, acid soil correspond mainly to Oxisols and Ultisols, which represent more than 60% of tropical soils. Acidity originates from the natural intense weathering of rocks under high drainage and temperature. Such conditions cause silicate mineral hydrolysis, leaching of bases, formation of low activity clays (kaolinite, oxides, hydroxides), and desaturation of the clays. In northern countries, the Origins of acidity are more diverse with both natural biogeochemical and anthropogenic causes. Problems associated with soil acidification are multiple. These include degradation of land quality, reduction of crop production, contamination of food chains, and threat to human health.

Soil air. The soil atmosphere; the gaseous phase of the soil, being that volume not occupied by solid or liquid.

Soil association. A kind of map unit used in soil surveys comprised of delineations, each of which shows the size, shape, and location of a landscape unit composed of two or more kinds of component soils or component soils and miscellaneous areas, plus allowable inclusions in either case. The bodies of component soils and miscellaneous areas are large enough to be delineated individually at the scale of 1:24,000. Several to numerous bodies of each kind of component soil or miscellaneous area are apt to occur in each delineation and they occur in a fairly repetitive and describable pattern. The proportions of the components may vary appreciable from one delineation to another and all of the components need not occur in every delineation to another and all of the components need not occur in every delineation though they will be present in most delineations.

Soil auger. A tool for boring into the soil and withdrawing a small sample for field or laboratory observation. Soil augers may be classified into several types as follows: (1) those with worm-type bits, unenclosed; (2) those with worm-type bits enclosed in a hollow cylinder; and (3) those with a hollow cylinder with a cutting edge at the end.

Soil Building Crop rotation. The practice of alternating crops grown on a specific field in a planned pattern or sequence in successive crop years. Organic certification programmes require 'soil building' crop rotations, meaning that they must be specifically designed to steadily improve soil tilth and fertility while reducing nitrate leaching, week, pest and disease problems. IFOAM, for example, recommends specific rotations that include legumes and requires the rotation of non-perennial crops "in a manner that minimizes pressure from insects, weeds, diseases and other pests, while maintaining or increasing soil, organic matter, fertility, microbial activity and general soil health." Under limited cropping conditions (e.g., mushrooms, perennials) crop rotations may not be applicable, in such cases other methods that contribute to soil fertility may be required by certification programmes.

Soil classification. The systematic arrangement of soils into classes in one or more categories or levels of classification for a specific objective.

Soil compaction. Increasing the soil bulk density, and concomitantly decreasing the soil porosity, by the application of mechanical forces to the soil.

Soil complex. Much like a soil association but the different soils occur as smaller discreet units; it is not possible to separate the soils in a soil complex at a scale of 1:24,000, which is used for held mapping.

Soil conservation. (1) Protection of the soil against physical loss by erosion or against chemical deterioration; that is, excessive loss of fertility by either natural or artificial means. (2) A combination of all management and land use methods that safeguard the soil against depletion or deterioration by natural or by human-induced factors. (3) The branch of soil science that deals with soil conservation (1) and (2).

Soil conservation. A combination of all management and land-use methods that safeguard and soil against depletion or deterioration caused by nature and/or humans.

Soil creep. The slow movement of a soil mass downslope, usually in wet periods when clayey materials are lubricated with water.

Soil development. 1. The breakdown of rocks to unconsolidated materials called soil. 2. The changes in the soil profile brought about by natural processes of leaching, translocation of colloids, accumulation of organic materials, and continued mineral and rock weathering.

Soil Enzymes. The total biochemical activity of soil is comprised of a series of reactions catalyzed by enzymes. These may be intracellular within living or dead organisms or extracellular, sometimes distant from bound to soil colloids. More than 52 enzymes have been shown to be active in soil, including *oxidoreductases, hydrolases* and *transferases.* In some numbers and enzyme activity, but sometimes activity appears to be independent of microbial biomass, which is a better index of microbial activity, and there have been few attempts to *correlate biomass with enzyme activity*. A proportion of enzyme activity is independent of living cells because treatment of soil with toluene, sodium azide or gamma rays usually destroys microorganisms and extracellular unbound enzymes without preventing enzyme activity. The study of soil enzymes is clearly more difficult than studies of the pure enzymes that biochemists are familiar with and where the disappearance of substrate and appearance of products is monitored.

Soil Erosion. The word "erosion" literally means gnawing away. Soil erosion is the rapid blowing or washing of soil from its natural place in a landscape to another place, usually many feet or possibly even miles away.

Soil extract. The solution separated from a soil suspension or from a soil by filtration, centrifugation, suction, or pressure. (May or may not be heated prior to separation.)

Soil fertility evaluation. Use of fertilisers and manures needs to be properly planned. Selection of fertilisers to be used and the quantities to be applied depend upon the nutrient requirements of crop grown and the nutrient supplying power of the soil on which it is grown. A crop does not grow normally if some plant nutrients, which are essential for its normal growth, are lacking in the soil. The nutrients that are deficient should be determined and corrected by adding fertilisers and other sources of plant nutrients. There are several methods for determining the fertility status or plant nutrient supplying power of soils.

Soil fertility. Soil fertility has been considered in the past in a restricted sense as an index of available nutrients for plants, but the modern usage of the term connotes the capacity of the soil to produce crops of economic value and maintain the health of the soil for future use. Thus, any system of management of soil fertility will ultimately consider all aspects of soil-plant relationship and the pollution of environment as well. Under natural vegetation, as in the case of virgin forests, nutrients taken up by the plants from the soil are returned to it by leaf and fruit fall and there is no substantial drain of nutrients from the plant-soil system. The capacity of soil to supply essential nutrients to normal plants in adequate amounts and in a balanced proportion is thus maintained under natural conditions. However, when land is brought under cropping grain or fruit and sometimes the entire plants are removed from the land, the soil loses a considerable amount of its nutrients. If cropping is continued over a period of time, without nutrients being restored to the soil, its fertility will be reduced and crop yields will decline. Apart from removal by crops, nutrients may also be lost from the soil through leaching and erosion. Even to maintain soil productivity at the existing levels, it is necessary to restore to the soil the nutrients removed by crops as also those lost through leaching and erosion. Continued maintenance of a high level of soil fertility is an indispensable condition for profitable and use and sustained agricultural production.

Soil formation factors. The variable, usually interrelated natural agencies that are active in and responsible for the formation of soil. The factors are usually grouped into five major categories: parent rock, climate, organisms, topography, and time.

Soil genesis. (1) The mode of origin of the soil with special reference to the processes or soil-forming factors responsible for development of the solum, or true soil, from unconsolidated parent material. (2) The branch of soil science that deals with soil genesis.

Soil horizon. A layer of soil or soil material approximately parallel to the land surface and differing from adjacent genetically related layers in physical, chemical, and biological properties or characteristics, such as color, structure, texture, consistency, amount of organic matter, and degree of acidity or alkalinity.

Soil interpretations. Predictions of soil behavior under specific uses or management and based on inferences from the soil's characteristics as reported in a Soil Survey Report.

Soil loss tolerance. (1) The maximum average annual soil loss that will allow continuous cropping and maintain soil productivity without requiring additional management inputs. (2) The maximum soil erosion loss that is offset by the theoretical maximum rate of soil development which will maintain an equilibrium between soil losses and gains.

Soil management. (1) The sum total of all tillage and planting operations, cropping practices, fertilizer, lime, irrigation, herbicide and insecticide application, and other treatments conducted on or applied to a soil for the production of plants. (2) The branch of soil science that deals with the items listed in (1).

Soil mapping unit. A kind of soil, a combination of kinds of soil, or miscellaneous land type or types that can be shown at the scale of mapping for the defined purposes and objectives of the survey. Soil mapping units are the basis for the delineations of a soil survey map. A soil survey identification legend lists all mapping units for the survey of an area. Mapping units contain inclusions of soils outside the limits of the taxonomic name, r names, used as the name for the mapping unit.

Soil Mineral. (1) Any mineral that occurs as a part of or in the soil. (2) A natural inorganic compound with definite physical, chemical, and crystalline properties (within the limits of isomorphism) that occurs in the soil.

Soil moisture regimes. Moisture regimes indicate the time periods during which the soil "control section" (loam soil between about 10 and 30 cm deep) is either moist or dry (wilting percentage). In addition to details below, to simplify this tabulation, the data given are not precise in all particulars. All temperatures are at the 50 cm depth in soils; the "parts" refers to parts of the "control section" (CS).

Soil monolith. A vertical section of a soil profile removed from the soil and mounted for display or study.

Soil morphology. (1) The physical constitution, particularly the structural properties, of a soil profile as exhibited by the kinds, thickness, and arrangement of the horizons in the profile, and by the texture, structure, consistency, and porosity of each horizon. (2) The structural characteristics of the soil or any of its parts.

Soil organic matter. The organic fraction of the soil that includes plant and animal residues at various stages of decomposition, cells and tissues of soil organisms and substances synthesized by soil organisms.

Soil organic residue. Animal and vegetative materials added to the soil and which are recognizable as to their origin.

Soil productivity. Productivity is the present capacity of a soil to produce crop yield under a defined set of management practices. It is measured in terms of the yield in relation to the input of production factors.

Soil Reaction (Soil pH). The term "soil reaction" refers to the concentration of hydrogen ions (H^+) in the soil solution. This concentration is expressed on a hydrogen potential (pH) scale.

Soil science. That science dealing with soils as a natural resource on the surface of the earth, including soil formation, classification and mapping, and the physical, chemical, biological, and fertility properties of soils per se; and these properties in relation to their management for the growth of plants and to cleanse the environment.

Soil solution. The aqueous liquid phase of the soil and its solutes.

Soil survey. The systematic examination, description, classification, and mapping of soils in an area. Soil surveys are classified according to the kind and intensity of field examination. Also, the program of the National Cooperative Soil Survey that includes developing and implementing standards for describing, classifying, mapping, writing, and publishing information about soils of a specific area.

Soil test. A chemical, physical, or biological procedure that estimates a property of the soil pertinent to the suitability of the soil to support plant growth. Sometimes used as an adjective to define fractions of soil components, e.g., soil test phosphorus.

Soil variant. A kind of soil that differs enough from recognized series to justify a new series name but is so limited in area that creation of a new series is not justified at a given time.

Soil water characteristic or characteristic curve. The relationship between the soil-water content (by mass or volume) and the soil-water matric potential.

Soil. Soil, in its traditional meaning, is the natural medium for the growth of land plants, whether or not it has developed discernible soil horizons. In this sense soil has a thickness that is determined by the depth of rooting of plants. Soil is defined as the *unconsolidated mineral material on the immediate surface of the earth that serves as a natural medium for the growth of land plants*. Soils are that portion of the earth's crust in which land plants can grow, if water and temperature are adequate, at least the minimum nutrients are available, and toxic substances are in low concentration. Soils are often deposits of weathered, loosened, and transported particles. These materials, called *parent materials*. They are deposited by water, glaciers (ice), wind, or gravity into land areas called landforms. Many landforms are named for their method of transport and final shape: Alluvial is from flowing water; moraines are from glacier movement and melting; dunes are from wind movement of sands; and colluvium is from gravity's action. All of these various deposits, and others, become soils when exposed to weathering at the earth's surface. The soils, and their landform materials, vary enormously in the sizes of particles, slopes, uniformity of properties, and extensiveness across the landscape. The soils vary widely in quality and productivity. Sand dunes do not ordinarily support plant life, but with stabilization (retarded mobility) and water, if needed, they could do so, and so can be considered soils. Organic deposits, such as are found as peat or muck bogs, are very different from the majority of soils because they sometimes contain almost no mineral materials, which are the major components of most soils; nevertheless, they are soils.

Soils of Indian arid ecosystem. The arid zone of India located in north-western part of the country largely covers western Rajasthan, southern Punjab and Haryana and northern Gujarat. The soil cover ranges from fine aeolian sandy to medium textured alluvial and often dark reddish brown fine textured colluvial soils. This variation in soils is mainly attributed to the nature of parent material and also the history of landscape evolution. The soils ave low organic matter, high salinity, segregation of lime and aridic moisture regime are their peculiar characteristics which distinguish them from the rest of azonal soils.

Soil-testing. This refers to chemical tests of the soil that can be made rapidly and with low cost, as compared to conventional methods of soil chemical analysis which are more accurate but more time-consuming and expensive. Soil testing covers both rapid analysis in the field and in the laboratory.

Solifluction. The slow downhill flowage or creep of saturated soil and other loose materials.

Spatial variability. The variation in soil properties (1) laterally across the landscape, at a given depth, or with a given horizon, or (2) vertically downward through the soil.

Specific gravity. The relative weight of a given volume of any kind of matter (volume occupied by solid phase, pore space excluded) compared with an equal volume of distilled water at a specified temperature. The average specific gravity for soil is about 2.65 Mg/m^3. Contrast with bulk density.

Splash erosion. The spattering of small soil particles caused by the impact of raindrops on soils. The loosened and separate particles may or may be subsequently removed by surface run-off.

Spoil bank. Rock waste, banks, and dump depositions resulting from the excavation of ditches and strip mines.

Stemflow. That portion of precipitation or irrigation water that is intercepted by plants and then flows down the stem to the ground.

Stomata. (Singular = stoma) Minute openings in the surface (epidermis) of leaves through which oxygen, water vapor, and other gases leave the leaf and carbon dioxide is taken into the leaf.

Stones. Rock fragment > 25 cm in diameter if rounded, and > 38 cm along the greater axis if flat.

Stony. Containing sufficient stones to interfere with or to prevent tillage. To be classified as stony, > 0.01% of the surface of the soil must be covered with stones. Used to modify soil class, such as stony clay loam or clay loam, stony phase.

Straight fertilizer. Qualification generally given to a nitrogenous, phosphatic or potassic fertilizer having a declarable content of one primary nutrient only.

Stratospheric Changes. Losses of N from the soil-plant system takes place by way of ammonia volatilization, denitrification and chemodenitrification. Loss of applied fertilizer N takes place as NH_3 (ammonia), N_2O (nitrous oxide), N_2 (dinitrogen), NO (nitric oxide) and NO_2 (nitrogen dioxide). Dissimilatory denitrification is carried out by a limited number of aerobic bacteria that can grow in the soil in the absence of molecular oxygen while reducing nitrates and nitrites to N_2O and N_2. Chemodenitrification is a term that encompasses processes for gaseous losses of N from soils through chemical reaction of nitrite. Many research studies have been involved in view of the concern that N_2O released into the atmosphere may increase the rate of reactions in the stratosphere that lead to the destruction of ozone layer. The stratosphere ozone layer shields the biosphere from the harmful ultra violet (UV) radiation and also influences the earth's temperature. The overall global source of N_2O from nitrogenous fertilizer applications is probably smaller than a few Tg N per year. Thus, the impact of increasing use of nitrogenous fertilizers on the stratosphere ozone content shall not invite much concern. The global increase in atmospheric concentration of N_2O is about 0.2 per cent per year which may be attributed to 3.5 per cent per year increase in N_2O emission from fossil fuel combustion.

Strip cropping. Planting of alternate strips of grasses, or grains with other crops on contour in order to conserve moisture and decrease erosion.

Strip cropping. The practice of growing crops for controlling erosion that requires different types of tillage, such as row and sod, in alternate strips along contours or at right angles to the prevailing direction of erosive winds.

Strip tillage. Tillage operations performed in isolated bands separated by bands of soil essentially undisturbed by the particular tillage equipment.

Structural charge. The charge (usually negative) on a mineral caused by isomorphous substitution within the mineral layer. (Expressed as moles of charge per kilogram of clay.)

Stubble mulch. The stubble of crops or crop residues left essentially in place on the land as a surface cover before and during the preparation of the seedbed and at least partly during the growing of a succeeding crop.

Stubble mulching. The use of stubble of crops or crop residues left essentially in place on the land as a surface cover before and during the preparation of the seedbed and at least partly during the growing of succeeding crop.

Subsidence. A lowering of the land elevation caused by solution and collapse of fluid pressures within an aquifer or petroleum reservoir, or decomposition of organic soils (Histosols).

Subsoiling. Any treatment to loosen soil below the tillage zone without inversion and with at least partly during the growing of a succeeding crop.

Substrate. (1) That which is laid or spread under; an underlying layer, such as the subsoil. (2) The substance, base, or nutrient on which an organism grows. (3) Compounds or substances that are acted upon by enzymes or catalysts and changed to other compounds in the chemical reaction.

Substratum. Any layer beneath the soil solum, either conforming or unconforming.

Subsurface tillage. Tillage beneath the surface, cutting plant roots and loosening the soil, without inverting it and without appreciably incorporation surface residues.

Sulphidic material. Waterlogged material or organic material that contains 7.5 g kg^{-1} or more of sulfur.

Sulphur. It is absorbed from soils as divalent sulfate ions (SO_4^{2-}). Sulphur deficiency is usually encountered in light-textured high-pH soils. Sulfur deficiency symptoms resemble somewhat those of nitrogen deficiency. There is general chlorosis followed by the production of anthocyanin pigments in some species. Unlike nitrogen deficient plants, deficiency of sulfur causes chlorosis in the young leaves first. Under severe conditions, however, all of the leaves may undergo some loss of green color. Most of the sulfur in plants occurs in proteins, specially in the amino acids cysteine and methionine that are building blocks for proteins. Other essential compounds that contain sulfur are the vitamins thiamine and biotin and coenzyme A, a compound essential for respiration and for synthesis and breakdown of fatty acids.

Sulphur-bacteria. Sulphur-bacteria convert sulphide into sulphuric acid. These compounds act on insoluble soil-compounds like Calcium carbonate, Magnesium carbonate, Calcium silicate, Tricalcium phosphate and bring them in soluble state. These soluble compounds are then absorbed by plants. The conversion of hydrogen sulphide into sulphur is purely a chemical reaction. Sulphur produced by these reactions is then further oxidized by the bacteria (*Thiobacteria*) into sulphuric acid.

Sulphuric horizon. A diagnostic horizon with both a pH < 3.5 and yellow jarosite (sulfate) mottles.

Summit. The highest point of any landform remanant, hill, or mountain.

Surface area. The area of the solid particles in a given quantity of soil or porous medium.

Surface sealing. The orientation and packing of dispersed soil particles in the immediate surface layer of the soil, thus greatly reducing it permeability.

Surface soil. The uppermost part of the soil, ordinarily moved in tillage, or its equivalent in uncultivated soils and ranging in depth from 7 to 20 cm. frequently designated as the surface layer, Ap layer, or Ap horizon.

Swarm ions. Soil cations are sometimes called "swarm ions" because they resemble a swarm of bees around a hive, with the greatest concentration of bees close to it. In a soil colloidal system the ions with two (or more) positive charges (like calcium) and a small diameter are attracted close to the colloid, while monovalent cations (such as sodium with one positive charge) with large diameters tend to migrate farther. All the attracted cations are in constant motion, but attraction holds them tightly enough so they are not readily lost to water that is moving through the soil. This action is very important to plant life because it keeps many nutrients within the root zone of the crops.

Sweep. In tillage, an implement with a point and flat or curved "wings" going back from the point at 30-60° angles.

Symbiosis. The living together in intimate association of two dissimilar organisms, the cohabitation being mutually beneficial, such as *Rhizobia* legume bacteria with the host leguminous plant.

Synergism. The action of two or more substances, organs, or organisms to achieve an effect of which each alone is incapable. In nutrient uptake, the presence of one nutrient aids in the uptake of another nutrient.

Synthetic Inputs. NO single definition of "synthetic" exists, although the various material lists of allowed and prohibited inputs for organic production, developed in different countries and by different certification programmes, are fairly consistent, reflecting an implicit agreement on a definition. The few legal definitions of synthetic reflect the common understanding of the term in organic trade. For example, British Columbia, Canada defines "synthetically compounded" as "a process which chemically changes a material extracted from naturally occurring plant, animal or mineral sources, excepting microbiological, mechanical and heat processes." The USA defines synthetic as "a substance that is formulated or manufactured by a chemical process or by a process that chemically changes a substance extracted from naturally occurring plant, animal, or mineral sources, except that such term shall not apply to substances created by naturally occurring biological processes."

Synthetic organic nitrogenous fertilizer. Nitrogenous fertilisers in which the nitrogen is combined with carbon by organic synthesis.

Talud. A short, steep slope formed gradually at the downslope margin of a field by deposition against a hedge, a stone wall, or similar barrier.

Talus. Fragments of rock and other soil material accumulated by gravity at the foot of cliffs or steep slopes.

Taxon. In the context of soil survey, a class at any categorical level in the U.S. system of soil taxonomy.

Tensiometer. A device for measuring the soil-water matric potential (or tension, or suction) of water in soil *in situ*; a porous, permeable ceramic cup connected through a water-filled tube to a manometer, vacuum gage, pressure transducer, or other pressure-measuring device.

Terrestrial environments. The soil which covers much of the land surface is a complex of water, air, mineral matter and organic matter. The dominant microorganisms are chemoheterotrophs. Microbial numbers in the top surface (3 to 8 cm) of a temperate soil as (numbers per gram of soil). Overall the abundance of the microorganisms relates directly to the amount of available fixed carbon provided by primary and secondary producers. The dominant microbial events are those associated with organic degradation. Physico-chemical features of importance include the soil particulate matter and the soil pore space. Attachment to the particulate material involves cation bridges, van der Waals forces, extracellular polysaccharides, fimbriae, etc. The soil pore space affects oxygenation, redox potential, water-holding capacity, etc. Excess moisture may limit oxygen diffusion, produce an anaerobic environment and set in train a series of anaerobic events including denitrification, sulphate reduction, methanogenesis and the release of adsorbed nutrients, just as in freshwater sediments.

Thermic. A soil temperature regime averaging between 15 and 22°C.

Thermophilic organisms. Organisms that grow readily at temperatures of about 113°F (45°C).

Thermosequence. A sequence of related soils that differ, one from the other, primarily as a result of temperature as a soil-formation factor. These products are mainly methylene ureas in the form NH_2-CO-($NHCH_2NHCH)_n$ NH_2, with low molar mass ($1<n<8$).

Threshold moisture content (biology). The minimum moisture condition, measured either in terms of moisture content or moisture stress, at which biological activity just becomes measurable.

Throughfall. That portion of precipitation that falls through or drips off of a plant canopy.

Tidal flats. Areas of nearly flat, barren mud periodically covered by tidal waters. Normally, these materials have an excess of soluble salt. A miscellaneous area.

Tile drain. Concrete or ceramic pipe placed at suitable depths and spacings in the soil or subsoil to provide water outlets from the soil.

Tillage action. The specific form or forms of soil manipulation performed by the application of mechanical forces to the soil with a tillage tool, such as cutting, shattering, inversion, or mixing.

Tillage equipment (tools). Field tools and machinery which are designed to lift, invert, stir, and pack soil, reduce the size of clods and uproot weeds (i.e., plows, harrows, disks, cultivators, and rollers).

Tillage objective. A desired soil, soil surface, or soil residue cover condition that is to be produced by one or more tillage actions.

Tillage operation. Act of applying one or more tillage actions in a distinct mechanical application of force to all or part of the soil mass.

Tillage requirements. The soil physical conditions, which after complete evaluation of fundamental utilitarian and economic requirements, is deemed necessary.

Currently, often based on practical experience and judgment.

Tillage. The manipulation, generally mechanical, of soil properties for any purpose; but in agriculture it is usually restricted to modifying soil conditions for crop production.

Tilth. The physical condition of soil as related to its ease of tillage, fitness as a seedbed, and its impedance to seedling emergence and root penetration.

Top dressing. An application of fertilizer to a soil surface, without incorporation, after the crop stand has been established.

Toposequence. A sequence of related soils that differ, one from the other, one from the other, primarily because of topography as a soil-formation factor.

Topsoil. (1) The layer of soil moved in cultivation. (2) The A horizon. (3) The all horizon. (4) Presumably fertile soil material used to top-dress roadbanks, gardens, and lawns.

Torric. A dry soil water regime defined like aridic but used in a different category level of the U.S. soil taxonomy than is "aridic."

Tortuosity. The nonstraight nature of soil pores.

Traditional agriculture. Agriculture which has been practiced for generations based on transfer of knowledge within the community. This transfer is often associated with cultural and religious practices. Such systems were rather sustainable. For different reasons (population pressure, introduction of agrochemicals, colonization, new religious and cultures introduced) these have been disrupted in most cases.

Transpiration. The process by which water vapor is released from plants to the atmosphere primarily through the leaf stomata.

Trioctahedral. An octahedral sheet that has all of the sites filled, usually by divalent ions such as magnesium or ferrous iron.

Truncated soil. Having lost all or part of the upper soil horizon or horizons.

Udic. A soil water regime usually having adequate water during all of the growing periods of the year.

Umbric horizon. A dark-colored, high-organic-matter, diagnostic apipedon similar to mollic, but it is more acidic.

Underground runoff (seepage). Water that seeps toward stream channels after infiltration into the ground.

Unique agricultural lands. These are soils aniquely suited to the high production of certain desired crops. The uniqueness may be aided by a lack of frost, a needed wetness or acidity, a peculiar climate, or a particular location relative to markets.

Universal soil loss equation (USLE). An equation for predicting A, the average annual soil loss in mass per unit area per year, expressed as A = RKLSPC, where R is rainfall factor, K is soil erodibility factor, L is length of slope, S is per cent slope, P is conservation practice factor, and C is cropping and management factor.

Ustic. A soil water regime with deficient water in the year but with most of the water coming in the warm summer period.

Vergin soil. A soil essentially undisturbed from its natural environment; never cultivated.

Vermiculite. A clay similar to hydrous mica and having 2:1 layers of 2 tetrahedral sheets to 1 octahedral sheet. Instead of the interlayer potassium found in illite, vermiculite has the layers held together by hydrated cations, and has less swelling than montmorillonite.

Vermitechnology. Vermitechnology (VT) is the application of earthworms in following areas: 1. for development of arable soils, turn over of soil, break down of plant organic matter, aeration and drainage. 2. For production of useful products like vermi fertiliser, worm tissue for animal feed etc. 3. In monitoring of the environment for soil fertility, organic and heavy metal, non-biodegradable toxic material pollution etc. 4. For maintenance of environmental quality. 5. Earthworms on soil improves the physico-chemical properties of the soil and help in enhancing organic manure production through earthworm consumption. In this way they are helpful in managing waste biomass. They also help in reclamation or conservation of waste lands by managing their anecic and endogeic category.

Vertical mulching. A subsoiling operation in which a vertical band of mulching material is placed into a vertical slit in the soil immediately behind a special chisel.

Void ratio. The ratio of the volume of soil pore (or void) space to the solid particle volume.

Volume water percentage. Volume water content multiplied by 100 to convert to percentage.

Volumetric water content. The soil-water content expressed as the volume of water per unit bulk volume of soil.

Wastelands. Any type of land which because of neglect, over use or degradation by climate, edaphic and or anthropic factors is not being used to its fullest potential. Such lands also termed as *derelict*, *divasted*, *destroyed*, *poor* or *neglected*, are all covered under wastelands. The technical Task Group Report of National Wastelands Development Board in 1985 had defined the wastelands as `the land which is presently lying unutilized due to different constraints'. It comprises of culturable and unculturable wasteland.

Waste-lime (by-product lime). An industrial waste or by-product containing calcium or calcium and magnesium in forms that will neutralize acids. It may be designated by prefixing the name of the industry or process by which it is produced; for example gas house lime, tranners' lime, lime kiln ashes etc.

Water availability. Soil contains two sorts of water: that which is unavailable to plants and that which is available. Not all water entering a soil can be used by growing vegetation. Two circumstances make soil water unavailable. One is the rapid percolation of rainwater through large pores, quickly bypassing the roots. When this fast-moving gravitational water has drained away within two to three days after a rain, the soil is said to be at field capacity. The other kind of unavailable water is in thin films that are held too tightly on the surfaces of soil particles for plant roots to extract. The wilting point has been reached when this is the only water left in a soil. The soil water between the field capacity and the wilting point is available. Loamy soils have more available water than sandy soils (which let water drain on through) and clay soils (which hold water films tightly).

Water harvesting. The term used to describe the process of collecting and storing water from an area that has been modified or treated to increase precipitation run-off. A water-harvesting systems is a complete facility for collecting and storing run-off for productive purposes. The system consists of a catchment or water-collecting area, a water storage facility and various auxiliary components such as sediments or trash traps, fencing, and evaporation control.

Water mass content. The soil-water content expressed as the mass of water per unit mass of oven-dry soil.

Water table, perched. The surface of a local zone of saturation held above the main body of groundwater by an impermeable layer, usually clay or rock, and separated from the main body of groundwater by an unsaturated zone.

Water table. The upper surface of groundwater or that level in the ground where the water is at atmospheric pressure.

Water tension. The equivalent negative pressure in the soil water. It is equal to pressure needed to bring the water into hydraulic equilibrium.

Water use efficiency. Crop production per unit of water used, irrespective of water source, expressed in units of water depth per unit weight of crop; for example, 25 cm (10 in.) of water used per ton of alfalfa hay produced.

Water-retention curve. A graph showing the soil-water content versus applied tension or suction.

Water-stable aggregate. A soil aggregate that is stable to the action of water, such as falling drops, or agitation as in wet-sieving analysis.

Wing strip cropping. The production of crops in relatively narrow strips placed perpendicular to the direction of prevailing erosive winds.

Worms. Earthworms the world over perform an important function in mixing organic matter with mineral matter. In a sense, they are *soil factories*. Among the many kinds of earthworms, the *night crawler*, *Lumbricus terrestris,* was brought to the United States from Europe by settlers. The tuft is called an earthworm midden. In moist summers, vast quantities of leaves on the forest floor are pulled into the soil in this manner, enriching it and making it more porous. As a result of this activity the thick mats of tree leaves that accumulate on the forest floor in the autumn may be almost completely encorporated into the soil be the following year. Many other kinds of earthworms do not make middens but in their own way incorporate organic matter into soil. Worms perform an important aeration and mixing function by burrowing through the soil, consuming organic matter, and bringing the residue to the surface as castings. It is estimated that worms bring 7 to 18 tons of soil per acre annually to the surface in this way.

Xenobiotic. A compound foreign to biological systems. Often refers to human-made (anthropogenic) compounds that are resistant or recalcitrant to biodegradation and/or decomposition.

Yield. The amount of a specified substance produced (e.g., grain, straw, total dry matter) per unit area.

Appendix

1. INTERNATIONAL FEDERATION OF ORGANIC AGRICULTURE MOVEMENTS

Ökozentrum Imsbacb
66636 Tholey-Theley, Germany
Tel: +49-6853-5190
Fax: +49-6853-3010
Email: headquarter@ifoam.org

IFOAM represents the worldwide movement of organic agriculture, and provides a platform for global exchange and cooperation

Appendix

2. PUBLICATIONS OF IFOAM

1. **Organic agriculture worldwide:** IFOAM directory of the member organizations and associated 1998/1999. 1997. IFOAM. 64 p. US$ 16.00.

   The most recent update IFOAM-directory of member organizations and associates contains the addresses, email addresses, web page phone and fax numbers, and general information of about 650 organic agriculture contacts, grouped per country. Proof of a continuously spreading (in number and geographically) organization and a very useful networking tool! (IHG)

   Guide to regulatory requirements for exporting organic foods into international markets. Suzanne Vaupel and Ken Commins. IFOAM.
2. **Trust in organics:** A guide to setting up organic certification programmes. Gunnar Rundgren (ed.). IFOAM
3. 'Ecology and Farming' is IFOAM's English language magazine.

   Editorial address: Sue Stolten, Equilibrium

   Consultants, 23 Bath Buildings, Bristol BS65PT, UK.

   Tel/Fax: +44-117-9428-674

   Mail: euilibrium@compuserve.com.

   Subscription handling: IFOAM, Subscription rate (6 tissues/two years) US$ 30, postage for airmail outside Europe and US$ 18 (only bank transfers accepted). ISSN 1016-5061.

   The magazine acts as in international forum for the exchange of information and experiences, and facilities a dialogue about organic agriculture and agro-ecology within the IFOAM network.

   The future agenda for organic trade Proceedings from the 5th IFOAM International Conference on Trade in Organic Products. The Soil Association and IFOAM, US$ 30.
4. **Certification of organic foodstuffs in emerging countries.** by J Neuendorff; R Rosskamp, U Sabel-Koschella, 1997. 21 p. Deutsche Gesellschaft fuer Technische Zusammenarheit (GTZ), PO Box 5180, D-65726 Eschborn, Germany.

   The market for organic products in North America and Europe are growing, Legal regulation guarantee organic standards and require certification by independent inspection bodies. International inspection bodies may conduct supervision and inspection in 'third' countries, interested in penetrating these organic markets, but they do that at a high cost. The establishment of national inspection bodies in emerging countries is being promoted nowadays as a part of development cooperation. This booklet gives a brief overview of the basic regulations and the aspects of the inspection and certification of organically produced products in third world countries through local inspection bodies. Examples are given, such as INKA CERT in Peru. (IHG)

5. ***EU regulation 'organic farming':*** a legal and agro-ecological commentary on the EU's council regulation (EEC) No. 2092/91 by HP Schmidt, M Haccius. 1998. 417 p. ISBN 3 8236 1288 3. USD 45.00. Margraf Verlag, PO Box 1205, D-97985Weikersheim, Germany.

   In the European Union, production and marketing of organic food is regulated by law, and is subject to the EU-Regulation No. 2092/91. The regulation is based on the IFOAM Basic Standards and prohibits the use of organo-synthetic pesticides and synthetic nitrogen fertilizers, and also processing is bound to described restrictions. To guarantee truthful organic labeling, the access of farmers in non-EU countries is based on the concept of equivalence. This means that the organic agricultural production and food processing, their documentation and their inspection must be equivalent to requirements in the European Union. A third world country may apply for a place on a list of accredited organic importers or (more costly and time consuming) case-to-case import permits may be demanded. The very complicated structure of the European Union regulations makes this book necessary, as it explains to farmers' originations, government officials, importers and exporter or organic products how the legislative process in the EU functions and what possibilities exist that can influence these procedures. It contains the full text of the Council Regulation No. 2092/91 and its amendments and its relationship with the FAO/WHO Codex Alimentarius Guidelines on organic farming is set out. Reading the Question and Answer chapter is very useful start to getting into the book, as it presents this intricate matter in a tangible manner. Addresses of professional manner. Address of professional contact in EU countries and useful directories are also provided (IHG)

Appendix

3. JOURNALS AND NEWSLETTERS

1. *Acres U.S.A* (monthly/$20 yr.), from Acres U.S.A., P.O. Box 8800, Metairie LA 70011-8800, Phone 504-889-2100,
fax 504-889-2777
e-mail acresusa@aol.com.
2. *American Journal of Alternative Agriculture* (quarterly/$24 yr.), from Henry A. Wallace Institute for Alternative Agriculture 9200 Edmonston Road, Suite 117, Greenbelt MD 20770-1551, phone 301-441-8777, fax 301-220-1551.
3. *American Small Farm* (10 issues/$18 yr), from Magnet Communications, Inc. 9420, Topanga Canyon Blvd., Suite 202, Chatsworth CA 91311-5759, Phone 818-727-2236
fax 818-727-1358.
4. *Biodynamic* (bimonthly $14-$75 yr.), from Bio-Dynamic Farming and Gardening Association, Inc., P.O. Box 550, Kimberton PA 19442, phone 610-935-7797 (800-516-7797 for CSA/BDA farms information), fax 610-983-3196, URL: http://www/his.com/~claymont/bda.html
5. Common Sense Pest Control *Quarterly* (quarterly $30 yr.), from Bio-Integral Resource Center (BIRC), P.O. Box 7414, Berkeley CA 94707, phone 510-524-2567.
6. *Countryside & Small Stock Journal* (bimonthly/$18 yr.), from Countryside & Small Stock Journal; N2601 Winter Sports Rd., Withee WI 54498 (to subscribe: W11564 Hwy. 64, Withee WI 54498), phone 1-880-551-5691 (715-785-7979 for editorial, advertising).
7. *Countryside & Small Stock Journal* (bimonthly/$18 yr.), from Countryside & Small Stock Journal; N2601 Winter Sports Rd., Withee WI 54498 (to subscribe: W11564 Hwy. 64, Withee WI 54498), phone 1-800-551-5691 (715-785-7979 for editorial, advertising).
8. *Farmer to Farmer* (bimonthly $15 yr.), from Community Alliance with Family Farmers (CAFF) Foundation, P.O. Box 73674, Davis CA 95617, phone 916-756-7428, fax 916-756-7857, e-mail f2f@igc.apc.org.
9. *Farmer's Information Network Organic Market News* (formerly Organic Wholesale Market Report: Organic Market News and Information Service (OMNIS)) (bimonthly/mail $65/ fax or e-mail $75 yr.), from FIN, P.O. Box 2067, Santa Clara CA: 95055-2067, phone 408-247-2067, fax 408-247-5823, e-mail FarmerNet@aol.com.
10. *The Furrow* (8 to 10/ complimentary to qualified farmers), from Deere & Company, John Deere Rd., Moline IL 61265-8098, phone 309-765-5403, fax 309-765-4498

11. *Growing for Market* (monthly/$26 yr.), from Fairplain Publications P.O. Box 3747 Lawrence KS 66046, phone/ fax 913-841-2559 (also telephone 1-800-307-8949).
12. *Journal of Pesticide Reform (*quarterly/membership $15 or $25 yr.), from Northwest Coalition for Alternative to Pesticides (NCAP), P.O. Box 1393, Eugene OR 97440, (office: 1249 Willamette St., Eugene OR 97401) phone 541-344-5044, fax 541-344-6923, e-mail ncap@igc.apc.org url: http://www.efn.org/~ncap/.
13. *The Natural Farmer* (quarterly, $10 yr.), from Northeast Organic Farming Association (NOFA), 41 Sheldon Rd., Barre MA 01005-9252, phone 508-355-2853.
14. *Natural Foods Merchandiser* (monthly, includes Organic Times annual supplement/$48 yr.), from New Hope Communications, 1301 Spruce St., Boulder CO 80302-4832, phone 303-939-8440, fax 303-939-9559, URL: http://devel.newhope.com/public/nfm/NFM_home.html
15. *Organic Commodity Price Bulletin* (weekly fax service), from Organic Food Business News, Hotline Printing & Publishing, P.O. Box 161132, Altamonte Springs FL 32716-1132, fax 407-628-9935.
16. *Organic Food Business News* (monthly/$48 yr.), from Hotline Printing and Publishing, P.O. Box 161132, Altamonte Springs FL 32716-1132, phone 407-628-1377, fax 407-628-9935.
17. *Organic Food Matters* (quarterly/membership, sliding scale from $20), from Committee for Sustainable Agriculture (CSA), P.O. Box 838, San Martin CA 95046, phone 408-778-7366, fax 408-778-7186.
18. *Organic Gardening* (9 issues/$25 yr.), from Rodale Press, Inc., 33 E. Minor St., Emmaus PA 18098, phone 610-967-5171 (610-967-8715 for advertising).
19. *The Organic Report* (monthly/free with membership; non-members $50 yr.), from Organic Trade Association (OTA), P.O. Box 1078, Greenfield MA 01302, (office: OTA, 20 Federal St., Suite 3, Greenfield MA 01301), phone 413-774-7511, fax 413-774-6432, e-mail ota@igc.apc.org.
20. *The Packer* (weekly; semi-weekly at end of each quarter/$55 yr.) (special issues: "Packer's Produce Availability and Merchandising Guide" and "Fresh Trends: A Profile of the Fresh Produce Consumer" available separately), from Vance Publishing Crop., 10901 W. 84th Terr., Lenexa KS 66214-1631, phone 913-438-8700.
21. *The Permaculture Activist* (quarterly $16 yr.), from Permaculture Activist, P.O. Box 1209, Black Mountain NC 28711.USA
22. *Rural Heritage* (bimonthly/$19 yr.), from Allan Damerow, Publisher, 281 Dean Ridge Lane, Gainesboro TN 38562-5039, phone 612-268-0655. (subscription includes annual supplement "The Evener Work Horse, Mule, and Oxen Directory and Guide").
23. *Small Farm Today* (bimonthly $21), from Missouri Farm Publishing Inc., 3903 W. Ridge Trial Rd, Clark MO 65243, phone 800-633-2535 or 314-697-3525, fax 314-687-3148.

24. *Small-Scale Agriculture Today* (quarterly/complimentary) from Office for Small Scale Agriculture (OSSA), c/o Bud Kerr, Editor, USDA/CSREES, Ag Box 2244, Rm. 3349, South Bldg., Washington DC 20250-2244, phone 202-720-5245, fax 202-205-2448, e-mail bkerr@reeusda.gov.
25. *The Stockman Grass Farmer* (montly/$24.50 yr.) from Mississippi Valley Publishing Corporation, 5135 Galaxies Drive, Suite 300C, Jackson MS 39206, phone 601-981-4805, fax 601-981-8558, (to subscribe: SGF, P.O. Box 9607, Jackson MS 39286-9909, phone 800-748-9808, fax 601-981-8558).
26. *Sustainable Farming*: (quarterly/membership C$20 or C$17. 12 yr., US rates on request), from Resources Efficient Agricultural Production (REAP)-Canada, Box 125, Glenaladale House, Ste-Anne-de-Bellevue, Quebec H9X 3V9, Canada, phone 514-398-7743, fax 514-398-7972.

Appendix

4. ORGANIC COTTON DIRECTORY

Available from:
Nathan Boone, OFC Coordinator
Tel: +1-530-750-2265
Fax: +1-530-756-3122
email: *ofc@igc.org*

1. **Tropical commodities and their markets: a guide and directory**
By P Robbins. 1995. 276p. ISBN 0 7494 16270.
Twin Ltd, 5-11 Worship Street
London EC2A 2BH
UK. Kogan Page
120 Pentonville Road
London N1 9JN, UK.
2. **Fair trade yearbook towards 2000,** 1998
EFTA, 151 p. Available in Dutch, French, English, Spanish, Italian and German.
3. **Guide to organic cotton**
1998
The Pesticides Trust
Eurolink Centre
49 Effra Road
London SW2 IBZ, UK
Tel: +44-71-274-8895
Fax: +44-71-274-9084.
4. **Traditional marketing system**
by L Cammann (ed.). 1992. 204 p
German Foundation for International Development (DSE)
Hans-Boeckler-Strasse 5
D-5300 Bonn 3, Germany.
(DSE Dok No. 1662 A/a).
5. **Participatory market planning: facilitators' manual.** 1998. 22 p. The Marketing System and Enterprise Promotions Group, Small Islands Agricultural Support Services Programme (SMISLE), PO Box 1193, Cebu City 6000, Philippines.

Appendix

5. ECO FRIENDLY FARMING

Organic farming

1. **Rodale Institute Research Center**
   611 Siegfriedale Rd.
   Kutztown, PA 19530
   (610) 683-6383
2. **Organic Crop Improvement Association (OCIA)-International**
   3185 Township Rd. 179
   Bellefontane, OH 43311
   (513)592-4983

Biodynamic Farming

3. **Biodynamic Farming & Gardening Assen.**
   P.O. Box 550
   Kimberton, PA 19442 USA
   (215) 935-7797
4. **Michael Fields Agricultural Institute**
   West 2493 Country Rd. ES
   East Troy, WI 53120 USA
   (414) 642-3303
5. **Bio-Dynamic Research Institute**
   Main Road, Powelltown, 3797
   Victoria, Australia.
   Tel: +61-59-66 73 33,
   Fax:+61-59-66 74 33
   http://www.demeter.net/International/Betriebe/australien.htm
6. **Institute for Biodynamic Research**
   Brandschneise 5, D-64295
   Darmstadt, Germany.
   Tel: +49-6155 8421-0
   Fax: +49-6155 8421-25.
   http://www.ibdf.de/
7. **Biological Farming**
   2617-C Edenborn Ave
   Metairie, LA 70002 USA
   (504) 889-2100
8. **Biological Farming Systems Group**
   Wageningen University
   Marijkeweg 22, Wageningen
   Building nr. 527, 6709 PG
   Wageningen,
   The Netherlands.
   Tel: +31.(0)317.478201
   Fax: +31.(0)317.478213.
   http://www.dpw.wageningen-ur.nl/biob/

Nature Farming

9. **Mokichi Okada Association**
   c/o Pacific Culture Center
   1835 Vancouver Drive
   Honolulu, HI 96822
10. **Nature Farming Research & Develoopment Center**
    6495 Santa Rosa Rd.
    Lompac, CA 93436
    USA

Appendix

6. VERMICOMPOST RESOURCES

1. **Appropriate Technology Transfer for Rural Areas (ATTRA)**
   Senior Technical Specialist,
   ATTRA, P.O. Box 3657, Fayetteville,
   Arkansas 72702 USA
   Phone: (800) 346-9140
   Fax: (501) 442-9842
   E-mail: aliceb@ncatfyv. uark.edu
2. **Beaver River Associates**
   P.O. Box 94
   West Kingston
   RI 02982 USA
   Phone & Fax: (401) 782-8747
3. **Bhawalkar Earthworm Research Institute**
   A/3 Kalyani
   Pune, 411 0317 India
   Phone: 91 212 413 714
   Fax: 91 212 442 305
4. **CSIRO, Davies Laboratory**
   Contact: Dr. A.V. Spain
   CSIRO, Private Mail Bag
   Aitkenvale, Queensland 4814
   Australia
   Phone: 77-719511
   Fax: 77-251009
5. **Enviro-Ganics Ecological Systems Inc.**
   P.O. Box 20241
   Grantham Postal Statino
   St. Catharines
   Ontario L2M 7W7
   Canada
   Phone: (905) 646-8690
   Fax: (905) 937-0517
6. **Environmental Earthworm Products, Inc.**
   Contact: Frank Stevenson
   8114 Port Said St. Orlando
   FL 32813 USA
   Phone: (407) 678-6454
   Fax: (407) 677-0790
7. **The French Institute For Scientific Research For Cooperative Development (ORSTOM)**
   Contact: Dr. Patrick Lavelle
   ORSTOM
   Ceentre de Bondy
   72, Route d'Aulnay
   95143 Bondy Cedex France
   Phone: 33 1 48 02 55 01
   Fax: 33 1 48 47 30 88
8. **The National Agricultural Library (NAL)**
   10301 Baltimore Blvd.,
   Beltsville, MD 20705 USA
   Phone: (301) 504-5755
9. **Oregon Soil Corporation**
   7810 SSW Bunker Oak Rd.,
   Aloha, OR 97006, USA
   Phone: (503) 629-5933.
10. **Original Vermitech Systems, Ltd.**
    2338 Queen Street East
    Toronto, Ontario M4E 1G9
    Canada
    Phone: (416) 693-1027
    Fax: (416) 698-02
11. **Vermitechnology Unlimited**
    P.O. Box 130
    Orange Lake, FL 32681
    USA
    Phone: (904) 591-1111
    Fax: (904) 591-4550

12. **Hy Hunter Industries**
    2001 Highwater Raod
    Granada Hills
    CA 91344 USA
    Phone: (818) 363-0202
    Fax: (818) 360-8562
13. **Institute of Natural Organic Agriculture (INORA)**
    Contact: Professor M.R. Bhiday
    100, Kalpana Apartments
    Erandwan
    Pune – 411 004 India
    Phone: 336215
14. **Maharashtra Agricultural Bloteks**
    Contact: Dr. Hemangee Jambhekar
    B/9 Shivai Housing Society
    Near Sane Guruji Smark
    Pune, 411 030, India
    Phone: 91 212-431983
15. **Resource Conversion Corp.**
    7825 Fay Ave. Suite 380
    La Jola, CA 92037 USA
    Phone: (619) 551-4800
    Fax: (619) 459-2794
16. **Sound Resource Management Group**
    Seattle, WA 98101-1540 USA
    Phone: (206) 622-9454

Archives on the Internet

17. **Venaura Farm**
    7119 Old Greensboro Rd.
    Chapel Hill, N.C. 27516
    E-mail: london@sunsite.unc.edu
    gopher and www: gopher://sunsit.unc.edu:70/11/.pub/academic/agriculture/sustainable_agriculture/vermiculture
    Description: This gopher site contains information about sustainable agriculture and vermicomposting.
18. **Unco Industries**
    7802 Old Spring Street
    Racine, WI 53406
    USA
    Phone: (414) 886-2665
    Description: Unco Industries sells earthworms and vermicompost. They have a network of 32 producers around the united States and have worked with groups is Israel, Belgium and Mauritius.
19. **The Worm Connection**
    581 Camino Manzanas
    Thousand Oaks
    CA 91360 USA
    Phone: (805) 496-2872
    Description: The Worm Connection promotes sustainable agriculture and organic gardening. Morhar has 19 years experience in recycling organic materials. The Worm Connection diverts landfill waste for vermiculture. The vermicompost is sold by the pail or truckload. They also sell vermiculture materials.

Publications

20. **Worms Eat My Garbage**
    Flower Enterprises
    0332 Shaver Raod
    Kalamazoo, MI 49002 USA
    Phone: (616) 343-4505
21. **The Warmer Bulletin**
    The Journal of the World Resource Foundation
    Bridge house, High Street
    Tonbridge
    Kent, TN91DP UK
    Phone: 44 (0) 1732368333
    Fax: 44 (0) 1732368337
22. **Worm Digest**
    Contact: Stephen White
    P.O. box 544
    Eugene, OR 97440-998 USA
    Phone: (503) 896-9058

Appendix

7. ENVIRONMENTAL-RELATED TRADE FAIRS AND PUBLICATIONS IN EUROPE

1. **BIO FACH**
   The largest Trade Fair in the world for Organic Food and Natural Products. Germany
   Organizers: Ökowelt Veranstaltungs-GmbH
   Industriestrasse 12
   D-91186 Büchenbach
   Germany.
   Tel: +49-9171-9610-0
   Fax: +49-9171-4016
   Email: info@biofach.de
   Internet: http://www.biofech.de.
2. **The European Fair Trade Association**
   EFTA, Witmakersstraat 10
   6211 JB Maastricbt
   The Netherlands
   Tel: +31-43-3258433
   Fax: +31-43-3258433
   Email: efta@antenna.nl
   Website: http://www.antenna.apc.org/fair-trade/fair71.btml.
3. **BIOFOOD.NET BUSINESS**
   Lombardenstraat 8
   B-2000 Antwepen
   Belgium
   Tel/Fax: +32-3-2261870
   Email: office@biofood.net
   Homepage; http://www.biofood.net,
4. **SOIL ASSOCIATION**
   The Organic Food & Farming Centre
   86 Colston Street
   Bristol BS1 5BB
   Tel: +44-1179-290661
   Fax: +44-117-9252504
   Homepage: http://www.earthfoods.co.uk.
5. **MAX HAVELAAR**
   Lucas Bolwerk 7
   PO Box 1252
   3500 BG Utrecht
   The Netherlands
   Tel: +31-30-2334602
   Email: maxhavelaar@maxhavelaar.nl
   Homepage: http://www.maxhavelaar.nl
6. **Ecotop**
   Brussels International Trade Fair
   Place de Belgique
   B-1020 BRUSSELS
   Belgium
   Phone: +32 2 4770432 / +32 2 4770477
   Fax: +32 2 4770391
7. **IFEST**
   Maaltekouter 1
   B-9051 GENT
   Belgium
   Phone: +32 9 2419211
   Fax: +32 9 2419495
8. **DANMILJÖ**
   Messecenter Herning
   P.O. Box 524
   D-7400 HERNING
   Denmark
   Phone: +45 97126000
   Fax: +45 97223060
9. **Euro Environment**
   Deloitte & Touche
   Phone: +45 981 80133 / 993 55555
   Fax: +45 981 85233 / 993 55580
   E-mail: euro@akkc.dk
   Internet: http://www.akkc.dk

10. **Pollutec**
SEPFI Technoexpo
70 Rue Rivay
F-92532 LEVALLOIS PERRET CEDEX
France
Phone: +33 1 47 562115
Fax: +33 1 47 562120
International trade fair, held in Lyon (France), devoted to technologies and services to the environment. The emphasis lies on pollution prevention and clean-up. The exposition focuses on water, waste, air, noise, analysis-measurement-control, and energy management.
Frequency: Annual

11. **Econova**
Deutsche Messe AG
Messegelände
D-30521 HANNOVER
Germany
Phone: +49 9 511 890
Fax: +49 9 511 8932626

12. **Envitee**
Düsseldorfer Messegesellschaft mbH
NOWEA
P.O. Box 101006
D-40001 DÜSSELDORF
Phone: +49 2 11456001
Fax: + 49 2 114560668

13. **Entsorga**
Messe und Ausstellungsgesellschaft m.b.H. Cologne
P.O. Box 210760
D-5232 COLOGNE
Germany
Phone: +49 221 8210
Fax: + 49 221 8213411

14. **IFAT**
Münchener Messe – und Ausstellungsgesellschaft mbH
Messegelände
D-80325 MÜNCHIN
Germany
Phone: +49 89 51070
Fax: + 49 89 5107506

15. **ÖKO + ÖKO BAU**
ÖKOWELT Veranstaltungs-GmbH
Industriestr. 12
91186 BÜCHENBACH
Germany
Phone: +49 9171 96100
Fax: +49 9171 4016
E-mail: info@biofach.de

16. **ÖKOWELT**
ÖKOWELT VERANSTALTUNGS-GmbH
Industriestr. 12
91186 BÜCHENBACH
Germany
Phone: +49 9171 96100
Fax: +49 9171 4016
E-mail: info@biofach.de

17. **InterHOLZ**
Deutsche Messe AG
Messegelände
D-30521 HANNOVER
Germany
Phone: +49 511 89 0
Fax: +49 511 89 331 26
or

18. **Fachgemeinschaft Holzbearbeitungsmaschinen im VDMA e.V.**
Lyoner Straβe 18
D-60528 FRANKFURT AM MAIN
Germany
Phone: +49 69 66 03 13 40
Fax: +49 69 66 03 16 21

The Netherlands

19. **Ecotech**
Koninklijke Nederlands Jaarbeurs Ecotech
Jaarbeursplein
NL-3521 AL UTRECHT
The Netherlands
Phone: +31 30 2955911
Fax: +31 30 2940379

20. **MACHEVO**
Koninklijke Nederlands Jaarbeurs
MACHEVO
Jaarbeursplein
NL-3521 AL UTRECHT
The Netherlands
Phone: 31 30 2955911
Fax: +31 30 2940379

21. **Petrotech**
Amsterdam RAI International
Exhibition & Congress
P.O. Box 77777
1070 MS AMSTERDAM
The Netherlands
Phone: +31 20 5491212
Fax: +31 20 6464469
E-mail: mail@rai.nl

22. **Interclean**
Amsterdam RAI International
Exhibition & Congress
Organisers
P.O. Box 77777
1070 MS AMSTERDAM
The Netherlands
Phone: +31 20 5491212
Fax: + 31 20 6464469
E-mail: mail@rai.nl
Environment-related trade fairs outside Europe

Brazil

23. *Quimtec*
ISC KölnMesse GmbH
Deutz-Mulheimerstr. 30
D-50679 KÖLN
Germany
Phone: +49 0221 821 0

Canada

24. **Globe**
Globe Foundation of Canada
World Trade Centre
504-999 Canada Place
VANCOUVER (BC), V6C 3E1
Canada
Phone: +1 604 7751994
Fax: +1 604 6668123

People's Republic of China

25. **CIP**
CIEC Exhibition Co. (HK) Ltd.
Rm 3909, China Resources Building
26 Harbour Road
Wanchai
HK-HONKONG
Phone: +852 8275078
Fax: +852 8275535

26. **ITEP China**
Hannover Messe International GmbH
Messegelände
D-30521 HANNOVER
Germany
Phone: +49 511 8931400
Fax: +49 511 8932646

India

27. **Environmental India**
Exhibition India
C-390
Defence Colony
NEW DELHI 110 024
India
Phone: +91 11 462 2710
E-mail: exhibind@giasd101.vsnl.net.in

28. **Envirotech International**
Tafcon Projects (India) Pvt. Ltd.
C-60, Nizamuddin East
NEW DELHI 110 013
India
Phone: +91 11 4633889 / 4692182 / 4692184
Fax: +91 11 4635215
web: http://www.tafcon.com
or:
Royal Dutch Jaarbeurs
P.O. Box 8500
3503 RM UTRECHT
The Netherlands
Phone: +31 30 2955549
Fax: +31 30 2955709

Poland

29. **POLEKO**
Poznan International Fair
Glogowaks 14
60734 POZNAN
Poland
Phone: +48 61 8692592
Fax: +48 61 8665827
E-mail: MTP@pol.pl

30. **Wastecare**
ISWA
Bremerholm 1
DK-1069 COPENHAGEN K
Denmark
Phone: +45 33 914491
Fax: +45 33 919188

31. **Envirotek**
Natasha Berryman
ITE J/V Ltd.
112 A Shirland Rd.
LONDON W9 2EQ
United Kingdom
Phone: +44 171 306 0033

Singapore

32. **ENVIROASIA**
International Environmental Technologies Exhibition
Times Conferences and Exhibitions Pte. Ltd.
1 New Industrial Road
Times Centre
SINGAPORE 1953
Singapore
Phone: +65 3807428
Fax: +65 2865754

33. **Aquatech Asia**
Aquatech Singapore
1 Maritime Square, #09-01 WTC
099253
SINGAPORE
Phone: +65 2722250
Fax: +65 2726744
E-mail: raispore@singen.com.sg

South Africa

34. **Safety & Health in the Environment**
South Africa RAI
P.O. Box 652495
2010 BENMORE
South Africa
Phone: +27 11 444 3937
Fax: +27 11 444 7987
E-mail: raisa@iafrica.com

35. **Safety, Health & Environment**
South Africa RAI
P.O. Box 652495
2010 BENMORE
South Africa
Phone: +27 11 444 3937
Fax: +27 11 444 7987
E-mail: raisa@iafrica.com

Turkey

36. **PRO-PAK; International Packaging Exhibition**
Overseas Exhibition Services Ltd.
11 Manchester Square
GB-LONDON W 1M 5AB
United Kingdom
Phone: +44 171 4861951
Fax: +44 171 4868773
International trade fair for packaging technologies and equipment. Held in Istanbul (Turkey).
Frequency: Annual

Appendix

8. IMPORT PROMOTION ORGANIZATIONS IN EUROPE

1. **International Trade Centre (IFC)**
   P.O. Box 10
   CH-1211 GENEVA 10
   Switzerland
   Phone: +41 22 730 0111
   Fax: +41 22 7334439
   Internet: http://www.intracen.org
2. **Bundeskammer der Gewerblichen Wirtschaft Wiedner Hauptstrasse 63**
   A-1045 VIENNA
   Austria
   Phone: + 43 1 501050
   Fax: +43 1 50206250
   Børsen
   DK-1217 COPENHAGEN K
   Denmark
   Phone: +45 33 950500
   Fax: +45 33 325216
   Email: dipo@commerce.dk
   Internet: http://www.dipo.dk
3. **Finnish Import Promotion Office for Finnish import Products from Developing Countries (FINIPO)**
   P.O. Box 908
   SF-00100 HELSINKI
   Finland
   Phone: +358 4 69591
   Fax: +358 4 851037
4. **Protrade**
   P.O. Box 5180
   D-65726 ESCHBORN
   Germany
   Phone: +49 61 96 793169
   Fax: +49 61 96 797414
5. **Irish Trade Board**
   P.O. Box 203
   IRE-DUBLIN 4
   Ireland
   Phone: +353 1 2695011
   Fax: +353 1 2695820
6. **Instituto Nazionale per il Commercio Estero (ICE)**
   P.O. Box 10057
   I-00144 ROME-EUR
   Italy
   Phone: +39 6 59921
   Fax: +39 6 5910508
7. **Centre for the Promotion of Imports from developing countries (CBI)**
   P.O. Box 30009
   3001 DA ROTTERDAM NL-
   The Netherlands
   Phone: +31 10 2013434
   Fax: +31 10 4114081
   E-mail: cbi@cbi.nl
   Internet: http://www.cbi.nl
8. **Norwegian Agency for Development Co-operation (NORAD)**
   P.O. Box 8304, Dep
   N-0030 OSLO 1
   Norway
   Phone: +47 22 314400
   Fax: +47 22 314401

9. **Swedish International Development Co-operation Agency (Sida)**
S-10525 STOCKHOLM
Sveavägen 20
Sweden
Phone: +46 8 6985000
Fax: +46 8 208864
E-mail: sida@sida.se
Internet: http://www.sida.se

10. **Swiss Office for Trade Promotion (OSEC)**
P.O. Box 1128
CH-1001 LAUSANNE
Switzerland
Phone: +41 21 3203231
Fax: +41 21 3207337
E-mail: mdarbre@osec.ch
Internet: http://www.osec.ch

Other Governmental Developmental Co-Operation Organizations In Europe

1. **Danish International Development Assistance (DANIDA)**
2 Asiatisk Plads
DK-448 COPENHAGEN K
Denmark
Phone: +45 9 20000
Fax: +45 9 2540533

2. **Finnish International Development Agency (FINNIDA)**
Tehtaankatu la
SF-00140 HELSINKI
Finland
Phone: +358 0134151

3. **Deutsche Gesellschaft für Technische Zusammenarbeit (GTZ) GmbH**
P.O. Box 5189
D-65726 ESCHBORN
Germany
Phone: +49 61 96793169
Fax: +49 61 96797414

4. **Directorate General for International Co-operation (DGIS)**
Ministry of Foreign Affairs
P.O. Box 20061
NL-2500 EB THE HAGUE
The Netherlands
Phone: +31 70 3486486
Fax: +31 70 3484848

Non-Governmental Development Co-Operation Organizations in Europe

1. **Both Ends**
Damrak 28-30
NL-1012 LJ AMSTERDAM
The Netherlands
Phone: +31 20 6230823
Fax: +31 20 6208049.

2. **European Fair Trade Association**
Witmakersstraat 10
NL-6211 JB MAASTRICHT
The Netherlands
Phone: +31 43 3256917
Fax: +31 43 3258433

3. **Forum Syd**
P.O. Box 17510
S-118 91 STOCKHOLM
Sweden
Phone: +46 8 7027700
Fax: +46 8 7029099
E-mail: forum.syd@forumsyd.se
Internet: http://www.forumsyd.se

4. **COLEACP**
5, Rue de la Corderie
Centra 342
F-94586 RUNGIS CEDEX
France
Phone: +33 1 41800210
Fax: +33 1 41800219

Appendix

9. DIRECTORATES GENERAL AND AGENCIES OF THE EUROPEAN UNION

1. **Commission of the European Communities (CEC)**
   DG-I/C
   Wetstraat 200
   B-1049 BRUSSEL
   Belgium
   Phone: +32 2 2990739
   Fax: +32 2 2962316
   For development and regulation of the EU market; DG-I/C for foreign trade
2. **Commission of the European Communities (CEC)**
   DG-VIII Development and Co-operation
   Wetstraat 200
   B-1049 BRUSSELS
   Belgium
   Phone: +32 2 2999852
   Fax: +32 2 2993002
3. **Commission of the European Communities (CEC)**
   DG XI
   Wetstraat 200
   B-1049 BRUSSELS
   Belgium
   Phone: +32 2 2969515
   Fax: +32 2 2969560
4. **Commission of the European Communities (CEC)**
   DG XXIII – B/l -
   Euro Info Centra
   Wetstraat 200 (AN 80)
   B-1049 BRUSSELS
   Belgium
   Fax: +32 2 2957335
5. **European Environment Agency**
   Kongens Nytorv 6
   COPENHAGEN K
   Denmark
   Phone: +45 33 367100
   Fax: +45 33 146599
   EU's central point for collection and dissemination of environmental data; open for questions from the public
6. **European Foundation for the Improvement of Living and Working Conditions**
   House, Shankill
   IRC-Co. DUBLIN
   Ireland
   Phone: +353 1 2826888
   Fax: +353 1 2826456

Appendix

10. RELEVANT UNITED NATIONS ORGANIZATIONS

1. **United Nations Environment Programme (UNEP)**
   P.O. Box 30552
   NAIROBI
   Kenya
   Phone: +254 2 230800/520600
   Fax: +254 2 520711
2. **United Nations Environment Programme (UNEP)**
   Industry and Environment Programme Tour Mirabeau
   39-43, quai André Citröen
   F-75739 PARIS CEDEX
   France
   Phone: +33 1 40 588850
   Fax: +33 1 40 588850
3. **United Nations Environment Programme (UNEP)**
   Geneva Executive Center
   16, Avenue Jean Trembley
   CH-1209 CHATELAINE/GE
   Switzerland
   Phone: +41 22 9799111
   Fax: +41 22 7973420
4. **UNEP Working Group on Sustainable Product Development (UNEP-WG-SPD)**
   c/o University of Amsterdam
   Department Science Environment Science
   Plantage Muidergracht 14
   NL-1018 TV AMSTERDAM
   The Netherlands
   Phone: +31 20 5256237
   Fax: +31 20 5256272
   UNEP working group on sustainable product development.
5. **Food and Agriculture Organization of the United Nations (UN FAO)**
   Environment and Sustainable Development Programmes
   Via della Terme di Caracalla
   I-000100 ROME
   Italy
   Phone: +39 6 57971
   Fax: +39 6 57973152 / 57975155
6. **United Nations Conference on Trade and Development (UNCTAD)**
   Division of Trade and Environment Palais des Nations
   Greentrade
   CH-1211 GENEVA 10
   Switzerland
   Phone: +41 22 9171234
   Fax: +41 22 9070042
   Restructures the traditional trade patterns in order for developing countries to play their part in world trade; manages a database
7. **United Nations Development Programme (UNDP)**
   Environment and Natural Resource Unit
   One United Nations Plaza
   NEW YORK, N.Y. 10017
   USA
   Phone: +1 212 9065000
   Fax: +1 212 9066947

8. **United Nations Economic Commission for Europe (ECE) Palais des Nations**
   CH-1211 GENEVA 10
   Switzerland
   Phone: +41 22 9172893 (direct)
   +41 22 9171234
   Fax: +41 22 9170039
   +41 22 9170123
9. **United Nations Industrial Development Organization (UNIDO)**
   P.O. Box 300
   Vienna International Centre
   A-1400 VIENNA
   Austria
   Phone: +43 1 211310
   Fax: +43 1 232156
10. **United Nations International Labour Office (ILO)**
    4, Route des Morillons
    CH-1211 GENEVA 22
    Switzerland
    Phone: +41 22 7996111
    Fax: +41 22 7986358
11. **United Nations World Health Organization (UN WHO)**
    20, avenue Appia
    CH-1211 GENEVA 27
    Switzerland
    Phone: +41 22 7912111
    Fax: +41 22 7910746
12. **International Trade Centre UNCTAD/GATT**
    Palais des Nations
    CH-1211 GENEVA 10
    Switzerland
    Phone: +41 22 7300111
    Fax: +41 22 7334439

Appendix

11. SUBSIDY AND SUPPORT INSTITUTIONS

European Union

1. **Commission of the European Communities (CEC)**
   TACIS Information Office
   DG I AN 88 1/06
   Rue de la Loi 200
   Belgium
   Phone: +32 2 2952585
   Fax: +32 2 2310441
2. **Commission of the European Communities (CEC)**
   DG XI – Environment, Nuclear, Safety and Civil Protection
   Unite B.2 – LIFE programme
   Rue de la Loi 200
   B-1049 BRUSSELS
   Belgium
   Phone: +32 2 2969552
   Fax: +32 2 2969561
3. **Commission of the European Communities (CEC)**
   DG XIII-E
   Montoyerstraat 75
   B-1040 BRUSSELS
   Belgium
   Phone: +32 2 2990269
   Fax: +32 2 2969037
4. **European Bank for Reconstruction and Development (EBRD)**
   1 Exchange Square
   GB-LONDON EC24 2EH
   United Kingdom
   Phone: +44 71 3386000
   Fax: +44 71 3386100
   Internet: http://www.worldbank.org/html/extdr/backgrd/ibrd/ibrd.htm
5. **European Development Fund (EDF)**
   Rue de la Loi 200
   B-1049 BRUSSELS
   Belgium
   Phone: +32 2 235111
   Fax: +32 2 2362725
6. **European Development Fund (EDF)**
   100, bd Konrad Adenauer
   L-2950 LUXEMBOURG
   Luxembourg countries
   Phone: +35 2 43791
   Fax: +35 2 437704
   Internet: http://www.eib.org
7. **Directorate General for International Co-operation (DGIS)**
   Ministry of Foreign Affairs
   P.O. Box 20061
   NL-2500 EB THE HAGUE
   The Netherlands
   Phone: +31 70 3486486
   Fax: +31 70 3484848
   The Netherlands Ministry of Foreign Affairs; DG for international development co-operation; DPO/BL Division of DGIS manages the MILIEV and ORET subsidy programmess
8. **EDCS, Ecumenical Development Co-operative Society**
   P.C. Hooftlaan 3
   3818 HG AMERSFOORT
   The Netherlands
   Phone: +31 33 4633122
   Fax: +31 33 4650336
   E-mail: office@edcs.nl

9. **ICCO**
P.O. Box 151
3700 AD ZEIST
The Netherlands
Phone: +31 30 6927816 / 819
Fax: +31 30 6925614
E-mail: comm-unit@icco.nl

10. **NOVIB**
P.O. Box 30919
2500 GX THE HAGUE
The Netherlands
Phone: +31 70 3421621
Fax: +31 70 3614461
E-mail: admin@novib.nl

11. **Humanistisch Instituut voor Ontwilkkelings-samenwerking (Hivos)**
Rammweg 16
2596 HL THE HAGUE
The Netherlands
Phone: +31 70 3765500
Fax: +31 70 3734472
Supporting small-scaled projects in developing countries, particularly in the fields of environment, education, women's rights and human rights

12. **Bilance**
P.O. Box 77
2340 AB OESTGEEST
The Netherlands
Phone: +31 71 515950
Fax: +31 71 5175391
E-mail: bilance@antenna.nl
Internet: http://www.antenna.nl/bilance/index.html
Supporting small-scaled projects in developing countries, particularly in the fields of environment and women's rights

13. **Swiss Agency for Development and Cooperation**
(Environment, Forest, Energy Division)
Schwarztorstrasse 59
CH-3003 BERN
Switzerland
Phone: +41 31 325 92 82
Fax: +41 31 325 93 62

Subsidy and support Institutions Outside the European Union

1. **African Development Bank (ADB)**
01 BP 1387
ABIDJAN 01
Côte d'Ivoire
Phone: +225 20 4444
Fax: +225 20 4171
Internet: http://www.afdb.org

2. **Asian Development Bank (ADB)**
P.O. Box 789
1099 MANILA
Philippines
Phone: +632 711 3851
Fax: +632 741 7961 / 6816
Email: information@mail.asiandevbank.org
Internet: http://www.asiandevbank.org

3. **Caribbean Development Bank (CDB)**
P.O. Box 408
Widley
ST. MICHAEL
Barbados
Phone: +1 809 431 1600
Fax: +1 809 426 7269
Internet: http://www.cdbank.org

4. **Inter-American Development Bank (IDB)**
1300 New York Avenue NW
WASHINTON DC 20577
USA
Phone: +1 202 623 1000
Fax: +1 202 789 2835
Internet: http://www.iadb.org

5. **International Fund for Agricultural Development (IFAD)**
Via del Serafico 107
I-00142 ROME
Italy
Phone: +39 6 54591
Fax: +39 6 5043463
E-mail: ifad@ifad.org
Internet: http://www.ifad.org

6. **World Bank**
   1818 H Street NW
   USA
   Phone: +1 202 477 1234
   Fax: +1 202 477 6391
   Internet: http://www.
   worldbank.org

Appendix

12. INTERNATIONAL RESEARCH AND POLICY INSTITUTES

1. **Business Council for Sustainable Development**
   Route de Florissant 160
   CH-1231 CONCHES
   Switzerland
   Phone: +41 22 8393100
   Fax: +41 22 8393131
   E-mail: info@wbcsd.ch
   Internet: http://www.wbcsd.ch
2. **Caribbean Conservation Association (CCA)**
   Savannah Lodge
   Thc Garrison
   ST. MICHAEL
   Barbados
   Phone: +1 809 426 5373/9635
   Fax: +1 809 429 8483
   E-mail: CCA@caribsurf.com
   Internet: http://www.tidco.co.tt/local/seduweb/cca/cca.htm
3. **International Environmental Bureau ICC-OEE**
   c/o ICC
   38, Cours Albert 1
   F-75008 PARIS
   France
   Phone: +33 1 49532952
   Fax: +33 1 42258663
4. **International Institute for Sustainable Development**
   212 McDermot Avenue
   WINNIPEG, MANITOBA
   Canada, R3B OS3
   Phone: +1 204 958 7700
   Fax: +1 204 958 7710
   Internet: http://www.iisdl.ca/
5. **The International Network for Environmental Management**
   Hellgrud 92
   D-2000 WEDEL
   (Schleswig Holstein)
   Germany
   Phone: +49 4103 84019
   Fax: +49 4103 13699
6. **International Union for Conservation of Nature and Natural Resources**
   (IUCN)
   Rue Mauverney 28
   CH-1196 GLAND
   Switzerland
   Phone: +41 22 9990001
   Fax: +41 22 9990002
   Internet: http://www.uicn.org
7. **Organisation for Economic Co-operation and Development (OECD)**
   12, Avenue d'Eylon
   F-75116 PRRIS CEDEX 16
   France
   Phone: +33 1 44 053160
   Fax: +33 1 44 053160
   Internet: http://www.oecd.org
8. **World Resource Institute**
   709, New York Avenue, NW
   WASHINGTON, DC 20006
   United States
   Phone: +1 202 6622589/6622522
   Fax: +1 202 6380036

9. **World Trade Organization (WTO) (formerly GATT)**
   Division of Technical Barriers and Trade
   Centre William Rappard
   154, rue de Lausanne
   CH-1211 GENEVA 21
   Switzerland
   Phone: +41 22 7395111
   Fax: +41 22 7314206
   Internet: http://www.wto.org

Appendix

13. STANDARDIZATION INSTITUTES

1. **International Organisation for Standardisation (ISO)**
   P.O. Box 56
   CH-1211 GENEVA
   Switzerland
   Phone: +41 22 7490111
   Fax: +41 22 7333430
   E-mail: central@iso.ch
   Internet: http://www.iso.ch
2. Comité Européen de **Normalisation (CEN)**
   Third Countries Unit
   Rue de Strassart 36
   B-1050 BRUSSELS
   Belgium
   Phone: +32 2 5196825/5196847
   +32 2 5196854
   Fax: +32 2 5196819
3. **Dansk Standardiseringraad**
   Baunegaardsvej 73
   DK-2900 HELLFRUP
   Denmark
   Phone: +45 39 770101
   Fax: +45 39 770202
4. **Deutsches Institut für Normung**
   P.O. Box 1107
   D-1000 BERLIN 30
   Germany
   Phone: +49 30 26011
   Fax: +49 30 26011231
   Internet: http://www.din.de
   German institute for the development of national standards and the adoption of international standards
5. **Association Française de Normalisation**
   (AFNOR)
   Tour Europe
   Cedex 7
   F-92049 PARIS LA DEFENSE
   France
   Phone: +33 1 42915555
   Fax: 33 1 42915656
6. **Nederlands Normalisatie Instituut (NNI)**
   P.O. Box 5059
   NL-2600 GB DELFT
   The Netherlands
   Phone: +31 15 2690390
   Fax: +31 15 2690190
   Internet: http://www.nni.nl
7. **Norges Standardiserungs-forbund**
   P.O. Box 7020,
   Homansbyen
   N-0306 OSLO 3
   Norway
   Phone: +47 2 466094
   Fax: +47 2 464457
8. **Standardiserings kommissionen i Sverige**
   (The Swedish Standards Institution (SIS))
   P.O. Box 6455
   S-11382 STOCKHOLM
   Sweden
   Phone: +46 8 6103000
   Fax: +46 8 307757
   E-mail: info@sis.se
   Internet: http://www.sis.se

9. **Schweizerische**
Normenvereinigung (SNV)
Swiss Organisation
for Standardisation
Mühlebachstrasse 54
CH-8008 ZÜRICH
Switzerland
Phone: +41 1 2545454
Fax: +41 1 2545474
Internet: http://www.snv.ch

10. **British Standard Institution**
(BSI)
2 Park Street
GB-LONDON WI A 2BS
United Kingdom
Phone: +44 171 629 9000
Fax: +44 171 629 0506

Appendix

14. ACCREDITATION AND CERTIFICATION BODIES

1. **Netherlands Board for Accreditation (RvA)**
   Stationsweg 13F
   NL-3972 KA DRIEBERGEN
   The Netherlands
   Phone: +31 343 512604
   Fax: +31 343 518554
   Internet: http://www.rva.nl
2. **KEMA Nederland NV**
   P.O. Box 9035
   The Netherlands
   Phone: +31 26 3562477
   Fax: +31 26 3517362
   Internet: http://www/kema.nl
3. **Lloyd's Register Quality Assurance Ltd.**
   P.O. Box 701
   NL-3000 AS ROTTERDAM
   The Netherlands
   Phone: +31 10 2018400
   Fax: +31 10 4126806
4. **Bureau Verities Quality International Ltd.**
   Borough High Street
   GB-SE1 1XF LONDON
   United Kingdom
   Phone: + 44 1713 788113
   Fax: +44 1713 780309
5. **Det Norske Verities Industry B.V.**
   Certification Department
   P.O. Box 9599
   NL-3007 AN ROTTERDAM
   The Netherlands
   Phone: +31 10 4798700
   Fax: +31 10 4796768
   Internet: http://www.dnv.nl
6. **KPMG Certification**
   P.O. Box 74103
   NL-1070 BC
   AMSTERDAM
   The Netherlands
   Phone: +31 20 6568751
   Fax: +31 20 6568750
   Internet: http://www.kpmg.nl

Appendix

15. ENVIRONMENTAL PRESSURE GROUPS

1. **World Wildlife Fund for Nature (WWF)**
   Avenue du Mont Blane
   CH-1196 GLAND
   Switzerland
   Phone: +41 22 3649111
   Fax: +41 22 3645829
   Internet: http://www.wwf.org
   http://www.panda.org
2. **Greenpeace**
   Keizersgracht 174
   NL-1016 DW
   AMSTERDAM
   The Netherlands
   Phone: +31 20 4223344
   Fax: +31 20 6221272
   Internet: http://www.greenpeace.org
3. **Friends of the Earth**
   26-28 Underwood Street
   GB-LONDON N1 7JQ
   United Kingdom
   Phone: +44 171 4901555
   Fax: +44 111 4900881
   E-mail: foeint@antenna.nl
   Internet: http://www.xs4all.nl/~foeint
4. **Milieudefensie**
   Damrak 26
   1012 LJ AMSTERDAM
   The Netherlands
   Phone: +31 20 6221366
   Fax: +31 20 6275287
   E-mail: info@milieudefensie.nl
   Internet: http://www.milieudefensie.nl
5. **Stichting Natuur & Milieu**
   Donkerstraat 17
   NL-3511 KB UTRECHT
   The Netherlands
   Phone: +31 30 2331328
   Fax: +31 30 2331311
   Internet: http://www.dds.nl/~pdc/678/491/49103.htm
6. **Swedish Society for Nature Conservation (SNF)**
   P.O. Box 4625
   S-11691 STOCKHOLM
   Sweden
   Phone: +46 8 7026500
   Fax: +46 8 7020855
   E-mail: info@snf.se
   Internet: http://www.snf.se
7. **Miljöförbundet-Jordens** Vänner
   Fjällgatan 23A
   S-116 28 STOCKHOLM
   Sweden
   Phone: +46 8 6417579
   Fax: +46 8 7140154
8. **The Natural Step (Det Naturliga Steget)**
   Amiralitetshuset, Skeppsholmen
   S-11149 STOCKHOLM
   Sweden
   Phone: +46 8 54512500
   Fax: +468 54512599
   E-mail: info@detnaturligasteget.se
   Internet: http://www.naturalstep.se

Appendix

16. ENVIRONMENTAL LABELING CONTACT POINT

Competent Bodies for the EU Ecolabel

1. **Commission of the European Environment Communities**
   DG XI-C-6
   Rue de la Loi 200
   B-1049 BRUSSELS
   Belgium
   Phone: +32 2 2990324
   Fax: +32 2 2955684
2. **The Norwegian Foundation Environmental Product**
   Labeling
   Kristian August gate 5
   N-0164 OSLO
   Norway
   Phone: +47 22 360710
   Fax: +47 22 360729
3. **Unweltbundesamt (German Federal Environment Agency)**
   Bismarckplatx 1
   D-14193 BERLIN
   Germany
   Phone: +49 30 23145699
   Fax: +49 30 2993096
4. **Stichting Milieukeur (SMK)**
   Eisenhowerlaan 150
   NL-2517 KP THE HAGUE
   The Netherlands
   Phone: +31 70 3586300
   Fax: +31 70 3502517
5. **Ministry of Environment, Physical Planning and Public Works**
   17 Amaliados Street
   GR-EL-ATHENS
   Greece
   Phone: +30 1 6411717 / 6426531
   Fax: +30 1 6456758
6. **Danish Ministry of the National Agency of Environmental Protection**
   Strandgade 29
   DK-1401 COPENHAGEN
   Denmark
   Phone: +45 32 660100
   Fax: +45 32 660479
7. **SIS Environmental for Labeling**
   The Swedish Standards Institution
   P.O. Box 32955
   S-103 66
   STOCKHOLM
   Sweden
   Phone: +46 8 6135327
   Fax: +46 8 2130770
8. **RAL, Deutsches Institut für Gütesicherung & Kennzeichnung**
   Siegburger Strasse 39
   D-53757
   SANKT AUGUSTIN
   Germany
   Phone: +49 2241 160523
   Fax: +49 2241 160511

9. **UK Ecolabeling Board**
7th Floor,
Eastbury House
30-34 Albert
Embankment
GB-LONDON SE 1 7TL
United Kingdom
Phone: +44 171 8201199
Fax: +44 171 8201104

10. Association Espanola de **Normalizacion y Certification (AENOR)**
Fernandez de la Hoz 52
E-28010 MADRID
Spain
Phone: +34 1 3104851
Fax: +34 1 3104976

11. **Dirección General De Calidad Ambiental**
Departmento de Medio
Ambiente Generalitat
de Catalunya
C/Diagonal, 525
E-08029 BARCELONA
Spain
Phone: +34 3 4193085
Fax: +34 3 4197630

12. **Ministére de**
l'Environment
18 Montée de la Pétrusse
L-2918 LUXEMBOURG
Luxembourg
Phone: +35 2 4786816
Fax: +35 2 400410

13. **Association Francaise de Normalisation (AFNOR)**
Tour Europe Cedex 7
F-92049 PARIS
La Défense
France
Phone: +33 1 42915916
Fax: +33 1 42915911

14. **Direccao General da Industria**
Av. Conselheiro
Fernando de Sousa, 11
P-1000 LISBOA
Portugal
Phone: +351 1 659161
Fax: +351 1 691042

Contact Points for the EU Ecolabel

1. **Ministerie Volksgezondheid**
en Leefmilieu
Rijksadministratief Centrum
Vesalius 2/3
Pachécolaan 19 bus 5
B-10010 BRUSSELS
Belgium
Phone: +32 2 2104874
Fax: +32 2 2104873

2. **Il Ministro dell' Ambiente**
Via della Ferratella
in Laterano 33
I-00185 ROME
Italy
Phone: +39 6 50072170
Fax: + 39 6 50072048

National Ecolabeling Schemes

1. **Stichting Milieukeur**
Eisenhowerlaan 150
NL 2517 KP THE HAGUE
The Netherlands
Phone: +31 70 3586300
Fax: +31 70 3502517
Netherlands Milieukeur label

2. **SIS ECO-labeling**
The Swedish Standards Institution
P.O. Box 3295
S-10366 STOCKHOLM
Sweden
Phone: +46 8 613 5200
Fax: +46 8 213 070
Nordic Swan label (Sweden)

3. **Norwegian Foundation for Environmental Labeling**
Kristin Augusts gate 5
N-0614 OSLO
Norway
Phone: +47 22 36 0710
Fax: +47 22 36 0729
Nordic Swan label (Norway)

4. **Department of the Environment**
Custom House
IRL-DUBLIN 1
Ireland
Phone: +353 1 6793377
Fax: +353 1 8742423

5. **Finish Standards Association**
SFS. Environmental Labeling
P.O. Box 116
FIN-00241 HELSINKI
Finland
Phone: +358 0 149 9331
Fax: +358 0 1499 3320
Nordic Swan label (Finland)

6. **Umhverfismerki, Ministry of the Environment**
P.O. Box 8080
IS-128 REYKJAVIK
Iceland
Phone: +354 1 68 88 48
Fax: +354 1 68 18 96
Nordic Swan label (Iceland)

7. **Den Grøne Nøgle**
(The Green Key)
Vester Ferimagsgade 3, 5, Fl.
DK-1606 COPENHAGEN V
Denmark
Phone: +45 33 369910
Fax: +45 33 369919
Danish Green Key label

8. **RAL: Institut für Gütesicherung**
und-kennzeichnung E.V.
Siegburger Str. 39
D-53757 SANKT AUGSTIN
Germany
Phone: +49 224 1605 0
Fax: +49 224 1605 11
German Blaue Engel
(Blue Angel) label

9. **Stiftung Warentest**
P.O. Box 304141
D-10785 BERLIN
Germany
Phone: +49 30 2631 0
Fax: +49 30 2631 429
German Stiftung Warentest label

10. **Panda Fördergesellschaft**
für Umwelt GmbH
Hedderichstrasse 110
D-60591 FRANKFURT AM MAIN
Germany
Phone: +49 69 605003 72
Fax: +49 69 605003 66
German WWF Panda label

11. **Neckermann Versand AG**
Hanauer Landstrasse 360-400
D-60368 FRANKFURT AM MAIN
Germany
Phone: +49 69 404 01
Fax: +49 69 404 235
German Umweltprädikat label

12. **Karstadt**
Theodor Althoff-Strasse 2
D-45133 ESSEN
Germany
Phone: +49 201 7271
Fax: +49 201 7275216
German Umweltbewust
Einkaufen label

13. **OTTO Versand**
Umweltkoordination
D-22172 HAMBURG
Germany
Phone: +49 40 6461 8430
Fax: +49 40 6461 8571
German mail order company, in
charge of:
-Umweltbaum label
-Future collection label
-Hautfreundlich-
schadstoffgeprüft label

14. **IBR GmbH**
Heilig-Geist-Strasse 54
D-83022 ROSENHEIM
Germany
Phone: +49 8031 17091
Fax: +49 8031 37546
German Empfohlen vom IBR label

15. Swedish Society for Nature Conservation (SSNC)
P.O. Box 7005
S-40231 GÖTEBORG
Sweden
Phone: +46 31 7116450
Fax: +46 31 7116430
E-mail: naturskyddsforeningen.gbg@snf.se
Internet: http://www.snf.se
Swedish Falcon (Bra Miljöval) label

16. Association Francaise de Normalisation (AFNOR)
Tour Europe Cedex 7
F-92049 PARIS La Défense
France
Phone: +33 1 42915916
Fax: +33 1 42915911
French NF-Environmental label

Fair Trade and Labeling Organizations

1. **IFAT, International Federation for Alternative Trade**
P.O. Box 500
AKRON, PA 17501
United States of America (USA)
Phone: +1 717 7333006
Fax: +1 717 7330294
E-mail: ifat@success.net
Internet: http://www.ifat.org
International federation of producers and alternative trading organizations

2. **European Fair Trade Association (EFTA)**
Witmakersstraat 10
NL-6211 JB MAASTRICHT
The Netherlands
Phone +31 43 3256917
Fax: +31 43 3258433
E-mail: efta@antenna.nl
European organization for promotion of fair trade

3. **EZA Deitte Welt**
Plainbachstr. 8
A-5101 BERGHEIM
Austria
Phone: +43 662 52178
Fax: +43 662 52586
E-mail: eza3welt@magnet.at
Internet: http://www.or.alpin.at/eza
Austrian member of EFTA

4. **Magasins du Monde – Oxfam**
7a, Rue E. Michiels
B-1180 BRUSSELS
Belgium
Phone: +32 2 3320110
Fax: +32 2 3321888
Belgian member of EFTA

5. **Oxfam Wereldwinkels**
Nieuwland 35-37
B-9000 GENT
Belgium
Phone: +32 92 230161
Fax: +32 92 250478
Belgium member of EFTA

6. **Federation Artisans du Monde**
17 Place de l'Argonne
F-75019 PARIS
France
Phone: +33 1 40351722
Fax: +33 1 42096666
French member of EFTA

7. **Solidar'Monde**
86, rue Berthie Albrecht
F-94400 VITRY S/SEINE
France
Phone: +33 1 45736543
Fax: +33 1 45736542
French member of EFTA

8. **GEPA**
Talstrasse 20
D-58332 SCHWELM
Germany
Phone: +49 2336 91820
Fax: +49 2336 10966
German member of EFTA

9. **C.T.M**
   Via Macello 18
   I-39100 BOLZANO
   Italy
   Phone: +39 471 975333
   Fax: +39 471 977599
   E-mail: a_vaccaro@link-B2.comlink.apc.org
   Internet: http://www.citinv.it/equo/homefair.htm
   Italian member of EFTA
10. **Fair Trade Organization**
    P.O. Box 115
    4100 AC CULEMBORG
    The Netherlands
    Phone: +31345 545151
    Fax: +31 345 541423
    E-mail: post@fairtrade.antenna.nl
    Internet: http://www.antenna.nl/fairtrade
    Netherlands EFTA member, owner of the Netherlands Fair Trade label
11. **CLARO AG**
    P.O. Box 69
    CH-2552 ORPUND
    Switzerland
    Phone: +41 32 553155
    Fax: +41 32 553159
    Swiss member of EFTA
12. **Vereinigung Dritte Welt Läden**
    P.O. Box 104
    CH-5015 ND. ERLINGSBACK
    Switzerland
    Phone: +41 64 343078
    Fax: +41 64 343078
    Swiss member of EFTA
13. **Intermon**
    Marques de Dos Aguas 5
    E-46002 VALENCIA
    Spain
    Phone: +34 6 3524193
    Fax: +34 6 3525772
    Spanish member of EFTA
14. **IDEAS**
    Llano de las Maravillas s/n
    11130 CHICLANA
    Spain
    Phone: +34 56 230380
    Fax: +34 56 230380
    Spanish member of EFTA
15. **Oxfam Trading**
    Murdock Rd
    GB-BICESTER, Oxon OX6 7RF
    United Kingdom
    Phone: +44 1869 245011
    Fax: +44 1869 247987
    E-mail: oxbridge@gn.apc.org
    UK member of EFTA
16. **Traidcraft Plc**
    Kingsway, Gateshead
    GB-TYNE & WEAR NE11ONE
    United Kingdom
    Phone: +44 191 4910591
    Fax: +44 191 4822690
    E-mail: traidcraft@gn.apc.org
    traidcraft@globalnet.co.uk
    Internet: http://www.globalnet.co.uk/~traidcraft
    UK member of EFTA
17. **Max Havelaar Foundation Netherlands**
    P.O. Box 1252
    NL-3500 BG UTRECHT
    The Netherlands
    Phone: +31 30 23344602
    Fax: +31/30/2332992
    E-mail: maxhavelaar@maxhavelaar.nl
18. **Max Havelaar Belgium**
    Handelskaai 9
    B-100- BRUSSELS
    Belgium
    Phone: +32 2 2198440
    Fax: +32 2 2176078
    E-mail: chris@maxhavelaar.be
19. **Max Havelaar Switzerland**
    Malzgasse 25
    CH-4052 BASEL
    Switzerland
    Phone: +41 61 2717500
    Fax: +41 61 2717562
    E-mail: maxhavelaarstiftung@access.ch

20. **Max Havelaar Denmark**
c/o DanChurchAid
Nørregade 13
DK-1165 COPENHAGEN
Denmark
Phone: +45 33 152802 - 255
Fax: +45 33 935802
E-mail: havelaar@pip.dknet.dk

21. **Max Havelaar Norge**
C/o Kirkens Nødhjelp
Sandakerveien 74
0483 OSLO
Norway
Phone: +47 22 092711
Fax: +47 22 222420

22. **Fairtrade Foundation**
7th Floor, Regent House
89 Kingsway
GB-LONDON WC2B 6RH
United Kingdom
Phone: +44 171 405 5942
Fax: +44 171 405 5943
E-mail: fairtrade@gn.apc.org

23. **Max Havelaar France**
67 Rue Robespierre
F-93558 MONTREUIL CEDEX
France
Phone: +33 1 42877021
Fax: +33 1 48515273

24. **TransFair International**
Hinter der Burg 27
D-73614 SCHORNDORF
Germany
Phone: +49 7181 929822
Fax: +49 7181 929823
E-mail: coord@transfair.com

25. **TransFair Germany**
Remigiusstrasse 21
D-50937 KÖLN
Germany
Phone: +49 221 942040-0
Fax: ++49 221 942040-40
E-mail: transfair@project.fido.de

26. **TransFair Austria**
Wipplingerstrasse 32
A-1010 VIENNA
Austria
Phone: +43 222 5330956
Fax: +43 222 5330957
E-mail: transfair.a@magnet.a

27. **TransFair Minka**
3, Rue des Vergers
L-7338 HEISDORF
Luxembourg
Phone: +352 332276 / +352 331417
Fax: +352 405849 / +352 332276

28. **TransFair Italy**
Piazzetta Forzaté 1
I-35137 PADOVA
Italia
Phone: +39 49 8750 823
Fax: +39 49 8750 910
E-mail: ctmmag-s@protec.it

29. **Föreningen för Rättvisemärkning Sverige**
Peter Myndes Backe 12, 5tr
S-118 46 STOCKHOLM
Sweden
Phone: +46 8 7148990
Fax: +46 8 6422641
E-mail: handla@raettvist.se
Internet: http://www.raettvist.se

30. **Fair Trade Center**
Malmgårdsvägen 14
S-116 38 STOCKHOLM
Sweden
Phone: +46 8 6434364
Fax: +46 8 642 21 60
E-mail: info@fairtradecenter.a.se
Internet: http://www.fairtradecenter.a.se

31. **Världsbutikerna för Rättvis Handel**
P.O. Box 161
S-271 24 YSTAD
Sweden
Phone: +46 411 65220
Fax: +46 411 65220
E-mail: info@valdsbutikerna.m.se

Organizations for Advice, Trade and Labeling of Organic and Biodynamic Agriculture

1. **IFOAM: International Federation of Organic Agricultural Movements**
   Ökozentrum Imsbach
   D-66636 THOLEY-THELEY
   Germany
   Phone: +49 68535190
   Fax: +49 68533010
   E-mail: ifoam@T-online.de
2. **IFOAM Accreditation Programme**
   IOAS
   2041/2 First Avenue South, Suite 1
   JAMESTOWN, ND 58401
   United States of America (USA)
   Phone: +1 701 2524070
   Fax: +1 701 2524124
   E-mail: Kcom@daktel.com
3. **DEMETER International**
   Baumschulenweg 11
   D-64295 DARMSTADT
   Germany
   Phone: +49 6155 4061
   Fax: +49 6155 5774
4. **SKAL**
   P.O. Box 384
   NL-8000 AJ ZWOLLE
   The Netherlands
   Phone: +31 38 4226866
   Fax: +31 343 531700
5. **Agro Eco Consultancy**
   P.O. Box 176
   NL-3970 AD DRIEBERGEN
   The Netherlands
   Phone: +31 343 536137
   Fax: +31 343 531700
6. **EOSTA**
   P.O. Box 632
   NL-3740 AP BAARN
   The Netherlands
   Phone: +31 35 5423747
   Fax: +31 35 5417578
7. **Vereniging voor Biologisch**
   Dynamische Landbouw
   P.O. Box 17
   NL-3970 AA DRIEBERGEN
   The Netherlands
   Phone: +31 343 531740
   Fax: +31 343 516943
8. **Milieu Project Sierteelt (MPS)**
   Environmental Project Floriculture)
   P.O. Box 533
   NL-2675 ZT HONSELERSDIJK
   The Netherlands
   Phone: +31 174 615715
   Fax: +31 174 632059
9. **The Greenery International b.v.**
   P.O. Box 70588
   NL-5201 CZ DEN BOCH
   The Netherlands
   Phone: +31 73 6499999
   Fax: +31 73 6494400
10. **AGÖL: Arbeitsgemeinschaft**
    Ökologischer Landbau
    Brandschneise 1
    D-64295 DARMSTADT
    Germany
    Phone: +49 6155 2081
    Fax: +49 6155 2083
11. **Naturland-Verband**
    Kleinhadener Weg 1
    D-82166 GRÄFELFING
    Germany
    Phone: +49 89 545071
    Fax: +49 89 855974
12. **Forschungsring für Biologisch-Dynamische Wirtschaftsweise**
    Demeter-Bund e.V
    Baumschlenweg 11
    D-64295 DARMSTADT
    Germany
    Phone. +49 615 42674
    Fax: +49 6155 5774

13. **Gäa e.V. Vereingung Ökologischer Landbau**
Plauenscher Ring 40
D-01187 DRESDEN
Germany
Phone: +49 351 4012389
Fax: +49 351 4012389
14. **Bioland-Bioland Verband für Organisch Biologischen Landbau**
P.O. Box 208
D-73002 GÖPPINGEN
Germany
Phone: +49 7161 910120
Fax: +49 7161 910127
15. **BNN/Gh- Bundesverband Naturkost-Naturwaren**
Robert-Bosch-Str. 6
D-50354 HÜRTH-EFFEREN
Germany
Phone: +49 2233 63326
Fax: +49 2233 66911
German association of retailers, wholesalers and processors of organic products
16. **KEAV**
P.O. Box 1940
S-75149 UPPSALA
Sweden
Phone: +46 18 100290
Fax: +46 18 100366
E-mail: info@krav.se
Internet: http://www.krav.se
17. **KRAV International Certification**
GroLink AB
Torfolk
S-684 95 HÖJE
Sweden
Phone: +46 563 722345
Fax: +46 563 72066
E-mail: info@grolink.se
18. **Svenska Demeterförbundet**
Skillebyholm
S-15391 JÄRNA
Sweden
Phone: +46 8 551 57988
Fax: +46 8 551 57976
Swedish Demeer organization, certification body for biodynamic agriculture
19. **Ekologiska Lantbrukarna**
Sågargatan 10
S-75318 UPPSALA
Phone: +46 18 101006
Fax: +46 18 101066
E-mail: lennart.larsson@ekolantbruk.se
Internet: http://home6.swipnet.se/~w-60294/index.html
20. **Future Earth (Framtidsjorden)**
Tideliusgatan 36
S-11869 STOCKHOLM
Sweden
Phone: +46 8 7206568
Fax: +46 8 7206581
E-mail: f-earth@algonet.se
Internet: http://www.framtidsjorden.se
Swedish organization which support projects on urban and rural development and organic production in developing countries
21. **ASOAB/BIO SUISSE**
Missionstrasse 60
CH-4055 BASEL
Switzerland
Phone: +41 61 3859610
Fax: +41 61 3859611
22. **Demeter Produzentenverein**
Grabenacker 15
CH-4141 MÜNCHENSTEIN
Switzerland
Phone: +41 61 4160643
Fax: +41 61 4160644

23. **Migros-Genossenschaftsbund**
Service Migros-Sano
P.O. Box 203
CH-1010 LAUSANNE
Switzerland
Phone: +41 21 6517300
Fax: +41 21 6537271
Managing the Swiss MIOPLANT-Natural label for organically grown cut flowers and plants

24. **United Kingdom Register of Organic Food Standards (UKROFS)**
c/o Ministry of Agriculture, Fisheries and Food (MAFF)
Room 320C Nobel House
17 Smith Square
London SW 1P3JR
Phone: +44 171 2385915
Fax: +44 171 2386553

25. **Bio-Dynamic Association**
Woodman Lane
GB-DY9 9PX
CLENT STOURBRIDGE
United Kingdom
Phone: +44 1562 884933
Fax: +44 1562 886219

26. **French Ministry for agriculture, fishery and food**
Ministére de l'Alimentatino
Direction Générale de l'Alimentation
Service de la Protection des Végétaux
175, rue du Chevaleret
F-75013 PARIS
France
Phone: +33 1 49558155

27. **Forum de ' Agriculture Raisonnée Respectueuse de l' Environment (FARRE)**
1, Rue Gambetta
F-92100 BOULOGNE
France
Phone: +33 1 46050714
Fax: +33 1 48255065

28. **Association Demeter France**
P.O. Box 286
F-67606 SELESTAT CEDEX
France
Phone: +33 88 828878
Fax: +33 88 922450

29. **Demeterforbundet I Denmark**
Norregard
Nedermarken 8
DK-5560 ARUP
Denmark
Phone: +45 64 431225
Fax: +45 64 431226

Appendix

17. LOGO OF ECO-LABELS

The Netherlands Milieuker Lable

Germany's Blue Angel Lable

Stiftung Warentest Label

WWF Panda Lable

Neckermann Umweltpradikat
(Eco Seal)

Umweltbaum Label

Organisation of the Nordic Swan

The Swedish Bra Milijoval (Good Environmental Choice) Label

The Max Havelaar Label

The TransFair Label

Case Example: Bananas

The EKO Quality Symbol

Organic Labels in Germany

Bioland

Organic Label is Sweden

Bio. Mit der Knospe.

Bio. Avec le bourgeon.

Bio. Con la gemma.

Organic Labels in Switzerland

German Oko-Tex Standard 100 Label

German Toxproof/Ecoproof Label

The German SG-
Schadstoffgepruft Label

Arbeitskreis
Naturtextil e.V.

German Arbeitskreis Naturtextil Label

eco-tex®

German Ecotex Label

Indo-German Rugmark Label

Green Cotton

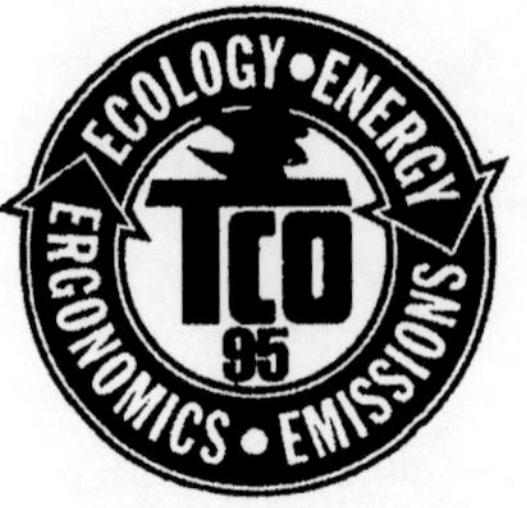

Swedish Label for Computers
and Computer Screens

International Interest

Green Dot Label

Appendix

18. RESOURCES IN ELECTRONIC MEDIA

General

1. Asia Pacific Seed Association – APSA
   http://www.apasseed.com/
2. Association of Official Seed Analysis – AOSA
   http://www.aosaseed.com/
3. Association of Seed Certifying Agencies – AOSCA
   http://www.aosa.org/
4. Commercial Seed Analysts of Canada Inc. – COSCA
   http://www.seedanalysis.com/
5. Community Plant Variety office – CPVO
   http://www.cpvo.fr/
6. Consultative Group on International Agricultural Research – CGIAR
   http://www.cgiar.org/
7. European Plant Protection Organization – EPPO
   http://www.eppo.org/
8. Federation International Seed Trade FIS
   http://www.worldseed.org/
9. Food and Agricultural Organizatoin – FAO
   http://www.fao.org/
10. International Centre for Agricultural Research in Dry Areas – ICARDA
    http://www.icarda.cgiar.org/
11. International Centre for Tropical Agriculture – CIAT
    http://www.ciat.cgiar.org/
12. International Crop Research Institute for Semi Arid Tropics – ICRISAT
    http://www.icrisat.org/
13. International Maize and Wheat Improvement Centre – CIMMYT
    http://www.cimmyt.org/
14. International Plant Genetic Research Institue – IPGRI
    http://www.ipgri.cgiar.org/
15. International Plant Protection Convention – ICPP
    http://www.fao.org/WAICENT/FAOINFO/AGRICULT/AGP/AGPP/PQ/
16. International Rice Research Institue – IRRI
    http://www.cgiar.org/irri/
17. International Seed Testing Association – ISTA
    http://www.seedtest.org/

18. International Society for Seed Science – ISSS
http://www.css.cornell.edu/ISSS/
19. International Union of Forestry Research Organizations – IUFRO
http://www.iufro.boku.ac.at/
20. National Institute of Agricultural Botany – NIAB
http://www.niab.com/
21. New Zealand National Seed Laboratory – NZNSL
http://www.seedlab.co.nz/
22. North American Plant Protection Organization – NAPPO
http://www.nappo.org/
23. Organization for Economic Co-operation and Development – OECD
http://www.oecd.org/
24. Seed and Plant Genetic Resources Service – AGPS
http://www.fao.org/WAICENT/FAOINFO/AGRICULT/AGP/AGPS/
25. Seed Quest – any information on seed
http://www.seedquest.com/
26. Society of Commercial Seed Technologists – SCST
http://www.seedtechnologists.com/
27. Syngenta Seeds
http://www.syngenta.com
28. System-Wide Information Network for Genetic Resources – SINGER
http://www.singer2.cgiar.org/
29. The Association of American Seed Control Officials – AASCO
http://www.seedcontrol.org/
30. The Germplasm Resources Information Network – GRIN
http://www.ars-grin.gov/npgs
31. The International Union for the Protection of New Varieties of Plants – UPOV
http://www.upov.int/eng
32. Appropriate Technology Transfer for Rural Areas (ATTRA)
http://www.attra.org/index.html
33. Comprehensive database and information service, covers almost all topics of organic farming
34. Food and Agriculture Organisation
http://www.fao.org/organicag/
35. Biopuglia
http://www.biopuglia.iamb.it
36. Danish Organic
http://www.ecoweb.dk/english/
37. Documentation of Ecological Agriculture (DOCEA)
http://www.bib.wau.nl/docea/
38. Ecological Agriculture Projects (EAP)
http://eap.mcgill.ca/general/home_frames.htm

39. Land, Livestock and Environment
http://www.geocities.com/RainForest/Canopy/3770/aslam.htm
40. Multinational Exchange for Sustainable Agriculture (MESA)
http://www.went.net/~mesa/
41. Alternative Farming System Information Center (AFSIC)
http://www.nal.usda.gov/afsic/afslinks.htm
42. Danish Institute of Agricultural Sciences
http://www.sp.dk/~jph/Areas/Organic_Farming/organic_farming.html
43. CABI Publishing
http://www.organic-research.com
44. CABI Bioscience
http://www.cabi-bioscience.org/
45. Organic Resource Database
http://www.wye.ac.uk/bioscienice/soil/
46. Center for Agroecology and Sustainable Food Systems
47. http://www.ucsc.edu/casfs
48. European Network for Scientific Research Coordination in Organic Farming (ENOF)
http://www.cid.csic.es/enof/
49. European Initiative for Agricultural Research for Development InfoSys
http://www.dainet.de/eiard/infosys/europa/frameset_hot_topics.htm
50. Institute of Organic Agriculture (University of Bonn, Germany)
http://www.uni-bonn.de/iol/english.htm
51. Organic Farming Research Foundation (OFRF)
http://www.ofrf.org
52. Research Institute of Organic Agriculture (FiBL)
http://www.fibl.ch/
53. Henry Doubleday Research Association
54. http://www.hdra.org.uk/
55. Scottish Agricultural College (SAC)
http://sac004.ed.sac.ac.uk/cropsci/external/ORGANIC/dfault.htm
56. Welsh Institute of Rural Studies
http://www.wirs.aber.ac.uk/research/organic.shtml

Marketing and Trade

1. Beta Pure Foods
http://www.betapure.com/
2. BioFach – The World Organic Trade Fair
http://www.biofach.de/english/index.html
3. BioFood.Net Business
http://www.biofoodnet.com/
4. Green Trade Net
http://www.green-tradenet.de

5. Information Network For Organic Produce
   http://www.sources.de/infop/
6. Organic Trading and Information Centre
   http://www.organicfood.com/
7. Organic Farmers Marketing Association (OFMA)
   http://web.iquest.net/ofma/
8. Organic Trade Association (OTA)
   http://www.ota.com
9. SANA – The Fair of Qualities
   http://www.sana.it
10. Syndicat European des transformateurs et distributeurs de l'agriculture biologique (SETRAB)
    http://www.setrabio.com/setrab/default.htm

Standards and Regulations

1. Australian National Standards
   http://www.bfa.com.au/frames/national.htm
2. California Certification Standards
   http://www.ccof.org/certification_standards.htm
3. Codex Alimentarius Commission
   http://www.fao.org/WAICENT/FAOINFO/ECONOMIC/ESN/codex/STANDARD/standard.htm
4. EU Organic Regulation
   http://europa.eu.int/comm/sg/consolid/en/391r2092/artm.htm
5. US National Organic Program
   http://www.ams.usda.gov/nop/index.htm

Organic Commodities

1. Organic Grapes into Wine Alliance (OGWA)
   http://www.isgnet.com/ogwa
2. Organic Cotton
   http://www.sustainable.org/
3. Organic Seed
   http://www.biogene.org

Selected National and International NGOs

1. Biodynamic Farming and Gardening Association (BDA)
   http://www.biodynamics.com/
2. Canberra Organic Growers Society (COGS)
   http://www.netspeed.com.au/cogs/cogsabt.htm
3. Canadian Organic Growers (COG)
   http://www.gks.com/cog/
4. Henry A. Wallace Institute for Alternative Agriculture
   http://www.hawiaa.org

5. Organic Consumers Association (OCA)
   http://www.organicconsumers.org/
6. Soil Association
   http://www.soilassociation.org

Organic Farming General

1. agAccess: http://www.mother.com/agaccess
   Includes comprehensive directory of US Organic Certification organizations, "Internships, Apprenticeships, and Working Farms Programs in the US" from ATTRA, in addition to their publications catalog/information.
2. Aggie Horticulture (Texas A&M): http://aggie-horticulture.tamu.edu
   Cooperative Extension information on crop production, etc. plus "Sustainable Practices for the South", IPM Network, and related topics.
3. Agricultural Network Information Center (AgNIC): http://www.agnic.org
   The National Agricultural Library's cooperative effort to compile and link agricultural databases (AgDB), Directories of Expertise, and more.
4. Alternative Farming Systems Information Center (NAL/ARS/USDA): http://www.nal.usda.gov/afsic
   Full text AFSIC bibliographic and reference publications, links.
5. Aquatic Ecology at the Earth Centre: http://www.shef.ac.uk/~ec/aetec/index.html
   One of the few places to find information on "Sustainable Freshwater Aquaculture" and "Sustainable Wetland Production".
6. Cascadian Farm: http://www.cfarm.com
   Commercial site with a Health food Store database and a National Organic Standards Board (NOSB) update section.
7. Center for Sustainable Agricultural Systems (University of Nebraska, Lincoln):
   http://ianrwww.unl.edu/ianr/csas
   Newsletters, research reports, curriculum materials, etc.
8. Conservation Technology Information Center: http://www.ctic.purdue.edu/ctic.html
   National Crop Residue Survey database and map,
   National watershed Network, newsletters.
9. Cooperative Extension System Information Serves: http://www.esusda.gov/partners/ces-locs.htm
   Linked directory of US and Canadian Extension On-line resources, including publications, by state/province.
10. EcoNet's Agriculture & Trade Directory: http://www.igc.apc.org/igc/www.agritrade.html

11. IFOAM '96 Book of Abstracts: http://ecoweb.dk/english/ifoam/conf96/index.html
    Hundreds of abstracts of papers, posters, etc. presented at the International Federation of Organic Agricultural Movements Conference in Denmark, August 1996. Author and subject indexes included.
12. Global Entomology Agriculture Research Server (ARS/USDA): http://gears.tucson.ars.ag.gov/
    Text, pictures, and sounds of insects, especially bees.
13. Midwest Organic Alliance: http://www.midwestorganic.org
    Background information for consumers, producers and marketers.
14. National Integrated Pest Management Network: http://www.reeusda.gov/ipm/ipm-home.htm
15. New Crops Information System: http://newcrop.hort.purdue.edu
    Searchable databases CropSEARCH, CropExpert, CropREFERENCE, plus a U.S.
    FarmMARKET directory, newsletters, and calendar.
16. New Hope Group On-line: http://devel.newhope.com/public/Home_page.html
    Commercial site, full-text articles from their publications, commercial links, etc.
17. Noah's Ark: An Organic Farmer's Homepage: http://www.rain.org/~sals/my.html
18. Not Just Cows
    A Guide to Internet/Bitnet Resources in Agriculture and Related Sciences by Wilfred Drew: http://www.snymor.edu/~drewwe/njc/njcbeta/njcmain.htm
    Just what it says; especially good for listing of agriculture-related Libraries.
19. Organic Agriculture and Permaculture at Sunsite (University of North Carolina): http://sunsite.unc.edu/london/index.html
    Very useful page of links: composting, biodynamic, permaculture, vermiculture, community supported agriculture, livestock, and more.
20. Organic Farming Marketing Association (OFMA): http://www.iquest.net/ofma/
    Organic laws including NOSB documents, and text of National Organic Marketing Cooperative Feasibility Study.
21. Organic Trading and Information Center: http://www.organicfood.com/welcome.html
    Legislative issues, resource groups, certifying organization as well as consumer and commercial information and links.
22. Permaculture International: http://nornet.nor.com.au/environment/perma/
    Global directory, mail order resources, more.
23. RotWeb: http://net.indra.com/~topsoil/Compost_Menu.html
    Composting how-to and links.

24. Sustainable Agriculture Network (SAN): http://www.ces.ncsu.edu/san/index.html
    SARE research and grant information, SAN Directory of Expertise, sanet-mg archives, more.
25. Sustainable Agriculture Research and Education Program (SAREP) (University of California, Davis): http://www.sarep.ucdavis.edu/
    Two downloadable databases: Cover Crop (PC & Mac) and Dairy Debate (Mac only).
26. Urban Agriculture Notes (City Farmer, Canada Office of Urban Agriculture): http://www.cityfarmer.org/
27. Water Quality Information Center (ARS/USDA/NAL); http://www.nal.usda.gov/wqic
    Full text bibliographic publications, searchable WQ databases including Agricultural Waste Database from Auburn University.

Appendix

19. DISCUSSION GROUPS/ MAILING LISTS

Receive newsletters and calendar; converse interactively with others. To subscribe generally requires a request to a computer listserv or listproc or majordomo address. For example, to subscribe to Graze-1, below, you would send in the message portion of an e-mail message the phrase "subscribe graze-1 your first name your last name" to the subscription address. Use only these words (substituting your own first name and last name), without the quotation marks. Subsequent message to the group are addressed to the list name and address.

There are several directions of discussion lists on the Internet:

Try Clark Consulting International, Inc. for a good one on agriculture: http://www.interaccess.com/consulting/maillist.html. Here is a sampling.

1. CSA-1 (Community Supported Agriculture networking): Subscriptions to: listproc@prairienet.org
   Discussion mailings to: csa-1@prairienet.org
2. Forage-mg (Forage production/management): Subscriptions to: almanac@ oes.orst.edu
   Discussion messages to: forage-mg@oes.orst.edu
3. Graze-1 (Rotational grazing and seasonal dairying): Subscriptions to: listserv@taranaki.ac.nz
   Discussion messages to: graze-1@taranaki.ac.nz
4. Permaculture-mg (US permaculture discussion): Subscriptions to: almanac@ces.ncsu.edu
   Discussion messages to: permaculture-mg@ces.ncsu.edu
5. Smallfruit-mg (berries, grapes, etc. for growers and researchers): Subscriptions to:
   almanac@esusda.gov
   Discussion messages to: smallfruit-mg@esusda.gov
6. Sanet-mg (Sustainable Agriculture Network-SAN): Subscriptions to: almanac@ces.ncsu.edu
   Discussion messages to: sanet-mg@ces.ncsu.edu (An informational pamphlet, *Getting Started Electronically with the Sustainable Agriculture Network is available through SAN's NAL office.
   (Contact: SAN Coordinator, National Agricultural Library, 10301 Baltimore Ave., Room 304, Baltimore Ave., Room 304, Beltsville MD 20705-2331, phone 301-504-6425, e-mail san@nal.usda.gov)

Appendix

20. ELECTRONIC JOURNALS AND NEWSLETTERS:

1. Full-text publications, selected articles, and abstracts of many periodicals are available. You may also subscribe to hard copy publications via the Internet. Two good directories: Online Agricultural Journals and Magazines: http://www.interaccess.com/consulting/agpubs.html and Oxbridge Communication's Media Finder Directory of Newsletters, Directory of Catalogs, and Directory of Magazines: http://www.oxbridge.com/index.html
2. A selection of current and back issues of environment-related journals, including several on alternative/sustainable agriculture is available via the Sustainable Earth Electronic Library (SEEL), URL http://www.envirolink.org/pubs/index.html

Appendix

21. RESEARCH PERSONALS

1. **Dr. Alore Hugo**
Danish Institute of Agricultural Sciences Research Centre Foulum
Tjele, DK-8830, Denmark
E.mail hugo.Alroe@agrsci.dk

2. **Dr. Alroe Hugo F.**
Research Centre for Organic Farming Foulum
P.O. Box 50
Tjele, DK-8830, Denmark
E.mail hugo.Alore@agrsci.dk

3. **Dr. Ansaloni Francesco**
DIPROVAL Sez. Economia
University of Bologna
Faculty of Agriculture, Via F. lli Rosselli 107
Reggio Emilia, 1-42100
Italy
E.mail ansaloni@stpa.unibo.it

4. **Dr. Baars Ton**
Louis Bolk Instituut Hoofstraat 24
Driebergen
NL-3972 LA
The Netherlands
E.mail louis.bolk@pobox.accu.uu.nl

5. **Dr. Bucarelli Francesco M**
Istituto della Nutrizione Via Ardeatina 546
Roma, 1-00178, Italy
E.mail quaglia@inn.ingrm.it

6. **Dr. Colazza Stefano**
Istituto di Entomologia
University of Perugia
Borgo XX Giugno 74
Perugia, 1-06121, Italy
E.mail colazza@unipg.it

7. **Dr. David Cristophe**
ISARA 31 Place Bellecour
Lyon Cedex 02 F-64288 France
E.mail david@isara.fr

8. **Dr. Dubikovsky Georgy P**
28 Tereshkova str.
Grodno 230600, Belarus
E.mail niwa@selhoz.belpak.grodno.by

9. **Dr. Dubois David**
FAL
Zurich
Switzerland
E.mail david.dubois@fal.ad

10. **Dr. Eltun Ragnar**
The Norwegian Crop Research Institute
Apeslvoll Research Centre
Kapp, N-2858, Norway
E.mail ragnar.eltun@planteforsk.no

11. **Dr. Fersino Vincenzo**
I.A.M.B. – Via Ceglie 9
Valenzano (BA), 1-70019, Italy
E.mail fersino@iamb.it

12. **Dr. Fried Padruot**
FAL-Zurich
Zurich, Switzerland
E.mail padruot.fried@fal.ad

13. **Dr. Gautronneau Yvan**
ISARA 31, Place Bellecour
Lyon Cedex 02
F-69288, France
E.mail yvan.gautronneau@isara.fr

14. **Dr. Granstedt Artur**
The Biodynamic Research Institute
Skilleby Farm, S-15391
Sweden
E.mail artur.granstedt@jdb.se

15. **Dr. Halberg Niels**
Danish Institute of Agricultural Sciences Research Centre Foulum
Tjele, DK-8830, Denmark
E.mail niels.halberg@agrsci.dk

16. **Dr. Hansen Sissel**
Norwegian Centre for Ecological Agriculture
Tingvoll N-6630
Norway
E.mail nsosh@post.nlh.no

17. **Dr. Hess Jürgen**
FG Ecological Agriculture
University of Kassel
Witzenhausen
D 37213
Germany
E.mail jhesz@wiz.uni-kassel.de

18. **Dr. Kabourakis Emmanouil**
Institute of Viticulture National Agricultural Foundation
P.O. Box 1841
Heraklion
GR-711100, Greece
E.mail agapi@mir.forthnet.gr
or
E.mail kabourakis@her.forthnet.gr

19. **Dr. Keatinge Ray**
ADAS Redesdale Rochester
Otterbum
Newcastle-upon-Tyne
NE19 1SB, UK
E.mail Ray_Keatinge@adas.co.uk

20. **Dr. Kilcher Lukas**
FIBL, Postfach
Frick CH-5070 Switzerland
E.mail Lukas.kilcher@fibl.ch

21. **Dr. Kloen Henk**
Research Institute for Agrobiology and Soil Fertility (AB-DLO)
P.O. Box 14
Wageningen
NL-6700 AA, The Netherlands
E.mail h.kloen@ab.dlo.nl

22. **Dr. Koepke Ulrich**
Institut für Organischen Landbau
Katzenburgweg 3
Bonn, D-53115, Germany
E.mail iol@uni-bonn.de

23. **Dr. Krell Rainer**
Regional Office for Europe
Food and Agriculture Organization (FAO)
Viale delle Terme di Caracalla
Rome 1-00100, Italy
E.mail Rainer.Krell@fao.org

24. **Dr. Kristensen Lb Sillebak**
Danish Institute of Agricultural Sciences
Dept. of Agricultural Systems Research Centre
Foulum, P.O. Box 50, Tjele
DK-8830
Denmark
E.mail lbs.Kristensen@agrsci.dk

25. **Dr. Lacko-Bartosova**
Magdalena, Dept. of Agricultural Systems
Slovak Agric University
Nitra, 94976, Slovak Republic
E.mail bartosov@afnet.uniag.sk

26. **Dr. Lammerts Van Bueren**
Edith Louis Bolk Institute
HoofdStraat 24, Driebergen, NL-3972 LA
The Netherlands
E.mail louis.bolk@pobox.ruu.nl

27. **Dr. Lockeretz William**
School of Nutrition Science and Policy Tuffs University
Medford, MA 02155, USA
E.mail wlockeretz@infonet.tufts.edu

28. **Dr. Lund Vonne**
Dept. of Animal Environment and Health
P.O. Box 234
Skara, S-53223, Sweden
E.mail vonne.lund@hmh.slu.se

29. **Dr. Mader Paul**
FiBL, Postfach
Frick
CH-5070, Switzerland
E.mail paul.maeder@fibl.ch

30. **Dr. Maurer Veronika**
FiBL, Postfach
Frick, CH-5070, Switzerland
E.mail Veronika.maurer@fibl.ch

31. **Dr. Mavrogianpoylos George**
Agricultural University of Athens
Str. lera Odos 75
GR-11850, Greece
E.mail Naxos@auadec.aua.gr

32. **Dr. McKinlay Rod G.**
SAC West Mains Rd.
Edinburgh
EH9 3JG, UK
E.mail r.mckinlay@ed.sac.ac.uk

33. Dr. Morel Jaques
BLW, Mattenhoffstr.5
Berm
CH-3003
Switzerland

34. **Dr. Müller-Haye Berndt**
Fuldastr-6
Rastede-Wahnbek
D-26180
Germany
E.mail mullerhaye@aol.com

35. **Dr. Niggli Urs**
FiBL, Postfach
Frick
CH-5070, Switzerland
E.mail urs.niggli@fibl.ch

36. **Dr. Padel Susane**
WIRS Leanbadarn Campus
Aberystwyth
SY23AL, UK
E.mail sxp@aber.ac.uk

37. **Dr. Paraskekov Plamen Pavlov**
Agro-ecology Centre Agricultural University
12 Mendeleev Str.
Plovdiv, BG-4000, Bulgaria
E.mail plamenp@au-plovdiv.bg

38. **Dr. Philipps Lois**
Elm Farm Research Centre
Hamstead Marshall
RG20 OHR, UK
E.mail efrc@compuserve.com

39. **Dr. Popov Vladislav H.**
Agro-ecology Centre Agricultural Univ.
Plovdiv 12 Mendeleev Str.
Plovdiv, BG-4000, Bulgaria
E.mail vpopov@au-plovdiv.bg

40. **Dr. Quaglia Giovanni Battista**
Istituto della Nutrizione
Via Ardeatine 546
Roma, 1-00178, Italy
E.mail quaglia@inn.ingrm.it

41. **Dr. Radics Làszlò**
University of Horticulture and Food Industry
Dept. of Agricultural Systems
Ecological Farming Division
Villànyl ùt 29-35, Budapest,
H-1518, Hungary
E.mail mezg@hoya.kee.hu

42. **Dr. Rasmussen llse**
Danish Institute of Agriculture Sciences Research Centre
Flakkebjerg, Dept of Crop Protection
Slagelse, DK-4200, Denmark
E.mail llseA.Rasmussen@agrsci.dk

43. **Dr. Raupp Joachim**
Institute for Biodynamic Research
Darmstadt
D-64295, Germany
E.mail raupp@ibdf.de

44. **Dr. Reents Hans Jurgen**
Technical University Munich-Weihenstephan
Freising, D-85350, Germany
E.mail reents@serv.agrinf.agrar.tu-muenchen.de

45. **Dr. Roderick Stephen**
Veterinary Epidemiology & Economics
Research Unit, Dept. of Agric.
University of Reading
Reading, RG6 6AT, UK
E.mail veeru@reading.ac.uk

46. **Dr. Ruissen Theo**
Norsok, Tingvoll Gard
Tingvoll, N-6630, Norway
E.mail nsotr@post.nlh.no

47. **Dr. Rutkovienè Vida Marija**
Lithuanian University of Agriculture
Kaunas
LT-4324, Lithuania
E.mail ek@nora.lzua.it

48. **Dr. Santucci Fabio Maria**
Istituto di Economia e Politica Agraria Università di Perugia
Borgo XX Giugno 74
Perugia, 1-06 121, Italy
E.mail fmsant@unipg.it

49. **Dr. Sarapatka Borivoj**
Palacky University Dept. of ecology
tr. svobody 26
Olomolic, 77146, Czech Republic
E.mail sar@risc.upol.cz

50. **Dr. Schaub Daniel M.**
FiBL: Postfach
Frick, CH-5070, Switzerland
E.mail daniel.schaub@fibl.ch

51. **Dr. Schmid Otto**
FiBL: Postfach
Frick, CH-5070, Switzerland
E.mail otto.schmid@fibl.ch

52. **Dr. Seppanen Laura Elina**
Dept. of Plant Production
Univ. of Helsinki
P.O. Box 27
Helsinki, SU-00014, Finland
E.mail Laura.Seppanen@helsinki.fi

53. **Dr. Sisto Luigi**
Mediterranean Institute
CIHEAM
Valenzano (BA), 1-70010, Italy
E.mail sisto@iemb.it

54. **Dr. Soltysiak Ursula**
Warsaw University of Agriculture
Nowoursynowska 166
Warsaw, PL-02-787, Poland
E.mail ekoland2@pol.pl

55. **Dr. Spranger Jörg**
FiBL: Postfach
Frick, CH-5070, Switzerland
E.mail joerg.spranger@fibl.ch

56. **Dr. Srinivasan Ancha**
Regional Science Institute
Sapporo, 001-0024, Japan
E.mail srinivas@vtt.co.jp

57. **Dr. Stehlo Pavol**
Research Institute of Agricultural and Food Economics
Nitra, SK-97901, Slovak Republic
E.mail vuepp1@wriag.sk

58. **Dr. Stolze Matthias**
Universität Hohenheim Institut 410 A, Stuttgart, D-70593
Germany
E.mail stolze@uni-hohenheim.de

59. **Dr. Stopes Christopher**
Wood End Garden Close Lane
Newbury, RG14 6PP, UK
E.mail c.stopes@dial.pipex.com

60. **Dr. Toncea I Ion**
Research Institute for Cereals and Industrial Crops
N. Titulescu 1, Fundulea
8264, Romania

61. **Dr. Troxler Jakob**
Station Fèdèrale de recherché en production vègètale Changins
Nyon, CH-1260, Switzerland
E.mail Jakob.troxler@rac.adm.ch

62. **Dr. Von Fragstein Peter**
University of Kassel Dept. of Ecological
Agriculture Nordbahnhofstrasse 1a
Witzenhausen, D-37213, Germany
E.mail pvf@wiz.uni-kassel.de

63. **Dr. Weibel Franco**
FiBL: Postfach,
Frick, CH-5070, Switzerland
E.mail franco.weibel@fibl.ch

64. **Dr. Wijnands F.G.**
Applied Research for Arable Farming & Filed Production of Vegetables
P.O. Box 430
Lelystad
NL-8200 AK, Netherlands
E.mail F.G.Wijnands@apv.agro.nl

65. **Dr. Willeke-Wetstein** Christiane
Dept. of Livestock Ecology
University
of Giessen Ludwigstrasse 21
Giessen, D-35390, Germany
E.mail ghi5@agrar.uni-giessen.de

66. **Dr. Willer Helga**
Stiftung Ökologie und Landbau
Weinstrasse Süd 51
Bad Dürkheim
D-67098, Germany
E.mail willer@soel.de

67. **Dr. Wolfe Martin S.**
Elm Farm Research Centre
Hamstead
Marshall, Newbury
RG20 OHR, UK
E.mail wolfe@wakelyns.demon.co.uk

68. **Dr. Zanoli Raffaele**
DIBIAGA
University of Ancona
Via Brecce Bichone
Ancona, 1-60100, Italy
E.mail zanoli@agrecon.unian.it

69. **Dr. Zerger Uli**
Stiftung Ökologie und Landbau
Weinstrasse Süd 51
Bad-Dürkheim
D-67098, Germany
E.mail zerger@soel.de

70. **Dr. Znaor Darko**
Wageningen Agricultural University
Dept. of Ecological Agriculture
Haarweg 333, Wageningen
NL-6709 RZ, The Netherlands
E.maildarko.znaor@users.eco.wau.nl

Appendix

22. VARIOUS INDIAN AGRICULTURAL RESEARCH INSTITUTES

Central University

1. CAU
   Central Agricultural University, Imphal (Manipur) – 795001

Demand to be Universities

1. CIFE
   Central Institute of Fishers Education, Versova, Mumbai (Maharashtra) – 400061
2. IARI
   Indian Agricultural Research Institutes, Pusa (New Delhi) – 110012
3. IVRI
   Indian Veterniary Research Institutes, Izatnagar (Uttar Pradesh) – 243122
4. NDRI
   National Dairy Research Institue, Karnal (Haryana) – 132001

State Agricultural Universities

1. AAU
   Assam Agricultural University, Jorhat (Assam) – 785003
2. ANGRAU
   Acharaya N.G. Ranga Agricultural University (APAU), Rajendrangar, Hyderabad (Andhra Pradesh) – 500030
3. BAU
   Birsa Agricultral Universtiy, Ranchi (Jharkhand) – 834006
4. BCKVV
   Bidhan Chandra Krishi Vishwa Vidyalaya, Mohanpur (West Bengal) – 741252
5. CCS HAU
   Chaudhary Charan Singh, Haryana Agricultural University (HAU), Hisar (Haryana) – 125004
6. CSAUA &T
   Chandra Shekhar Azad University, of Agriculture and Technology, Kanpur (Uttar Pradesh) – 208002
7. DPDKV
   Dr. Paunjabro Deshmukh krishi Vidyapeeth, Akola (Maharashtra) – 444104
8. DYSPUHF
   Dr. Yahwant Sing Parmar University of Horticultural and Forestry, Nauni, Solan (Himachal Pradesh) – 173230
9. GAU
   Gujarat Agricultural University, Banaskantha (Gujarat) – 385506
10. GBPUA &T
    Govind Ballabh Pant University of Agriculture and Technology, Pantnagar (Uttranchal) – 263145
11. HPKVV
    Himachal Pradesh Krishi Vishwa Vidyalaya, Palampur (Himachal Predesh) – 176062
12. IGKVV
    Indria Gandhi Krishi Vishwa Vidyalaya, Raipur (Chattisgarh) – 492012

13. JNKVV
    Jawahar Lal Nehru Krishi Vishwa Vidyalaya, Jabalpur (Madhya Pradesh) – 482004
    KAU
    Kerala Agricultural University, Vellanikkara, Thrissur (Kerala) – 680654
14. KKV
    Konkan krishi Vidyapeeth, Depoli (Maharashtra) – 431712
15. MAU
    Marathwada Agricultural University, Parbhani (Maharashtra) – 431402
16. MPKV
    Mahatma Phule Krishi Vidyapeeth, Rahuri (Maharashtra) – 413722
17. NDUA&T
    Narendra Deva University of Agriculture and Technology, Faizabad (Uttar Pradesh) – 224229
18. OUA&T
    Orissa University of Agriculture and Tecnology, Bhubaneshwar (Orissa) – 751003
19. PAU
    Punjab Agricultural University, Ludhiana (Punjab) – 141004
20. RAU
    Rajasthan Agricultural University, Bikaner (Rajasthan) – 334002
21. RAU
    Rajendra Agricultural University, Samastipur (Bihar) – 8481125
22. SKUAS&T
    Sher-E Kashmir University of Agriculture Science and Technology, Gandhinagar, Jammu Tawi, (Jammu & Kashmir) – 180004
23. SKUAS&T
    Sher-E Kashmir University of Agriculture Science and Technology, Srinagar (Jammu & Kashmir) – 191121
24. TNAU
    Tamil Nadu Agricultural University, Coimbatore (Tamil Nadu) – 641003
25. TNV&ASU
    Tamil Nadu Veterinary and Animal Sciences University, Chennai (Tamil Nadu) – 600007
26. UAS
    University of Agricultural Sciences, Bangalore (Karnataka) – 560065
27. UAS
    University of Agricultural Sciences, Dharwad (Karnataka) – 580005
28. WBUA&FS
    West Bengal University of Animal and Fishery Sciences, Belgachia, Kolkata (West Bengal)

Institutes of Indian Council of Agrucultural Research

1. CARI (A&N)
   Central Agricultural Research Institue for Andman and Nicobar groups of Islands, Port Blair (Andman and Nicobar) – 744101
2. CARI
   Central Avian Research Institute, Izatnagar (Uttar Pradesh) – 243122
3. CAZRI
   Central Arid Zone Research Institute, Jodhpur (Rajasthan) – 342003
4. CICFRI
   Central Inland Capture Fisheries Research Institue, Barrackpore (West Bengal) – 743101
5. CIRB
   Central Institute for Research on Buffaloes, Hisar (Haryana) 125001
6. CICR
   Central Institute of Cotton Research, Nagpur (Maharashtra) – 440001

7. CIRCOT
Central Institute for Research on Cotton Technology, Matunga, Mumbai (Mahatrashtra) – 440019
8. CIRG
Central Institute for Research on Goats, Mathura (Uttar Pradesh) – 281122
9. CIAE
Central Institute of Agricultural Engineering, Nabi Bagh, Berasia Road, Bhopal (Madhya Pradesh) – 462039
10. CIBA
Central Institute of Brackishwater Aquaculture, 160 Mahalingapuram, Main Road, Nungambakkam, Chennai (Tamil Nadu) – 600034
11. CIFA
Central Institue of Freshwater Aquaculture, Kaushalyaganj, Bhubaneshwar (Orissa) – 751002
12. CIFT
Central Institute of Fisheries Technology, Willingdon Island, Matsyapuri PO, Cochin (Kerala) – 682014
13. CIPHET
Central Institute of Post Harvesting Engineering and Technology, PAU Post Office, Ludhiana (Punjab) – 141004
14. CISH
Central Institute of Sub-tropical Horticulture, Rai Bareli Road, P.O. Dilkusha, Lucknow (Uttar Pradesh) – 226002
15. CITH
Central Institute of Temperature Horticulture, Santanagar, Srinagar (Jammu and Kashmir) – 190005
16. CMFRI
Central Marine Fisheries Research Institue, Ernakulam, Cohin (Kerala) – 682014
17. CPCRI
Central Plantation Crops Research Institute, Kasaragod (Kerala) – 671124
18. CPRI
Central Potato Research Institute, Simla (Himachal Pradesh) – 171001
19. CRIDA
Central Research Institue for Dryland Agriculture, Santoshnagar, Hyderabad (Andhra Pradesh) – 500049
20. CRIJAF
Central Research Institute for Jute and Allied Fibres, Dist. 24, Parganas, Barrackpore (West Bengal) – 734201
21. CRRI
Central Rice Research Institute, Cuttack (Orissa) – 753006
22. CSWRI
Central Sheep and Wool Research Institute, Avikanagar (Rajasthan) – 304501
23. CSWCRTI
Central Soil and Water Conservation Research and Training Institute, 218, Kaulagarh Road, Dehradun (Uttranchal) – 248195
24. CSSRI
Central Soil Salinity Research Institute, Karnal (Haryana) – 132001
25. CTRI
Central Tobacco Research Institute, Rajahmundry (Andhra Pradesh) – 533105
26. CTCRI
Central Tuber Crops Research Institue, Shreekariyam, Thiruvanantha-puram (Kerala) – 695017
27. DBC
Directorate of Biological Control, Bangalore (Karnataka) – 560024

28. DMR
Directorate of Maize Research, Cummings Lab, IARI, Pusa Campus, New Delhi – 110012
29. DOR
Directorate of Oilseed Research, Rajendranagar, Hyderabad (Andhra Pradesh) – 500030
30. DRR
Directorate of Rice Research, Rajendranagar, Hyderbad (Andhra Pradesh) – 500030
31. DWMR
Directorate of Water Management Research, Walmi Complex, P.O. Phulwari Shariff, Patna (Bihar) – 801505
32. DWR
Directorate of Wheat Research, P.B. 158, Kunjpura Road, Karnal (Haryana) – 132001
33. IASRI
Indian Agricultural Statistical Research Institute, Library Avenue, Pusa Campus, New Delhi – 110012
34. ICAR-RCG
ICAR Research Complex for Goa, Ela Old Goa (Goa) – 342402
35. ICAR-RCNEHR
ICAR-Research Complex for North-Eastern Hills Region, Umroi Road, Shillong (Meghalaya) – 793103
36. IGFRI
Indian Grassland and Fodder Research Institute, Pahuj Dam, Jhansi (Uttar Pradesh) – 284003
37. IIHR
Indian Institute of Horticulture Research, P.O. Hessaraghatta Lake, Bangalore (Karnataka) – 560089
38. IIPR
Indian Institute of Pulse Research, Kalyanpur, Kanpur (Uttar Pradesh) – 208024
39. IISS
Indian Institute of Soil Science, 2-6, Zone I, Nabi Bagh, Bersaia Road, Bhopal (Madhya Pradesh) – 462001
40. IISR
Indian Institue of Species Research, P.B. No. 1701, Marikunnu P.O Calicut (Kerala) – 673012
41. IISR
Indian Institute of Sugarcane Research, P.O. Dilkusha, Lucknow (Uttar Pradesh) – 226002
42. IIVR
Indian Institute of Vegetable Research I-Gandhi Nagar (Naria), P.B. 5002, BHU Varanasi (Uttar Pradesh) – 221005
43. NAARM
National Academy of Agricultural Research Management, Rajendranagar, Hyderbad (Andhra Pradesh) – 500030
44. NBAIM
National Bureau of Agriculturally Important Micro-organisms, NBPGR-Old Buildings, IARI, Pusa Campus, New Delhi – 110012
45. NBAGR
National Bureau of Animal Genetic Resources, P.O. No. 129, Karnal (Haryana) – 132001
46. NBFGR
National Bureau of Fish Genetic Resources, Canal Ring Road, P.O. Dilkusha, Telibagh, 251/28, Dariyapur Talkatora Road, Lucknow (Uttar Pradesh) – 226002
47. NBPGR
National Bureau of Plants Genetic Resources, Pusa, New Delhi - 110012
48. NBSS&LUP
National Bureau of Soil Survey and Land Use Planning, Shankar Nagar, Amravati Road, Nagpur (Maharashtra) 440110

49. NCIPM
National Centre for Integrated Pest Management, Lal Bahadur Shastri Bhavan, Wing L-1, M1, Block F, IARI Campus, Pusa, New Delhi – 110012
50. NIANP
National Institute of Animal Nutrition and Physiology, C/o-Campus Adigudi, Hosur Road, Bangalore (Karnataka) – 560030
51. NIRJ&AFT
National Institute for Research on Jute and Allied Fiber Technology, 12, Reagent Park, Kolkata (West Bengal) – 700040
52. NCAP
National Research Centre of Agricultural Economics and Policy Research, IASRI Campus, Library Avenue, Pusa, New Delhi – 110012
53. NRCAF
National Research Centre for Agroforestry, IGFRI Campus, Pahuj Dam, Gwalior – Jhansi Road (Uttar Pradesh) – 284003
54. NRCAH
National Research Centre for Arid Horticulture, Near 10th K.M. Milestone National Highway, Sri Ganga Nagar Road, Bikaner (Rajasthan)
55. NRCB
National Research Centre for Banana, 44, Ramalingam Nagar, South Extension, Vayihur Road, Tiruchirapalli (Kerala) – 620017
56. NRCC
National Research Centre for Cashew, Dakshina Kannada, Puttur (Karnataka) – 574202
57. NRCC
National Research Centre for Citrus, P.B. 464, Shankar Nagar P.O., Nagpur (Maharashtra) – 440010
58. NRCC
National Research Centre on Camel, Jorbeer, P.B. No. 701, Bikaner (Rajasthan) – 334001
59. NRCCWF
National Research Centre for Coldwater Fisheries, Saurabh Cottage, Thandi Sarak, Brimtal, Nainital (Uttranchal) – 263136
60. NRCDF
National Research Centre on DNA Fingerprinting, NBPGR Campus, IARI, Pusa, New Delhi – 110012
61. NRCE
National Research Centre for Equines, Sirsa Road, Hisar (Haryana) – 125001
62. NRCG
National Research Centre for Grapes, PO No. 3 Manjeri Farm Post, Sholapur Road, Pune (Maharashtra) – 412307
63. NRCG
National Research Centre for Groundnut, Ivnagar Road, PB No. J, Junagadh (Gujarat) – 362001
64. NRCMAP
National Research Centre for Medicinal and Aromatic Plants, Bonavi Taluka, Anand (Gujarat) – 387310
65. NRCM
National Research Centre for Mithun, ICAR Research Complx, Jhamapani, Kohima (Nagaland) – 797106
66. NRCM
National Research Cenre for Mushroom Chambaghat, Solan (Himachal Pradesh) – 173213
67. NRCOP
National Research Centre for Oil-Palm, Door No. 1-66-D, Shalvahana Nagar, Gooravarms, Elurn (Andhra Pradesh) – 534003
68. NRCOG
National Research Centre for Onion and Garlic, Rajgurunagar, Pune (Maharashtra) – 410505

69. NRCO
National Research Centre for Orchids, Tadang, Gangtok (Sikkim) – 737102
70. NRCPB
National Research Centre on Plant Biotechnology, IARI, Pusa Campus, New Delhi – 110012
71. NRCRM
National Research Centre on Rapessed and Mustard, Sewar Farm, Bharatpur (Rajasthan) – 321303
72. NRCSS
National Research Centre for Seed Species, Tabiji, Ajmer, Rajasthan
73. NRCS
National Research Centre for Sorghum, Rajendranagar, Hyderabad (Andhra Pradesh) - 500030
74. NRCS
National Research Centre for Soybean, Indore (Madhya Pradesh) – 452001
75. NRCWS
National Research Centre for Weed Science, 215 Rajendra Nagar, JNKVV, Jabalpur (Madhya Pradesh) – 482004
76. NRCY
National Research Centre on Yak, Dirang, West Kameng (Arunachal Predesh) – 790101
77. PDCSR
Project Directorate for Cropping Systems Research, Modipuram, Merrut (Uttar Pradesh) – 250110
78. PDADMAS
Project Directorate on Animal Disease Monitoring and Surveillance, IAH and VB Campus, Hebbal, Bangalore (Karnataka) – 560024
79. PDC
Project Directorate on Cattle, PH-7, Pallavpuram, Phase-II, Modipuram, Merrut (Uttar Pradesh) – 250110
80. PDFMD
Project Directorate on Foot and Mouths Diseases, IVRI Campus, Mukteshwar, Kumaon (Uttranchal) – 263138
81. PDP
Project Directorate on Poultry, ANGRAU Campus, Rajendranagar, Hydreabad (Andhra Pradesh) – 500030
82. SBI
Sugarcane Breeding Institute, Coimbatore (Tamil Nadu) – 641007
83. VPKAS
Vivekananda Parvatiya Krishi Anushandhan Shala, Almora (Uttaranchal) – 263601
84. WTCER
Water Technology Centre for Eastern Region, No. VII – H/176-180, Sailshree Vihar, Chandra Shekharpur, Bhubaneshwar (Orissa) - 751102

Appendix

23. AGENCIES/ INDIVIDUALS WORKING ON ORGANIC FARMING IN INDIA

IFOAM- India
Institute for integrated Rural Development (IIRD)
IIRD is the Asia Regional Office for International Federation of Organic Agricultural Movements (IFOAM)
Postal address
54, Kanchan Nagar, Nakshatrawadi, Aurangabad – 431 002.,
Maharashtra, India.
Tel. +91-240-332336, 322828
FAX +91-240-322866
Electronic mail -General Information: Project Office (iirdind@bom4.vsnl.net.in

1. **Acharya, S. A. R. and Madhwaraj, K.B.**
Vishvaneedam International Sarvodaya Centre
13 K.M. Magadi Rd.
P.O. Vishraneedam
Bangalore – 560 091
2. **Agarwal, Anil**
Centre for Science and Environment
41, Tughlakabad Institutional Area
New Delhi – 110 062
3. **Agriculture & Processed Food Product Export Development Authority (APEDA)**
3rd Floor, Ansal Chambers II,
Bhikaji Cama Place
New Delhi – 110 066
4. **Agriculture Man and Ecology Programme**
P.O. Box 7836
Bangalore – 560 078
5. **Ali, Farhat**
Sana Adarsh Sewa Samiti
Masauli
Barabanki – 225 204
6. **Babu, Lucas**
Rural Integrated Development Association
Hqsevagram
Morappur – 635 305
Dharmapuri Dist. Tamil Nadu
7. **Babu, T. H.**
Indian Institute of Forest Management
10-IIFM Colony
Kottrasultanabad
Bhopal – 462 003
M.P. (India)
8. **Balasubramanian**
Moses Compound
Sunoora Patti
Segii Patti (PO) 613 402
Dist. Tanjavore (India)
9. **Behera, Shanti Ranjan**
Society For Development Action
P.O. Box 16
Baripada – 757 001
Mayurbhanj
Orissa (India)

10. **Chakrabarty, D.N.**
CSIR (Council of Scientific & Industrial Research)
PO Box No. 6
Palampar – 176 061
H.P. (India)

11. **Chandrashekar Kote & Co.**
Charted Accountants
M.G. Road
Chikmagalur – 577 101

12. **Chauhan, G.C.**
Pudumjee Plant Laboratories
Jatia Chambers, 60
Dr. V.B. Gandhi Marg,
Mumbai – 400 023
India

13. **De, Hansi**
Joint Co-ordinator
Friends Rural Centre
Rasulia
Hoshangabad
M.P. – 461 001 (India)

14. **Gable, Thomas**
Nelliampathy Hills Conservation Society
Padagiri PO
Palakkad Dist. – 678 509

15. **Gloria Land**
Aurobrindavan
Auroville – 605 101

16. **Gram Manglam-Sultanpur**
At PO Bhudasan
via Akarund – 383 260
Ta Bayad. Dist. Sabarkantha
Gujarat

17. **Gupta, Anil K.**
Chairperson
R.M. Centre for Educational Innovation,
Indian Institute of Management,
Ahmedabad – 380 015 (India)

18. **Gururaja, B.**
E-1 Maithree Apt
6th Main, 15th Cross
Malleshiraram
Bangalore
Karnataka State (India)

19. **Haranath Tadepally**
Academy of Gandhian Studies
2-2-1133/4/5
New Nallakunta
Hyderabad – 500 044

20. **Haria, M.D.**
Kachch Sanjivani Farm
Bidada (Mondvi)
Kachch – 370 435
Gujarat

21. **Hegde, N.G.**
BAIF Development Research Foundation
BAIF Bhavan,
Dr. Manibhai Desai Nagar
Warje
Pune – 411 029

22. **Helen, S.**
Justus Bhavan
Taruvai
Manavalakurichi PO
Kanayakumari
Dt. 629 252
Tamil Nadu (India)

23. **James Rajasekaran E.**
Project Director
PARD
PO Box 87
Madurai – 625 020
Tamil Nadu

24. **Jenkins, Brian**
Shanthi Vedanayakam
Centre for Learning, Agriculture & Appropriate Technology
PO Box 57
Kodaikanal – 624 101
Tamil Nadu (India)

25. **Jeste, Milind**
Caprico Enterprises
Pearl Apts – 3/47
Sidharth Nagar – 1
Goregaon (W)
Mumbai – 400 104 (India)

26. **Jindal, Preeti**
Gorakhpur Environmental Action Group
244 Purdilpur
M.G. College Rd.
P.O. Box 60
Gorakhpur – 273 001 (U.P.)

27. **Joseph, Maria**
SEEDS
Thachankurichy (P.O.)
Gandharvakottai
Pudukkottai DT – 613 301 (India)

28. **Kannikeswaran, S.V.**
Annamalai Reforestation
5 Manakulavinayagar St.
Sri Ramanasramam
P.O. Tiruvanamalai
Tamil Nadu (India)

29. **Kumar, Arbind**
Lok Jagriti Kendra
Madhupur – 815 353
Deoghar Bihar

30. **Kumar, Devendra**
Centre of Science for Village
Dattapur,
Vardha – 442 001 (M.H.)

31. **M.G. Mony**
The Director
Society of Health, Education & Development (SHED)
PO Box No. 18
Rayagada
Orissa – 765 001 (India)

32. **Madhhaik, A.S.**
Kupvi Herbarium Workers Association
Kupvi – 171 217
Sub. tch Kupvi
Distt. Simla H.P.

33. **Margosan Exports**
P.O. Box 1122,
Bhavani Ganesh Complex,
Bejai Church Road,
Mangalore – 575 004

34. **Mehta, Kisan**
President
Prakruthi
620 Jame Jameshed Road
4th Floor, Dadar (East)
Mumbai – 400 014 (India)

35. **Mitra, Amit**
Kassar Trust
Mankote
P.O. Udera
Dist. Almora
Uttar Pradesh (India)

36. **Mohan, T.**
Greenpeace
J-15, Saket
New Delhi – 110 017

37. **Nair, K. Harikrishnan**
Association Professor & Co-ordinator
Tribal Area Research Centre (Ambouri)
College of Agriculture
Vellayani – 695 522
Thiruvananthapuram (India)

38. **Nambi, V. Arivudai**
Save the Easter Ghats Association
Y109, Raja Street
Valaijam Battu
Chengam – 606 701 (T.N.)

39. **Narayan Bhai**
Sarbik Gram Bikash Kendra (Gandhi Mission)
PO Daspur
Dist. Midnapore
W.B. (India)

40. **Nazareth, Jagdish**
H-3 Rushika Appartments
Near Kiran Park
Nava Wadaj
Ahmedabad – 380 013 (India)

41. **P.S. Mani**
"Ramakripa"
7/386 Tharakad
Palgat 678 001
Kerala State (India)

42. **Padama**
Prarambha
18/1 All Asker Road
Bangalore – 560 052
India
43. **Pandhripande, N.D.**
Kumarappa Go Samvardhan Kendra
PUSAD,
Yavatmal – 445 204 (M.H.)
44. **Patel, Gunvant R.**
Sarsa – 388 365
Gujarat
45. **Pattanayak, Parikshita**
Gania Unnayan Committee (GUC)
At/P.O. Gania
Dist. Nayagarh
Orissa – 752 085 (India)
46. **Ponnusamy, K.**
M.Sc. (Agri)
P.M. Kovil
Kallankattuvalasu (PO)
Salem (DT),- 638 183
Tamil Nadu (India)
47. **Prajapati, N.D.**
Sonamukhi Nagar
Snagaria Phanta, Basni
Jodhpur – 342 004
48. **Prakash, R.**
Associate Professor
Kerala Agricultural University
College of Agriculture
Vellayani PO
Thiruvananthapuram – 695 522
Kerala (India)
49. **Punater, Dilip**
Honorary Secretary
Prakruti
123 Mahatma Gandhi Road
Opp. Bombay Univesity Fort
Mumbai – 400 023 (India)
50. **Raghavendra, K. Rao & Rao, Shylaja R.**
Plot No. 13, Srinagar Colony,
Hyderabad – 500 873
Andhra Pradesh (India)
51. **Raju, T.**
Regional Coffee Research Station
Thandigundi
Kodaikamal Tuluk
Tamil Nadu (India)
52. **Ramachandra, T.V.**
Centre for Ecological Sciences
Indian Institute of Science
Bangalore – 560 012
India
53. **Ramprasad, Vanja**
Green Foundations
839, 23rd Main, J.P. Nagar
2nd Phase,
Bangalore – 560 078
54. **Reddy, L. Narayana**
Srinivasapura
Via Maralenahalli
Hanabe P.O.
Doddaballapur Taluk
Karnataka – 561 203
55. **Reddy, S.T. Somashekhara**
Research Fellow
Indian Institute of Management
Bannerghatta Road
Bangalore – 560 076
56. **Save, Bhaskar H.**
Kalpavruksha
At & Post Deheri
Ta. Umbergaon – 396 170
Dist. Valsad (Gujarat)
57. **Setty, H.C. Ashoka**
Setty's Farm and
Gopal T. Kulkami
Vasundhara Group of Organic Farmers
Rattihalli – 581 116
KTKst (India)
58. **Sherief, A.K.**
Assistant Professor
Dept. of Agricultural Extension
College of Agriculture
(PO) Vellayani
Thiruvananthapuram – 695 522
Kerala (India)

59. **Shiva, Vandna**
Research Foundation for Science
Technology and Ecology
A-60 Hauz Khas,
New Delhi – 110 016

60. **Singh, H.B.**
Central Institute of Medicinal and
Aromatic Plants
Post Bag No. 1
PO Ram Sagar Mispa Nagar
Lucknow – 226 016 (India)

61. **SKAL (Certification & Exports)**
3rd Floor – Monarch Chambers
122 Infancy Road,
Bangalore

62. **Society for Human Integrety and Prosperity**
9-81 Gandhinagar
Nuzvid – 521 201
Krishna Dist. (India)

63. **Thomson, Jhansy**
President/Project Director
CAFARD (Community Action for
Food & Rural Development)
Second Seven Wells Street
St. Thomas Mount
Butt Road
Chennai – 600 016 (S. India)

64. **Todiara, N.P.**
Dept. of Forestry
P.O. Box 59
H.N.B. Garwhal University
Srinagar

65. **Venugopal, R.**
Bodi Bio Farms
Rajamani Cloth Centre Upstairs
Kamaraj Bazaar
Bodi Nayakanur – 626 513
Tamil Nadu (India)

66. **Verma, Jeevan Lal**
Director
Rural Organization on Social
Elevation (Rose)
K.S.S. – Social Awareness Centre
KANDA
PO Kanda
Almora
U.P. – 263 631 (India)

67. **Vijay Shah**
Nu Tech Farm,
Rayan
Dist: Kutch-Mandivi
Gujarat – 370 465 (India)

Appendix

24. COMMON ABBREVIATIONS

Abbr.	Description
2,4 D	2,4 Dichlorophenoxy Acetic Acid
AAASA	Association for the Advancement of Agricultural Science in Africa
AA-N	Action Aid-Nepal
AASCO	Association of American Seed Control Officials
AATG	Actino Aid The Gambia
ABA	Abscisic acid
ABA	Abscisic acid
ABS	Agricultural Bank of Sudan
ACMV	African cassava mosaic virus
ACORD	Agency for Cooperation and Research in Development
ADA (SCN)	Assistant Director of Seed Certification
ADP	Adinosine Di Phosphate
AFC	Agricultural Finance Corporation
AFGP	Antifreeze Glycoproteins
AFLP	Amplified Fragment Length Polymorphism
AFLP	Amplified Fragment Length Polymorphism
AFPs	Antifreeze Proteins
AIC	Agricultural Inpus Corporation (Nepal)
AIMV	Alfalfa mosaic virus
AMCEN	African Ministerial Conference on the Environment
AMP	Adinosine Mono Phosphate
AO (SCN)	Agricultural Officer (Seed Certification)
AO (SI)	Agricultural Officer (Seed Inspection)
AOSA	Association of Official Seed Analysis
AOSCA	Association of Official Seed Certifying Agencies
ARC	Agricultural Refinance Corporation

Abbr.	Description
ARD	Associates in Rural Development (US)
ASC	Agricultural Service Centre (Nepal)
ASIA	American seed Trade Association
ASS	African Seeds of Survival (Ethiopia)
ASSINEL	International Association for the Protection of Plant
ASTEC	Australian Seed Technology and Education Centre
ATP	Adinosine Tri Phosphate
AVRDC	Asian Vegetable Research and Development Center
AVRDC	Asian Vegetable Research and Development Centre, China
BAP	6-Benzylaminopurine
BARC	Bhaba Atomic Research Centre Bangladesh Agriculture Research Council
BAU	Bangladesh Agricultural University
BMV	Brome mosaic virus
BP	Between Paper method of germination
BR	Breeders Right
BS	Breeder seed
CAD	Cinnamyl-CoA-reductase and Cinnamyl alcohol dehrydogenase
CaMV	Cauliflower mosaic virus
CAT	Chloramphenicol amino transferase
CBN	The cassava biotechnology network
CCC	Chloro Choline Chloride
cDNA	complementary DNA
CEAO	Communaute Economique de 1'Afrique de 1'Ouest

Abbr.	Description
CEC	Commission of the European Communities
CEPGL	Communaute aaeconomique des Pays des Grands Lacs
CESA	Ecuadorian Centre for Agricultural Services
CG	Canadian Genebank
CGIAR	Consultative Group on International Agricultural Research
CGIAR	Consultative Group for International Agricultural Research
CGMS	Cytoplasmic Genetic Male Sterility
CHA	Chemical Hybridizing Agent
CIAT	International Center for Tropical Agriculture
CIAT	International Centre for Tropical Agriculture
CIC	Centro Internazionaale Crocevia
CIDA	Canadian International Development Authority
CILSS	Permanent Interstate Committee for Drought Control in the Sahel
CIMMYT	International maize and wheat improvement center
CIP	International potato center
CIP	International Potato Centre
CITES	Convention on International Trade in Endangered Species of Wild Fauna and Flora (IUCN), Washington D.C.
CLADES	Latin American Consortium for Agroecology and Development
CLRV	Cherry leafroll virus
CMV	Cuoumbor mocaio viruc
CP	Cross Pollination
CP	Coat Protein
CS	Certified Seed
CSB	Community Seed Bank (Philippines)
CSC	Central Seed Committee
CSCB	Central Seed Certification Board
CS-I	Certified Seed stage one
CS-II	Certified Seed stage two
CSSTD	Central Seed Science and Technology Division (Nepal)
CSTA	Canadian Seed Trade Association
CSTL	Central Seed Testing Laboratory
CSVT	Combined Stress Vigour Test
CVPO	Certification Based on Verietal Purity Only
CVRC	Central Variety Release Committee
CyRSV	Cymbidium ringspot virus
D (SCN)	Director (Seed Certification)
DAR	Department of Agricultural Research (The Gambia)
DAS	Department of Agricultural Services (The Gambia)
DD (SCN)	Deputy Director (Seed Certification)
DFSC	Danida Forest Seed Centre
DMP	Dry Matter Production
DOA	Department of Agriculture (Nepal)
DSSTS	Danish State Seed Testing Station
DUS	Distinct, Uniform and Stable
DUST	Distinctness, Uniformity and Stability Testing
EC	Energy Charge/Electrical Conductivity
EC	European Community
ECOWAS	Economic Community of West African States
EDB	Ethylene Di Bromide
ELC	Environmental Liaison Centre International
ELISA	Enzyme Linked Electrophoresis Assay
EMI	Embu-Meru-Isiolo project (Kenya)
ENDA	Environemtn and Development Activities, Zimbabwe
ENS	Empresa Nacional de Sementes (Mozambique)
EPPO	European and Mediterranean Plant Protection Organisation
EPSP	5-enol pyruvyl shikimate 3-phosphate synthase
ESA	Edinburg School of Agriculture
ESC	Ethiopia Seed Croporation
EU	European Union
EUCARPIA	European Association for Research on Plant Breeding

Abbr.	Description
F_1	Filial generation one or First Filial generation
F1	First filial generation hybrid
FAO	Food and Agricultural Organization
FAO	UN Food and Agriculture Organization
FFHC	Freedom From Hunger Organization
FFHC	Freedom From Hunger Campaign
FIS	Federation Internationals du Semances (Federation of International seed trade)
FS	Foundation Seed
FS-I	Foundation Seed – Stage One
FS-II	Foundation Seed – Stage Two
FV	Farmer's variety
GA	Gibberellic Acid
GA3	gibberellic acid
GADA	Gulatmic Acid Decarboxylase Activity
GATT	General Agreements on Tariff and Trade
GATT	General Agreement on Tariffs and Trade
GCMV	Grapevine chrome mosaic virus
GCU	The Gambia Cooperative Union
GDP	Gross Domestic Product
GEMS	Global Environment Monitoring System
GIFT	Genetic Improvement of Farmed Tilpia.
GIS	Geographic Information System
GM	Genetically Modified (Transgenics)
GMS	Genetic Male Sterility
GNP	Gross National Product
GOI	Government Of India
GOT	Grow Out Test
GPMB	Gambia Produce Marketing Board
GPSN	Niassa Seed Production Board (Mozambique)
GRAIN	Genetic Resource Action International
GRAIN	Genetic Resources Action International
GRID	Global Resource Information Database
GSM	Good Seed Mission (The Gambia)
GTZ	Deutsche Gesellschaft fur Technische Zusammenarbeit
GUS	β-glucuronidase
HSS	Herbage Seed Scheme
HYV	High Yielding Variety
IA	Image Analysis (Method of cv. Characterization)
IAA	Indole Acetic Acid
IAA	Indole-3-acetic acid
IAPI	International Association of Plant Taxonomy
IAPSC	Inter African Phyto-Sanitary Commission
IARC	International Agricultural Research Centre
IARI	Indian Agricultural Research Institute
IBA	Indole-3-butyric acid
IBP	International Biological Programme
IBPGR	International Board for Plant Genetic Resources
ICAR	Indian Council of Agricultural Research
ICARDA	International Center for Agricultural Research in Dry Areas
ICDA	International Coalition for Development Action
ICRISAT	International Crop Research Institute in Semi Arid Tropics
ICRISAT	International Crops Research Institute for the Semi-Arid Tropics (India)
IDESSA	Institute des Savanes
IDRC	International Devleopment Research Centre, Canada
IFOAM	International Federation of Organic Agricultural Movements
IITA	International Institute for Tropical Agriculture
IITA	International Institute of Tropical Agriculture (Nigeria)
IJO	International Jute Organization

Abbr.	Description
ILCA	International Livestock Centre for Africa
INIAP	National Agricultural Research Institute (Ecuador)
Ip	Inorganic Phosphorous
IPB	Institute for Plant Breeding
IPGRI	International Plant Genetic Resources Institute
IPR	Intellectual Property Rights
IRAT	Institute for Research in Tropical Agriculture and Food Crops (France)
IRAZ	Institut de Recherche Agronomique et Zootechnique de la CEPGL
IRRI	International Rice Research Institute
IRRI	International Rice Research Institute (The Philippines)
ISO	International Standards Organization
ISST	Indian Society of Seed Technology
ISTA	International Seed Testing Association
ISTA	International Seed Testing Association
IUCN	World Conservation Union (formerly International Union for the Conservation of Nature and Natural Resources)
IUFRO	International Union for Forestry Research Organization
IVO	Development Research Centre (The Netherlands)
JICA	Japan International Cooperation Agency
KHAP	Koshi Hills Agriculture Project (Nepal)
Kin	Kinetin
KMP	Peasant Movement of The Philippines
KSHP	Kebkabiya Smallholder's Project (Nepal)
L	Light essential for germination
LEA	Late Embryogenesis Abundant
LEC	Livelihood Enhancement Committee (The Philippines)

Abbr.	Description
LFR	Liable For Rejection
MARs	Matrix attachment/associated regions
MASIPAG	Farmers-Scientists Partnership for Agricultural Development (The Philippines)
MC (%)	Moisture Content (%)
MCC	Mennonite Central Committee
MIND	Mindoro Institute for Development Inc (The Philippines)
MNC	Multi National Corporation
MNC	Multi-National Corporation
MONAP	Mozambique Nordic Agriculture Programme
MS	Male Sterile
MV	Modern Variety
NAA	Naphthalene acetic acid
NAAK	General Netherlands Inspection Service for Seeds and Seed Potatoes
Na-alginate	Sodium alginate.
NABARD	National Bank for Agriculture and Rural Development
NACGRAB	National Centre for Genetic Resources and Biotechnology
NADP	Nicotinamide Adenine Di Phosphate
NADPH	Nicotinamide Adenine Di Phosphate Hydrogen
NAIB	National Institute of Agricultural Botany
NARC	National Agricultural Research Centre
NARP	National Agricultural Research Project
NARS	National Agricultural Research Service
NATP	National Agriculture Technology Project
NBPGR	National Bureau of Plant Genetic Resources
NCA	National Commission on Agriculture
NEF	Near East Foundation
NG	Nordic Gene bank
NGO	Non-Governmental Organization
NIB	Replicase

Abbr.	Description
NIHORT	National Horticultural Research Institute of Nigeria
NMR	Nuclear Magnetic Resonance
NOS	Nopaline synthase
NPK	Nitrogen-Phosphorus-Potassium fertilizer
NPT-II	Neomycin phosphotransferase
NRI	Natural Resources Institute (UK)
NS	Nucleus seed
NSA	National Seed Administration (Sudan)
NSB	National Seed Board (Bangaldesh)
NSC	National Seeds Corporation
NSDC	National Seed Development Council
NSP	National Seed Project
NSSL	National Seed Storage Laboratory of USA
NSSTS	Norwegian State Seed Testing Station
NUV	Near Ultra Violet radiation
OCP	Other Crop Plants
OCS	Other Crop Seeds
ODA	Overseas Development Administration (UK)
ODI	Overseas Development Institute (UK)
ODV	Other Distinct Variety
OECD	Organization for Economic Cooperation and Development
OECD	Organization for Economic Cooperation and Development
OFSP	On-Farm Seed Production Project (The Gambia and Senegal)
OP	Open-pollinated variety
OPV	Open Pollinated Variety
ORSTOM	Office de la Recherche Scientifique et Technique d' Outer-Mer (France)
OWS	Other Weed Seed
PAC	Pakhribas Agricultural Centre (Nepal)
P-AF-P	Paper-Aluminium Foil – Polythene laminated pouch
PAGE	Poly Acrylamide Gel Electrophoresis
PAI	Paran Agronomy Institute (Brazil)
PARE	Partnership in Agricultural Research and Extension (Bangladesh)
PBR	Plant Breeder's Right
PBR	Plant Breeder's Rights
PCR	Polymerase Chain Reaction
PEBV	pea early browning virus
PEG	Poly Ethylene Glycol
PEG	polyethyleneglycol
Pfr	Phytochrome far-red
PG	Polygalaturonase
PGI	Plant Germplasm Institute
PGMS	Photo Genetic Male Sterility
PGR	Plant Genetic Resources
PGRC	Plant Genetic Resources Center
PGRC/E	Plant Genetic Resources Centre/ Ethiopia
PGS	Pure Germinating Seeds
PLRV	Potato leafroll virus
PLS	Pure Live Seed
PMRA	Pest Management Regulatory Agency
PP	Pleated Paper method of germination
PPS	Private Producer-Sellers Project (Nepal)
Pr	Phytochrome red
PRSV	Papaya Ring-spot Potyvirus
PTGS	Post-Transcriptional Gene Silencing
PVA	Plant Variety Protection
PVP	Plant Variety Protection
PVS	Potato virus S
PVX	Potato virus X
PVY	Potato virus Y
RAFI	Rural Advancement Foundation International
RAPD	Random Amplified Polymorphic DNA
RASE	Royal Agricultural Society of England
RCA	Royal Commission on Agriculture
RENAMO	Mozambique National Resistance
RFLP	Restricted Fragment Length Polymorphism
RQ	Resiration Quotient

Abbr.	Description
RSS	Research and Specialist Services
RST	Registered Seed Technologist
RSV	Rice stripe virus
RT	Reverse Transcriptase
S	Sand method of Germination
SAD	State Agricultural Department
SADC	Southern African Development Community
SADCC	Southern African Development Coordination Conference
SAM	Sahabat Alam Malaysia
SARs	Scaffold attachment regions
SAU	State Agricultural University
SBD	Seed Borne Disease
SCF	Save the Children Federation (US)
SCO	Seed Certification Officer
SCST	Society of Commercial Seed Technologist
SDS-PAGE	Sodium Dedocyl Sulphate-PolyAcrylamide Gel Electrophoresis
SEAN	Seed Entrepreneurs Association of Nepal
SEARICE	South East Asia Regional Institute for Community Education
SEM	Scanning Electron Microscope
SEMOC	Sementes de Mozambiuqe Lda.
SFCI	Stage Forms Corporation of India
SHMV	Sunn hemp mosaic virus
SI	Seed Inspector
SIBAT	Spring of Science and Technology (The Philippines)
SIDP	Seed Improvement and Development Programme
SP	Self Pollination
SPG	Seed Producer Group (Nepal)
SRGB	SADC Regional Genebank
SRR	Seed Replacement Rate
SRT	Seed Review Team
SSC	State Seed Corporation
SSTCI	Swedish Seed Testing and Certification Institute
SSTL	State Seed Testing Laboratory
STIP	Seed Technology and Improvement Program (Nepal)
STL	Seed Testing Laboratory
STO	Seed Testing Officer

Abbr.	Description
STU	Seed Technology Unit (The Gambia)
SWOT	Strength, Weakness, Opportunities and Threats
TAV	Tomato aspermy virus
TBSV	Tomato bushy stunt virus
TCV	Turnip crinkle virus
TDC	Tarai Development Corporation
TEMED	N,N,N, N – Tetra methyl ethylene diamine
TEV	Tobacco etch virus
TGMS	Thermo Genetic Male Sterility
TGMV	Tomato golden mosaic virus
TMO	Technology Mission on Oilseeds
TMV	Tobacco Mosaic Virus
TMV	Tobacco mosaic virus
TobRV	Tobacco ringspot virus
ToMV	Tomato mosaic virus
TP	Top of Paper method of germination
TPS	True Potato Seed
TRIPS	Trade Related Intellectual Property Rights
TRIPs	Trade Related Aspects of Intellectual Property Rights
TRIS	Tris-hydroxymethyl amino methane
TRV	Tobacco rattle virus
TS	Top of Sand method of germination
TSWV	Tomato spotted wilt virus
TTZ	Tri phenyl Tetrazolium Chloride
TYLCV	Tomato yellow leaf curl virus
TZ	Tetrazolium chloride
UN	United Nations
UNCED	UN Conference in Environment and Development
UNDP	United Nations Development Programme
UNESCO	Un Educational, Scientific and Cultural Organization
UPLB	University of The Philippines at Los Banos
UPOV	International Union for the Protection of new Plant Varieties
USAID	United States Agency for International Development

Abbr.	Description
USC	Unitarian Service Committee (Cananda)
USDA	United States Department of Agriculture
USDC	United States Department of Commerce
USNOAA	United States National Oceanic and Atmospheric Administration
UV	Ultra Violet radiation
VCU	Value for Cultivation and Use
VI	Vigour Index
VRC	Village Relief Committee (Sudan)
WCMC	World Conservation Monitoring Centre
WIPO	World Intellectual Property Organization
WRI	World Resources Institute
WTO	World Trade Organisation
WWF	Worldwide Fund for Nature

25. UNENDING NOTE

Cycle of Information

Philosophy of organic farming is very old but the technology and marketing in the present world are fast changing. Sharing of experience and knowledge of organic farming with other, start its further refinement, and again information reached to origin in more refined form. Thus, this is a cycle which needs to be continued. If you have experience and knowledge more than the available in the handbook, please do send it to me for propagation of your knowledge to the others, with due acknowledgement it will be incorporated in the future edition of the handbook.

Please provide- information in the following format

1. Name
2. Address
3. Type of information
   a) Technology
   b) Adoption/Motivation
   c) Marketing
4. Details of success e.g. constraints converted into opportunity
5. Suggestion

Please send information to:

Arun K. Sharma
Scientist (Agronomy)
Central Arid Zone Research Institute
Jodhpur – 342003, INDIA
E-mail aksharma@cazri.raj.nic.in
Phone: +91-291-2643547,
Fax: +91-291-2740706